CHAOS APPLIED TO FLUID MIXING

CHAOS, SOLITONS AND FRACTALS

Applications in Science and Engineering

An interdisciplinary journal of nonlinear science

Chaos, Solitons & Fractals is an interdisciplinary journal devoted to the rapid publication of full length original papers, short communications, reviews and tutorial articles in the following nonlinear sciences: bifurcation and singularity theory, deterministic chaos and fractals; stability theory, soliton and coherent phenomena; and formation of pattern, evolution, complexity theory, knot theory and neural networks.

Contributions on both fundamental and applied studies are welcome, but the emphasis of the journal is on applications in the following fields:

Physical sciences – classical mechanics, including fluid mechanics; quantum and statistical mechanics; lasers, optics and acoustics; plasma physics and fusion; condensed matter physics; chemistry and chemical physics; astronomy and astrophysics; materials science; geophysics; meteorology.

Biomedical and life sciences – biology; molecular biology; population dynamics; zoology; theorectical ecology.

Engineering – marine engineering; mechanical, aeronautical and astronautical engineering; electrical engineering; chemical engineering; structural and civil engineering.

Contributions may be of three kinds:

Invited review and papers
Length by agreement. Authors should consult the Executive Editor if they wish to prepare a review.

Contributed papers
Normally 2000–5000 words plus illustrations in black and white or colour. Authors should follow the instructions on the inside back cover.

Short communications
Less than 1500 words plus two illustrations. Short communications may give advance notice of important results.

General and informal papers
The journal will have a limited amount of space for informal articles of a general nature.

Special issues
The journal also publishes special issues which draw together outstanding collections of papers on topics of current importance.

CHAOS APPLIED TO FLUID MIXING

H. Aref *(Urbana, Illinois, USA, Guest Editor)*
M. S. El Naschie *(Cambridge, UK, Editor)*

Pergamon

U.K. Elsevier Science Ltd, The Boulevard, Langford Lane, Kidlington, Oxford OX5 1GB, U.K.

U.S.A. Elsevier Science, Inc., 660 White Plains Road, Tarrytown, New York 10591-5153, U.S.A.

JAPAN Elsevier Science Japan, Tsunashima Building Annex, 3-20-12 Yushima, Bunkyo-ku, Tokyo 113, Japan

---

First edition 1995

Library of Congress Cataloging in Publication Data
A catalogue record for this book is available from the Library of Congress

British Library Cataloging in Publication Data
A catalogue record for this book is available from the British Library

ISBN 0 08 042028 1

Reprinted from *Chaos, Solitons and Fractals*, **4**(6)

Printed and bound in Great Britain by BPC Wheatons Ltd, Exeter

Contents

Preface – H. Aref 1
Chaotic Advection in Perspective

D. Beigie, A. Leonard and S. Wiggins 5
Invariant Manifold Templates for Chaotic Advection

M. Liu, F. J. Muzzio and R. L. Peskin 125
Quantification of Mixing in Aperiodic Chaotic Flows

K. Bajer 151
Hamiltonian Formulation of the Equations of Streamlines in Three-dimensional Steady Flows

A. J. Szeri and L. G. Leal 169
Orientation Dynamics and Stretching of Particles in Unsteady, Three-dimensional Fluid Flows: Unsteady Attractors

S. W. Jones 185
Interaction of Chaotic Advection and Diffusion

P. E. Hydon 197
Resonant and Chaotic Advection in a Curved Pipe

Hsueh-Chia Chang and M. Sen 211
Application of Chaotic Advection to Heat Transfer

V. V. Meleshko and G. J. F. van Heijst 233
Interacting Two-dimensional Vortex Structures: Point Vortices, Contour Kinematics and Stirring Properties

S. P. Beerens, H. Ridderinkhof and J. T. F. Zimmerman 267
An Analytical Study of Chaotic Stirring in Tidal Areas

M. Tabor and I. Klapper 287
Stretching and Alignment in Chaotic and Turbulent Flows

K. B. Southerland, R. D. Frederiksen, W. J. A. Dahm and D. R. Dowling 313
Comparisons of Mixing in Chaotic and Turbulent Flows

R. T. Pierrehumbert 347
Tracer Microstructure in the Large-eddy Dominated Regime

R. T. Pierrehumbert, I. M. Held and K. L. Swanson 367
Spectra of Local and Nonlocal Two-dimensional Turbulence

Author Index 373

Subject Index 379

Chaotic Advection in Perspective

HASSAN AREF

Department of Theoretical and Applied Mechanics, University of Illinois at Urbana-Champaign, Urbana, IL 61801, USA

Abstract—As a preface to the papers in this volume I make some general philosophical remarks on chaotic advection and its emergence as a subtopic of fluid mechanics. Based on these remarks the following papers are placed in perspective.

The term 'chaotic advection' used—as far as I am aware*—for the first time in the archival literature in my 1984 paper [1], has spawned its own subtopic within the science of fluid mechanics as witnessed by many articles, chapters in several books, and a number of workshops and conferences dealing with the subject in the decade since then. A representative sample of recent research papers in this general area are collected in the present volume.

The basic idea is simple enough. If a particle is so light, so inert, so passive that it follows the fluid in which it finds itself embedded everywhere and for all times, then its equations of motion are simply the *advection equations*:

$$\frac{d\mathbf{x}}{dt} = \mathbf{V}(\mathbf{x}, t), \tag{1}$$

where $\mathbf{V}$ is the fluid velocity field that—in the present context—is thought of as having been obtained from other considerations. Now, from the standpoint of dynamical systems theory the format of the problem presented by equation (1)—D-coupled ordinary differential equations in a space of D dimensions—is sufficiently rich for $D \geqslant 2$ that, in general, one must assume the solutions to be chaotic (in the technical sense). Exceptions will, of course, occur for certain constrained systems, such as the familiar case of two-dimensional steady flow in which $\mathbf{V}$ has the simple form $(u(x, y), v(x, y), 0)$ and in similar situations.

The appearance of chaos in the solutions of equation (1) is hardly a surprise, and analogous situations that arise in other branches of physical science have been studied for many years. For example, the motion of a charged particle in a prescribed electric and magnetic field yields equations of motion of the general form (1) under suitable approximations, and the role of chaos in those equations has been studied in great detail in the context of plasma confinement device technology and particle accelerator physics. In these cases, as in the present advection problem, the new qualitative insight is that very simple functions $\mathbf{V}$ in equation (1) can lead to very complicated solutions $\mathbf{x}(t)$.

In spite of these obvious prior models, perusal of the fluid mechanics literature prior to 1984 shows essentially no appreciation for these simple facts from the qualitative theory of differential equations. Indeed, quite to the contrary, the journal, monograph and textbook literature spends a great deal of time and effort on the treatment of integrable cases. A

*The term was actually first used in print in a conference abstract published in 1982.

superficial reading of the literature may actually leave the reader with the impression that simple flows will lead to simple particle trajectories, and that complex particle trajectories require complex flows. The phenomenon of chaotic advection points out that such an inference is entirely false. The main reason that this is so important is that chaotic trajectories of advected particles, in general, signal enhanced mixing of the advectant with the surrounding fluid. Indeed, since efficient mixing is typically a desirable goal in applications, achieving chaotic advection is typically seen as a positive outcome (whereas chaos in plasma devices and accelerators leads to charged particle losses and is seen as undesirable). The realization that 'chaos theory' can contribute to a significant practical problem such as fluid mixing has done much to interest applied scientists and engineers in the phenomenon and to propel a field that might otherwise be viewed as somewhat mundane to the headlines of scientific inquiry. It has become clear that the tools of 'chaos theory' provide a language in which to express quantitatively aspects of efficient mixing systems design that were previously expressible only in a qualitative empirical manner.

In view of its apparent importance it seems appropriate to single out the non-integrable case of advection (which we now realize must be by far the predominant one) through the use of a term such as *chaotic advection* (reserving for the common textbook case the term integrable advection). More fanciful names have been suggested, but have not stuck. In particular, the term *Lagrangian turbulence* was used to describe the same phenomenon. However, while turbulence may be some form of multi-degree-of-freedom chaotic solution of the Navier–Stokes equations, there are many instances of chaotic solutions of those equations that have little to do with turbulence. The kinematics of fluid particles is one such example. Even in very simple, viscosity-dominated flows at essentially vanishing Reynolds number one can see chaotic advection. Indeed, some of the most spectacular experimental confirmations of the theory have been produced in such flows. It seems utterly inappropriate to attach the word turbulence to these situations. Nevertheless, the spatial intricacy of advection patterns produced by chaotic advection often reminds one of what is seen in a snapshot of turbulent mixing, and the geometric correlation of data from the two regimes is a theme of current research interest.

It is not entirely accurate to say that the notion of chaos in equation (1) took until 1984 to mature in fluid mechanics. It was, in fact, introduced for the case of steady *three-dimensional* flow in two brief papers by Arnold [2] and Hénon [3] in 1965–66. One can only speculate as to why these papers were not seized upon with greater vigor by the community, appearing as they did at essentially the same time as Lorenz's seminal paper on strange attractors [4]. Such, I assume, is the nature of the history of science. Possibly the papers by Arnold and Hénon, to the extent they became known at all to the fluid mechanics community, were seen as studies of certain special cases, and the generality of the phenomenon was not appreciated. A brief paper by Arter [5], published in 1983 (but actually submitted a bit later than ref [1]), pursued chaotic advection in three-dimensional steady convection.

I like to see the emergence of interest in chaotic advection as part of a broader interest in what we may call the 'Lagrangian view' of fluid mechanics. I should probably say the 'particle-based point of view' instead of Lagrangian, as opposed to the 'field-based point of view' usually associated with the name of Euler. As Truesdell has pointed out this terminology of 'Eulerian' and 'Lagrangian' representations of fluid mechanics is accidental and quite inappropriate, but the terminology seems to have established itself, and it may be too late to change it. Some years ago Stuart and Tabor edited a special volume on precisely this topic of the 'Lagrangian view' of fluid mechanics [6]. Certain subjects in fluid mechanics have always been much clearer conceptually, if not simpler analytically, from the particle point of view. I am thinking of large portions of vortex dynamics, for example, where Helmholtz's equation for the evolution of vorticity has a clear, mechanistically appealing interpretation from a 'Lagrangian' point of view, but is just another field equation from an 'Eulerian' perspective. The 'Lagrangian' interpretation in the case of vortex dynamics leads to all manner of intriguing models and computational methods. In the case of passive advection, I should like to argue, a particle perspective applied to (1) opens up the panoply of dynamical systems theory methods, and leads to new qualitative insight, new physics and, in due course, will produce new technology.

A very important feature of equation (1) is the link between 'phase space' and 'real space' that it embodies. Dynamical systems live in phase space. For a Hamiltonian system, for example, we are used to writing equations of motion in the well-known canonical form:

$$\dot{q} = \frac{\partial H}{\partial p}; \qquad \dot{p} = -\frac{\partial H}{\partial q}. \tag{2}$$

Here H is the Hamiltonian and (q, p) is a pair of conjugate, generalized coordinate-momentum variables. Such a pair corresponds to one degree of freedom of the system in question. In an early paper of remarkable vision Welander [7] explored potential links between the motion of a fluid in real space and the development of an ensemble in phase space as dictated by Gibbsian statistical mechanics. At that time expectations presumably were that in order to get complex behavior in phase space (i.e. satisfy the 'ergodic hypothesis') one needed a complex system with many degrees of freedom. Today we know that time-dependent systems with one degree of freedom (sometimes referred to as one-and-a-half degree-of-freedom systems) are capable of complex phase space motion. And we understand, of course, that the connection between (1) and (2) (at least in two dimensions) is given by the *kinematic* relation

$$\dot{x} = \frac{\partial \psi}{\partial y}, \qquad \dot{y} = -\frac{\partial \psi}{\partial x}, \tag{3}$$

where ψ is the streamfunction of the flow, and (x, y) is the position of an advected particle. It takes a bit of adjustment to think of the x-coordinate of an advected particle as its generalized coordinate, and the y-coordinate as its generalized momentum! It does, of course, drive home the point of the equivalence of coordinates and momenta, and their interchangeability as witnessed in the theory of canonical transformations. More significantly, it shows why chaotic advection is immediately visualizable and has led to so many very beautiful images.

On the basis of these rather general remarks I can already place in perspective several of the papers to be found in this volume. The first paper, a major review by Beigie, Leonard and Wiggins of some of the mathematical tools that have evolved to describe and quantify chaotic advection, is sure to become a valued reference for workers in the field for several years to come. Some of the more recent developments include the detailed exploration of 'lobe dynamics' whereby fluid is exchanged between various spatial regions, and the study of aperiodic systems where the concept of a Poincaré section, so extensively used in simulations and experiments just a few years ago, loses its meaning. The following paper by Liu, Muzzio and Peskin also deals with these issues. How does one quantify mixing when the underlying flow is spatially 'simple' but temporally complex?

We may see these studies of chaotic advection by aperiodic flows as steps on the way to understanding mixing by fully turbulent flows. In this volume we have organized the papers according to some general measure of 'flow complexity'. Issues of chaotic advection versus turbulent mixing will thus be found towards the end of the volume in the papers by Southerland, Frederiksen and Dahm on experimental comparisons of mixing in chaotic and turbulent flows—a paper that in many ways addresses the issues Welander [7] was trying to get at many years ago—and in the analytical study by Tabor and Klapper on stretching and alignment.

There has always been a sense in which 3D steady flows could be seen as a counterpart of 2D unsteady flows. This correspondence implicitly underlies much of the work on mixing in pipe flows. The paper by Bajer sets out a Hamiltonian formulation for 3D steady flows, a topic that has been pursued in the plasma dynamics literature from time to time. Application of chaotic advection ideas to pipe flow are pursued in the paper by Hydon.

Much of the chaotic advection literature has dealt with point particles and their dynamics. For particles of finite extent chaotic motion may be seen both in the trajectory followed by the center of mass and in the orientation dynamics of the particle. The paper by Szeri and Leal deals with this topic. Analogously, if one adds other effects to chaotic advection, such as diffusion of chemical reaction, new phenomena arise. The paper by Jones explores interactions of chaotic advection and diffusion, presumably the simplest case possible, but still surprisingly rich.

Much of the interest in chaotic advection stems from applications and the new ways in which this phenomenon allows one to interpret experimental results and field observations. In the present volume we include the exciting engineering application by Chang and Sen on heat transfer enhancement, and several applications with geophysical motivations, such as the paper by Pierrehumbert and collaborators on statistics of 'microstructure' in flows dominated by large eddies. Finally, as a 'synthetic' GFD study we include the laboratory experiments and numerical experiments on colliding vortex dipoles by Meleshko and van Heijst.

The papers in this volume are by no means inclusive in terms of the current state of the topic of chaos and its application to fluid mixing. Some individuals approached as potential authors were simply too busy to contribute. They do, however, provide a comprehensive picture of the state of development of the subject, and apart from individual achievements, will hopefully provide an inspirational collection of examples for workers in a wide variety of problems that involve fluid mixing processes.

REFERENCES

1. H. Aref, Stirring by chaotic advection, *J. Fluid Mech.* **143**, 1–21 (1984).
2. V. I. Arnold, Sur la topologie des écoulements stationnaires des fluides parfaits, *C. R. Acad. Sci. Paris* **A261**, 17–20 (1965).
3. M. Hénon, Sur la topologie des lignes courant dans un cas particulier, *C. R. Acad. Sci. Paris* **A262**, 312–314 (1966).
4. E. Lorenz, Deterministic non-periodic flow, *J. Atmos, Sci.* **20**, 130–141 (1963).
5. W. Arter, Ergodic stream-lines in steady convection, *Phys. Lett.* **A97**, 171–174 (1983).
6. J. T. Stuart and M. Tabor, The Lagrangian picture of fluid motion, *Phil. Trans. R. Soc. Lond.* **A333**, 261–400 (1990).
7. P. Welander, Studies of the general development of motion in a two-dimensional ideal fluid, *Tellus* **7**, 141–156 (1955).

Invariant Manifold Templates for Chaotic Advection

DARIN BEIGIE

Center for Applied Mathematics and Theory Center, Cornell University, Ithaca, NY 14853, USA

and

ANTHONY LEONARD

Graduate Aeronautical Laboratories, California Institute of Technology, Pasadena, CA 91125, USA

and

STEPHEN WIGGINS

Department of Applied Mechanics, California Institute of Technology, Pasadena, CA 91125, USA

Abstract—Invariant manifolds can serve as a geometric template for the study of chaotic advection. In particular, global stable and unstable manifolds of invariant hyperbolic or normally hyperbolic sets form the boundaries of chaotic tangles in physical space, and these manifolds criss-cross one another to form an intricate network of *lobes* in the tangles. Manifold invariance implies that these lobes of fluid evolve from one to another in a well-defined manner, and it is in the context of *lobe dynamics* in the tangles that one exploits the invariant manifold template as a skeletal backbone for the study of transport, stretching, and mixing of fluid under chaotic advection, all of which are intimately connected. More concretely, in the study of fluid *transport*, one can use segments of the invariant manifolds to partition the irregular flow into regions of qualitatively different fluid motion, such as open flow versus closed flow, or different rolls, and then perform a lobe-dynamic study of fluid transport into, and out of, these regions (e.g. *entrainment* and *detrainment*) in a geometrically exact context, specified in terms of images and pre-images of a set of *turnstile lobes*. Mixing can be thought of as a consequence of barrier destruction, and transport across partially destroyed barriers can be studied in a lobe-dynamic context, providing a basic measure of mixing. A practical example of a transport calculation is the use of Melnikov theory to obtain analytical expressions for lobe areas, and hence a measure of *flux* in the mixing region, which allows for efficient searches through parameter space in order to enhance or diminish a mixing process. The *stretching* of fluid elements can be studied in the context of the lobes repeatedly stretching, folding, and wrapping around one another in an approximately self-similar manner. Such a picture can be viewed as a generalization of the horseshoe map paradigm for stretching in 2D chaotic tangles, which restricts its interest to a Cantor set near hyperbolic fixed points. A symbolic description of the evolution of lobe boundaries provides a framework for studying the global topology of, and mechanisms for, enhanced stretching in chaotic tangles. In particular, it is found that, though stretching under chaotic flows can be viewed in terms of products of weakly correlated events, there is a range of stretch, spatial, and temporal scales associated with these products, leading to a range of stretch processes. Invariant manifold geometry plays a central role in these stretch processes. For example, the invariant hyperbolic or normally hyperbolic sets have special relevance as engines of good stretching, the turnstile lobes play a fundamental role of re-orienting line elements between each successive pass by the hyperbolic sets, and intersections of stable and unstable manifolds act as partitions between segments of lobe boundaries that experience qualitatively different stretch histories. A notable consequence of the range of scales in the stretch processes is the range of statistics, and hence multifractal properties, associated with the high-stretch tails of finite-time Lyapunov exponent distributions, which has significant impact on interfacial evolution and striation width in the small-scale limit. Regions in physical space of vanishingly small initial measure associated with these tails are thus shown to be able to play a significant role under the flow. The *mixing* of passive scalars or vectors as a result of such stochastic effects as molecular diffusion or chemical reactions tends to wash out the structure associated with invariant manifolds. However, one can study this interplay between advection and mixing in the context of lobe evolution. In particular, mixing efficiency is seen to depend not only on the stretch experienced by lobe boundaries, but also on the thickness of the lobes, and their separation from other lobes. Though essentially all chaotic advection studies have been in the context of 2D time-periodic velocity fields, we show how an invariant manifold template can apply to quite general circumstances, such as quasiperiodic and indeed aperiodic time dependences, and 3D velocity fields. In all cases, the invariant manifold templates are seen to have physical reality, and the unstable manifolds act as a dominant structure in the regions of irregular flow.

1. INTRODUCTION TO INVARIANT MANIFOLD TEMPLATES AND LOBE DYNAMICS

Over the last decade, growing attention has been given to the phenomenon of chaotic advection, and its relation to transport, stretching, and mixing properties of fluid flows (e.g. see [1–3, 6–9, 11, 12, 18, 26–28, 35–37, 41, 44, 48–53, 63, 69–77, 80, 83, 87–90, 92, 101]). Far less attention has been devoted, however, to an invariant manifold description of chaotic fluid flow, even though invariant manifold theory lies at the heart of conventional dynamical systems analysis. We thus wish to present an exposition on the use of invariant manifolds as a geometric template for the study of chaotic advection of fluids.

The notion of such a geometric template for chaotic advection is best introduced in the concrete setting of an elementary example. Consider the two-dimensional fluid flow induced by a pair of equal and opposite point vortices. In the frame comoving with the vortex pair the flow is steady, and the equations of motion for fluid particle trajectories can be written in the time-independent, or autonomous, form

$$\begin{aligned} \dot{x} &= \frac{\partial \psi_0(x, y)}{\partial y} \\ \dot{y} &= -\frac{\partial \psi_0(x, y)}{\partial x}, \end{aligned} \tag{1}$$

where (x, y) denotes the spatial coordinates of the fluid and $\psi_0(x, y)$ is the streamfunction of the steady flow (see Appendix). Equation (1) defines a one-degree-of-freedom Hamiltonian flow, where the streamfunction plays the role of the Hamiltonian and y the role of conjugate momentum, and thus gives rise to an *integrable*, area-preserving flow, since an integral of the motion is provided by the Hamiltonian, and the velocity field is divergence-free. The level sets of the Hamiltonian, $\psi_0(x, y) = \text{constant}$, define streamlines (see Fig. 1), which are *invariant manifolds*—any point starting on a given manifold remains on it—and these curves thus act as a *template* for studying the fluid motion. Though one could of course perform a variety of numerical experiments to study the evolution of the fluid, the *phase portrait* offered by these invariant manifolds affords a basic understanding of the flow. Of particular note in the phase portrait are the *hyperbolic fixed points*, where the velocity field vanishes and the eigenvalues associated with the linearized flow about the fixed points have nonzero real components (these eigenvalues are of course real for our fluid example, and have opposite sign as a consequence of area preservation). These fixed points are connected by *heteroclinic manifolds* that define *separatrices*, two of which join to divide bounded from unbounded fluid motion (another class of separatrices are formed out of *homoclinic manifolds* when the separatrix forms a closed loop originating and terminating at the same fixed point). Inside the separatrix fluid particles move along closed streamlines (1-tori), and outside the separatrix fluid particles translate from right to left

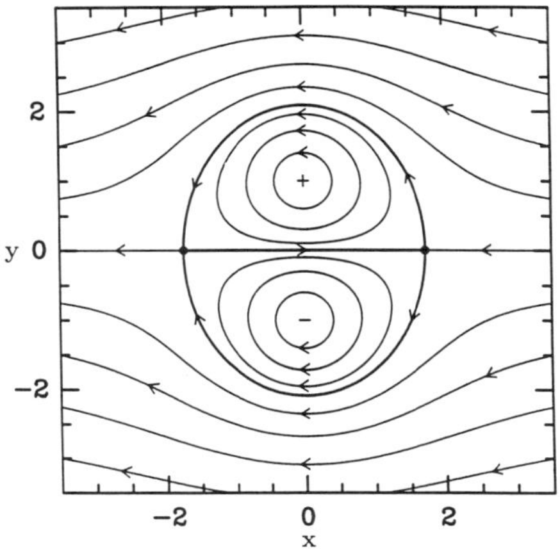

Fig. 1. The streamlines of the unperturbed vortex pair flow in the frame comoving with the vortex pair (the vortices are situated at $(x, y) = (0, \pm 1)$).

along open streamlines. No particle can cross the separatrix, which thus acts as a *complete barrier* to fluid transport. Integrable flows such as this are well-understood and pretty mundane. For example, here there is no *transport* of fluid across the separatrix, or any streamline for that matter, *stretching* is generically asymptotically linear in time, and if we choose to add molecular diffusion to the problem, the resulting *mixing* is poor.

Suppose we now add to the velocity field a time-dependent perturbing straining field of the form

$$\varepsilon\tilde{\psi}(x, y, \omega t + \theta_0; \varepsilon) = \varepsilon xy \sin(\omega t + \theta_0) + O(\varepsilon^2), \tag{2}$$

where ε is the perturbation amplitude, ω is the perturbation frequency, θ_0 is the initial $(t = 0)$ phase of the perturbation, and the $O(\varepsilon^2)$ term is understood to be time-periodic with frequency ω. This straining field oscillates periodically in time, and can arise, for example, if one embeds the vortex pair in a wavy-walled channel. The vortices respond to the external perturbation by oscillating periodically in time, and the net result is a time-periodic perturbation to the original velocity field, so that the equations governing particle motion now have the form

$$\begin{aligned} \dot{x} &= \frac{\partial \psi_0(x, y)}{\partial y} + \varepsilon \frac{\partial \tilde{\psi}(x, y, \omega t + \theta_0; \varepsilon)}{\partial y} \\ \dot{y} &= -\frac{\partial \psi_0(x, y)}{\partial x} - \varepsilon \frac{\partial \tilde{\psi}(x, y, \omega t + \theta_0; \varepsilon)}{\partial x}, \end{aligned} \tag{3}$$

where $\tilde{\psi}(x, y, \omega t + \theta_0; \varepsilon) = \tilde{\psi}(x, y, \omega t + \theta_0 + 2\pi; \varepsilon)$. In order to obtain invariant manifolds, we recast this time-dependent, or *nonautonomous*, 2D velocity field in 3D autonomous form (or, equivalently, as a $1\frac{1}{2}$-degree-of-freedom (d.o.f.) Hamiltonian system)

$$\begin{aligned} \dot{x} &= \frac{\partial \psi_0(x, y)}{\partial y} + \varepsilon \frac{\partial \tilde{\psi}(x, y, \theta; \varepsilon)}{\partial y} \\ \dot{y} &= -\frac{\partial \psi_0(x, y)}{\partial x} - \varepsilon \frac{\partial \tilde{\psi}(x, y, \theta; \varepsilon)}{\partial x} \\ \dot{\theta} &= \omega, \end{aligned} \tag{4}$$

where $\theta \in T^1$, the 1-torus. In order to simplify the underlying geometrical structure associated with the invariant manifolds, we define a Poincaré section associated with fixing the phase of the perturbation

$$\Sigma^{\theta_0} \equiv \{(x, y, \theta) | \theta = \theta_0\} \tag{5}$$

and the resulting Poincaré map generated by the autonomous flow (4) is given by

$$P_\varepsilon: \Sigma^{\theta_0} \to \Sigma^{\theta_0}$$
$$(x(0), y(0)) \to \left(x\left(\frac{2\pi}{\omega}\right), y\left(\frac{2\pi}{\omega}\right)\right). \tag{6}$$

Studying the autonomous flow via the 2D Poincaré map is equivalent to sampling fluid particle trajectories at time intervals equal to $2\pi/\omega$. Hence, one can think of the Poincaré section as the 2D fluid at any one of the sample times $t = (2\pi/\omega)n$, $n \in Z$, i.e. the fluid represented via a strobing process. It is thus by periodic sampling that one reduces the analysis of an ODE system to one of a 2D map, which considerably simplifies the analysis, as we shall see. This reduction is unique to time-periodic velocity fields, and it is precisely this reduction scheme that has made periodically forced systems exceedingly popular, subject to vast amounts of literature, and one of the canonical classes of dynamical systems out of which many of the initial paradigms and constructs of the field have been built.

The seemingly innocuous addition to the velocity field of a small time-periodic perturbation entails drastic changes in the fluid motion, as portrayed in the Poincaré section of Fig. 2. In particular, the resulting $1\frac{1}{2}$-d.o.f. Hamiltonian system is not integrable. In some regions of phase space regular motion gives way to *irregular* or *chaotic* motion, as indicated by the 'fuzzy' regions of the Poincaré section. These chaotic zones come in two classes.

(i) *Resonance bands*. When the forcing frequency ω is *commensurate* with (rationally related to) a closed level set rotation frequency of the unperturbed flow, the level set breaks up under the perturbation into a resonance band.

(ii) *Heteroclinic and homoclinic tangles*. For any forcing frequency ω the heteroclinic and homoclinic manifolds will break up into a heteroclinic and homoclinic tangle, respectively (in this example there are only heteroclinic tangles).

Figure 2 shows a substantial heteroclinic tangle associated with separatrix break-up, one that has in fact 'eaten up' a significant amount of nearby resonance bands. At the resolution of the portrayal in Fig. 2, resonance bands outside the heteroclinic tangle are too thin to decipher, although a more highly resolved study indeed shows their presence. The heteroclinic tangle is the overwhelmingly dominant chaotic region, and though both classes of chaotic zones are studied by similar techniques, we will be focusing our attention primarily on homoclinic/heteroclinic tangles. The 'fuzzy' portrayal of these chaotic zones is generated by the orbit under successive iterates of the Poincaré map of one or more initial conditions inside the chaotic zone. In the chaotic domains of resonance bands a single orbit will, instead of being confined to a curve, map to successive locations in a rather erratic fashion, filling up an available *area*. Orbits in the heteroclinic tangle exhibit similar behavior, except for the fact that they will asymptotically approach $x = -\infty$, since this particular heteroclinic tangle is unbounded; hence, one initializes more than one orbit to obtain a sufficiently dense portrayal of the 'fuzz' (see the caption for further details on this point).

These chaotic zones correspond to radically different fluid motion from the integrable flow, accompanied by greatly enhanced transport, stretching, and mixing processes. For example, the destruction of the unperturbed separatrix entails that open flow fluid can be

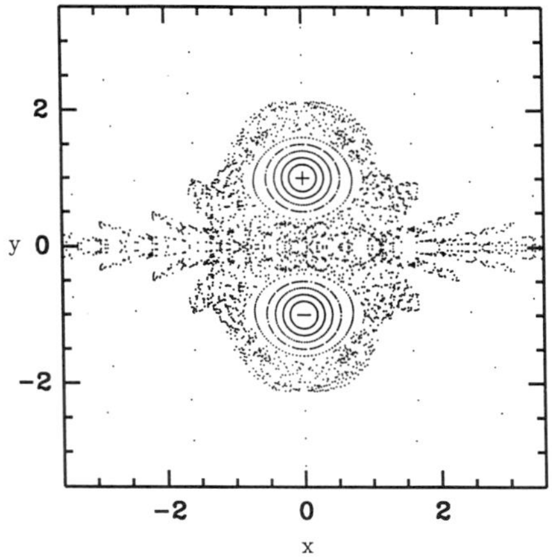

Fig. 2. The two-dimensional Poincaré section associated with the oscillating vortex pair flow exhibits regions of regular and irregular (i.e. chaotic) motion (system parameters are given later in the text, Figs 36 and 43). A substantial chaotic zone is associated with separatrix break-up. The 'fuzzy' portrayal of this zone is the result of evolution forwards and backwards in time of a set of fifty points initialized in a particular subset of the zone. This subset is the *entraining turnstile lobe* of the chaotic tangle, whose meaning will become quite clear in later sections.

entrained into the bounded region, and closed flow fluid can be *detrained from* the bounded region. The entrained fluid is stretched and folded in a violent manner until it is eventually detrained, and the result is a *stirring* process or (when we later add molecular diffusion) a *mixing process* in the heteroclinic tangle. This stirring process is inherently related to the chaotic advection of the fluid, as we shall discuss in great length. At the outset we make two observations about the appearance of chaotic regions under the presence of the perturbing straining field.

(i) The appearance of chaos is *immediate* for an *arbitrarily small* perturbing straining field. No *route to chaos* consisting of progressively complicated motions as one increases some perturbation parameter is needed here, although this latter scenario seems to have been more widely disseminated among the general scientific community.

(ii) The recognition that extremely simple, non-turbulent fluid velocity fields can evolve exceedingly complicated, indeed chaotic, fluid motion, a phenomenon commonly referred to as *chaotic advection*, has stirred much excitement and activity in the fluid mechanics community since the seminal paper by Aref [1]. Though such a phenomenon might seem paradoxical against the backdrop of complicated (i.e. turbulent) velocity fields, it is important to recognize that a central point of dynamical systems analysis since its inception has been that simple vector fields can evolve complicated dynamics.

Between the chaotic zones lie regions of regular motion, as indicated by the persistence of invariant tori in the Poincaré section of Fig. 2. Though trajectories may wander freely throughout individual chaotic zones, they may not pass through one of these invariant curves, which thus act as *complete barriers* between distinct regions of irregular motion. The entire phase portrait is thus filled with interspersed regions of regular and irregular motion, and there is a well-developed theory describing the intertwining of these regions. Most notably, the response of tori to external forcing is described by KAM (Kolmogorov–Arnold–Moser) theory (see Berry [17] for a nice introduction). There has also been considerable numerical study portraying these intertwining regular and irregular regions, offering rich and intricate pictures (for example, there is a rich set of *islands* of regular motion nested in the resonance bands). These numerical portrayals of chaos are extremely popular in the literature even to this day. A major limitation of such 'fuzzy' portrayals of the chaotic zones, however, is that they basically offer for chaotic zones just a *big mess*. One gains little understanding of the dynamics inside regions of chaos, e.g. of the stirring or mixing process in the oscillating vortex pair flow.

The principal interest of this paper is in the dynamics inside the chaotic regions, in particular, in the idea that underlying the chaotic regions is a *skeletal backbone* of *invariant manifolds* which act as a *geometric template* with which to study the dynamics, as shown in Fig. 3. These manifolds in our oscillating vortical flow are the global stable and unstable manifolds of hyperbolic fixed points of the perturbed Poincaré map, as will be discussed in more detail throughout the rest of the text. The basic idea to convey for introductory purposes is the following. It is an implicit function theorem result that the hyperbolic fixed points of the unperturbed Poincaré map persist under small perturbations of the velocity field, and the global stable and unstable manifolds of each hyperbolic fixed point are the curves in the Poincaré section Σ^{θ_0} representing the collection of fluid particles which asymptotically approach the fixed point forwards and backwards in time, respectively (the existence and smoothness of these manifolds is discussed later). By definition these manifolds are *invariant* under the flow, since a fluid particle with the above asymptotic property retains that property under the flow, and hence remains on the manifold. Under the unperturbed ($\varepsilon = 0$) flow, the stable and unstable manifolds of the two hyperbolic fixed points coincide to form the separatrices. The time-periodic perturbation, however, causes the stable and unstable manifolds to split apart, a phenomenon referred to as a *global bifurcation*. As a consequence of manifold invariance, a single transversal intersection of stable and unstable manifolds implies infinitely many of these intersections, and hence these invariant manifolds will criss-cross each other *ad infinitum* to form the boundary of a *chaotic tangle* and to form *lobes* in phase space (see Fig. 4). Figure 5 shows a finite number of these lobes more explicitly (the shading, labeling, and dynamics associated with these lobes is explained in more detail in subsequent sections). The lobes are divided into two classes, one associated with an entraining process (shaded lobes $\{E(m); m \in Z\}$), and one associated with a detraining process (unshaded lobes $\{D(m); m \in Z\}$), as will become clear in the upcoming transport discussion, where we will see that the parameter m specifies the number of iterates of the Poincaré map for the entraining or detraining

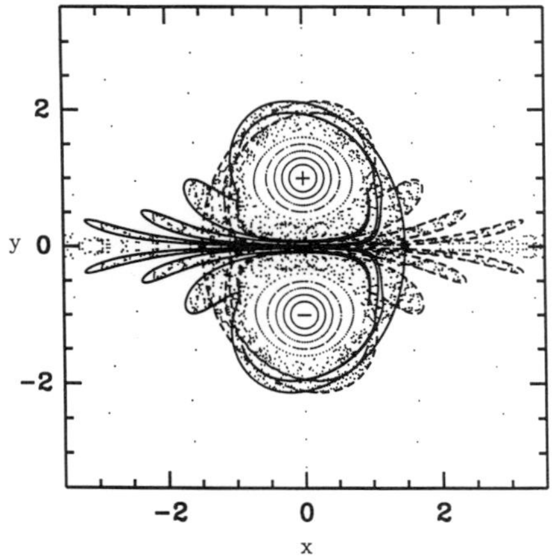

Fig. 3. Underlying the 'fuzzy' portrayal of the chaotic zones is a skeletal backbone formed out of intersecting global stable (dashed lines) and unstable (solid lines) manifolds of persisting hyperbolic fixed points. These invariant manifolds offer a *template* with which to study the chaotic motion. We are of course showing only finite segments of these manifolds.

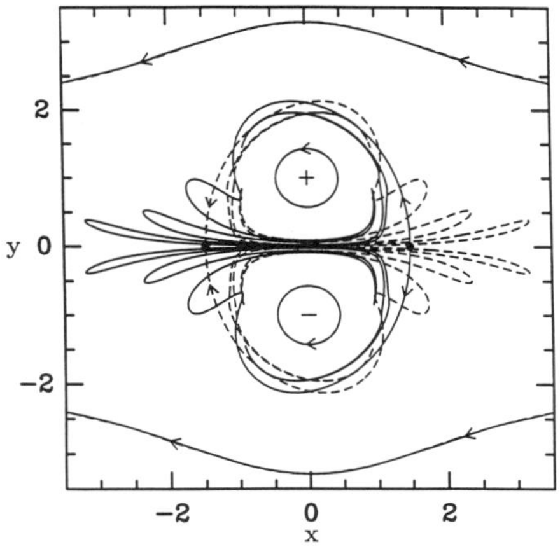

Fig. 4. The skeletal backbone without the 'fuzz'. The two solid dots denote hyperbolic fixed points of the Poincaré map which persist under the perturbation, the solid lines denote *invariant global unstable manifolds* of the right fixed point, and the dashed lines denote *invariant global stable manifolds* of the left fixed point. These manifolds criss-cross each other *ad infinitum* to form the boundary of an *invariant heteroclinic tangle*.

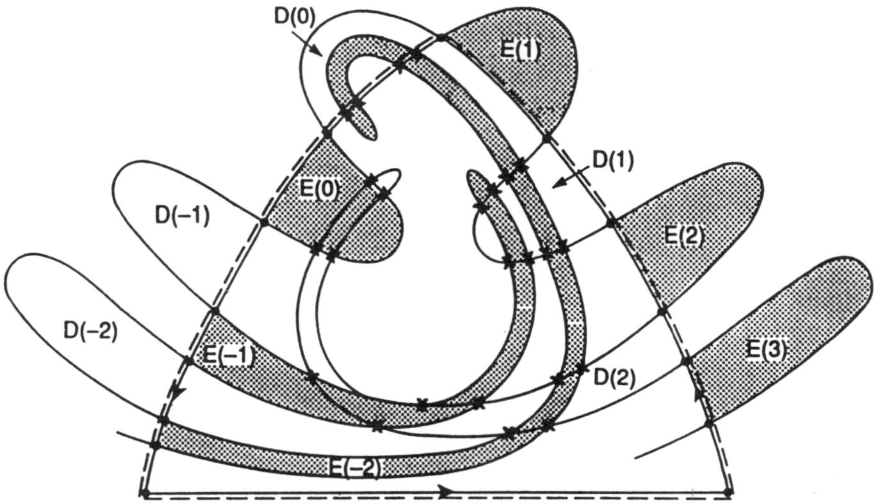

Fig. 5. The stable and unstable manifolds criss-cross each other *ad infinitum* to form an *invariant lobe structure*. A finite number of lobes of the upper heteroclinic tangle are portrayed here. The lobes are labelled according to the dynamics they undergo, as will be explained in detail in later sections: the shaded lobes are those associated with *entrainment*, the unshaded lobes with *detrainment*, and under the Poincaré map they evolve according to $E(m) \to E(m-1)$ and $D(m) \to D(m-1)$.

process to occur. It is a consequence of the invariance of the stable and unstable manifolds that these lobes map from one to another under iterates of the Poincaré map. For this particular flow the map is described by

$$P_\varepsilon(E(m)) = E(m-1)$$
$$P_\varepsilon(D(m)) = D(m-1) \qquad m \in Z \tag{7}$$

(see Fig. 6). It will be within this context of *lobe dynamics* within the invariant chaotic tangle that we exploit the geometric template of invariant manifolds to study features of

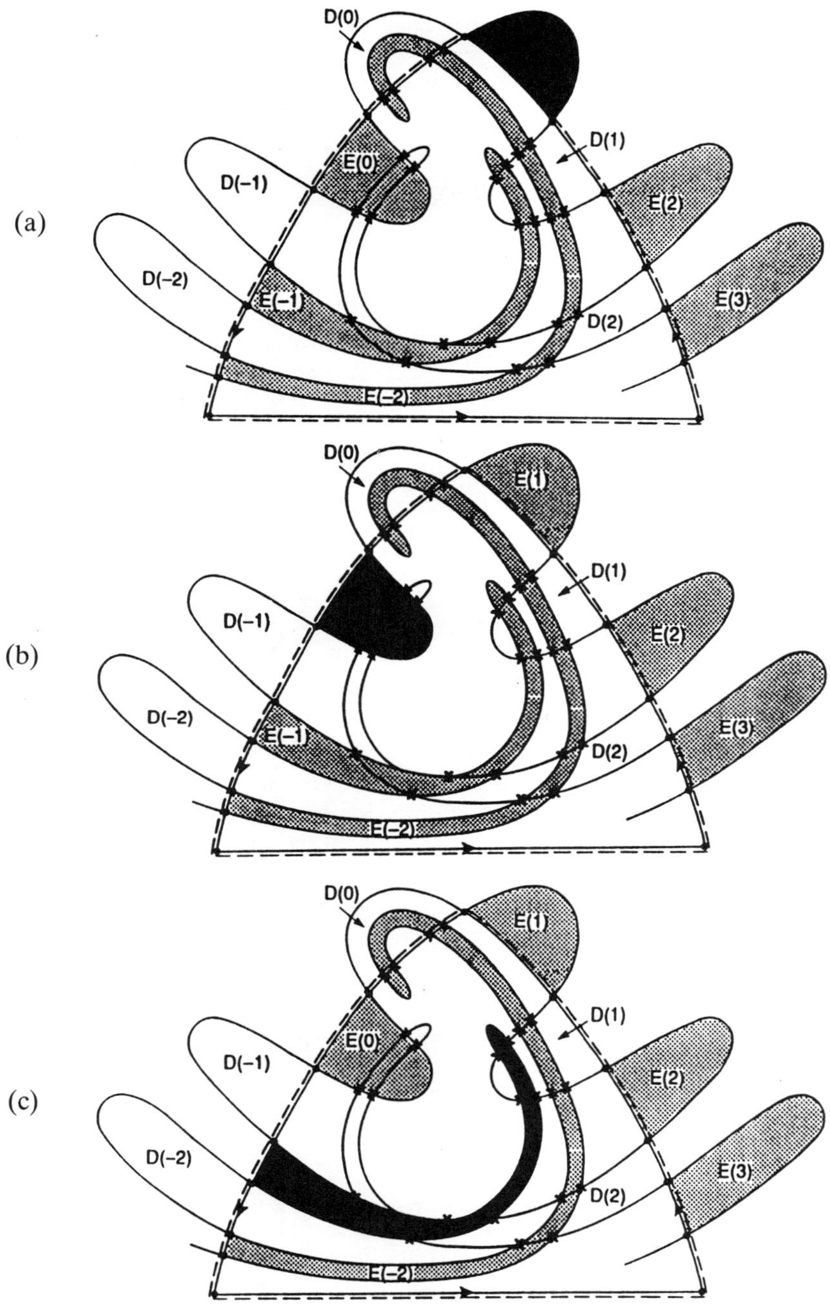

Fig. 6. A *lobe* (shown in black) mapping within the invariant lobe structure. The lobe is portrayed at (a) $t = 0$, (b) $t = 2\pi/\omega$, and (c) $t = 2(2\pi/\omega)$.

the dynamics within the tangle associated with *transport*, *stretching*, and *mixing* of fluid under chaotic advection.

We note that the existence of global stable and unstable manifolds in chaotic tangles dates all the way back to Poincaré at the turn of the century [78]; however, using these manifolds as a template for systematic study of chaotic dynamics and to study basic nonlinear physical processes is a relatively recent pursuit, active primarily since the mid-eighties. Indeed, the first such exploitation of invariant manifold theory in the context of chaotic advection appears in the little-known article by Holmes [41] on chaotic particle paths in time-periodic three-dimensional swirling fluid flows. Both resonance bands and homoclinic/heteroclinic tangles can in principal be studied in this context, though the tangles in resonance bands can be geometrically more complicated. Our attention will focus on homoclinic/heteroclinic tangles. In particular, we focus on two oscillating vortical flows—the opposite-signed vortex pair flow we just introduced (henceforth referred to as the *open flow*)—and a same-signed vortex pair flow illustrated in Fig. 7, and described in the Appendix (henceforth referred to as the *closed flow*). We note that both oscillating vortex pair flows consider a chaotic tangle in a *near-integrable* setting, meaning we add a small perturbation to the velocity field of an initially integrable flow. Our presentation will be described primarily in this setting, since it simplifies the description of the dynamics and allows for additional results, such as perturbative analysis of manifold geometry via *Melnikov theory* (to be discussed later). *We stress, however, the idea of a geometric template formed out of invariant manifolds does not necessitate small perturbations, an unperturbed phase-portrait that contains fixed points, or even a perturbative setting.*

Having introduced the notion of an invariant manifold template and its exploitation via lobe dynamics, we address in Section 2 the range of applicability of these templates. Though the overwhelming attention in chaotic advection has been devoted to 2D time-periodic velocity fields, we stress that invariant manifold theory applies to a much wider class of velocity fields, including 2D velocity fields with multi-frequency and aperiodic time dependences, as well as 3D velocity fields. Section 3 impresses upon the reader the physical reality and dominant role of these invariant manifolds in chaotic flows. In particular, passive scalars and vectors are rapidly stretched out in one direction and contracted in another such that they tend to be stretched out along unstable manifolds, which thus play a role in area- and volume-preserving flows analogous to attractors in dissipative flows. The remaining sections then turn attention towards exploiting the invariant manifold templates for practical understanding of irregular fluid flow. Section 4 introduces the analytical technique of Melnikov theory for obtaining a scalar measurement of stable and unstable manifold separation resulting from the velocity field perturbation.

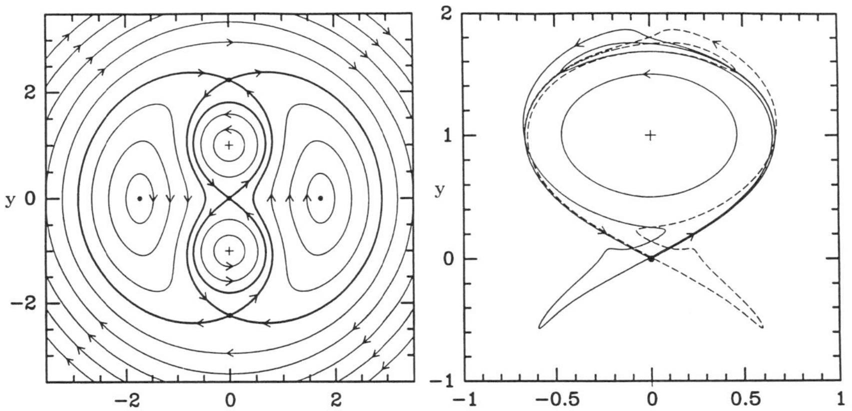

(a) (b)

Fig. 7. (a) The streamlines of the unperturbed closed flow in the frame corotating with the vortex pair (the vortices are situated at $(x, y) = (0, \pm 1)$). (b) The *invariant homoclinic tangle* associated with break-up of the upper homoclinic loop for the perturbed Poincaré map; in particular, the global stable (dashed) and unstable (solid) manifolds of the persisting hyperbolic fixed point, which remains at the origin.

Such an analytical tool will have great practical value for uncovering the geometry associated with invariant manifold tangling, and measuring basic physical properties of the flow. Section 5 studies fluid *transport* in a geometrically exact setting of invariant manifold constructs such as *partitions*, *partial barriers*, *turnstiles*, and *images and pre-images of the turnstiles*. More concretely, one can use segments of the invariant manifolds to partition the irregular flow into regions of qualitatively different fluid motion, such as open flow versus closed flow, or different rolls, and then perform a lobe-dynamic study of fluid transport into and out of the regions (e.g. *entrainment* and *detrainment*) in a geometrically exact context, specified in terms of images and pre-images of turnstile lobes. A practical example of a transport calculation is the use of Melnikov theory to obtain analytical expressions for lobe areas, and hence a measure of *flux* in a mixing region, which allows for efficient searches through parameter space in order to enhance or diminish a mixing process. Section 6 studies *stretching* of fluid elements in chaotic tangles, with the principal theme being the study of lobe evolution as a generalization of the horseshoe map paradigm in order to obtain a more global understanding of stretching in chaotic tangles. Stretch processes, stretch rates, and stretch statistics are studied in the context of underlying invariant manifold geometry, accompanied by detailed numerical computation for illustration. Section 7 closes with a discussion of *mixing* resulting from the introduction of such stochastic processes as molecular diffusion or heat transfer, and its relation to some aspects of stretching and transport. The Appendix gives the equations of motion for our two oscillatory vortical flows, as well as the procedure for numerical computation of invariant manifolds.

We emphasize that, though the dynamical systems community has given great attention to the study of *asymptotic behavior*, many of the above physical processes are indeed *finite-time* processes (e.g. for a mixing experiment one may be interested in 20–40 perturbation periods). Hence, asymptotics and long-time behavior may have little relevance to the problem, and much of our study will be explicitly concerned with finite-time or transient phenomena, relatively virgin territory from a dynamical systems perspective. In the context of stretching, for example, rather than be interested in conventional Lyapunov exponents, we are instead interested in the finite-time analogue, referred to as *finite-time Lyapunov exponents*.

We lastly would like to make clear the intent of our exposition. Though lengthy in presentation, our discussion has not the slightest pretense of a comprehensive treatment. Indeed, quite the contrary, our goal is merely to introduce and encourage the utility of invariant manifold theory in the study of chaotic advection, and we leave many basic avenues unpursued. Further, we certainly do not wish to take the imperialistic view that an invariant manifold approach, though steeped in tradition of mainstream dynamical systems theory, is the best way to study chaotic advection. Nonlinear physical phenomena are best studied by a variety of approaches, each of which will have some merit, but will also fall short of a 'complete solution' to the problem. A common complaint by engineers and physicists of an invariant manifold approach is that it can sometimes degenerate into a significant amount of geometry with few detailed practical results. Our philosophy, however, as is conventionally held in the dynamical systems community, is that a primary goal for nonlinear dynamics is to use phase space geometry to obtain a qualitative understanding of the dynamics, formulate paradigms of behavior, initiate partial answers to certain features of the nonlinear dynamics, and provide a setting for intelligently chosen numerical investigations.

For other references directly related to this work, please see [7–14, 18, 39, 41, 43, 47, 51, 66, 82–84, 87–90, 102–106]. In particular, for extensions past time-periodic velocity fields see [8–11], for studies of Melnikov theory and manifold geometry see [8–14, 43, 47,

51, 66, 83, 102–106], for the development of a lobe-dynamic phase space transport formalism see [7–14, 18, 51, 82–84, 104–106], and for studies of stretching and mixing related to lobe boundary evolution see [7, 11, 12].

2. APPLICABILITY TO INCREASINGLY COMPLICATED CLASSES OF VELOCITY FIELDS

We have introduced the notion of an invariant manifold template in the context of a 2D time-periodic velocity field. As we have mentioned, fluid flows generated by these velocity fields enjoy the convenient property of being reducible to a 2D map, upon which many of the basic constructs and paradigms of dynamical systems theory are based—the Smale horseshoe map, hyperbolic fixed points and their 1D stable and unstable manifolds, KAM tori as complete barriers in phase space, partial barriers and turnstiles, and so on. Consequently, essentially all analysis of chaotic advection to date has been in the context of this class of velocity fields. Though the phenomenon of simple fluid velocity fields generating complicated advection properties is of great interest in its own right, the obvious reality remains that most physically relevant fluid flows have quite complicated, indeed turbulent, velocity fields. There is thus strong motivation to extend the analysis of chaotic advection towards more complicated scenarios. Thus, prior to exploiting these invariant manifold templates to gain understanding of transport, stretching, and mixing properties of chaotic fluid flow, in this section we discuss the relatively robust nature of invariant manifold theory, and its application to such general circumstances as multi-frequency, and indeed aperiodic, time dependences, as well as 3D velocity fields. We nevertheless end the section with a warning about the limitations on the applicability of invariant manifold theory to the study of fluid flow.

2.1. *Two-dimensional single-frequency velocity fields*

Our introduction provided most of the essentials for our invariant manifold description of 2D time-periodic velocity fields, and we add here only a few remarks to set the stage for later generalizations in this section. For this class of velocity fields we can write the equations governing fluid particle motion as

$$\begin{aligned} \dot{x} &= \frac{\partial \psi}{\partial y}(x, y, \omega t + \theta_0) \\ \dot{y} &= -\frac{\partial \psi}{\partial x}(x, y, \omega t + \theta_0), \end{aligned} \tag{8}$$

where ψ is 2π-periodic in $\omega t + \theta_0$. Equation (3) gives these equations in the near-integrable context of an $O(\varepsilon)$ time-periodic perturbation of a time-independent velocity field. Recall we then rewrote this 2D time-periodic velocity field in 3D autonomous form in order to obtain invariant manifolds, followed by reduction to a 2D Poincaré section Σ^{θ_0} in order to simplify the underlying structure associated with the invariant manifolds, as illustrated in Fig. 8(a, b). Such a reformulation of the problem will be essential to our attack on more general classes of velocity fields. The starting point for an invariant manifold description of global dynamics is typically some sort of *local* existence and smoothness result. For a 2D map, such a local result is in the context of a hyperbolic fixed point and its local 1D stable and unstable manifolds. As mentioned earlier, in the near-integrable context of an $O(\varepsilon)$ perturbation of a map containing a hyperbolic fixed point, the peristence of such a fixed point, $q_{\varepsilon=0}$, is guaranteed for small perturbations by a straightforward implicit function theorem argument (in practice persistence can be found for quite sizable perturbations).

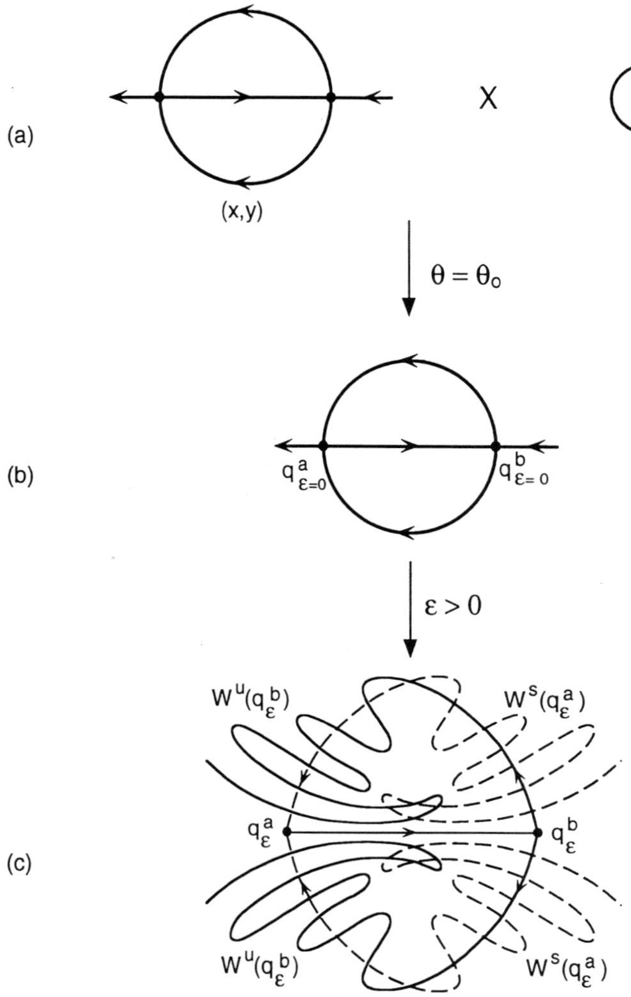

Fig. 8. The unperturbed separatrices of the single-frequency open flow in (a) the 3D autonomous representation, and (b) the 2D Poincaré section Σ^{θ_0}. (c) The chaotic tangle of the perturbed Poincaré map, whose boundaries are formed by global stable (dashed) and unstable (solid) manifolds of persisting hyperbolic fixed points.

Though it is convenient to pursue an invariant manifold study in this near-integrable context, we stress that the existence of a hyperbolic fixed point need not be tied to persistence under an $O(\varepsilon)$ perturbation of the map (e.g. an invariant manifold analysis can apply to hyperbolic fixed points created in a resonance, or that exist in a setting which is not near-integrable). The existence of *local* 1D stable and unstable manifolds of a hyperbolic fixed point, as smooth as the map and denoted by $W^s_{loc}(q_\varepsilon)$ and $W^u_{loc}(q_\varepsilon)$, respectively, is guaranteed by the stable manifold theorem (e.g. see [39]). In the previous section we provided a verbal description of stable and unstable manifolds as the collection of points which asymptotically approach a hyperbolic fixed point forwards and backwards in time, respectively. More formally we write

$$\begin{aligned} W^s_{loc}(q_\varepsilon) &\equiv \{(x, y) \in U \mid P^n_\varepsilon((x, y)) \to q_\varepsilon \text{ as } n \to \infty,\ P^n_\varepsilon((x, y)) \in U,\ \forall n \geq 0\} \\ W^u_{loc}(q_\varepsilon) &\equiv \{(x, y) \in U \mid P^{-n}_\varepsilon((x, y)) \to q_\varepsilon \text{ as } n \to \infty,\ P^{-n}_\varepsilon((x, y)) \in U,\ \forall n \geq 0\}, \end{aligned} \tag{9}$$

where $U \in \mathbb{R}^2$ is a small neighborhood of the hyperbolic fixed point q_ε. *Global* stable and unstable manifolds (again as smooth as the map) are continued from the local ones via evolution backwards and forwards in time, respectively, under the map:

$$W^{s}(q_{\varepsilon}) = \bigcup_{n\leq 0} P_{\varepsilon}^{n}(W_{\text{loc}}^{s}(q_{\varepsilon}))$$
$$W^{u}(q_{\varepsilon}) = \bigcup_{n\geq 0} P_{\varepsilon}^{n}(W_{\text{loc}}^{u}(q_{\varepsilon})). \tag{10}$$

These global manifolds possess three properties that are the key ingredients to their exploitation in a global study of the dynamics.

(i) *Critical asymptotic behavior*. Points that lie on these manifolds are associated with a critical asymptotic behavior that plays a fundamental role in the study of phase space transport and stretching phenomena.

(ii) *Invariance*. By the definition of global stable and unstable manifolds, a point which starts on either of these manifolds will remain on that manifold under iterates of the map (or alternatively under the flow) both forwards and backwards in time. In particular, a point which lies on *both* stable and unstable manifolds, thus corresponding to a point of manifold intersection, must remain on both manifolds and hence retain its identity as a point of stable and unstable manifold intersection.

(iii) *Non-self-intersecting property*. Though stable and unstable manifolds (of the same hyperbolic fixed point or of different hyperbolic fixed points) can intersect one another, each manifold cannot intersect itself, since this would entail a violation of uniqueness of solutions. The non-self-intersecting property of stable and unstable manifolds in tangles implies that they must wrap around themselves in a highly constrained manner, which plays an essential role in understanding the global geometry of the invariant manifolds and the tangle boundaries they define.

As discussed in the introduction and as follows from the manifold invariance property (ii), a single point of intersection of stable and unstable manifolds implies infinitely many of these intersections, entailing that the manifolds criss-cross *ad infinitum* to form the boundary of a *chaotic tangle*, as introduced in Section 1 and discussed more carefully here. We have mentioned how criss-crossing manifolds form *lobes*, and to firm up what is meant by a lobe, we need to distinguish between two types of manifold intersections, a distinction that plays a fundamental role in a variety of features of tangle dynamics. Let $\bar{q}$ be a point of intersection between stable and unstable manifolds of a hyperbolic fixed point q_{ε}, and let $S[q_{\varepsilon}, \bar{q}]$ and $U[q_{\varepsilon}, \bar{q}]$ denote the segment of the stable and unstable manifold, respectively, joining q_{ε} and $\bar{q}$. Then if $S[q_{\varepsilon}, \bar{q}]$ and $U[q_{\varepsilon}, \bar{q}]$ intersect *only* at $\bar{q}$ (discounting q_{ε}) we refer to $\bar{q}$ as a *primary intersection point* (PIP); otherwise we refer to $\bar{q}$ as a *secondary intersection point* (SIP) (see Fig. 9 for an illustration of the distinction). A lobe is then defined as the 2D region enclosed by segments of stable and unstable manifolds joining adjacent PIPs. More formally, if $\bar{q}_a$ and $\bar{q}_b$ denote *adjacent* PIPs (meaning $S[\bar{q}_a, \bar{q}_b]$ and $U[\bar{q}_a, \bar{q}_b]$ contain no other PIPs than $\bar{q}_a$ and $\bar{q}_b$), then the boundary of a lobe $\mathcal{L}$ is defined by

$$\partial\mathcal{L} \equiv S[\bar{q}_a, \bar{q}_b] \cup U[\bar{q}_a, \bar{q}_b]. \tag{11}$$

In this manner one can conceptualize the tangle as an invariant lobe structure consisting of a bi-infinite sequence of lobes, as shown in Figs 8(c) and 9. It is a consequence of manifold invariance that these lobes map from one to another under iterates of the Poincaré map, and we thus exploit the invariant manifold template in the context of a *lobe dynamics* within the tangle, specified by a map such as equation (7) in the context of our open flow. Such a specification of lobe dynamics is best offered in the context of a transport-based

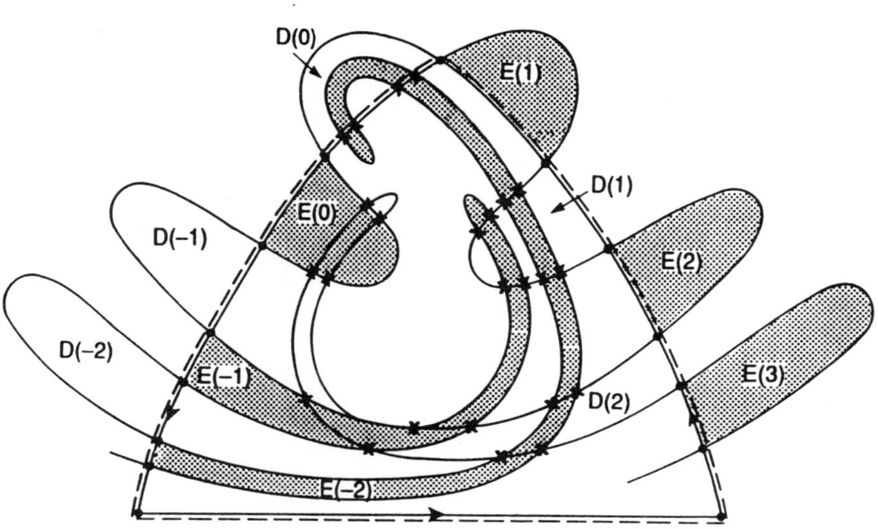

Fig. 9. Illustrating the difference between *primary intersection points* (PIPs, marked by dots) and *secondary intersection points* (SIPs, marked by crosses). In this context it should be clear the notion of a lobe boundary formed by segments of stable and unstable manifolds joining adjacent PIPs.

description, whose details are thus deferred to Section 5. We note at the outset, however, that the lobe dynamics given by equation (7) for the open flow, involving *entraining lobes* $\{E(m)|m \in Z\}$ and *detraining lobes* $\{D(m)|m \in Z\}$, applies to the special case of transport into, and out of, a particular region. In general, transport can occur between any number of distinct regions (e.g. several rolls in Rayleigh–Bénard convection, as studied in Camassa and Wiggins [18]), and in this general case the lobe dynamics will be described by

$$P_\varepsilon(L_{ij}(m)) = L_{ij}(m - 1), \tag{12}$$

where $L_{ij}(m)$ will denote the lobe that maps from the ith to the jth region upon the mth iterate of P_ε. Once the lobe-dynamic description is made, manifold properties (i)–(iii) will be essential to translating lobe dynamics into practical understanding of chaotic flows. For example, the property of a critical asymptotic behavior (i) will allow one to use segments of invariant manifolds to partition physical space into regions of qualitatively different fluid motion (e.g. bounded versus unbounded, or different rolls) and study in a *geometrically exact* context transitions across the partitions (e.g. entrainment and detrainment, or roll-to-roll transport). The non-self-intersecting property (iii), and the consequent highly constrained manner with which the lobes must wrap around themselves, will provide the basis for a symbolic description of lobe boundary evolution in order to study global stretch processes in chaotic tangles.

2.2. *Two-dimensional multi-frequency velocity fields*

Let us now consider a multi-frequency generalization of 2D time-periodic velocity fields. In particular, consider a streamfunction of the form

$$\psi(x, y, \omega_1 t + \theta_{1_0}, \omega_2 t + \theta_{2_0}, \ldots, \omega_l t + \theta_{l_0}), \tag{13}$$

where ψ is 2π-periodic in each of $\omega_i t + \theta_{i_0}$, $i = 1, 2, \ldots, l$, and hence exhibits a quasiperiodic dependence on time. Let us again focus on a near-integrable flow, and restrict our interest to the two-frequency problem. We can thus write the equations governing fluid particle motion as

$$\begin{aligned}\dot{x} &= \frac{\partial \psi_0}{\partial y}(x, y) + \varepsilon \frac{\partial \tilde{\psi}}{\partial y}(x, y, \omega_1 t + \theta_{1_0}, \omega_2 t + \theta_{2_0}; \varepsilon) \\ \dot{y} &= -\frac{\partial \psi_0}{\partial x}(x, y) - \varepsilon \frac{\partial \tilde{\psi}}{\partial x}(x, y, \omega_1 t + \theta_{1_0}, \omega_2 t + \theta_{2_0}; \varepsilon),\end{aligned} \tag{14}$$

where $\tilde{\psi}$ is 2π-periodic in each of $\omega_1 t + \theta_{1_0}$ and $\omega_2 t + \theta_{2_0}$. Our restriction to the two-frequency problem is merely for ease of presentation and concreteness of discussion, and we refer the reader to Beigie *et al.* [8–10] for a detailed discussion of the l-frequency problem, $l \geqslant 2$. For single-frequency velocity fields, we have seen that time-periodic sampling of fluid particle trajectories reduces the analysis to a 2D map, which simplifies the underlying geometrical structure with which to study the fluid flow. The addition of even one more frequency to the velocity field time dependence requires a fundamental departure from the time-periodic analysis, however, since any discrete-time sampling procedure will in general reduce the analysis now to a *bi-infinite sequence of nonautonomous 2D maps*. In particular, if we sample fluid particle trajectories at, say, the second frequency ω_2, then the evolution of fluid from time $t = (2\pi/\omega_2)n$ to time $t = (2\pi/\omega_2)(n + 1)$ defines a map on the plane

$$T_\varepsilon(\cdot\,; n):\quad \mathbb{R}^2 \mapsto \mathbb{R}^2$$

$$\left(x\left(\frac{2\pi}{\omega_2}n\right), y\left(\frac{2\pi}{\omega_2}n\right)\right) \mapsto \left(x\left(\frac{2\pi}{\omega_2}(n + 1)\right), y\left(\frac{2\pi}{\omega_2}(n + 1)\right)\right). \tag{15}$$

The goal then is to study the evolution of the fluid under the bi-infinite sequence of 2D nonautonomous maps $\{T_\varepsilon(\cdot\,; n); n \in Z\}$. To perform such a study in the context of invariant manifold theory, we use the same procedure employed in the time-periodic case of recasting the equations of motion in autonomous form in order to obtain invariant manifolds, followed by reduction via a Poincaré section in order to simplify the geometrical structure associated with the invariant manifolds. We then use this invariant template to define for each time $t = (2\pi/\omega_2)n$ a *time-dependent* geometric template for the 2D fluid. Hence, an invariant manifold template will govern the evolution of the 2D fluid, but this template will be seen to exist in a higher-dimensional setting. The two-frequency system rewritten in autonomous form is given by

$$\begin{aligned}\dot{x} &= \frac{\partial \psi_0}{\partial y}(x, y) + \varepsilon \frac{\partial \tilde{\psi}}{\partial y}(x, y, \theta_1, \theta_2; \varepsilon) \\ \dot{y} &= -\frac{\partial \psi_0}{\partial x}(x, y) - \varepsilon \frac{\partial \tilde{\psi}}{\partial x}(x, y, \theta_1, \theta_2; \varepsilon) \\ \dot{\theta}_1 &= \omega_1 \\ \dot{\theta}_2 &= \omega_2,\end{aligned} \tag{16}$$

where $(\theta_1, \theta_2) \in T^2$, the two-torus, and (ω_1, ω_2) are referred to as the *fundamental frequencies*. The autonomous system phase space, $\mathbb{R}^2 \times T^2$, is 4D, with the added two dimensions associated with the phases of the perturbation, as shown in Fig. 10(a). In the spirit of sampling fluid particles at the second frequency ω_2, we then define a Poincaré section associated with fixing the second phase of the perturbation

$$\Sigma^{\theta_{2_0}} \equiv \{(x, y, \theta_1, \theta_2) | \theta_2 = \theta_{2_0}\} \tag{17}$$

as shown in Fig. 10(b), and the associated Poincaré map generated by the autonomous flow

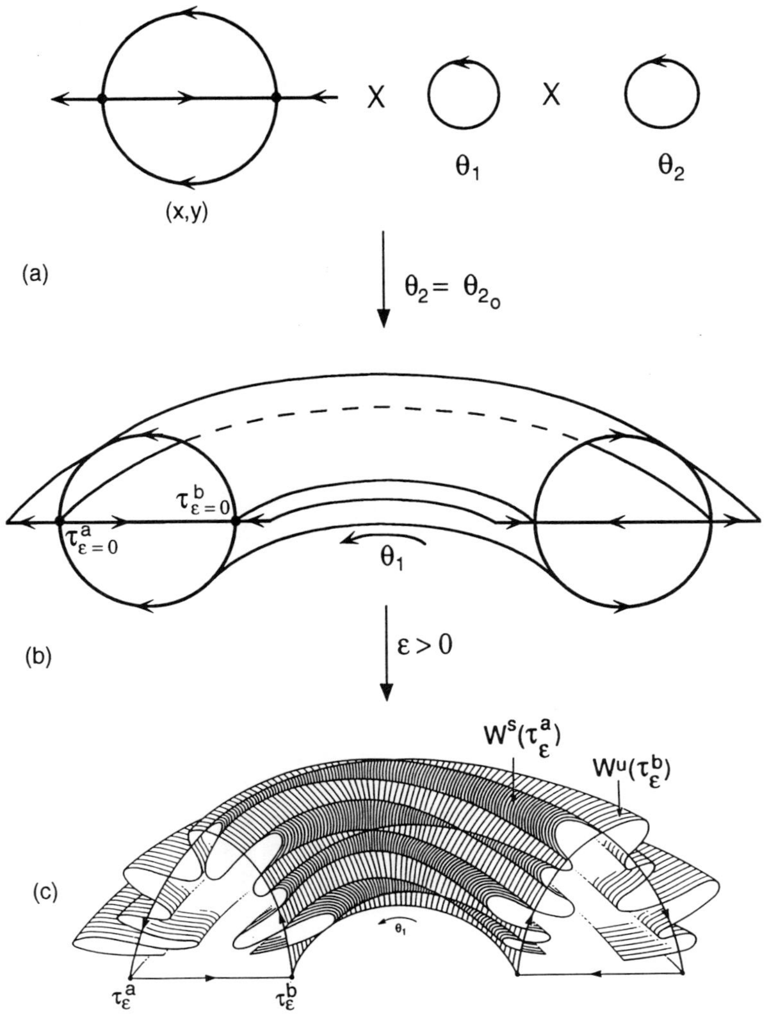

Fig. 10. The unperturbed separatrices of the two-frequency open flow in (a) the 4D autonomous representation, and (b) the 3D Poincaré section $\Sigma^{\theta_{2_0}}$. The 1-tori $\tau^a_{\varepsilon=0}$ and $\tau^b_{\varepsilon=0}$ denote *normally hyperbolic invariant manifolds*. (c) The chaotic tangle of the perturbed Poincaré map, whose boundaries are formed by global stable and unstable manifolds, $W^s(\tau^a_\varepsilon)$ and $W^u(\tau^b_\varepsilon)$, of persisting normally hyperbolic invariant manifolds, τ^a_ε and τ^b_ε (here we show just the upper chaotic tangle). As will be the case for all two-frequency portrayals, we show a half-cut of the full geometric objects in 3D.

(16) is given by

$$P_\varepsilon\colon\ \Sigma^{\theta_{2_0}} \mapsto \Sigma^{\theta_{2_0}}$$

$$(x(0),\ y(0),\ \theta_{1_0}) \mapsto \left(x\left(\frac{2\pi}{\omega_2}\right),\ y\left(\frac{2\omega}{\omega_2}\right),\ \theta_{1_0} + 2\pi\frac{\omega_1}{\omega_2}\right). \tag{18}$$

Studying the autonomous flow (16) via this 3D Poincaré map is equivalent to sampling the trajectories of (16) at time intervals equal to $(2\pi/\omega_2)$. Though our ultimate goal is to relate this 3D Poincaré map back to our 2D fluid, for the moment we remain suspended in the 3D Poincaré section to obtain invariant manifolds.

Let us first examine the phase space structure associated with the unperturbed Poincaré map $P_{\varepsilon=0}$, considering for concreteness the open flow (to be subjected now to a two-frequency perturbation, as described in the Appendix). The unperturbed phase portrait in the 3D Poincaré section $\Sigma^{\theta_{2_0}}$ is a cross product of the 2D phase portrait of the unperturbed fluid flow with a 1-torus, as shown in Fig. 10(b). In this higher-dimensional setting we encounter a fundamental generalization in the geometric constructs upon which invariant manifold theory is based. The cross product of a hyperbolic fixed point $q_{\varepsilon=0}$ with a 1-torus defines a *normally hyperbolic invariant 1-torus*

$$\tau_{\varepsilon=0} \equiv \{(x, y, \theta_1) | (x, y) = q_{\varepsilon=0}\}, \tag{19}$$

as shown in Fig. 10(b). The term *normal hyperbolicity* refers to the property that linear expansion and contraction rates *normal* to the 1-torus dominate over linear expansion and contraction rates *along* the 1-torus. This asymptotic rate comparison is made precise in the context of *generalized Lyapunov-type numbers* (e.g. see Fenichel [34], and Wiggins [103]). The 1-tori $\tau^a_{\varepsilon=0}$ and $\tau^b_{\varepsilon=0}$ are easily seen to satisfy the property of normal hyperbolicity from the fact that the fixed point is hyperbolic, while the phase variable $\theta_1(t)$ evolves only linearly in time. In a similar manner, the cross product of the 1D separatrix with a 1-torus defines a 2D separatrix in $\Sigma^{\theta_{2_0}}$, as shown in Fig. 10(b), and the cross product of local 1D stable and unstable manifolds with a 1-torus defines 2D local stable and unstable manifolds, denoted by $W^{\mathrm{s}}_{\mathrm{loc}}(\tau_{\varepsilon=0})$ and $W^{\mathrm{u}}_{\mathrm{loc}}(\tau_{\varepsilon=0})$, respectively (where we note that $\tau_{\varepsilon=0}$ can represent $\tau^a_{\varepsilon=0}$ or $\tau^b_{\varepsilon=0}$). The relevant feature of the normally hyperbolic property of $\tau_{\varepsilon=0}$ is the guarantee of a persistence and smoothness result analogous to that of hyperbolic fixed points of a 2D map. In particular, if $P_{\varepsilon=0}$ is a C^r map that contains a normally hyperbolic invariant 1-torus, $\tau_{\varepsilon=0}$, then for ε sufficiently small P_ε possesses a C^r normally hyperbolic invariant 1-torus, τ_ε, whose local 2D C^r stable and unstable manifolds, $W^{\mathrm{s}}_{\mathrm{loc}}(\tau_\varepsilon)$ and $W^{\mathrm{u}}_{\mathrm{loc}}(\tau_\varepsilon)$, are C^r ε-close to $W^{\mathrm{s}}_{\mathrm{loc}}(\tau_{\varepsilon=0})$ and $W^{\mathrm{u}}_{\mathrm{loc}}(\tau_{\varepsilon=0})$, respectively (see Fenichel [34] and Wiggins [103]). These local stable and unstable manifolds are continued to global manifolds by evolving them backwards and forwards in time, respectively, under the Poincaré map:

$$\begin{aligned} W^{\mathrm{s}}(\tau_\varepsilon) &= \bigcup_{n\leqslant 0} P^n_\varepsilon(W^{\mathrm{s}}_{\mathrm{loc}}(\tau_\varepsilon)) \\ W^{\mathrm{u}}(\tau_\varepsilon) &= \bigcup_{n\geqslant 0} P^n_\varepsilon(W^{\mathrm{u}}_{\mathrm{loc}}(\tau_\varepsilon)), \end{aligned} \tag{20}$$

and these 2D invariant manifolds can criss-cross one another *ad infinitum* to form the boundary of 3D chaotic tangles in $\Sigma^{\theta_{2_0}}$, as explained in detail in [8–10], and portrayed in Fig. 10(c) along with a more complete portrayal in Fig. 11. Indeed for the l-frequency problem one has, in an $(l+1)$-dimensional Poincaré section $\Sigma^{\theta_{l_0}}$, l-dimensional manifolds criss-crossing one another to form the boundaries of $(l+1)$-dimensional tangles [8–10]. Since the invariant manifolds now generically intersect in $(l-1)$-dimensional manifolds, rather than isolated points, the tangle boundaries and the regions they enclose can take on a far richer geometry than in the time-periodic case, and the description of tangle dynamics is correspondingly more complicated. Though one can always uncover global geometry via numerical computation of the invariant manifolds (as described in the Appendix, and indeed performed in later sections), such an approach quickly becomes prohibitively computationally intensive in the higher-dimensional setting of multi-frequency or 3D velocity fields, especially in the presence of several system parameters to be searched through. It is for these types of flows that the analytical technique of Melnikov theory and its recent generalizations [14, 103] for uncovering manifold geometry is particularly useful, and remains a state-of-the-art tool for understanding multidimensional flows [8–10, 104, 106].

For a particular type of intersection manifold geometry, namely when the 2D manifolds

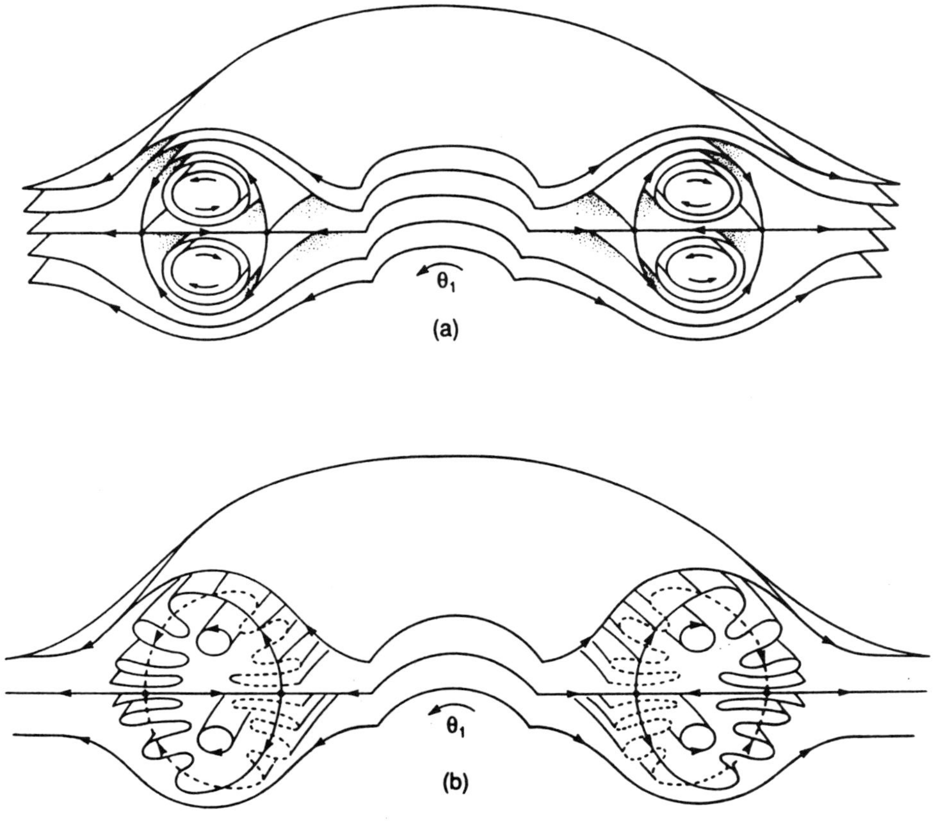

Fig. 11. A more complete portrayal of the two-frequency open flow in the 3D Poincaré section $\Sigma^{\theta_{2_0}}$ for (a) the unperturbed case and (b) the perturbed case.

intersect in the form of isolated 1-tori, the generalization of the time-periodic description of tangle dynamics is immediate and straightforward. Rather than intersection *points*, we now have intersection *tori*, and as in the time-periodic case we can distinguish between *primary intersection tori* and *secondary intersection tori* (in general we will refer to these as *primary intersection manifolds* (PIMs) and *secondary intersection manifolds* (SIMs)). A lobe $\mathscr{L}$ in the 3D Poincaré section now consists of the 3D region bounded by segments of stable and unstable manifolds, $S[\bar{\tau}_a, \bar{\tau}_b]$ and $U[\bar{\tau}_a, \bar{\tau}_b]$, respectively, joining adjacent PIMs $\bar{\tau}_a$ and $\bar{\tau}_b$; i.e. the lobe boundary $\partial\mathscr{L}$ is given by

$$\partial\mathscr{L} = S[\bar{\tau}_a, \bar{\tau}_b] \cup U[\bar{\tau}_a, \bar{\tau}_b]. \tag{21}$$

In this manner one can describe the tangle dynamics in $\Sigma^{\theta_{2_0}}$ in terms of 3D lobes mapping from one to another under P_ε. The intersection manifolds certainly can take on a richer geometry than isolated tori, however (e.g. isolated spirals or intersecting tori, as we shall see in Section 4). In these instances the same picture of 3D lobes and a lobe dynamics goes through, but their description is best formulated in the more natural setting of 2D *phase slices* $\chi(\bar{\theta}_1)$ of the 3D Poincaré section

$$\chi(\bar{\theta}_1) \equiv \{(x, y, \theta_1) | \theta_1 = \bar{\theta}_1\} \tag{22}$$

(see Fig. 12, which will be explained more fully momentarily). The basic point to appreciate here is that, regardless of higher-dimensional manifold geometry, it is possible

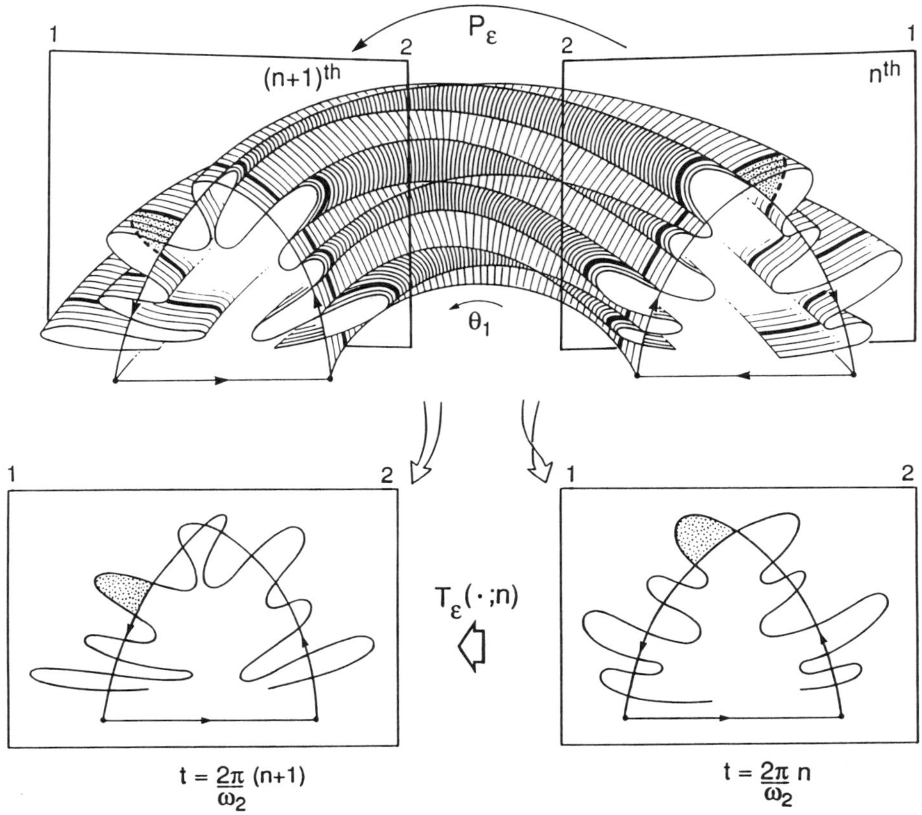

Fig. 12. The invariant manifold template (the upper heteroclinic tangle) for the two-frequency open flow: a sequence of time-dependent 2D tangles derived from a 3D tangle embedded in the Poincaré section $\Sigma^{\theta_{2_0}}$. The lobe dynamics is expressed in terms of 2D lobes mapping from one 2D tangle to the next, and here we show a single lobe of fluid (shaded) evolving between two 2D tangles.

to define higher-dimensional constructs (e.g. lobe boundaries, and then later partitions, partial barriers, turnstiles, etc.) as a union of 2D constructs associated with these phase slices, as described in detail in [8–10]. Not only is such a union the natural setting for geometrical descriptions of near-integrable multi-d.o.f. Hamiltonian flows [13, 14], we shall additionally see how these 2D slices take on a particular dynamical significance in the context of time-quasiperiodic velocity fields.

We thus have for the two-frequency problem an invariant tangle and a tangle dynamics in the context of a 3D Poincaré section. The obvious task that remains is to relate this higher-dimensional Poincaré section back to the original 2D fluid. This relation is made in a straightforward manner, as described heuristically here and rigorously in [8–10]. At each sample time $t = (2\pi/\omega_2)n$, the phase associated with the first forcing frequency is $\theta_1(n) = \theta_{1_0} + 2\pi(\omega_1/\omega_2)n$; hence one can think of the original 2D fluid at the nth sample time as a *2D slice* of the 3D Poincaré section $\Sigma^{\theta_{2_0}}$, defined by the *time slice*, or equivalently the *phase slice*,

$$\chi(\theta_{1_0} + 2\pi(\omega_1/\omega_2)n) \equiv \{(x, y, \theta_1) | \theta_1 = \theta_{1_0} + 2\pi(\omega_1/\omega_2)n\}. \tag{23}$$

Hence the sequence of 2D nonautonomous maps (15) on $\mathbb{R}^2$ can be understood in terms of P_ε acting on a sequence of phase slices of $\Sigma^{\theta_{2_0}}$, as shown in Fig. 12 (for incommensurate frequencies the phase slices will visit $\theta_1 \in [0, 2\pi))$ uniformly and densely, and for commensurate frequencies they will visit a finite number of θ_1 values). The intersection of each time slice with $W^s(\tau_\varepsilon)$ and $W^u(\tau_\varepsilon)$ (recalling τ_ε represents τ_ε^a or τ_ε^b) defines *time-dependent* 1D manifolds in (x, y) at the appropriate sample time (the manifolds vary with the time slice since $W^s(\tau_\varepsilon)$ and $W^u(\tau_\varepsilon)$ vary with θ_1). The result for each time slice is a time-dependent 2D chaotic tangle with the same topological constraints as in the familiar case of periodic forcing: the stable and unstable manifolds criss-cross *ad infinitum*, intersecting each other but never intersecting themselves, to define a countable infinity of 2D lobes. More precisely, the set of PIMs intersect the time slice $\chi[\theta_{1_0} + 2\pi(\omega_1/\omega_2)n]$ in a countable infinity of PIPs, and the segments of $W^s(\tau_\varepsilon)$ and $W^u(\tau_\varepsilon)$ in that time slice between adjacent PIPs define the boundary of a 2D lobe in that slice. The geometric template for chaotic advection under multi-frequency velocity fields is thus the sequence of time-dependent 2D tangles derived from invariant manifolds embedded in a 3D Poincaré section, as illustrated in Fig. 12. It is a straightforward consequence of the invariance of $W^s(\tau_\varepsilon)$ and $W^u(\tau_\varepsilon)$ that each lobe in the nth time slice maps under $T_\varepsilon(\cdot\,; n)$ to a lobe in the $(n + 1)$th time slice, as indicated in Fig. 12. The geometric template is thus exploited in the context of 2D lobes mapping from one to another within a sequence of 2D tangles, as illustrated in Fig. 13 in the context of a numerical simulation of the open flow. Since the lobe boundaries vary with the time slice $t = (2\pi/\omega_2)n$, a specification of lobes and lobe dynamics needs an additional parameter, n, over the time-periodic case. For each time slice $t = (2\pi/\omega_2)n$, however, the lobes associated with a quasiperiodic generalization of the open flow are still divided into two classes, one associated with an entraining process $\{E(m, n);\ m \in Z,\ n \in Z\}$ and one associated with a detraining process $\{D(m, n);\ m \in Z,\ n \in Z\}$, as will become clear in the upcoming transport discussion of Section 5, where we will see that, as in the time-periodic case, the parameter m specifies the number of time samples until the entraining or detraining process occurs. These lobes then satisfy the sequence of 2D maps

$$\begin{aligned} T_\varepsilon(E(m, n); n) &= E(m - 1, n + 1) \\ T_\varepsilon(D(m, n); n) &= D(m - 1, n + 1) \qquad m \in Z,\ n \in Z. \end{aligned} \tag{24}$$

In general, for transport between several regions, the lobe dynamics will be described by the sequence of 2D maps

$$T_\varepsilon(L_{ij}(m, n); n) = L_{ij}(m - 1, n + 1) \qquad m \in Z, \qquad n \in Z, \tag{25}$$

where $L_{ij}(m, n)$ denotes the lobes in (x, y) at the nth time sample that map from an ith region to a jth region between the $(m + n - 1)$th and $(m + n)$th time sample. In keeping with the time-dependent nature of the tangle sequence, these regions will be seen to be *time-dependent*. We also note that the lobe dynamics can be specified in terms of the 3D Poincaré map P_ε acting on a 2D time slice, or equivalently a 2D phase slice, of the Poincaré section, $\chi[\theta_{1_0} + 2\pi(\omega_1/\omega_2)n]$:

$$P_\varepsilon\{L_{ij}[m, \theta_{1_0} + 2\pi(\omega_1/\omega_2)n]\} = L_{ij}[m - 1, \theta_{1_0} + 2\pi(\omega_1/\omega_2)(n + 1)], \tag{26}$$

where $L_{ij}[m, \theta_{1_0} + 2\pi(\omega_1/\omega_2)n]$ denotes the lobes in the nth time slice $\chi[\theta_{1_0} + 2\pi(\omega_1/\omega_2)n]$ of the Poincaré section $\Sigma^{\theta_{2_0}}$ that map from an ith region to a jth region in Σ^{2_0} under the mth iterate of P_ε.

In this more general circumstance of a sequence of 2D maps derived from time-quasiperiodic velocity fields, the geometric entities $L_{ij}(m, n)$ or $L_{ij}[m, \theta_{1_0} + 2\pi(\omega_1/\omega_2)n]$ can represent now a set of lobes with in general a varying number of lobes for each m. A complete specification of the lobe dynamics thus needs a determination of these different number of lobes. The method for such a determination is presented and used in detail in

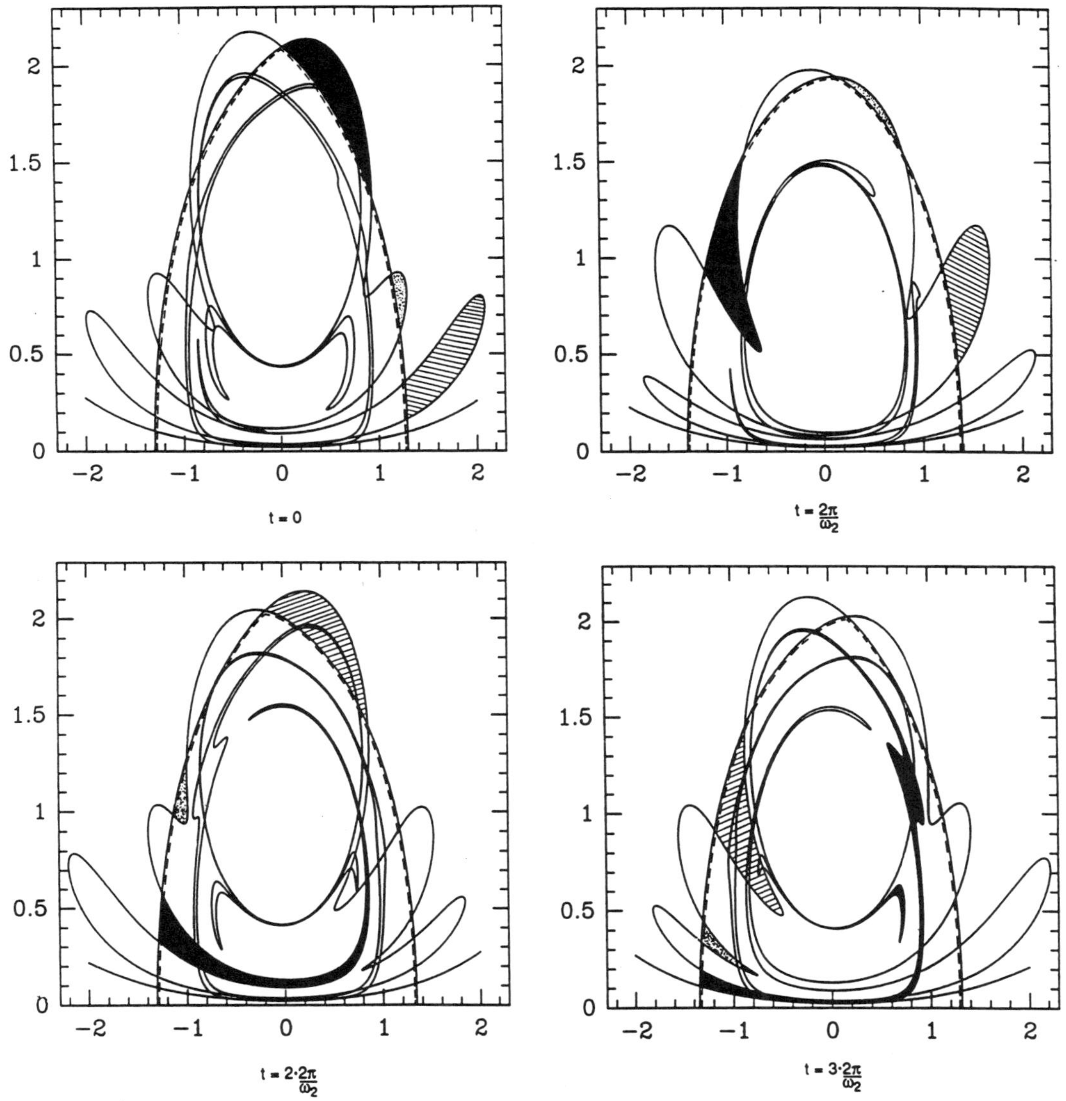

Fig. 13. A sequence of four time samples of the two-frequency open flow, whose equation of motion is given in the Appendix, with $\varepsilon = 0.12$, $f_1 = 0.4$, $f_2 = 0.95$, $\omega_1 = 2$, $\omega_2 = 2g^{-1}$, $\theta_{1_0} = 2\pi[8g - 4]$, and $\theta_{2_0} = 0$, where $g = (\sqrt{5} - 1)/2$ is the golden mean. Four material lobes (entraining lobes) are shaded so that we can monitor their dynamics. The dashed line indicates a time-dependent partitioning surface, whose meaning will become clear in the upcoming Transport Section.

[8–10], which we will not describe here. We do remark, however, that although in principle such a determination needs to address a bi-infinite sequence of 2D maps, one can exploit the underlying periodicity properties of the velocity field to obtain a compact description of the lobe dynamics in the context of the Poincaré map P_ε acting on phase slices $\chi(\bar{\theta}_1)$ of the Poincaré section $\Sigma^{\theta_{2_0}}$.

The scenario in the time-periodic case of a lobe dynamics within a 2D tangle is thus basically robust, and generalizes to multi-frequency velocity fields. Such a generalization entails interesting new features associated with chaotic dynamics and phase space structure. In particular, we stress two of these features.

2.2.1. *Time-dependent constructs*. In contrast to the typical single-frequency case, the geometric constructs in the (x, y) phase space associated with a discrete-time sampling of the fluid are now *time-dependent*. As a prime example, it is important to appreciate that there are no hyperbolic fixed points associated with the sequence of 2D maps, but rather points which move around, and yet live on an invariant normally hyperbolic torus in a higher-dimensional setting. In a similar manner, all other geometric constructs we will encounter associated with the sequence of non-autonomous 2D tangles—partial barriers, turnstiles, attracting-like objects, Smale-like horseshoes, etc.—will take on a time-dependent nature in the (x, y) phase space of the fluid.

2.2.2. *Greater freedom to vary features of the dynamics for optimization and control*. In contrast to the typical single-frequency case, for multi-frequency velocity fields the lobes can exhibit great variation that can be exploited for optimization and control. For example, the size, shape, and number of lobes associated with transport from one region to another can vary greatly from one time sample to the next, and such factors as relative amplitudes and relative phases associated with multi-frequency time dependences in the velocity field contribute to significant variation in the dynamics, such as flux relative to a partitioning surface (as we will see in Section 5).

2.3. *Two-dimensional aperiodic velocity fields*

Suppose we wish to consider 2D velocity fields with yet more complicated time dependences, and hence place no restriction on the time dependence of the perturbing streamfunction $\varepsilon\tilde{\psi}(x, y, t; \varepsilon)$. We can still always recast the problem in autonomous form

$$\begin{aligned} \dot{x} &= \frac{\partial \psi_0(x, y)}{\partial y} + \varepsilon \frac{\partial \tilde{\psi}}{\partial y}(x, y, \lambda; \varepsilon) \\ \dot{y} &= -\frac{\partial \psi_0(x, y)}{\partial x} - \varepsilon \frac{\partial \tilde{\psi}}{\partial x}(x, y, \lambda; \varepsilon) \\ \dot{\lambda} &= 1, \end{aligned} \tag{27}$$

where the enlarged phase space is now $\mathbb{R}^2 \times \mathbb{R}^1$ (see Fig. 14(a) for the unperturbed phase portrait). In this 3D setting, a 1D normally hyperbolic invariant curve is defined by a cross product of the hyperbolic fixed point $q_{\varepsilon=0}$ with the real axis

$$\tau_{\varepsilon=0} \equiv \{(x, y, \lambda) | (x, y) = q_{\varepsilon=0}\}, \tag{28}$$

and similarly a 2D separatrix and 2D local stable and unstable manifolds, $W^{s}_{loc}(\tau_{\varepsilon=0})$ and $W^{u}_{loc}(\tau_{\varepsilon=0})$, are defined by cross products of the 1D constructs in the 2D phase portrait with the real axis. The added complication from single- and multi-frequency time dependences is the non-compact property of $\lambda \in \mathbb{R}^1$. In this instance, however, the same persistence and smoothness results quoted earlier in the quasiperiodic case for normally hyperbolic invariant tori, τ_ε, and their local stable and unstable manifolds, $W^{s}_{loc}(\tau_\varepsilon)$ and $W^{u}_{loc}(\tau_\varepsilon)$, in the quasiperiodic case applies here for all perturbations that are bounded and uniformly continuous in time (e.g. see Meyer and Sell [65]). These local manifolds continue to global manifolds, $W^{s}(\tau_\varepsilon)$ and $W^{u}(\tau_\varepsilon)$, which thus can define boundaries of a 3D chaotic tangle in $\mathbb{R}^2 \times \mathbb{R}^1$, as illustrated in Fig. 14(b). One might wish to study the evolution of the fluid in the context of a 3D flow. Alternatively, a bi-infinite sequence of monotonically increasing real numbers $\{t_n | n \in Z\}$ defines a bi-infinite sequence of time slices $\{\chi(t_n) | n \in Z\}$, where

$$\chi(t_n) \equiv \{(x, y, \lambda) | \lambda = t_n\}, \tag{29}$$

and the evolution under (27) from one time slice to the next defines a bi-infinite sequence

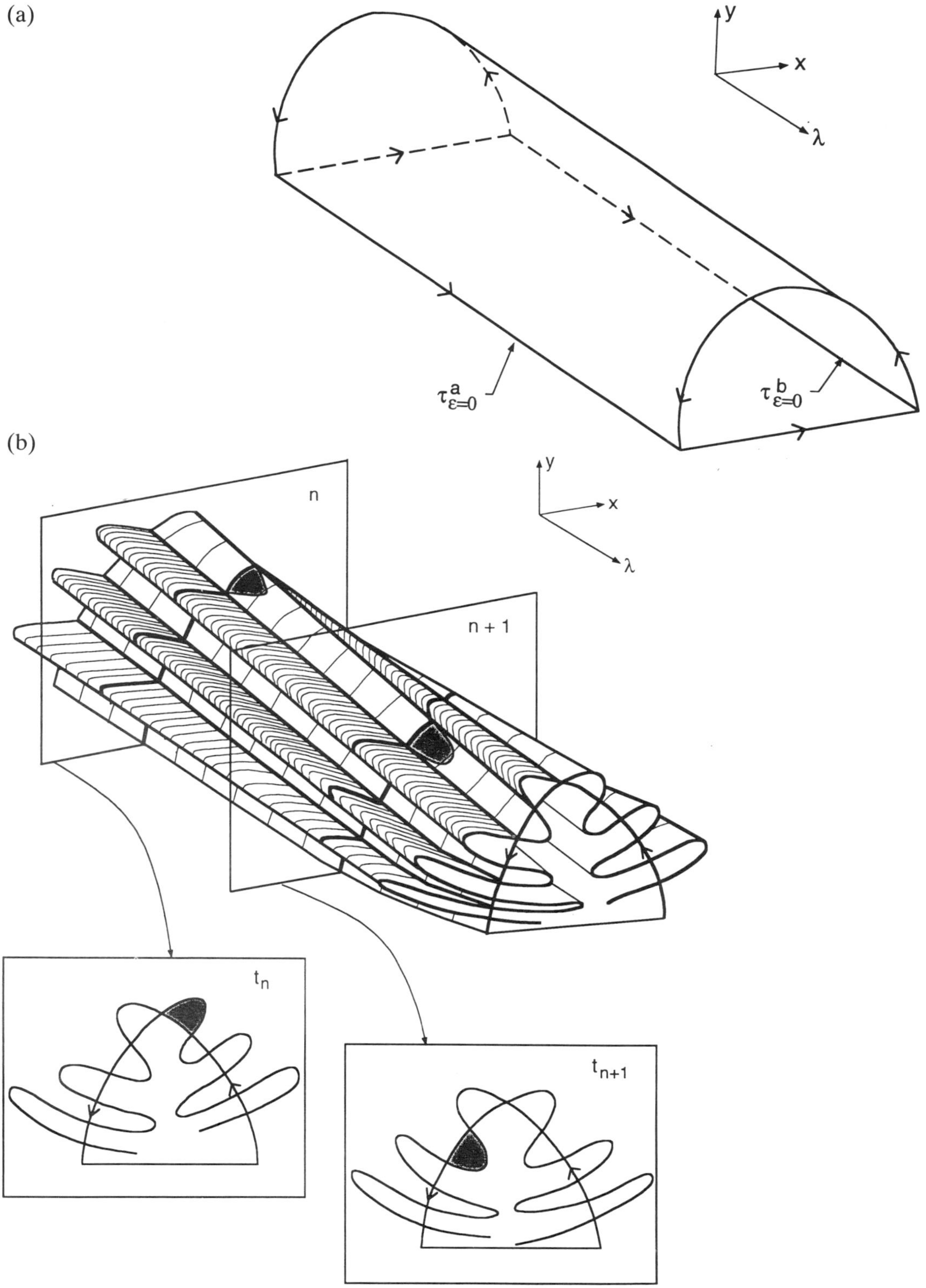

Fig. 14. (a) The unperturbed open flow in the autonomous representation $\mathbb{R}^2 \times \mathbb{R}^1$; (b) an invariant lobe structure in $\mathbb{R}^2 \times \mathbb{R}^1$ for the perturbed case, and a lobe mapping between two successive time slices. Here we show only the $y > 0$ portion.

of 2D nonautonomous maps

$$\begin{aligned} T_\varepsilon(\cdot\,; n)\colon \quad \mathbb{R}^2 &\rightarrow \mathbb{R}^2 \\ (x(t_n), y(t_n)) &\rightarrow (x(t_{n+1}), y(t_{n+1})). \end{aligned} \tag{30}$$

In this instance as well, a Melnikov function allows one to determine whether the manifolds criss-cross *ad infinitum* to form a chaotic tangle in one of the time slices, and hence (by manifold invariance) in all of the time slices (e.g. see Stoffer [95, 96]). If such a tangle exists, then we have the same scenarios as in the multi-frequency case of 2D lobes mapping from one to another within a sequence of nonautonomous 2D chaotic tangles, as shown in Fig. 14(b). The principal difference from the multi-frequency situation is possibly a less apparent global picture of the dynamics as a consequence of the λ variable being non-compact (for example, we will need a truly bi-infinite sequence of specifications to determine the lobe dynamics, rather than having a compact means of expression as in the multi-frequency case).

We point out two notable classes of aperiodic velocity fields, both of which can exhibit the phenomena of lobes which do *not* tangle around one another *ad infinitum* in a manner qualitatively similar to the single- and multi-frequency case. We stress the importance of these scenarios to fluid flow, for though the Smale horseshoe map paradigm for conventional 2D tangles is an elegant and illustrative paradigm, few realistic fluid flows enjoy the opportunity to undergo a repeated stretch and fold process *ad infinitum*.

2.3.1. *Velocity fields with a decaying or transient component in the time dependence*. This class of velocity fields is associated with a variety of fluid mixing phenomena, such as certain types of vortex collisions (e.g. see Meleshko *et al.* [63]). As a concrete, albeit contrived, example one could picture our open flow in a channel that had wavy walls with decaying amplitude of oscillations, or with nonzero amplitudes for only a finite length of the channel. In these instances, once has a tangle that consists of lobes of decaying area as one approaches the hyperbolic fixed point, or a tangle that consists of a finite number of lobes. As will be clear from forthcoming discussion, when a tangle has a finite number of lobes, the phenomena of transport across partially destroyed barriers and repeated stretching and folding will occur over only a finite time interval, and we refer to such a scenarios as one of *transient chaos*. The use of such terminology is in contrast to its use in the context of open chaotic tangles, such as found in our open flow, where chaotic behavior is transient except for a measure zero subset of phase space. In the case of a finite number of lobes, for both open and closed tangles, the transient chaotic behavior is found truly everywhere in phase space.

2.3.2. *Velocity fields whose time dependence contains a monotonically varying component*. Let us consider this class of velocity fields in the specific context of the following equations of motion with a slowly varying parameter

$$\begin{aligned} \dot{x} &= \frac{\partial \psi_0(x, y, \lambda)}{\partial y} + \varepsilon \frac{\partial \tilde{\psi}}{\partial y}(x, y, \lambda; \varepsilon) \\ \dot{y} &= -\frac{\partial \psi_0(x, y, \lambda)}{\partial x} - \varepsilon \frac{\partial \tilde{\psi}}{\partial x}(x, y, \lambda; \varepsilon) \\ \dot{\lambda} &= \varepsilon. \end{aligned} \tag{31}$$

In the unperturbed ($\varepsilon = 0$) case, the third dimension λ plays the role of a parameter for the 2D steady fluid flow described by $\psi_0(x, y, \lambda)$. For example, the area enclosed by heteroclinic separatrices might increase monotonically with λ. As λ increases monotonically in time under the perturbed flow, the area enclosed by the *instantaneous separatrices* of the unperturbed streamfunction (i.e. the separatrices associated with $\psi_0(x, y, \lambda(t = \bar{t})))$ will thus grow in time. Though typically perturbations with a time-oscillatory component cause separatrices of area-preserving flows to break into a tangle, in the present scenario it is possible for the stable and unstable manifolds to never intersect—even in the presence of a time-oscillatory component in the velocity field—entailing spiral-like behavior rather than tangle behavior. Though such spiral-like behavior is more typical of non-area-preserving flows, it is illuminating to note the possibility of this behavior for area-preserving flows. Since many fluid flows involve lobes exhibiting spiral-like behavior rather than horseshoe-like behavior, the present scenario is important to keep in mind when trying to apply chaotic advection ideas to more realistic fluid flows (e.g. see Meiburg and Newton [60] in the context of the shear layer).

2.4. *Three-dimensional velocity fields*

Invariant manifold descriptions of 3D fluid flows remain relatively unexplored. It is nevertheless important to appreciate the existence of invariant manifold tangle boundaries for certain classes of 3D flows. Such boundaries are formed by codimension-one manifolds eminating from either hyperbolic fixed points or normally hyperbolic invariant manifolds. Since these manifolds are codimension-one, i.e. of one less dimension than the phase space in which they are embedded, they are candidates for tangle boundaries analogous to ones associated with 2D maps. One would expect, however, the features of transport, stretching, and mixing associated with 2D maps to be discussed in depth in the remainder of this paper to have rich extensions in the setting of 3D flows.

As a simple example, we mention the delightful paper by Holmes [41], which appeared the same year as Aref's seminal paper [1], has received little attention, and yet showed a far-sighted vision of an invariant manifold description of chaotic advection, including such features as a higher-dimensional Melnikov theory and the notion of fluid entrainment studied via an invariant manifold template of a 3D heteroclinic tangle. The example considers 3D swirling flows in the context of an axisymmetric bubble type vortex breakdown, as shown in Fig. 15. The basic idea is to take an axisymmetric 3D flow which contains a recirculating region, with stagnation points, and study the effects of non-axisymmetric time-periodic perturbations to the velocity field on particle paths for the fluid. From an invariant manifold perspective the problem is phrased in terms of an unperturbed 3D flow containing a 2D (hence codimension-one) separatrix connecting two hyperbolic fixed points, a and b, as shown in Fig. 15(a), which is then subjected to a time-periodic perturbation motivated by the presence of a weakly growing non-axisymmetric mode. The net result of this perturbation is a splitting of the separatrix into the boundaries of a 3D heteroclinic tangle, allowing for fluid to be entrained from upstream, as indicated in Fig. 15(b). The perturbed situation is not unlike the open vortical flow example of this paper, but with richer dynamical features due to the extra dimension in the flow.

For more recent studies of 3D chaotic flows, we refer the reader to Feingold *et al.* [33], Lau and Finn [49], MacKay [58], and Mezić and Wiggins [66]. We note, however, that invariant manifold studies of 3D chaotic flows remain a virtually wide-open field with rich possibilities.

2.5. *Applicability and limitations*

The framework presented in this paper applies to flows that contain some sort of invariant hyperbolic set—a hyperbolic fixed point or normally hyperbolic invariant manifold of an associated Poincaré map, or a normally hyperbolic invariant manifold of the flow

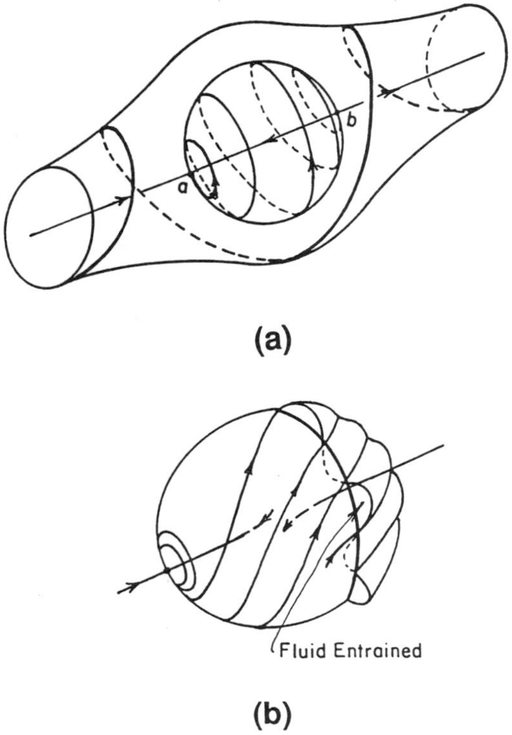

Fig. 15. Axisymmetric bubble type vortex breakdown. (a) The unperturbed manifolds, and (b) the resulting manifold splitting as a consequence of the perturbing velocity field. Taken from Holmes [41].

itself. From a dynamical systems perspective the condition of the existence of some invariant hyperbolic set is not very restrictive, since the basic scenarios associated with chaotic dynamics—e.g. homoclinic/heteroclinic tangles and resonance bands—contain such an invariant set. From a fluid mechanics perspective, such a condition can nevertheless be highly restrictive. Though the applicability of a dynamical systems approach to certain elementary mixing flows is well-established (e.g. see the work by Ottino and collaborators with cavity flows, flows between eccentric cylinders, and so on [74, 75]), possible relevance to pre-turbulent and turbulent flows is far less apparent. The single- and multi-frequency analysis should hold some promise for various realistic fluid flows (e.g. unsteady vortex breakdown [53], unsteady flow past a cylinder [89], Langmuir circulation [50], Rayleigh–Bénard convection [18, 92], and so on). Further, an interesting possibility is the utility of certain dynamical systems paradigms and constructs towards turbulent flows with coherent structures (e.g. the mixing layer or wall layer), such as the possible relevance of a repeated stretching and folding mechanism towards mixing phenomena. One must nevertheless exhibit extreme caution when attempting to draw any parallels between non-turbulent chaotic flows and turbulent chaotic flows, or to formulate any conclusions about the utility of low-dimensional dynamical systems techniques towards a robust understanding of turbulence.

We lastly stress that, as remarked earlier, our study is presented in the context of near-integrable flows. Some of the analysis, e.g. the global perturbation technique of Melnikov theory for uncovering manifold geometry, is valid only in the near-integrable regime; however, much of the analysis only needs the existence of invariant manifolds, which does not require a near-integrable setting. Our motivation for presenting the analysis in a near-integrable setting is primarily for ease of presentation and concreteness of discussion.

3. PHYSICAL REALITY AND DOMINANT ROLE OF INVARIANT MANIFOLDS

3.1. *Unstable manifolds as dominant structures*

Given the existence of invariant manifold templates for a wide class of flows, we now wish to exploit these templates for practical understanding of chaotic advection. Perhaps the best starting point is to recognize the physical reality and dominant role of these invariant manifolds. In particular, we discuss here how unstable manifolds of hyperbolic or normally hyperbolic invariant sets can play the role of a dominant structure for the patterns taken on by tracer elements under chaotic flows. We shall see how streaklines formed in certain flow visualization experiments are, in fact, nearly images of segments of unstable manifolds.

To understand the nature of this phenomenon, let us consider a simple example in the context of our time-periodic open flow. Figure 16 shows for seven perturbation periods snapshots of a material curve initially on the upper unperturbed separatix (the perturbed situation is symmetric about $y = 0$). With each snapshot a new lobe of fluid is *entrained* into the region originally bounded by the unperturbed separatrices (for the moment we refer to this region heuristically as the *bounded core*, with the understanding that a more precise distinction between regions and description of entrainment and detrainment with respect to these regions will be given in Section 5). The lobe stretches as it winds around the tangle, and then it folds across another entrained lobe, part of it escaping to be advected steadily to the left, part of it remaining in the bounded core to repeat the stretching and folding process. This process continues *ad infinitum* to give chaotic advection. The material curve is being essentially 'attracted' to the upper unstable manifold of the Poincaré map's right hyperbolic fixed point. Since the dynamical system is Hamiltonian (the fluid is incompressible), the unstable manifold is of course not an attractor; however, the strong stretching and contraction of the fluid lobes implies that, though area elements do not shrink to zero, they tend to be stretched in one direction and contracted in another, such that the unstable manifold dominates the evolution of material curves in the tangle region (see Fig. 17 for a heuristic portrayal). The unstable manifolds of the Poincaré map's hyperbolic fixed points are thus understood to be a dominant structure by which the dynamics in the tangle regions can be studied. Hence for area-preserving flows, breaking integrability causes the separatix to give way to, instead of an attractor, a *dominator*, so to speak. Attractors for dissipative systems, and in a similar manner dominators for Hamiltonian systems, are typically conceptualized as a time-independent construct in the context of a map or autonomous flow. For the elementary case of a time-periodic velocity field, reducible to a 2D-map, it is natural to conceptualize an attractor or dominator as a time-independent object. For more complicated fluid velocity fields, however, such as those exhibiting a quasiperiodic or aperiodic time dependence, attractors or dominators indeed must take the form of a time-dependent object in physical space (an immediate consequence of the time-dependent nature of the tangle boundaries, described in Section 2). One has the interesting phenomenon, then, of the evolution of tracer elements being dominated by an underlying geometrical object that itself evolves in time.

The observation of the dominant role of unstable manifolds is of course to be understood with major qualifications.

(i) Given an initial tracer distribution, a time interval must elapse before one can conclude the appearance of the tracer is dominated by the unstable manifold. Fig. 16 shows a conveniently chosen initial tracer distribution whose correspondence with the unstable manifold rapidly becomes apparent. In general, of course, a much longer time interval may be required before significant resemblance to an unstable manifold is apparent.

(ii) When invoking the dominant role of the unstable manifold in area-preserving flows, it is important to remember that area elements are *not* shrinking to zero. For example, a tracer line that extends, say, from two PIPs to only halfway into a lobe, will effectively retain that property with increasing time.

Given these and other qualifications, much of course remains to be learned of the relationship between passive scalars and unstable manifolds, but the basic idea of a dominant role played by the unstable manifolds should be clear.

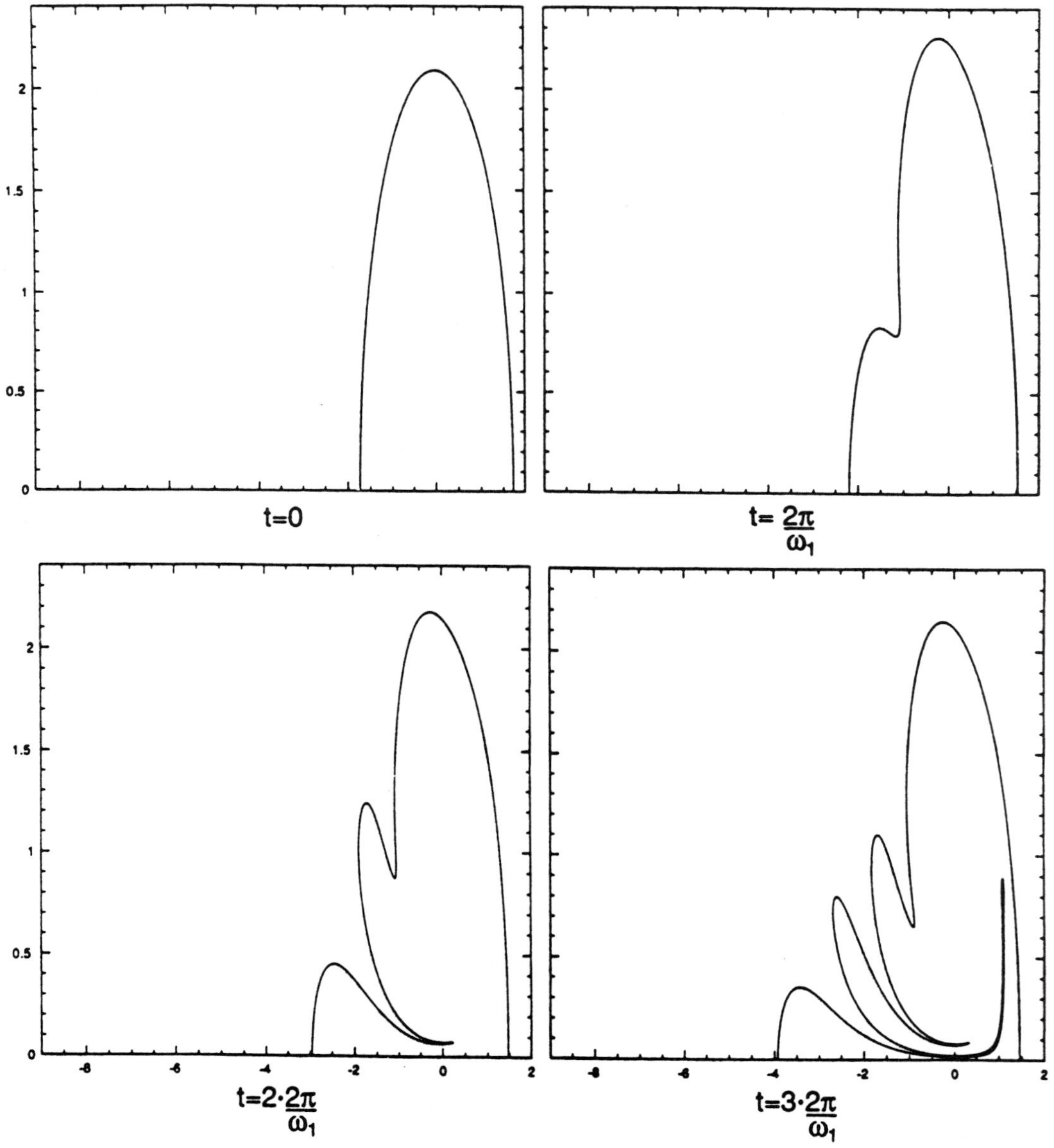

Fig. 16.

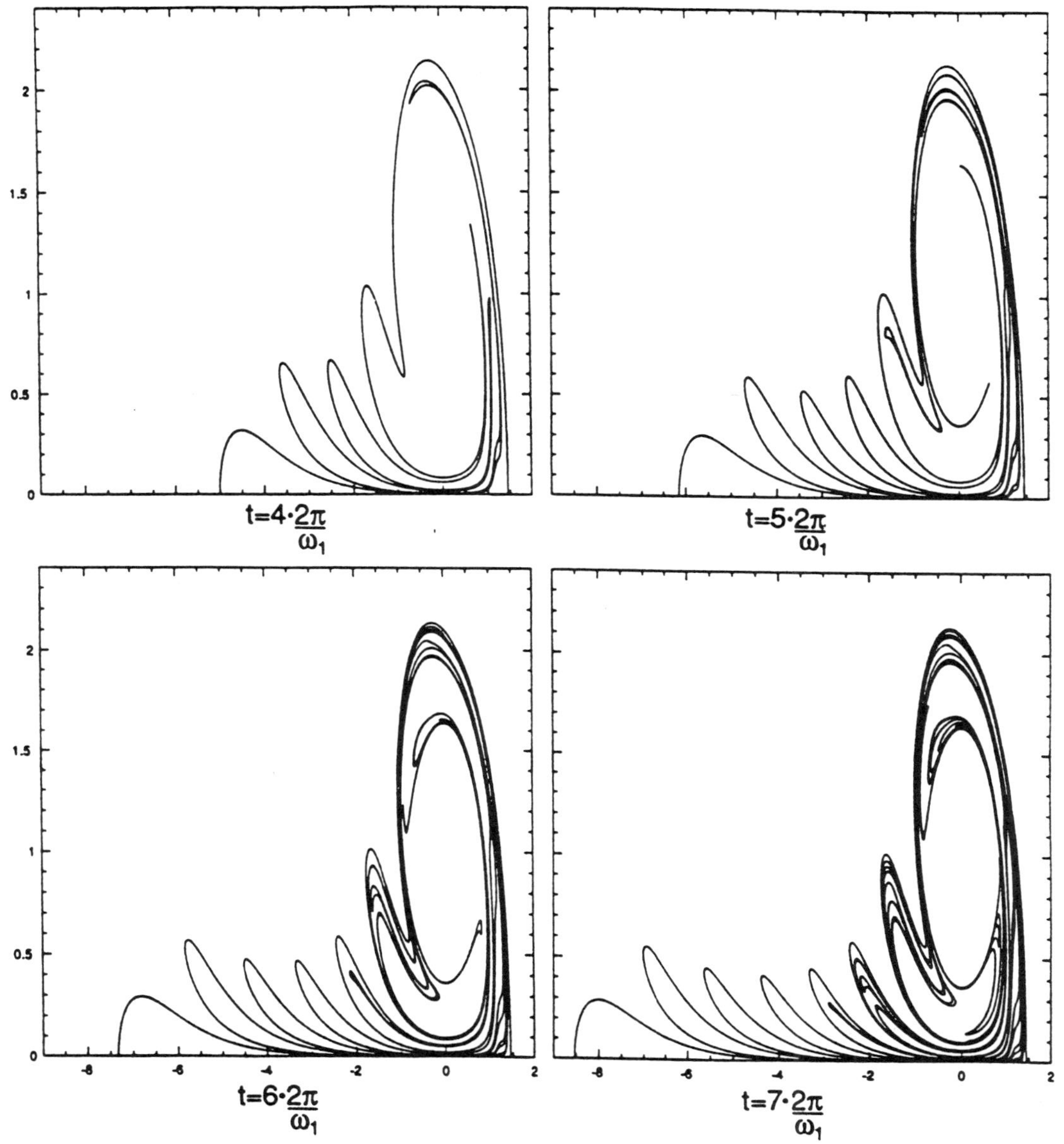

Fig. 16. Evolution, according to the single-frequency open flow, of a material curve that initially lies on the upper unperturbed separatrix. The system parameters are given in the Appendix (in the context of the single-frequency limit of the multi-frequency open flow streamfunction).

3.2. *Flow visualization experiments*

To help convey the physical reality and dominant role of unstable manifolds, let us consider some flow visualization experiments.

3.2.1. *Leapfrogging vortex rings.* Two ideal coaxial vortex rings, having the same sense of rotation, will leapfrog one another. Because of mutual induction, the forward vortex increases in diameter and slows down while the rearward vortex shrinks in diameter and speeds up. Once the vortices trade places, the process repeats. If one moves with the mean speed of the pair, one observes a periodic, axisymmetric flow.

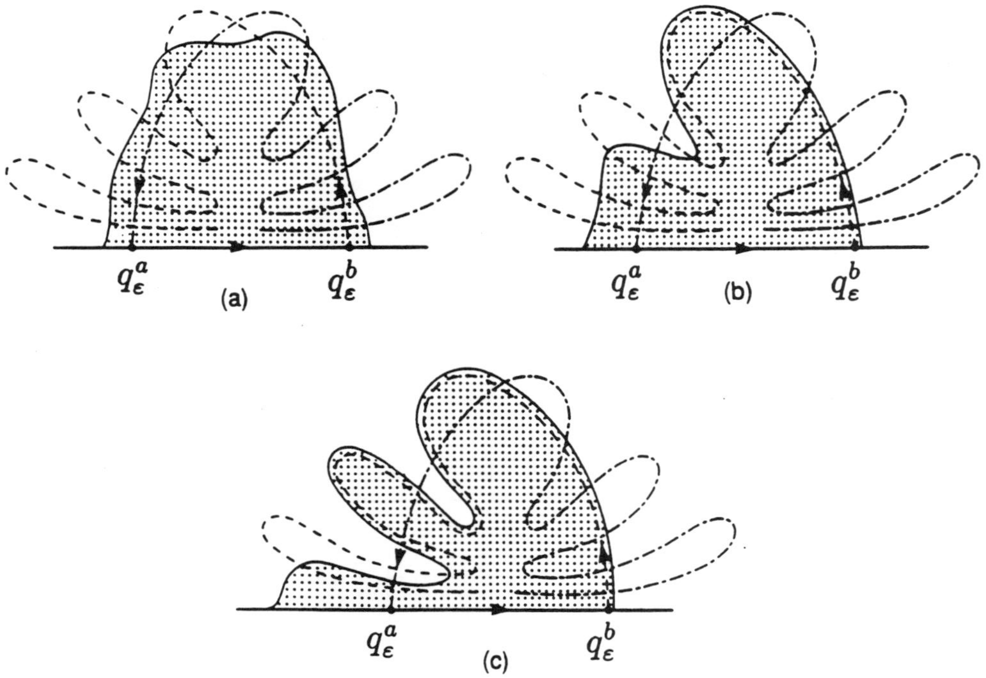

Fig. 17. Three successive time samples showing a material curve being 'attracted' to the unstable manifold. The dashed-dotted line is the stable manifold, the dashed line is the unstable manifold, and the solid line is the material curve.

Shariff *et al.* [87, 88], using an idealized model of vortex ring dynamics, have computed the unstable manifold of a fixed point in the Poincaré map that lies on the axis just forward of the leading vortex. The result is shown in Fig. 18(b) for three phases of the period. In this flow one has portions of lobes that escape as in our open vortical flow, but the basic flow has two centers of rotation instead of one. In Fig. 18(a) we show a flow visualization of two interacting, coaxial vortex rings from the experiment of Yamada and Matsui [108]. In this experiment the flow is visualized by ejection of smoke from a wire at a fixed axial location. The agreement of two patterns is quite remarkable, and extends even to the fine-scale features, even though the experimental rings would certainly have a more diffuse field of vorticity due to the formation process and subsequent advection and diffusion; see Shariff and Leonard [90] for further discussion. Apparently, fine details in the advected scalar can be determined by gross features of the advecting flow field. Certainly the dominating property of the unstable manifold is unmistakable.

3.2.2. *Unsteady vortex breakdown.* In this example we consider flow in a cylindrical container with a rotating endwall. A dominant feature of this flow is an axial vortex which also has an axial flow towards the rotating endwall. Within a certain range of Reynolds numbers, the axial vortex will form one or more local recirculation bubbles (each sometimes called a vortex breakdown) caused kinematically by the presence of ring-like vortices superposed on the axial vortex. Furthermore, and again depending on *Re*, the vortex rings may oscillate periodically in axial position and shape.

The flow has been studied experimentally by Escudier [31] and computationally by Lopez and Perry [53] using the full Navier–Stokes equations. For the case of periodic oscillations, Lopez and Perry also computed the relevant unstable and stable manifolds of the two

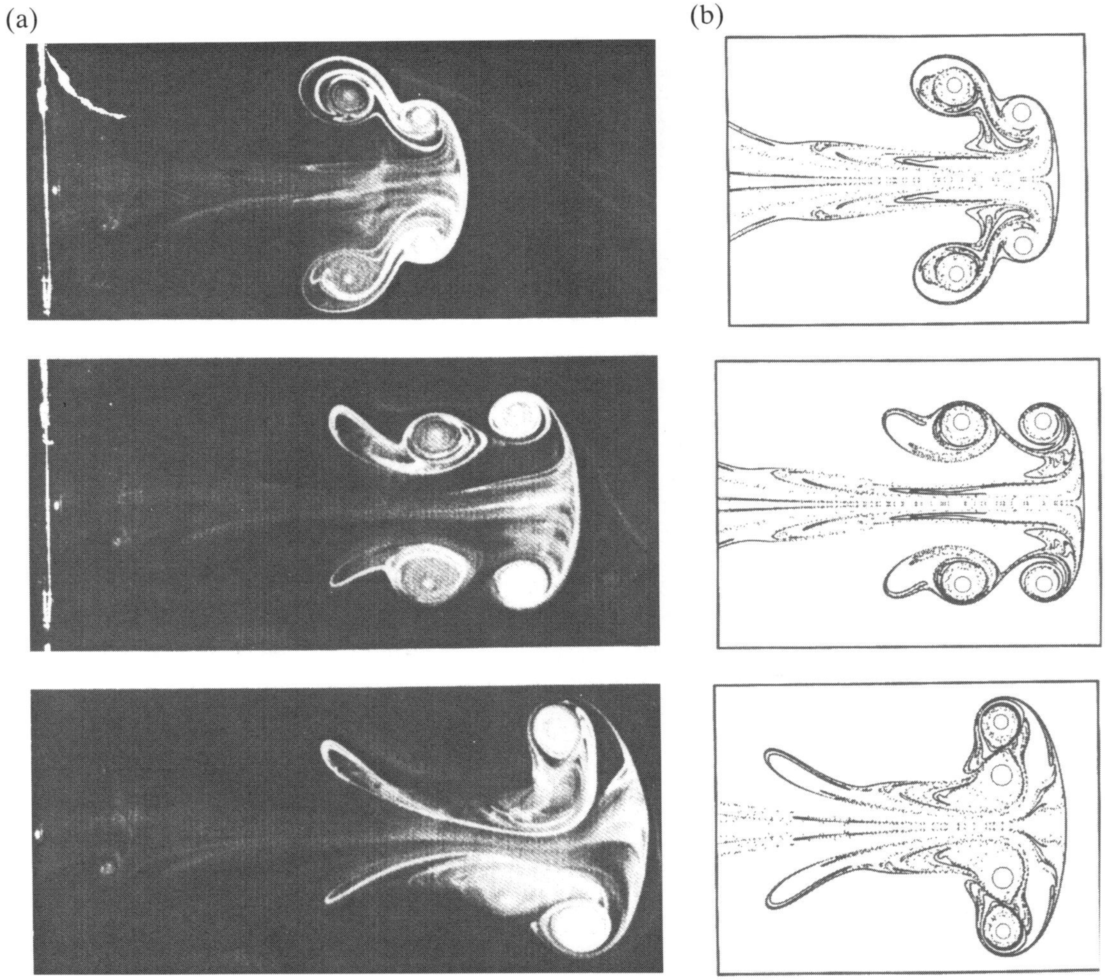

Fig. 18. Leapfrogging vortex rings: (a) photographs from an experiment by Yamada and Matsui [107], and (b) unstable manifold computed by Shariff *et al*. [88]. The circles shown are the vortex cores in the idealized model.

hyperbolic fixed points on the axis and, for higher Reynolds numbers, the manifolds of the hyperbolic point off the axis, thus establishing the transport of fluid by lobe dynamics as discussed in Section 5. In Fig. 19(b) we show the unstable manifold of the fixed point located on the axis just upstream of the first bubble or breakdown. Excellent agreement is seen with the flow visualization by Escudier shown in Fig. 19(a). In this case however the agreement may be more anticipated, because in the experiment the dye is released just on the axis near the top endwall.

3.2.3. *Unsteady flow past a cylinder.* Flow past a circular cylinder at moderately low *Re* (approximately 10 to 40) shows a steady symmetric separation zone at the rear of the cylinder. There is a hyperbolic stagnation point in the flow field at the rear of the separation bubble on the axis, two flow attachment points, also on the axis, and two separation points symmetrically located off the axis. These features can be seen in the flow visualization shown in Fig. 20, although a slight unsteadiness in the flow is detectable by the small oscillations in what would nominally be the separating streamline as it approaches the hyperbolic stagnation point.

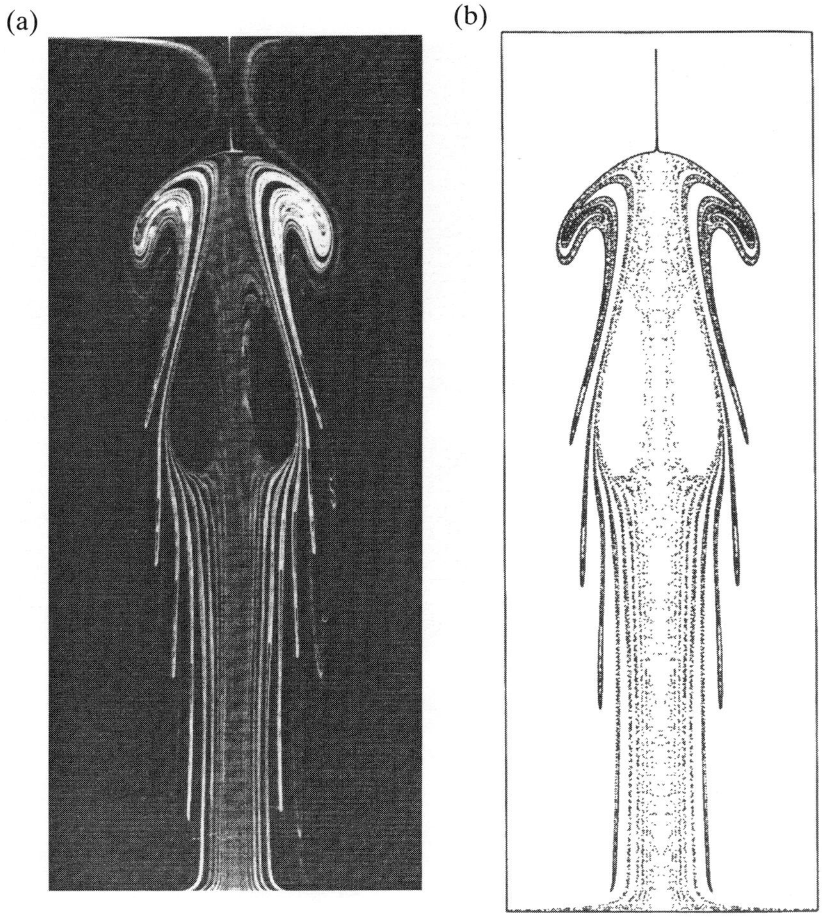

Fig. 19. Flow in a cylindrical cavity with a rotating endwall, $Re = 2765$: (a) streakline visualization of an experiment by Escudier [31], and (b) unstable manifold computed by Lopez and Perry [53].

Between $Re = 40$ and 250 flow past a cylinder oscillates periodically in time. Vortices shed from the body (two per period) form a wake known as a Kàrmàn street. Shariff *et al.* [89] have investigated the time-periodic flow fields generated numerically for $Re = 100$ and $Re = 180$. They identified the fixed points in the Poincaré map and have computed the lobe structures that govern transport in the wake of the cylinder.

In their analysis they had to pay special attention to the neighborhood of the surface of the cylinder. The fluid velocity at the surface is zero so that all points there are nonhyperbolic fixed points. Nevertheless they show that an unstable manifold does emanate from a point on the surface where the time-average of the wall shear stress (proportional to the normal derivative of the tangential velocity) is zero. The location of this point is therefore independent of phase, but the angle of the separation line or manifold at the surface may depend on phase.

We show in Fig. 21 the unstable manifold of the upper separation point on the cylinder as computed by Shariff *et al.* [89] for $Re = 180$, and a visualization of an experimental flow by S. Taneda (Van Dyke [108]) using smoke produced at the surface of the cylinder by electrolytic precipitation. The experiment, of course, shows roughly the unstable manifolds of both the upper and lower separations points.

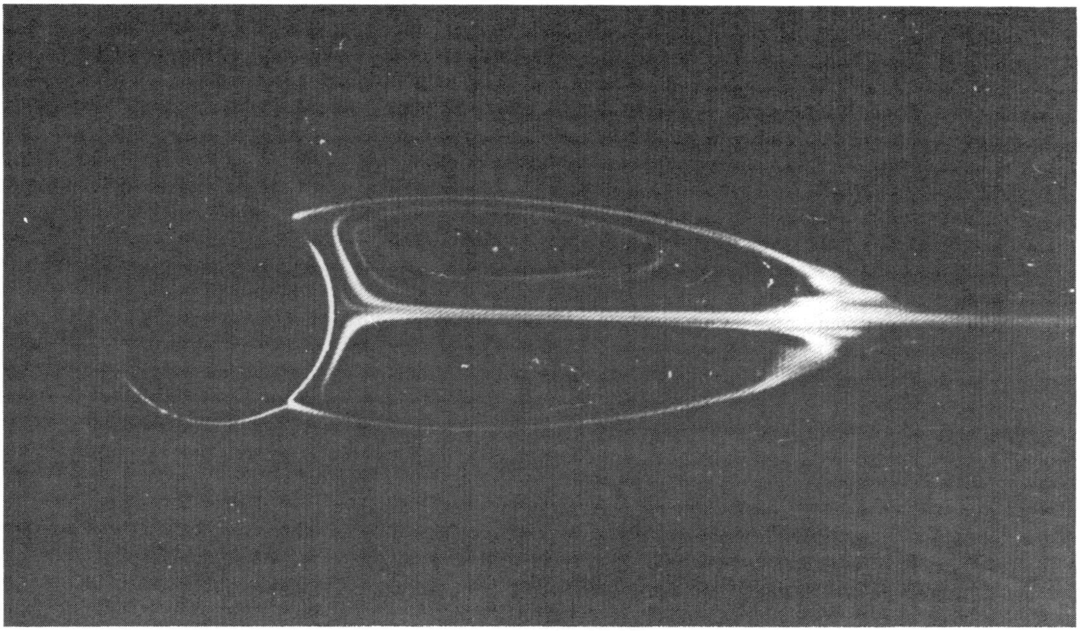

Fig. 20. Flow past a circular cylinder at $Re = 41$. Visualization by S. Taneda (Van Dyke [108], p. 30).

4. MANIFOLD GEOMETRY AND MELNIKOV'S METHOD

Thus far we have seen that a wide class of flows exhibit chaotic tangles whose boundaries are defined by the global stable and unstable manifolds of invariant hyperbolic or normally hyperbolic sets. Furthermore, we have had impressed upon us the physical reality and dominant role of these invariant manifolds. As a first step towards using the manifolds to obtain practical understanding of chaotic advection, we would like to study their geometry, i.e. the way in which the stable and unstable manifolds criss-cross one another to form lobes in the tangle. Melnikov's method [64], and generalizations of his original method [14, 103], provide an analytical technique for studying manifold geometry for near-integrable flows. For 2D time-periodic velocity fields, tangle geometry is straightforward, and Melnikov's original theory for uncovering this geometry has been in existence for over 30 years. However, in the higher-dimensional settings needed to describe more realistic fluid flows and other physical phenomena, tangle geometry can indeed be quite complicated, and these geometries are only now being uncovered by the relatively recent generalizations of Melnikov's original theory. These generalized Melnikov methods are thus truly state-of-the-art tools for studying invariant manifold geometry in these more complicated settings, and (as we shall see) for measuring practical physical quantities, e.g. transport quantities such as flux. Though one could determine these manifold geometries and physical measurements by direct numerical computation, for higher-dimensional flows with dependence on several parameters, such computations quickly become prohibitively intensive. A great practical advantage of generalized Melnikov theories will thus be a convenient and efficient means to search through phase space and parameter space in order to *control* or *optimize* various physical processes (examples of which are given in later sections).

We proceed to describe Melnikov theory and its use for increasingly complicated classes of flows, following the discussion of Section 2. The features of the theory common to all these flows are the following. Melnikov's method is a global perturbation technique for

(a)

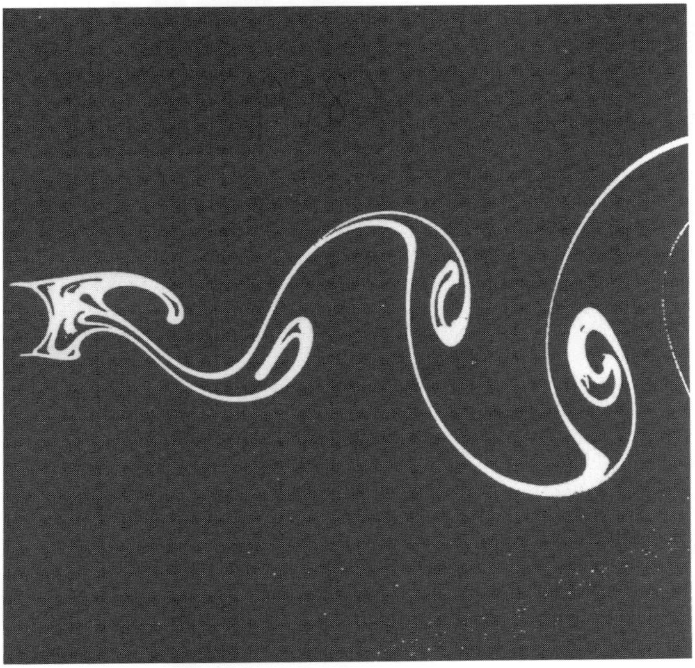

(b)

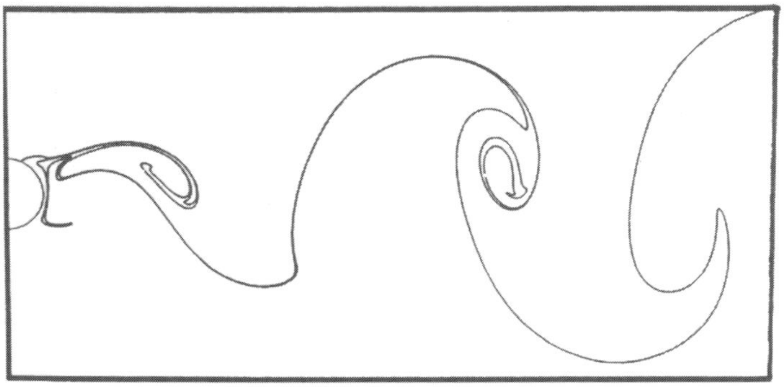

Fig. 21. Flow past a circular cylinder at $Re = 180$: (a) visualization by S. Taneda (Van Dyke [108], p. 56), and (b) unstable manifold computation by Shariff *et al*. [89].

measuring the separation, and hence geometry, of stable and unstable manifolds associated with near-integrable flows whose unperturbed phase portrait contains some sort of homoclinic or heteroclinic separatrix (e.g. refer back to the open and closed vortical flows of Section 2). (Melnikov theory can also be applied to the study of destroyed tori, but for concreteness we focus on destroyed separatrices.) These homoclinic/heteroclinic orbits can be thought of as global stable and unstable manifolds which, for the unperturbed flow or map, are identical to one another, and hence lie on top of each other to form the separatrix. The perturbation typically destroys the separatrix, causing the initially identical stable and unstable manifolds to split apart and intersect one another in some complicated fashion to give a tangling of the manifolds. In order to study this splitting process, one constructs a distance function, d, that measures the *separation of these manifolds normal to the unperturbed separatix* (see Fig. 22). The arguments of this distance function will thus be parameters that specify location on the unperturbed separatix. It is a consequence of the

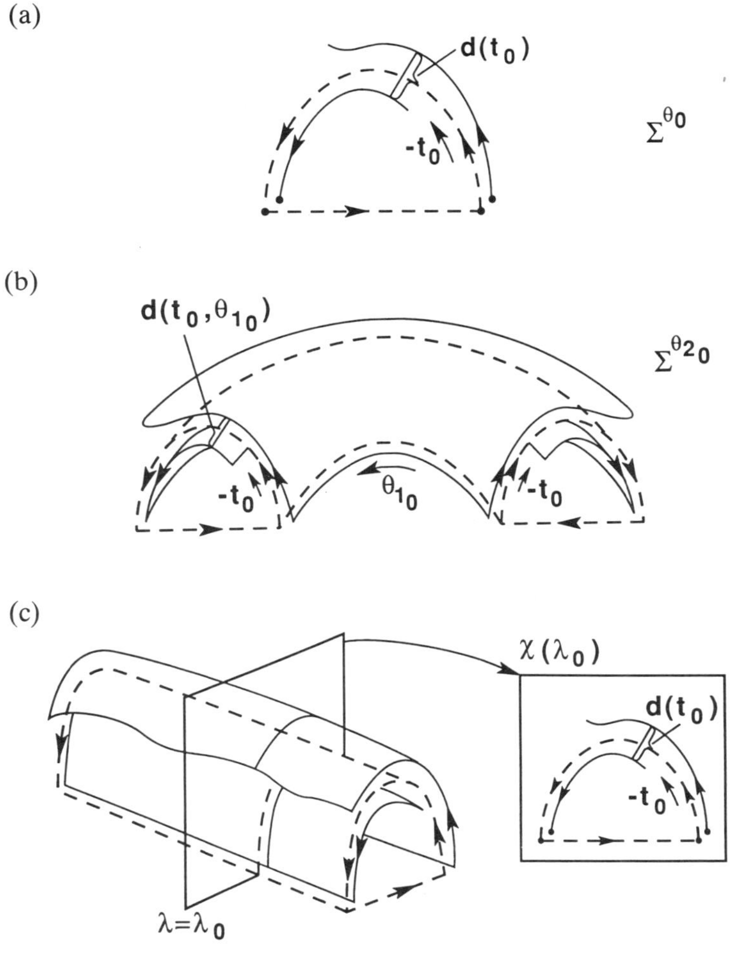

Fig. 22. The distance function d for (a) near-integrable 2D maps derived from 2D single-frequency velocity fields, (b) near-integrable 3D maps derived from 2D two-frequency velocity fields, and (c) near-integrable sequences of 2D nonautonomous maps derived from 2D aperiodic velocity fields. This function measures the separation of the stable and unstable manifolds normal to the unperturbed separatrix, and is thus expressed as a function of the parameters specifying location on the separatrix. The portrayal here is in the context of the upper separatrix of the open flow.

local persistence and smoothness theorems described in Section 2 that the invariant manifolds, and hence the distance function, will be as smooth as the velocity field that defines the flow or map. Hence we can write this distance expression as an expansion in the perturbation parameter, ε, and Melnikov's method provides an analytical expression for the $O(\varepsilon)$ contribution to this distance expansion. (We note parenthetically that the $O(\varepsilon)$ term in the distance expansion remains bounded only over a finite domain of the separatix, i.e. the term diverges as one approaches the invariant hyperbolic or normally hyperbolic set, but this need not concern us here.) In particular, one obtains for the distance expansion

$$d = \frac{\varepsilon M}{\|D\psi_0\|} + O(\varepsilon^2), \tag{32}$$

where M denotes the Melnikov function and $\|D\psi_0\|$ denotes the norm of the gradient of the unperturbed streamfunction, i.e. the magnitude of the unperturbed velocity field. Since arguments of this distance function, and hence of the Melnikov function, are the parameters specifying location on the unperturbed separatrix, these arguments will depend on the class of flows at hand, and are thus suppressed for the moment. The $O(\varepsilon)$ term in the distance expansion (32) is derived by solving an ODE for a time-dependent distance function constructed from trajectories that lie on the stable and unstable manifolds, and the resulting Melnikov function will be expressed, as we shall see, as a time-like integral involving the unperturbed streamfunction ψ_0 and the perturbing streamfunction $\tilde{\psi}$. This Melnikov function then gives, via the distance expression (32), the $O(\varepsilon)$ measure of the stable and unstable manifold separation normal to the unperturbed separatrix.

The key to uncovering tangle geometry will be to determine the geometry of the *intersections* of the stable and unstable manifolds, from which lobe geometry in the tangle will follow. From the distance expression (32) it is clear that a vanishing of the Melnikov function corresponds to an $O(\varepsilon)$ vanishing of the stable and unstable manifold separation. Indeed, implicit function theory implies that a zero of the Melnikov function continues to an actual transversal intersection of the stable and unstable manifolds nearby (i.e. within $O(\varepsilon)$). (Such a continuation result requires a mild transversality condition involving a non-vanishing of derivatives of the Melnikov function, which need not concern us here.) Hence a study of Melnikov zero set geometry translates into a study of stable and unstable manifold intersection geometry (e.g. a torus in the Melnikov zero set will continue to a torus in the set of stable and unstable manifold intersections, a spiral will continue to a spiral, and so on). The intersections continued from the Melnikov zero set correspond to the *primary intersection points* and *primary intersection manifolds* referred to in Section 2, and are thus instrumental in the definition of lobes in the tangles. A primary concern in the use of Melnikov theory will thus be a study of the zero sets of the Melnikov function.

4.1. *Melnikov theory for 2D single-frequency velocity fields*

Let us first consider near-integrable flows generated by the 2D time-periodic velocity fields of the form (3), and assume that the integrable 2D flow defined by the unperturbed velocity field contains one or more hyperbolic fixed points with a homoclinic or heteroclinic trajectory $\{(x_h(t), y_h(t)) | t \in \mathbb{R}^1\}$ (e.g. see the open and closed vortical flows of Section 1). Recall from Sections 1 and 2 that these periodically forced systems can be recast as a 2D Poincaré section of a 3D autonomous flow. The 1D unperturbed separatrix in the 2D Poincaré section Σ^{θ_0} is thus expressed as

$$\Gamma_{\varepsilon=0} \equiv \{(x_h(t), y_h(t)) | t \in \mathbb{R}^1\}. \tag{33}$$

A trajectory on the separatrix of the 3D autonomous flow can be written as $(x_h(t - t_0), y_h(t - t_0), \omega t + \theta_0)$, and thus the $t = 0$ solution parameterizes an initial condition on the unperturbed separatrix of the Poincaré section Σ^{θ_0}, $(x_h(-t_0), y_h(-t_0))$. Hence, for a fixed sampling phase θ_0, location on the 1D separatrix of the unperturbed Poincaré map can be specified by the time-like parameter t_0 (see Fig. 22(a)). The single-frequency Melnikov function, and the resulting distance expression, defined along the unperturbed separatrix of the Poincaré section $\Sigma^{\theta_0=0}$ is thus written as a function of the parameter t_0:

$$M(t_0) = \int_{-\infty}^{\infty} \{\psi_0, \tilde{\psi}\}(x_h(t - t_0), y_h(t - t_0), \omega t)\, dt, \tag{34}$$

where $\{\psi_0, \tilde{\psi}\}$ denotes the Poisson bracket

$$\{\psi_0, \tilde{\psi}\} \equiv \frac{\partial \psi_0}{\partial x}\frac{\partial \tilde{\psi}}{\partial y} - \frac{\partial \psi_0}{\partial y}\frac{\partial \tilde{\psi}}{\partial x}, \tag{35}$$

which is evaluated at $(x_h(t - t_0), y_h(t - t_0), \omega t)$, i.e.

$$\psi_0 = \psi_0(x_h(t - t_0), y_h(t - t_0))$$
$$\tilde{\psi} = \tilde{\psi}(x_h(t - t_0), y_h(t - t_0), \omega t).$$

A coordinate change involving translation in the time variable over which the integral is performed typically gives a more convenient expression

$$M(t_0) = \int_{-\infty}^{\infty} \{\psi_0, \tilde{\psi}\}(x_h(t), y_h(t), \omega(t + t_0))\,dt. \tag{36}$$

Hence, the Melnikov function is solved from a time-like integral of a Poisson bracket of the unperturbed streamfunction ψ_0 and the perturbing streamfunction $\tilde{\psi}$. For a perturbing streamfunction that involves a sinusoidal time dependence, one can conveniently use a trigonometric expansion to separate the t_0 dependence from the Melnikov integral, resulting in an integral for the Melnikov *amplitude* alone. For example, the time-periodic open vortical flow (see Section 1 and the Appendix) has the Melnikov function

$$M(t_0) = A(\omega)\sin(\omega t_0), \tag{37}$$

where $A(\omega)$ denotes the Melnikov amplitude, whose solution is given in Fig. 23. The two points to appreciate here are the following.

(i) The sinusoidal time dependence in the perturbing streamfunction translates into sinusoidal oscillations of the stable and unstable manifold separation. Hence the Melnikov zeros, and transversal intersections of stable and unstable manifolds, correspond to isolated points, and lobe geometry is plain.

(ii) For a given amplitude in the perturbing streamfunction's sinusoidal time dependence, one obtains a *frequency-dependent* amplitude in the Melnikov function. Hence, an opportunity for optimization in this elementary context is afforded by varying the perturbation frequency in such a way as to maximize the Melnikov amplitude for a fixed perturbation amplitude, thus maximizing manifold splitting, and (as we shall later see) lobe area.

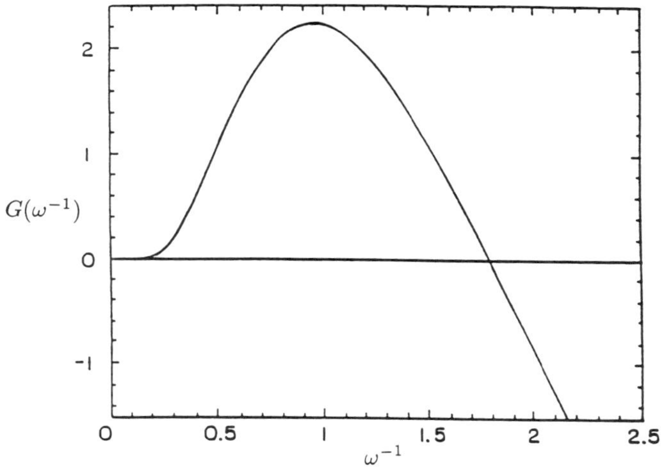

Fig. 23. The Melnikov amplitude $A(\omega)$ is given by $A(\omega) = \omega^{-1}G(\omega^{-1})$, where $G(\omega^{-1})$ is as shown. This amplitude expression is given with respect to the single-frequency limit of the two-frequency equations of the Appendix.

4.2. *Melnikov theory for 2D multi-frequency velocity fields*

Let us now consider near-integrable flows generated by 2D multi-frequency velocity fields of the form (14), again assuming the unperturbed 2D flow contains one or more hyperbolic fixed points with a homoclinic or heteroclinic trajectory $\{(x_h(t), y_h(t)) | t \in \mathbb{R}^1\}$. For simplicity of discussion we consider the two-frequency case, and recall from Section 2.2 that these quasiperiodically forced systems can be recast as a 3D Poincaré section of a 4D autonomous flow. The 2D unperturbed separatrix in the 3D Poincaré section $\Sigma^{\theta_{2_0}}$ is expressed as

$$\Gamma_{\varepsilon=0} \equiv \{(x_h(t), y_h(t), \theta_1) | t \in \mathbb{R}^1, \theta_1 \in [0, 2\pi)\}. \tag{38}$$

A trajectory on the separatrix of the 4D autonomous flow can be written as $(x_h(t - t_0)$, $y_h(t - t_0)$, $\omega_1 t + \theta_{1_0}$, $\omega_2 t + \theta_{2_0})$, and thus the $t = 0$ solution parameterizes an initial condition on the unperturbed separatrix of the Poincaré section $\Sigma^{\theta_{2_0}}$, $(x_h(-t_0)$, $y_h(-t_0)$, $\theta_{1_0})$. Hence, for a fixed sampling phase θ_{2_0}, location on the 2D separatrix of the unperturbed Poincaré map can be specified by the *two* parameters (t_0, θ_{1_0}). The additional parameter for the two-frequency problem, θ_{1_0}, thus specifies the phase associated with the unsampled frequency. The two-frequency Melnikov function, and the resulting distance expression, defined along the unperturbed separatrix of the Poincaré section $\Sigma^{\theta_{2_0}=0}$ is thus written as a function of the parameters (t_0, θ_{1_0}):

$$M(t_0, \theta_{1_0}) = \int_{-\infty}^{\infty} \{\psi_0, \tilde{\psi}\}(x_h(t), y_h(t), \omega_1(t + t_0) + \theta_{1_0}, \omega_2(t + t_0))\, \mathrm{d}t_0 \tag{39}$$

(see Fig. 22(b)). Again the Melnikov function is solved from a time-like integral of a Poisson bracket of the unperturbed streamfunction ψ_0 and the perturbing streamfunction $\tilde{\psi}$. For a perturbing streamfunction that has the form of a sum of sinusoidal dependences on time, again we can employ a trigonometric expansion to separate out the parameter dependences from the Melnikov integral. For example, the open flow subjected to the two-frequency perturbing streamfunction

$$\varepsilon\tilde{\psi}(x, y, \omega_1 t + \theta_{1_0}, \omega_2 t + \theta_{2_0}; \varepsilon) = \varepsilon xy\{\omega_1 f_1 \sin(\omega_1 t + \theta_{1_0}) + \omega_2 f_2 \sin(\omega_2 t + \theta_{2_0})\} \tag{40}$$

(see the Appendix) has for the Poincaré section $\Sigma^{\theta_{2_0}=0}$ a Melnikov function of the form

$$M(t_0, \theta_{1_0}) = A_1(\omega_1) \sin(\omega_1 t_0 + \theta_{1_0}) + A_2(\omega_2) \sin(\omega_2 t_0), \tag{41}$$

where $A_i(\omega_i)$, $i = 1, 2$, denote the frequency-dependent Melnikov amplitudes, each of whose dependence on perturbation frequency follows from the time-periodic relation $A(\omega)$ shown in Fig. 23, in particular $A_i(\omega_i) = f_i A(\omega_i)$. The two points to appreciate for this two-frequency Melnikov function are the following.

(i) Much as in the time-periodic case, the relative amplitude for each sinusoidal component in the perturbation, $F_i \equiv f_i \omega_i$, $i = 1, 2$, translates into a frequency-dependent Melnikov amplitude, A_i, $i = 1, 2$. Hence a basic measure for understanding the relative effectiveness of each frequency at producing manifold splitting is given by *relative scaling factors*, $\mathcal{RSF}(\omega_i)$, defined as the ratios of the Melnikov amplitude over the corresponding relative perturbation amplitude

$$\mathcal{RSF}(\omega_i) \equiv A_i(\omega_i)/F_i. \tag{42}$$

These two factors, $i = 1, 2$, are determined by a single *relative scaling function* $\mathcal{RSF}(\omega)$, whose functional dependence on frequency follows from the single-frequency relation of Fig. 23. This scaling function, and the resulting scaling factors, give a measure of the relative effectiveness of each frequency in the multi-frequency problem at producing manifold splitting.

(ii) Since the Melnikov function is now defined over a 2D domain, parameterized by (t_0, θ_{1_0}), Melnikov zero sets, and the consequent transversal intersections of stable and unstable manifolds, will now typically be 1D curves, rather than points. The result is a set of manifold intersections with much richer geometrical possibilities than the time-periodic case. A detailed study of these possibilities in the two-frequency case is given in [8–10], and Fig. 24 gives an indication of some of the geometries one can encounter. In particular, these intersection curves can take on the form of isolated tori, isolated spirals, and even intersecting spirals and tori. These richer intersection geometries entail richer lobe geometries, but in each case it is possible to construct well-defined lobes in the tangle, as indicated in Sections 2 and 5. We note that the way to interpret lobe geometry from

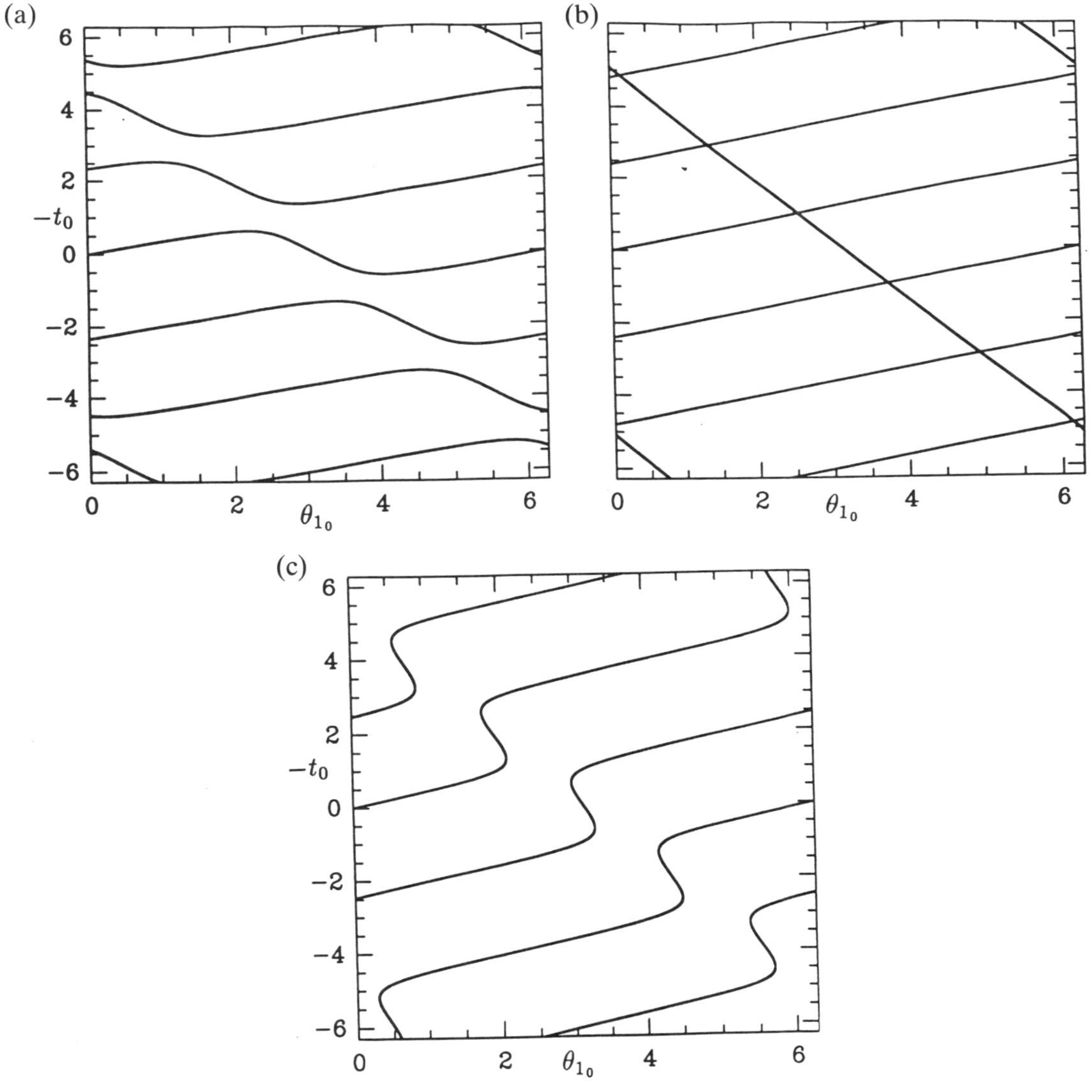

Fig. 24. Zero sets of the two-frequency Melnikov function $M(t_0, \theta_{1_0})$ for $\omega_1 = g\omega_2 = 1$, where g is the golden mean $(\sqrt{5}-1)/2$, and (a) $A_1(\omega_1)/A_2(\omega_2) = 5/6$, (b) $A_1(\omega_1)/A_2(\omega_2) = 1$, and (c) $A_1(\omega_1)/A_2(\omega_2) = 6/5$. The zero sets are portrayed on the 2D domain of the unperturbed separatrix, parameterized by (t_0, θ_{1_0}).

invariant manifold geometry is to recognize that the stable and unstable manifolds criss-cross one another, meeting at the intersections, as illustrated in Fig. 25.

4.3. *Melnikov theory for 2D aperiodic velocity fields*

Let us now consider near-integrable flows generated by 2D aperiodic velocity fields of the form (27), again assuming the unperturbed 2D flow contains one or more hyperbolic fixed points with a homoclinic or heteroclinic trajectory $\{(x_h(t), y_h(t))|t \in \mathbb{R}^1\}$. Recall from Section 2.3 that 2D systems subjected to forcing with general time dependences can be recast as a sequence of 2D nonautonomous maps derived from a 3D autonomous flow. The 1D unperturbed separatrix in the 2D slice $\chi(\lambda = \lambda_0)$ is expressed as

$$\Gamma_{\varepsilon=0} \equiv \{(x_h(t), y_h(t), \lambda)|t \in \mathbb{R}^1, \lambda = \lambda_0\}. \tag{43}$$

A trajectory on the separatrix of the 3D flow can be written as $(x_h(t - t_0), y_h(t - t_0), t + \lambda_0)$, and thus the $t = 0$ solution parameterizes an initial condition of the unperturbed separatrix of the 2D slice $\chi(\lambda = \lambda_0)$, $(x_h(-t_0), y_h(-t_0))$. Hence, for a fixed 2D slice $\chi(\lambda = \lambda_0)$, location on the 1D unperturbed separatrix is specified by the parameter t_0. The Melnikov function for forcing with general time dependences, and the resulting distance expression, defined along the unperturbed separatrix of the slice $\chi(\lambda_0 = 0)$ is thus written as

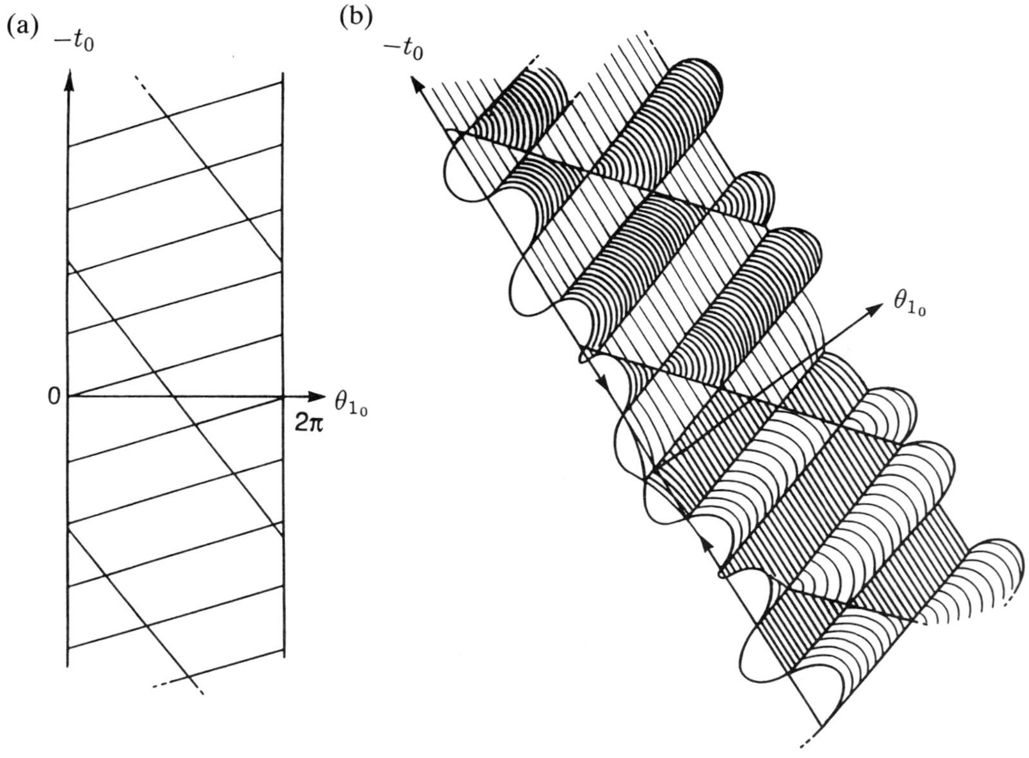

Fig. 25. Visualizing lobes by showing the dimension normal to the unperturbed separatrix. The stable and unstable manifolds criss-cross one another, meeting at intersections continued from the Melnikov zero sets, to form the boundaries of lobes in the tangle. The Melnikov zero sets are shown in (a), and the criss-crossing manifolds are shown in (b).

a function of the parameter t_0:

$$M(t_0) = \int_{-\infty}^{\infty} \{\psi_0, \widetilde{\psi}\}(x_h(t), y_h(t), t + t_0)\,dt \tag{44}$$

(see Fig. 22(c)). Given the fixation of the chaotic advection community on time-periodic velocity fields, it is illuminating to note that the Melnikov function for general time dependences has the same functional form as the time-periodic Melnikov function (36). The only new feature associated with the aperiodic Melnikov function will be the more complicated time dependence in the perturbing streamfunction translating into more complicated variation in manifold geometry with the parameter t_0. We lastly note that since the Melnikov function offered here applies to general time dependences (with the qualifications of Section 2.3 understood) this function also applies to the multi-frequency case. The choice between different representations of a multi-frequency system depends primarily on one's preference towards a higher-dimensional, but compact, representation (Section 2.2) versus a lower-dimensional non-compact representation (Section 2.3). For the two-frequency case the preference towards the former case should be clear.

4.4. *Melnikov theory for more complicated flows*

After having introduced Melnikov's method in the context of a few classes of flows, we end our discussion by pointing out that the method applies to many other, indeed a continually growing number of, classes of flows: systems with slowly varying parameters, 3D flows, various types of multi-d.o.f. flows, dissipative systems, and so on [14, 103]. The generalized Melnikov functions associated with these flows can involve not only a dependence on different flow parameters than seen thus far, but also different functional forms in the Melnikov integrand than a Poisson bracket of an unperturbed and perturbing streamfunction (or Hamiltonian).

5. TRANSPORT

5.1. *Partially destroyed barriers, partitions, turnstiles, and lobe dynamics*

Having conveyed the dominant role and physical reality of invariant manifolds, and presented an analytical technique for uncovering their geometry, we now turn in earnest to practical use of invariant manifold templates. In this section we discuss the rich and active dynamical systems field of *phase space transport* [7–11, 13–15, 29, 30, 54–58, 61, 62, 82–84, 104, 106], which in the context of fluid flows describes the transport of fluid in physical space. Efficient stirring under chaotic advection entails the destruction of barriers to fluid motion, allowing for fluid to be efficiently transported among various regions of physical space. Such transport phenomena are manifested, for example, as entrainment and detrainment processes in flows with coherent vortical structures [82], and roll-to-roll transport in pre-turbulent flows that contain an array of vortical rolls [18]. At first glance, it may seem prohibitively complicated to understand such transport processes in the context of non-integrable flows. Indeed, lacking such integrable structures as streamlines, it might not seem even possible to offer a reasonable, practical definition of 'entrainment' or a 'roll' of vorticity. The basic success of phase space transport theory, however, is to provide a rigorous and geometrically exact criterion not only for distinguishing between regions of qualitatively different fluid motion (e.g., bounded versus unbounded, or different rolls), but also for quantifying the transport of fluid amongst these various regions.

For convenience and concreteness of discussion, let us describe this transport theory in the context of the near-integrable flows of Section 2, with the recognition that such a theory need not require near-integrability, only the existence of invariant manifolds. The basic elements of the theory common to all classes of flows described in Section 2 are the following. In the integrable case (e.g. our unperturbed vortical flows) one has *complete barriers* in physical space, across which no fluid can be transported in the absence of stochastic processes such as molecular diffusion. Breaking integrability as a consequence of perturbing the velocity fields entails the destruction of these barriers, allowing for fluid to be transported across what was once a complete barrier. The barriers to fluid motion are not completely destroyed, however, leaving remnants referred to as *partial barriers*. Fluid transport across these partial barriers can be described in a geometrically exact context of images and pre-images of *turnstile lobes* [7–11, 13–15, 29, 30, 54–58, 61, 62, 82–84, 104, 106]. More precisely, the invariant manifolds of the chaotic flow allow for the following description of transport, as shall be made explicit in the context of individual classes of flows momentarily.

(i) A union of segments of stable and unstable manifolds, joined at an intersection of the two manifolds, defines a *partition*, i.e. a curve or surface that completely divides physical space into distinct regions. It will be a consequence of the critical asymptotic behavior of trajectories on the global stable and unstable manifolds that this partitioning surface will be seen to represent the true criterion for distinguishing between qualitatively different fluid motion under the non-integrable flow.

(ii) The partitioning surfaces are not complete barriers, however, but partial barriers across which fluid can be transported. As we shall see, this transport process can be quantified in the geometrically exact context of *turnstile lobes*. These lobes are the region bounded by segments of the partitioning surface and its pre-image under the Poincaré map. This region, and only this region, will be seen to cross the partition under the Poincaré map, and hence act as the sole mechanism for transport across the partition.

(iii) Transport of fluid over some finite time interval, i.e. under some n iterates of the Poincaré map, is then expressed in terms of images and pre-images of the turnstile lobes.

5.1.1. *Transport under velocity fields with periodic time dependences.* Let us first consider the invariant manifold formulation of transport theory in the context of time-periodic velocity fields, illustrated by our open vortical flow (refer back to Section 1). For the unperturbed flow a complete barrier is formed by a union of any of the three separatrices. In particular, a union of the upper and lower separatrices forms a complete barrier between bounded and unbounded motion, across which no fluid can pass in the absence of such stochastic phenomena as molecular diffusion. Breaking integrability as a result of the perturbing streamfunction destroys the upper and lower separatrices (but preserves the middle one due to the symmetry of the perturbation), allowing for fluid to be *entrained into* and *detrained from* the region originally bounded by the upper and lower separatrices. In order to understand this transport process, let us focus on the $y>0$ domain, with the understanding that a similar transport process occurs in the $y<0$ domain. A partition of the $y>0$ domain, shown by the dashed line in Fig. 26(a), is defined by the union of curves

$$\Gamma_\varepsilon \equiv U[q^b_\varepsilon, q_p] \cup S[q_p, q^a_\varepsilon] \cup X[q^a_\varepsilon, q^b_\varepsilon], \tag{45}$$

where q_p denotes the central PIP indicated in the figure and $X[q^a_\varepsilon, q^b_\varepsilon]$ denotes the unbroken separatrix along the x-axis from the left fixed point q^a_ε to the right fixed point q^b_ε (the meaning of $U[q^b_\varepsilon, q_p]$ and $S[q_p, q^a_\varepsilon]$ as segments of unstable and stable manifolds should be clear from Section 2). One can formally choose any PIP to play the role of q_p in

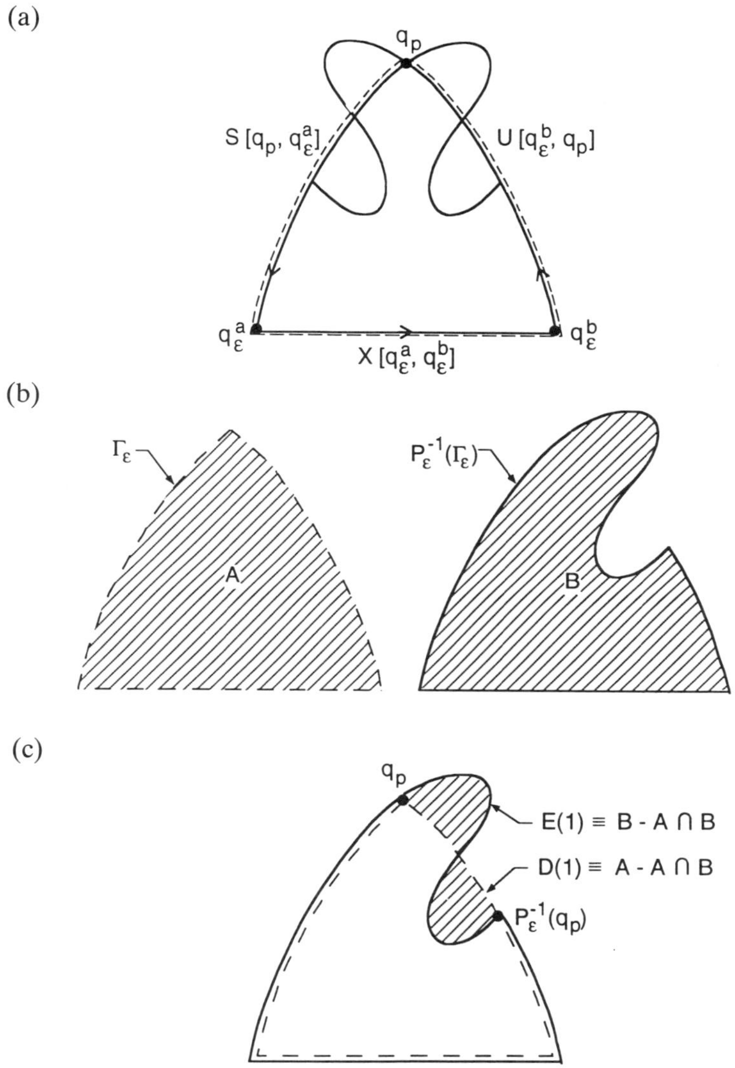

Fig. 26. (a) Defining a partition (dashed line) from segments of stable and unstable manifolds. (b) The region enclosed by Γ_ε and $P_\varepsilon^{-1}(\Gamma_\varepsilon)$, A and B, respectively. (c) The turnstile boundaries are the segments of Γ_ε and $P_\varepsilon^{-1}(\Gamma_\varepsilon)$ that differ.

defining the partition; however, q_p is chosen here to give a partition that best approximates the unperturbed separatrix. As we shall see, the non-uniqueness in the choice of partition construction is a reality of studying dynamics under non-integrable flows that in no way detracts from the power and utility of the transport theory. The role and interpretation of this partitioning surface shall become clear only after transport across this partition is described, which we now address. The key to understanding transport across the partition lies with the pre-image of the partitioning surface $P_\varepsilon^{-1}(\Gamma_\varepsilon)$. Since $P_\varepsilon^{-1}(\Gamma_\varepsilon)$ is diffeomorphic to Γ_ε, and hence is also a partitioning surface, it is a simple matter of set theory and invariant manifold theory to define the turnstiles and construct a lobe dynamic formalism. Denote the collection of points enclosed by Γ_ε ($P_\varepsilon^{-1}(\Gamma_\varepsilon)$) by A (B), as shown in Fig. 26(b). That P_ε sends $P_\varepsilon^{-1}(\Gamma_\varepsilon)$ in an orientation-preserving manner to Γ_ε then implies

$$
\begin{aligned}
B - A \cap B &= \text{points } \textit{captured into } \Gamma_\varepsilon \text{ under } P_\varepsilon, \\
&\quad \text{i.e. mapped from outside to inside the partition} \\
A - A \cap B &= \text{points that } \textit{escape from } \Gamma_\varepsilon \text{ under } P_\varepsilon, \\
&\quad \text{i.e. mapped from inside to outside the partition.}
\end{aligned} \tag{46}
$$

Hence, $B - A \cap B$ is referred to as the *entraining turnstile lobe*, denoted by $E(1)$, and $A - A \cap B$ is referred to as the *detraining turnstile lobe*, denoted by $D(1)$, and the sum of these two regions defines the turnstile

$$\mathscr{L}(1) \equiv E(1) \cup D(1) \tag{47}$$

(the motivation for this turnstile notation will become clear momentarily). Equation (46) implies the turnstile is thus formed by the region contained in $A \cup B$ minus the region contained in $A \cap B$, and is thus bounded by the segments of the partitioning surface Γ_ε and its pre-image, $P_\varepsilon^{-1}(\Gamma_\varepsilon)$, that differ (plus a subset of the intersection manifolds associated with the meeting of these differing segments), as portrayed in Fig. 26(c). By partition construction, the turnstile region is enclosed by the boundary

$$\partial\mathscr{L}(1) = S[q_p, P_\varepsilon^{-1}(q_p)] \cup U[q_p, P_\varepsilon^{-1}(q_p)]. \tag{48}$$

Delineation between the individual boundaries $\partial E(1)$ and $\partial D(1)$ depend on the relative orientations of $W^{\mathrm{s}}(q_\varepsilon^a)$ and $W^{\mathrm{u}}(q_\varepsilon^b)$, which vary from case to case, depending on invariant manifold geometry.

Once the turnstile lobe boundaries have been specified, all other lobes in the tangle are specified by images and pre-images of the turnstiles

$$
\begin{aligned}
&\{\mathscr{L}(m) \equiv P_\varepsilon^{-(m-1)}(\mathscr{L}(1)) \mid m \in Z\} \\
&\{E(m) \equiv P_\varepsilon^{-(m-1)}(E(1)) \mid m \in Z\} \\
&\{D(m) \equiv P_\varepsilon^{-(m-1)}(D(1)) \mid m \in Z\},
\end{aligned} \tag{49}
$$

as illustrated in Fig. 5. One thus obtains a lobe-dynamic description of phase space transport introduced in Sections 1 and 2

$$
\begin{aligned}
P_\varepsilon(E(m)) &= E(m-1) \\
P_\varepsilon(D(m)) &= D(m-1),
\end{aligned} \tag{50}
$$

which follows by the lobe definition (49). Hence, for $m > 0$, $E(m)(D(m))$ denotes the collection of points in phase space that are entrained into (detrained from) Γ_ε upon the mth iterate of P_ε (a similar interpretation follows for $m \leq 0$). In this manner we see how the lobe-dynamic formalism introduced and described in prior sections follows from a transport-based description of lobe evolution in the tangles. Figure 27 illustrates the entrainment and detrainment process associated with the turnstiles under one iterate of the Poincaré map. As mentioned in Section 2, and seen explicitly here, the lobe-dynamic equations (50) pertain to a rather special circumstance of two regions separated by a single partition. In general transport can occur between several regions (e.g. roll-to-roll transport in Rayleigh–Bénard convection [18]), in which case $E(m)$ and $D(m)$ are replaced by the more general description of $L_{ij}(m)$, which denotes the lobe which will map from the ith region to the jth region upon the mth iterate of the Poincaré map.

At this point it is essential to make clear the role and interpretation of a partition, and transport across the partition, under non-integrable flows. Points within the open flow partition are destined to oscillate within the partition *ad infinitum*, or until they land on the detraining turnstile lobe, $D(1)$, upon which they exit the partition under the next iterate of

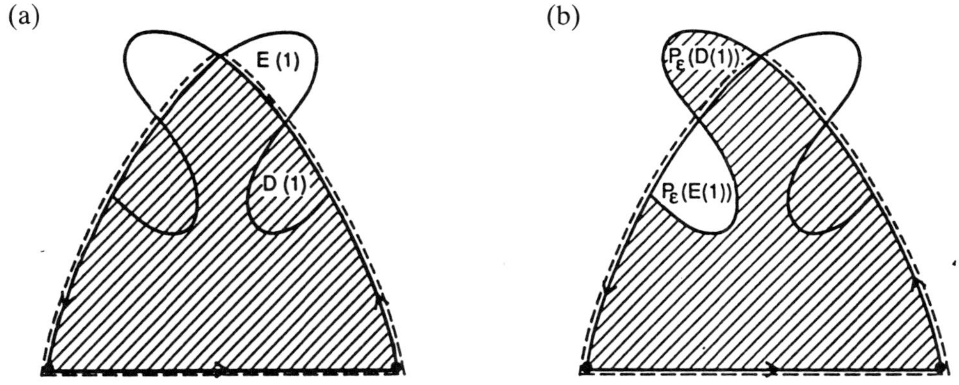

Fig. 27. Entrainment and detrainment process associated with the turnstiles of the time-periodic open flow. (a) The partition (dashed line) and the fluid it encloses (shaded), and (b) the image of this fluid under one iterate of the Poincaré map. $P_\varepsilon(E(1))$ has been captured into the partition and $P_\varepsilon(D(1))$ has escaped the partition.

the Poincaré map, henceforth undergoing unbounded motion. It is thus meaningful to make the following interpretation: the partition divides between qualitatively different motion (bounded versus unbounded in this instance), and it is through the turnstiles, and only the turnstiles, that points make a *transition* across the partition and into a region of qualitatively different motion. *Hence, in the context of non-integrable flows it is possible to determine a geometrically exact distinction between regions of qualitatively different motion, and offer a geometrically exact description of transitions between these regions.* The key to such an interpretation is the critical asymptotic behavior of trajectories on segments of stable and unstable manifolds, which implies that once a transition is made, the partition acts as the true dividing line between qualitatively different motion. The non-uniqueness of partition choice does not compromise this geometrically exact description, and can be thought of as akin to a mere shift in initial time.

5.1.2. *Transport under velocity fields with quasiperiodic and aperiodic time dependences.* A truly exciting prospect for description of transport processes in fluid mechanical situations is the applicability of an invariant manifold description to velocity fields with more complicated time dependences that periodic. Since the time-periodic analysis enjoys the convenient property of being reducible to a 2D map, transport under a non-integrable flow can be phrased in the context of fixed, well-defined regions in physical space. Such a theory appears rather quaint in the face of realistic fluid situations, where the more typical scenario is one of fluid being deformed and stirred in some highly irregular fashion whose description is not reducible to fixed regions in physical space. In this typical case, one would like to describe transport, if possible, relative to some sort of well-defined moving boundary, as illustrated in Fig. 28. There would appear to be a problem, however, with attempting to determine what this boundary is, as indicated in Fig. 29(a). Recall from Section 2 that for the class of 2D velocity fields in that section with a quasiperiodic or aperiodic time dependence, one studies the evolution of the fluid in the context of a *sequence* of 2D nonautonomous maps, defined by a discrete-time sampling of the fluid (e.g. for a two-frequency velocity field, sample the fluid at one of the frequencies). One studies this sequence of 2D maps in the context of a sequence of 2D slices of invariant constructs embedded in a higher-dimensional, autonomous setting (e.g. a 3D Poincaré section for the two-frequency problem). The goal then is to construct the partitioning and transport formalism in the context of the invariant manifolds of the higher-dimensional

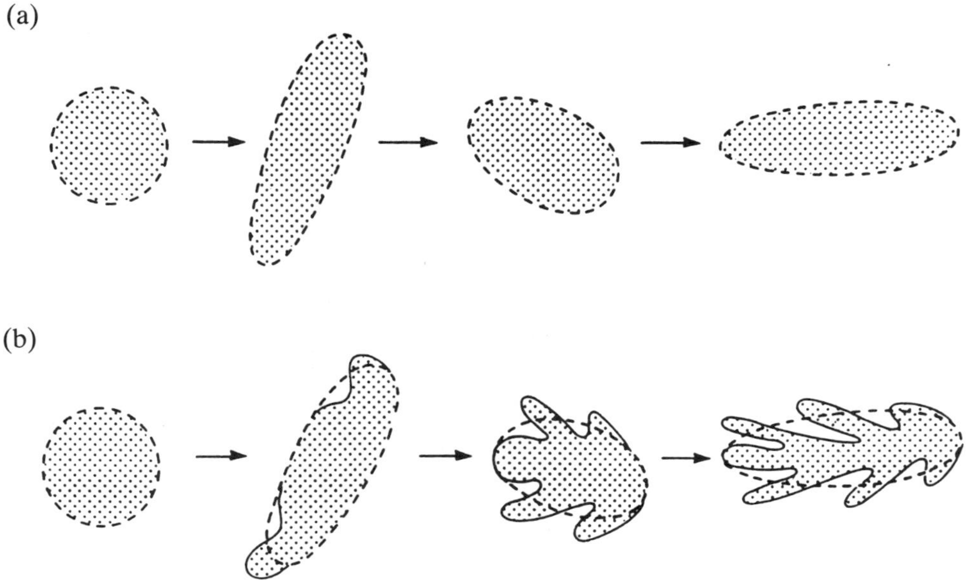

Fig. 28. (a) A pulsating material blob and (b) a pulsating and lobe-forming material blob. We would like to characterize the lobe formation, and not the pulsation, as entrainment and detrainment. Hence a natural partition is shown by the dashed line.

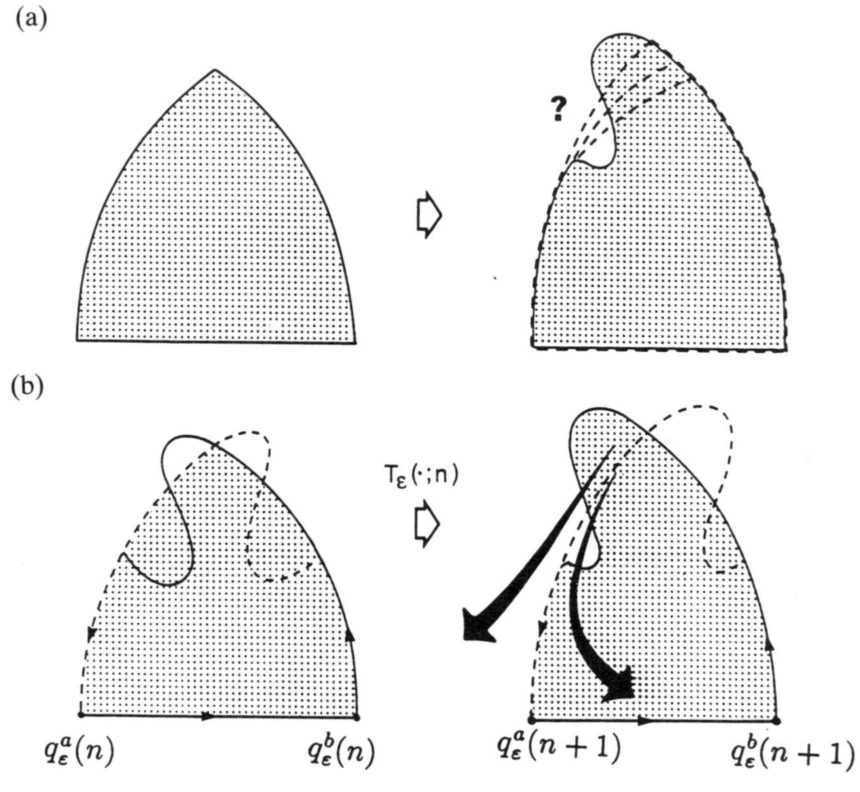

Fig. 29. (a) A material blob deforms; the dashed lines show possible dividing lines between 'inside' and 'outside'. (b) The stable manifold (the dashed line) provides a natural dividing line where the lobes form between 'inside' and 'outside'. Here $q^i_\varepsilon(n) \equiv \tau^i_\varepsilon \cap \chi[\theta_{1_0} + 2\pi(\omega_1/\omega_2)n]$ for $i = a, b$.

setting, from which the constructs and transport formalism for 2D slices of this setting will immediately follow. A formal description of transport in these more complicated settings can be much more involved and notationally intensive, but the basic idea follows the three elements described in the introductory part of this transport section, and in spirit follows much the same description given in detail for the time-periodic case. We thus defer the formal description to [8–10], and convey the essential idea for the case of two-frequency velocity fields. Recall from Section 2.2 that in this instance one has, rather than 1D invariant manifolds in a 2D Poincaré section intersecting in points, 2D manifolds in a 3D Poincaré section intersecting in *curves*. The key point to appreciate is that the invariant manifolds, namely global stable and unstable manifolds of normally hyperbolic invariant manifolds, are still of one less dimension than the Poincaré section, i.e. *codimension-one*, and hence a union of segments of these manifolds can serve as a dividing surface in the Poincaré section. In particular, if the upper stable and unstable manifolds of the open vortical flow, $W^s(\tau_\varepsilon^a)$ and $W^u(\tau_\varepsilon^b)$, intersect in a 1-torus τ_p, then a partition of the Poincaré section can be defined by forming a union of segments of 2D manifolds

$$\Gamma_\varepsilon \equiv U[\tau_\varepsilon^b, \tau_p] \cup S[\tau_\varepsilon^a, \tau_p] \cup X[\tau_\varepsilon^a, \tau_\varepsilon^b], \tag{51}$$

as illustrated in Fig. 30. For more complicated intersection geometries than toroidal a slightly more involved partition construction is needed [8–11, 13], but that need not

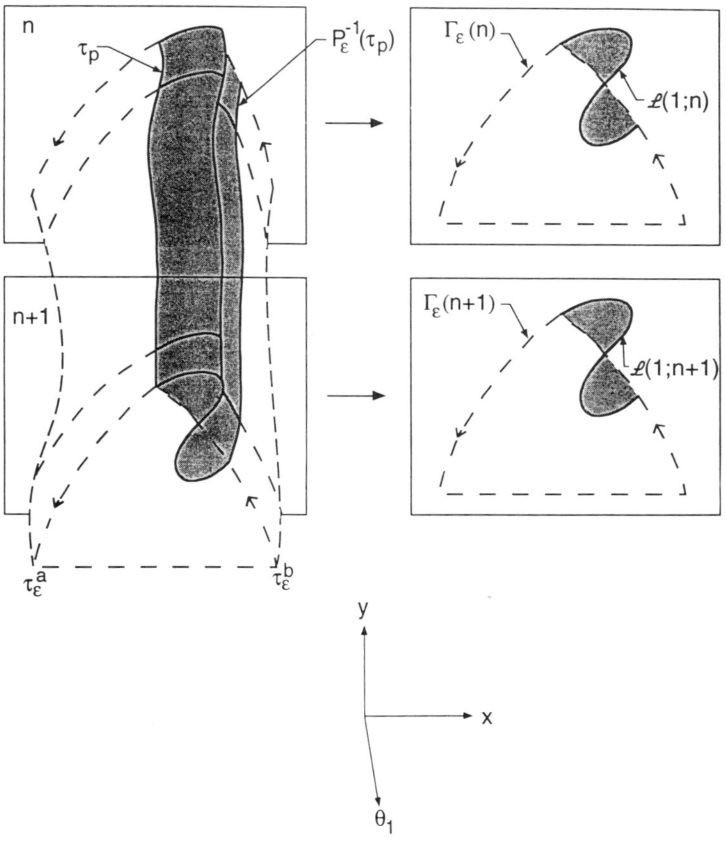

Fig. 30. Defining a partition (dashed line joining τ_ε^a, τ_ε^b, and τ_p) and turnstiles (shaded region) in the 3D Poincaré section $\Sigma^{\theta_{2_0}}$ for the two-frequency open flow, and the resulting sequence of time-dependent partitions and turnstiles for the 2D slices, which represent the fluid at a sequence of sample times.

concern us here. Much as in the time-periodic case, having defined a partition, transport across the partition is then described via turnstiles, denoted by $\mathscr{L}(1)$ (or by lobes $\mathscr{L}_{ij}(1)$ in the more general setting of several regions), whose boundaries are determined from appropriate segments of the partition and its pre-image. For example, when τ_p is a 1-torus we have

$$\partial\mathscr{L}(1) = S[\tau_p, P_\varepsilon^{-1}(\tau_p)] \cup U[\tau_p, P_\varepsilon^{-1}(\tau_p)], \tag{52}$$

a result in direct analogy with the time-periodic case (equation (48)).

Once one has the necessary constructs in the autonomous setting of the 3D Poincaré section, for the 2D fluid at any time sample, $t = (2\pi/\omega_2)n$, a time-dependent partition $\Gamma_\varepsilon(n)$, and turnstile, $\mathscr{L}(1; n)$, is then defined by the intersection of the autonomous constructs with the appropriate 2D time slice

$$\begin{aligned} \Gamma_\varepsilon(n) &\equiv \Gamma_\varepsilon \cap \chi[\theta_{1_0} + 2\pi(\omega_1/\omega_2)n] \\ \mathscr{L}(1; n) &\equiv \mathscr{L}(1) \cap \chi[\theta_{1_0} + 2\pi(\omega_1/\omega_2)n] \end{aligned} \tag{53}$$

(see Fig. 30). In this manner one can study the transport of fluid under sequences of 2D nonautonomous maps in the context of a sequence of *time-dependent* partitioning surfaces and turnstiles, an example of which is shown in Fig. 31. It is a consequence of the construction of time-dependent partitions from segments of invariant manifolds in a higher-dimensional setting that these partitions are found to be the true, albeit time-dependent, dividing lines between regions of qualitatively different fluid motion, much as described in the time-periodic case, and as illustrated in Fig. 29(b).

As described in detail in [8–10], velocity fields with quasiperiodic or aperiodic time dependences can engender much richer and more varied transport phenomena than in the typical time-periodic case. In particular, the turnstiles can vary in shape, number, and size from one time sample to the next, allowing for greater variation in the nature of the transport processes.

5.2. *Features of transport*

There are a variety of features of fluid transport one can study in the context of the theory described in the prior section. We briefly discuss two features.

5.2.1. *Flux, optimization, and control via Melnikov theory.* Perhaps the most elementary transport quantity is the flux of fluid across the partition, one measure of the amount of stirring or mixing occurring in the tangle. For area-preserving 2D maps derived from time-periodic velocity fields, flux across the partition (in either direction) is defined as the area of fluid crossing the partition (in either direction) under one iterate of the Poincaré map, divided by the period associated with the map. For sequences of 2D maps derived from velocity fields with quasiperiodic or aperiodic time dependences, several types of flux, with more complicated definitions, are allowed: *instantaneous flux* associated with crossing the partition under any successive pair of time samples, *finite-time average flux* associated with crossing the partition over some finite number of time samples, n, and *infinite-time average flux* associated with crossing the partition under an infinitely long time interval. The instantaneous flux will typically vary from one time sample to the next, and differ for a capture process versus an escape process, as will be the case for finite-time average flux. However, as the number of time samples, n, involved in the finite-time average flux goes to infinity, the finite-time average flux will of course asymptotically approach the infinite-time average. These various types of flux and the possible variety of their

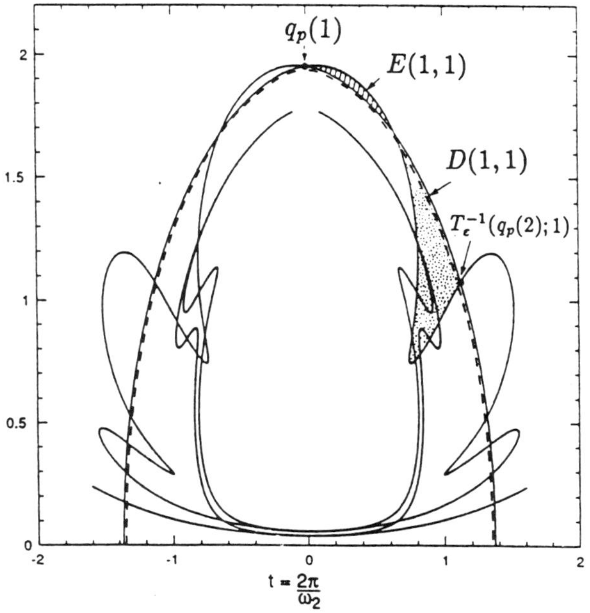

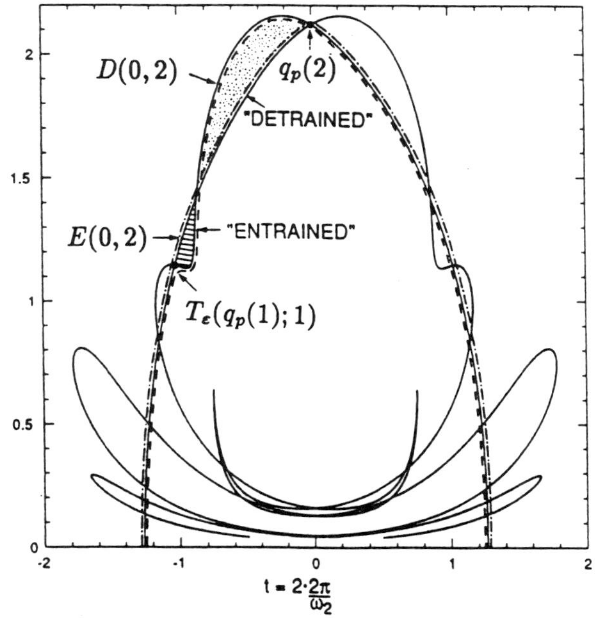

Fig. 31. Showing explicitly entrainment and detrainment between the $t = 2\pi/\omega_2$ sample and the $t = 2 \cdot 2\pi/\omega_2$ sample of the two-frequency open flow, whose equation of motion is given in the Appendix with $\varepsilon = 0.12$, $f_1 = 0.4$, $f_2 = 1.05$, $\omega_1 = 2$, $\omega_2 = 4$, $\theta_{1_0} = \theta_{2_0} = 0$. The dashed line is $\Gamma_\varepsilon(1)$ in the $t = 2\pi/\omega_2$ sample and $T(\Gamma_\varepsilon(1);\, 1)$ in the $t = 2 \cdot 2\pi/\omega_2$ sample. The dashed-dotted line is $\Gamma_\varepsilon(2)$, and $q_p(n) \equiv \tau_p \cap \chi[\theta_{1_0} + 2\pi(\omega_1/\omega_2)n]$.

magnitudes, are one example of the richer and more varied transport phenomena associated with velocity fields with quasiperiodic or aperiodic time dependences.

One could, of course, measure flux by a numerical computation of turnstile lobe area; however, a more convenient method of measurement for near-integrable flows is provided by the analytical technique of Melnikov theory described in the prior section. Given that Melnikov theory provides an $O(\varepsilon)$ measure of stable and unstable manifold separation normal to the unperturbed separatrix (refer back to Section 4), an area element of a lobe, δA, associated with an infinitesimal line element along the unperturbed separatrix, $\delta\lambda$, is given by

$$\delta A = \frac{\varepsilon|M(t_0)|}{\|D\psi_0(x_h(-t_0),\, y_h(-t_0))\|}\delta\lambda + O(\varepsilon^2). \tag{54}$$

The area of the turnstile lobes is given by an integral of this area element over the arclength of the unperturbed separatrix joining the two Melnikov zeros that define the turnstile lobe boundaries. Using that $\delta\lambda = (\mathrm{d}\lambda/\mathrm{d}t_0)\delta t_0 = \|D\psi_0(x_h(-t_0),\, y_h(-t_0))\|\delta t_0$ gives for the above area integral

$$\mu(\mathscr{L}(1)) = \varepsilon\int_{t_0^0}^{t_0^{-1}} |M(t_0)|\,\mathrm{d}t_0 + O(\varepsilon^2), \tag{55}$$

where $\mu(\mathscr{L}(1))$ denotes the area of the turnstile lobes $\mathscr{L}(1)$, and where t_0^0 and t_0^{-1} are the t_0 values of the chosen PIP for partitioning, q_p, and its pre-image, $P_\varepsilon^{-1}(q_p)$, respectively. For time-periodic velocity fields, the flux in either direction, Φ, is given by one-half the turnstile area divided by the sampling period,

$$\Phi = \frac{\omega}{2\pi}\,\frac{\varepsilon}{2}\int_{t_0^0}^{t_0^{-1}} |M(t_0)|\,\mathrm{d}t_0 + O(\varepsilon^2). \tag{56}$$

For velocity fields with quasiperiodic or aperiodic time dependences there are several flux expressions analogous to (56), as indicated previously. We refer the reader to [8–10] for most of these expressions, and give here merely one example, namely infinite-time average flux for two-frequency velocity fields:

$$\Phi(\theta_{1_0}) = \lim_{T\to\infty}\frac{\varepsilon}{2T}\int_0^T |M(t_0,\, \theta_{1_0})|\,\mathrm{d}t_0 + O(\varepsilon^2). \tag{57}$$

For ω_1 and ω_2 incommensurate, the above improper integral is found to be independent of θ_{1_0}; for ω_1 and ω_2 commensurate, the integral is found to depend on θ_{1_0}, and to be reducible to a proper integral over a finite time interval.

The attractive feature of a Melnikov-based measurement of flux is a convenient and efficient means for searching through parameter space in order to optimize flux. Such a control strategy could be helpful towards the enhancement of a mixing process. We give here a simple example of an optimization study in the context of the two-frequency open flow (see Section 2.2 and the Appendix). Recall that this flow arises from a perturbing streamfunction of the form

$$\varepsilon\tilde{\psi}(x,\, y,\, t;\, \varepsilon) = \varepsilon xy[\omega_1 f_1 \sin(\omega_1 t + \theta_{1_0}) + \omega_2 f_2 \sin(\omega_2 t + \theta_{2_0})]. \tag{58}$$

In particular, for some given perturbing frequency pair (ω_1, ω_2), let us study the dependence of infinite-time average flux on one of the amplitudes of the perturbing streamfunction, with the other amplitude determined by some normalization condition. Consider for example the normalization $F_1 + F_2 = 1$, where $F_i \equiv \omega_i f_i$ for $i = 1, 2$, and let us study how the flux varies with $F_1 \in [0, 1]$. The two endpoints of the F_1 interval, $F_1 = 0$ and $F_1 = 1$, then correspond to the two single-frequency limits of the perturbation (entirely ω_2 forcing and entirely ω_1 forcing, respectively). The solid line of Fig. 32(a) shows a case where the infinite-time average flux is identical at the two single-frequency limits. Everywhere between the two single-frequency limits the flux decreases as a consequence of interference effects. The dashed line of Fig. 32(a) shows a case where the flux is different

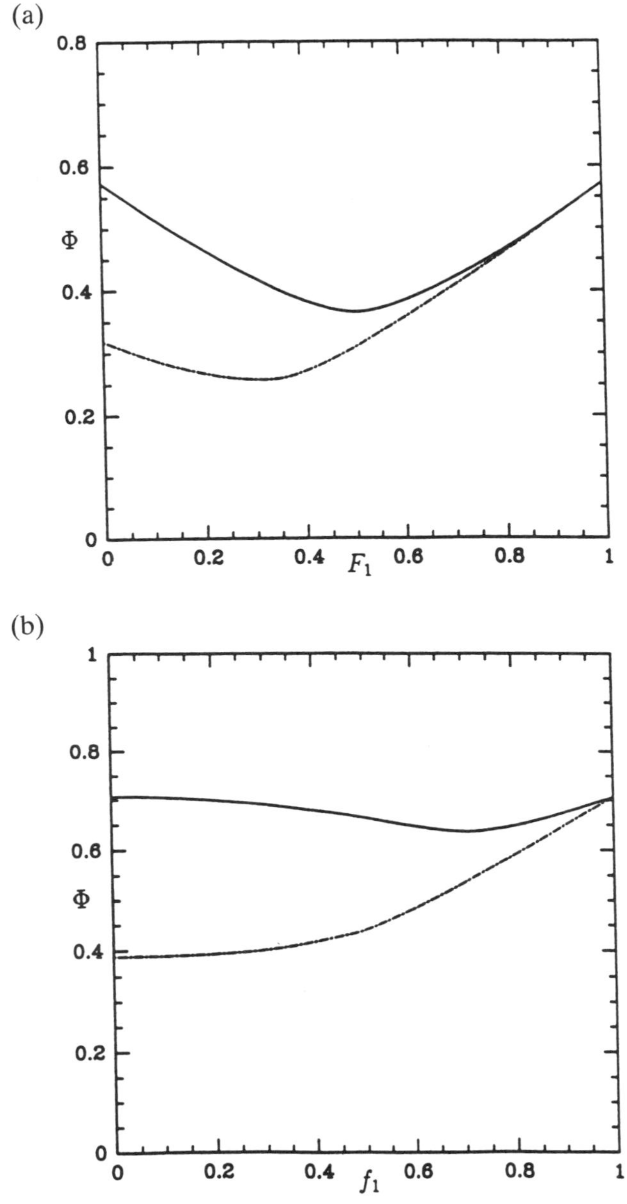

Fig. 32. (a) Infinite-time average flux for the two-frequency open flow as a function of F_1 for $F_1 + F_2 = 1$, with $(\omega_1, \omega_2) = (1.48, 0.78)$, $(A_1/F_1, A_2/F_2) = (1.80, 1.80)$ for the solid line and with $(\omega_1, \omega_2) = (1.48, 0.66)$, $(A_1/F_1, A_2/F_2) = (1.80, 1.00)$ for the dashed-dotted line, where A_1, A_2 are the Melnikov amplitudes of Section 4. (b) Infinite-time average flux for the two-frequency open flow as a function of f_1 for $f_1^2 + f_2^2 = 1$, with $(\omega_1, \omega_2) = (1.00, 1.94)$, $(A_1/f_1, A_2/f_2) = (2.22, 2.22)$ for the solid line and $(\omega_1, \omega_2) = (1.00, 0.75)$, $(A_1/f_1, A_2/f_2) = (2.22, 1.22)$ for the dashed-dotted line. In both plots the vertical scale is per unit ε and $\theta_{1_0} = \theta_{2_0} = 0$.

at the two single-frequency limits. As one varies F_1 from the lower to higher single-frequency limit, a dip away from a linear increase in the infinite-time average flux is again a consequence of interference effects. Hence an absolute maximum in the infinite-time average flux is obtained at one of the single-frequency limits. The optimal situation for the two-frequency problem is then the single-frequency limit with the greater infinite-time average flux; to include the second frequency only *detracts* from the infinite-time average flux. Such a phenomenon is typically found for other reasonable and physically motivated choices of normalization, an example of which is shown in Fig. 32(b).

The above example of an optimization search through parameter space involved the relative amplitudes of the perturbation; other searches could involve relative phases, frequency combinations, and so on. With so many system parameters to search through, numerical searches quickly become prohibitively computationally intensive, and the analytical searches afforded by Melnikov theory have great practical use.

5.2.2. *Lobe geometry and lobe intersections.* Lobe area, or lobe volume, and flux provide one measure of transport across partially destroyed barriers; however, most realistic questions pertaining to phase space transport necessitate a consideration of lobe geometry. For example, suppose we wish to compute the amount of fluid originally in the time-periodic open-flow partition that has been detrained after some n perturbation periods. To facilitate the discussion, let us assume there is originally ($t = 0$) fluid A in the partition and fluid B outside the partition. Clearly as time progresses there will be a decreasing amount of A remaining within the partition. Hence the detraining lobe $D(1)$ will tend to contain a smaller percentage of A with increasing time. Such an occurrence pertains to the issue of the *species content* of a lobe. As we shall see, the species content of a lobe is related to the geometry of the lobe within the tangle. The key to understanding lobe geometry within the tangles is to recognize that (i) lobes can, and indeed must, intersect with other lobes in the tangle as well as wrap around yet other lobes, and (ii) the non-self-intersecting property of each of the stable and unstable manifolds highly constrains how the lobes intersect and wrap around one another. Consider for example the tangle associated with the $t = 0$ sample of the two-frequency open flow shown in Fig. 13 and shown again in Fig. 33. Lobe $D(3,0)$ intersects with lobe $E(1,0)$; the fluid in $D(3,0) \cap E(1,0)$ will not only be entrained into the partition in the next time sample, it will then be detrained from the partition between the $n = 2$ and $n = 3$ time sample. Similarly $E(-2,0)$ intersects $D(1,0)$: the fluid in $E(-2,0) \cap D(1,0)$ will be detrained

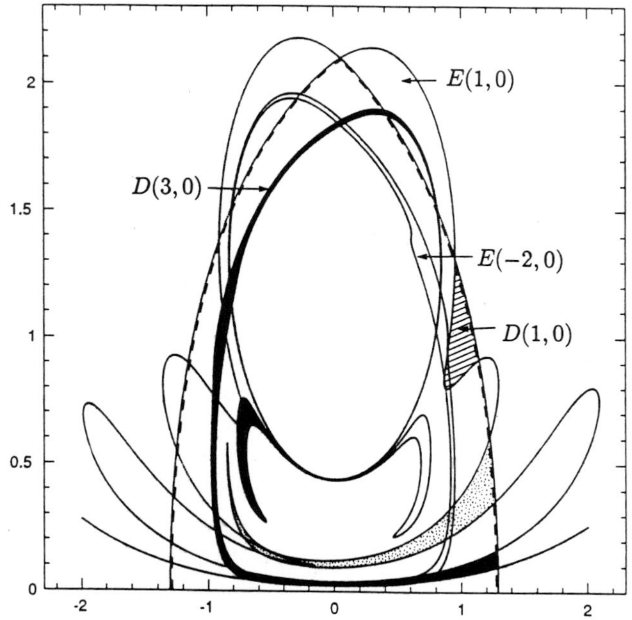

Fig. 33. The lobe structure of the $t = 0$ sample of Fig. 13, shown again with the necessary labeling.

from the partition in the next time sample and was entrained into the partition between the $n = -3$ and $n = -2$ time sample. These concerns relate to the *content* of those lobes about to be entrained/detrained in the next time sample: examining the intersections of these lobes with other lobes provides an understanding of the history of the fluid particles, i.e. when they might have been previously entrained/detrained or in the future when they may be entrained/detrained. Addressing concerns such as these allow one to answer the transport question we originally posed about the amount of fluid initially in a partition that exits the partition after some n perturbation periods. A systematic and detailed study of transport phenomena in the context of lobe geometry and lobe intersection formulas is given in [83, 105]. The development of the formulas, and indeed the formulas themselves, can be quite involved. The basic idea behind all development and use of lobe intersection formulas, however, is conveyed in Fig. 34. Consider lobe $D(3,0)$, which contains the fluid that will be detrained from the partition between the $n = 2$ and $n = 3$ time samples. The lobe $D(3,0)$ of course need not be contained entirely within the partition, and hence will contain a fraction of both fluid A and B. Given that the stable manifold $W^s(\tau_\varepsilon^a) \cap \chi(\theta_{1_0})$ cannot intersect itself, it is clear to see that (i) the portion of the lobe $D(3,0)$ that lies outside the partition (and hence contains fluid B) must *intersect* with one or more of the entraining lobes $E(1,0)$, $E(2,0)$ and $E(3,0)$, and (ii) the lobe $D(3,0)$ must *wrap around* one or both of the other detraining lobes $D(1,0)$ and $D(2,0)$. With considerations such as these, one can express transport processes in the context of highly constrained lobe-intersection formulas.

The lobe-intersection formulas are essentially statements of topological constraints based on the non-self-intersecting property of each invariant manifold. Though these and the lobe-dynamic formulas describe transport in a geometrically exact setting, the determination of transport rates of course requires explicit numerical computation. Some in the phase space transport community appear to dismiss a lobe-dynamic formalism because of its inability to offer an elegant and convenient expression for transport rates (e.g. see [62]). Indeed from the standpoint of a practical computation, Beigie *et al.* [8–10] point out how, in practice, one would typically avoid computing transport rates in terms of the highly convoluted intersections of lobes with each other; instead a more convenient, and no less fundamental, description of transport rates via species content of a turnstile involves the

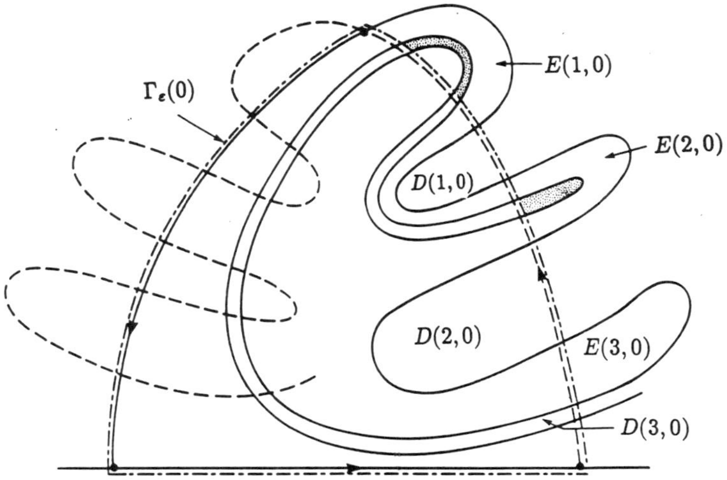

Fig. 34. A portrayal of the two-frequency open flow tangle at $t = 0$. The lobe $D(3,0)$ intersecting with the two entraining lobes $E(1,0)$ and $E(2,0)$, and wrapping around the detraining lobe $D(1,0)$.

intersections of the lobes with the partitioned regions themselves. Despite the merits or disadvantages of the lobe intersection formulas, the geometrically exact description of phase space transport in the context of lobe dynamics within invariant manifolds is a useful setting for understanding transitions across partially destroyed barriers. In order to appreciate the usefulness of this setting, it is important to recognize the following.

(i) Even the simplest transport issues associated with 2D maps derived from 2D time-periodic velocity fields can become quite challenging in the more complicated settings of higher-dimensional flows and velocity fields with quasiperiodic or aperiodic time dependences. Here even the mere establishment and determination of invariant manifolds, tangles, partial barriers, turnstiles, lobe area/volume, flux, and so on, is of great value.

(ii) It is typically unrealistic to expect a nonlinear theory phrased in the context of phase space geometry to provide easily obtained, comprehensive, and quantitative answers to the problem at hand. A more realistic expectation is to use phase space geometry to obtain a qualitative understanding of the dynamics, formulate paradigms of behavior, initiate partial answers to certain features of the nonlinear dynamics, and provide a setting for intelligently chosen numerical investigations.

Besides explicit numerical computation of phase space transport via turnstile images and pre-images, one can also seek approximate descriptions of transport using elementary models. A popular method, often employed in the early days of phase space transport theory, involves a Markov model approach [54, 61]. Here one solves for species content of a turnstile lobe at the nth time sample using the rather artificial assumption that, once a turnstile is sent under the Poincaré map to a new region, its contents then instantaneously diffuse throughout the new region. Such an unrealistic assumption has the advantage of giving convenient analytical expressions for phase space transport, which sometimes approximate the true transport process quite adequately. In other instances, however, it is possible for the approach to work poorly [10, 105]. The important point to appreciate about such a Markov model approach is that, once it addresses phase space geometry in the context of partial barriers and turnstiles, by design it then ignores any further geometrical considerations when invoking the instantaneous approximation. An interesting prospect, then, would be to take a modeling approach one step further and construct a model that attempts to respect geometrical considerations associated with the pre-images and images of the turnstiles as well. For example, in the upcoming section on stretching, we construct models by invoking the approximate self-similar behavior of lobe evolution as it repeatedly winds around the tangle. Beigie and Wiggins [10] mention the possible relevance of such an approach to a study of transport phenomena. Rom-Kedar [84] offers perhaps the first attempt at modeling transport processes in such a way as to respect geometrical considerations associated with images and pre-images of the turnstile lobes.

6. STRETCHING

Let us now turn our attention to stretching phenomena in chaotic tangles. A paradigm for chaotic dynamics under a 2D map is the Smale horseshoe map, illustrated in Fig. 35(a). The existence of such a map in homoclinic tangles (and in a similar manner heteroclinic tangles) is guaranteed by the Smale–Birkhoff Homoclinic Theorem (e.g. see [39]). The horseshoe map consists of a stretch and fold process, which upon repetition *ad infinitum* generates exponential stretching and sensitive dependence on initial conditions in an invariant subset of the domain (a Cantor set in the vicinity of a hyperbolic fixed point). One can represent horseshoe map dynamics on the invariant subset by a symbolic description that makes the essential properties of chaotic dynamics quite plain. This picture is essentially robust for the dynamics associated with sequences of 2D maps (obtained for example from 2D velocity fields with quasiperiodic or aperiodic time dependences), and generalizes to what we have called (and rigorously established [8–10]) a *traveling horseshoe map sequence*, shown heuristically in Fig. 35(b). This scenario consists of, rather than a single horseshoe map, a bi-infinite sequence S_H of different 'horseshoe maps' $H(\cdot\,; j)$: $\mathbb{R}^2 \rightarrow \mathbb{R}^2$,

$$S_H = \{\ldots, H(\cdot\,; -j), \ldots, H(\cdot\,; -1), H(\cdot\,; 0), H(\cdot\,; 1), \ldots, H(\cdot\,; j)\}, \tag{59}$$

and a bi-infinite sequence S_D of different domains $D(j) \in \mathbb{R}^2$,

$$S_H = \{\ldots, D(-j), \ldots, D(-1), D(0), D(1), \ldots, D(j)\} \tag{60}$$

such that $H(D(j); j)$ intersects $D(j+1)$ in the shape of a horseshoe. There is thus a sequence of formed horseshoes landing on different regions of $\mathbb{R}^2$; each time the horseshoe happens to land on the region that will next form a horseshoe, and land in such a way that the stretched direction 'aligns' with the direction about to be stretched. It is clear that this sequence of maps retains the essential ingredient of chaos—repeated stretching and folding, and hence sensitive dependence on initial conditions. A symbolic dynamics can be constructed in this instance relative to a time-dependent sequence of Cantor sets. For certain classes of 3D flows containing normally hyperbolic invariant manifolds, a horseshoe-like paradigm applies, where the flow exhibits a 3D horseshoe, i.e. a stretch and fold process in two dimensions, with minimal deformation in the third dimension. The stretching possibilities associated with 3D flows are much richer than this simple scenario, however, and will remain unexplored in this presentation.

For both cases of 2D maps and sequences of 2D maps the invariant subsets we speak of are Cantor sets in phase space near a hyperbolic fixed point or normally hyperbolic invariant manifold. Consequently, though the horseshoe map and traveling horseshoe map sequence, with their associated symbolic description, make quite plain the stretching properties of these Cantor sets, they give less information about the global stretching

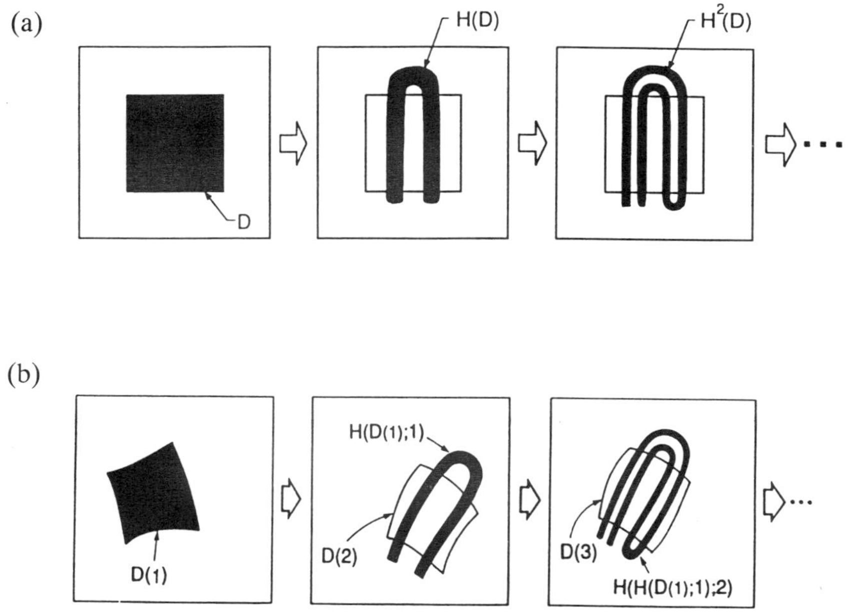

Fig. 35. (a) A horseshoe map, and (b) a traveling horseshoe map sequence.

properties in chaotic tangles. One would like to obtain a more detailed understanding of the stretching and folding process in chaotic tangles, which we thus explore in the context of the lobes of the invariant manifold template wrapping around one another in a highly constrained manner. Given that the time-periodic analysis robustly extends to more complicated time dependences (refer back to Section 2), for simplicity of discussion and concreteness of presentation we focus on the 2D time-periodic case in this section. In Section 6.1 we introduce the notion of lobe evolution as an extension of the horseshoe map paradigm in order to obtain a more global understanding of stretching in chaotic tangles. In particular, we construct a symbolic description for lobe boundary evolution, one that applies then to a one-dimensional curve bounding the lobe, rather than to a Cantor set in phase space. The symbolic description is based on a recursive partitioning of the lobe boundary by the *secondary intersection points* (SIPs) along the boundary, which divide between lobe boundary segments of qualitatively different behavior, namely, different outcomes of, and elapsed times for, successive revolutions around the chaotic tangle. These intersection points will be seen to be significant in their own right, since they asymptotically approach the hyperbolic fixed points. In the context of the symbolic description one can study the global topology of, and mechanisms for, enhanced stretching in chaotic tangles. We perform such a study in the context of numerical calculations of *stretch profiles* associated with the lobe boundaries, i.e. plots of stretch versus initial arclength along the lobe boundary. Loosely speaking, the basic picture we uncover in the near-integrable setting is one of an unperturbed stretch profile approximately repeating itself on smaller and smaller scales. A range of temporal, spatial and stretch scales are associated with this process, and the stretch profiles can have significant variation in their non-uniformities, which will have a basic impact on stretching and mixing, as we shall see later. In contrast, then, to the horseshoe map scenario of a repetition of an essentially uniform stretching and folding process in a small domain near a hyperbolic fixed point, our invariant manifold based analysis shows *approximate* repetition of highly *non-uniform* stretching and folding of lobe boundaries *throughout* the chaotic tangle. Section 6.2 turns our attention from stretch to a geometric understanding of stretch *rates*, i.e. rates of strain. Our principal interest here will be in the notion of irreversible rate of strain, the role of hyperbolic fixed points as engines for good irreversible straining, and the role of turnstiles as mechanisms for increasing stretch efficiency via *re-orientation* of line elements and *transport* of line elements to regions of superior straining. Having obtained some geometric understanding of stretch and stretch rates, we end in Section 6.3 with a study of one of many features of stretching one might wish to investigate in the context of an invariant manifold template, namely stretch *statistics* or, more precisely, probability distributions of finite-time Lyapunov exponents. In particular, we perform, via a point insertion scheme, a high-resolution numerical study of high-stretch tails of these distributions, and find a range of tail behavior, including significant non-Gaussian behavior as a consequence of non-uniformity in the stretch profiles.

6.1. *Lobe evolution as an extension of the horseshoe map paradigm*

Since the lobes in chaotic tangles map from one to another, let us consider the evolution of one of these lobes, in particular the entraining turnstile lobe (chosen since it is about to be entrained and hence undergo the first stretch and fold). Though for completeness we could also study the detraining turnstile lobe, we restrict our attention to the entraining lobe, and note that the images of the detraining lobe for the open flow undergo no repeated stretching and folding, which thus evolve non-chaotically. Given the dominant role of the unstable manifold (refer back to Section 3), we choose to study stretching associated with the unstable manifold segment of the lobe boundary. Since the stable manifold segment of the lobe boundary rapidly shrinks to negligible length, such a study effectively amounts to a study of the entire lobe boundary. Consider then a material curve that starts off ($t = 0$) on the segment of the unstable manifold that bounds the entraining turnstile lobe, as shown in Fig. 36. We denote this material curve by $r_0(t)$, with endpoints given by the PIPs $p1$ and $p2$. The evolution of r_0 is best understood by first making a comparison with an unperturbed case. Consider, for example, the evolution according to the *unperturbed* flow of a material curve $\tilde{r}_0(t)$ which at $t = 1$ lies on the segment of the *perturbed* unstable manifold that is essentially $P_\varepsilon(r_0(t = 0))$, except that its endpoints $\tilde{p}1$, $\tilde{p}2$ are the points which asymptotically approach the unperturbed fixed point rather than the perturbed fixed point (see Fig. 37). This material curve winds around the core *ad infinitum*, never intersecting the unperturbed separatrix, except at $\tilde{p}1$ and $\tilde{p}2$. Since the endpoints asymptotically approach a hyperbolic fixed point, the infinitesimal line elements of $\tilde{r}_0$ associated with the endpoints stretch asymptotically exponentially in time; however,

(a)

(b)

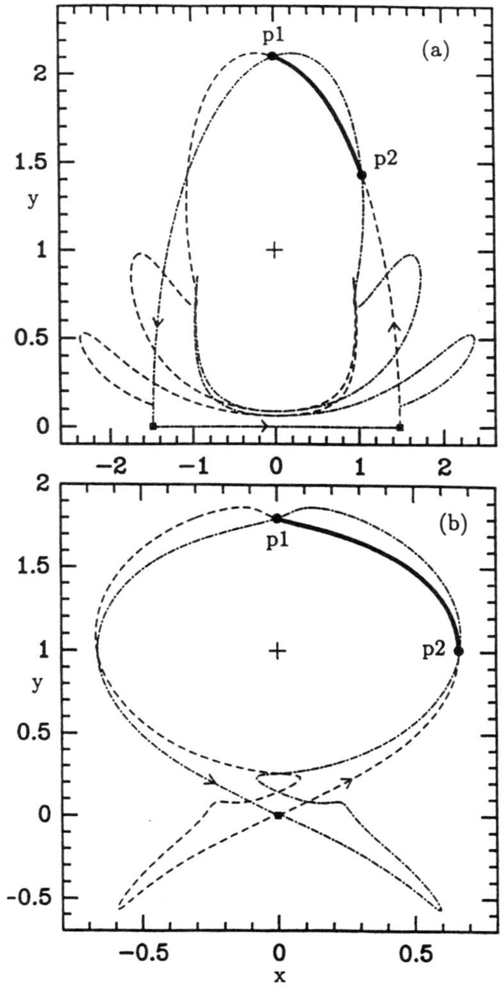

Fig. 36. Segments of stable (dashed-dotted) and unstable (dashed) manifolds of hyperbolic fixed points (solid squares) in Σ^{θ_0}, and our chosen interface (solid line) for the (a) open and (b) closed flow. The vortices are at $(x, y) = (0, \pm 1)$ at $t = n \in Z$, with magnitude of circulation $\Gamma = 0.4(2\pi)^2$ and (a) $\varepsilon = 0.085 \cdot 2\pi$, (b) $\varepsilon = 0.02 \cdot 2\pi$.

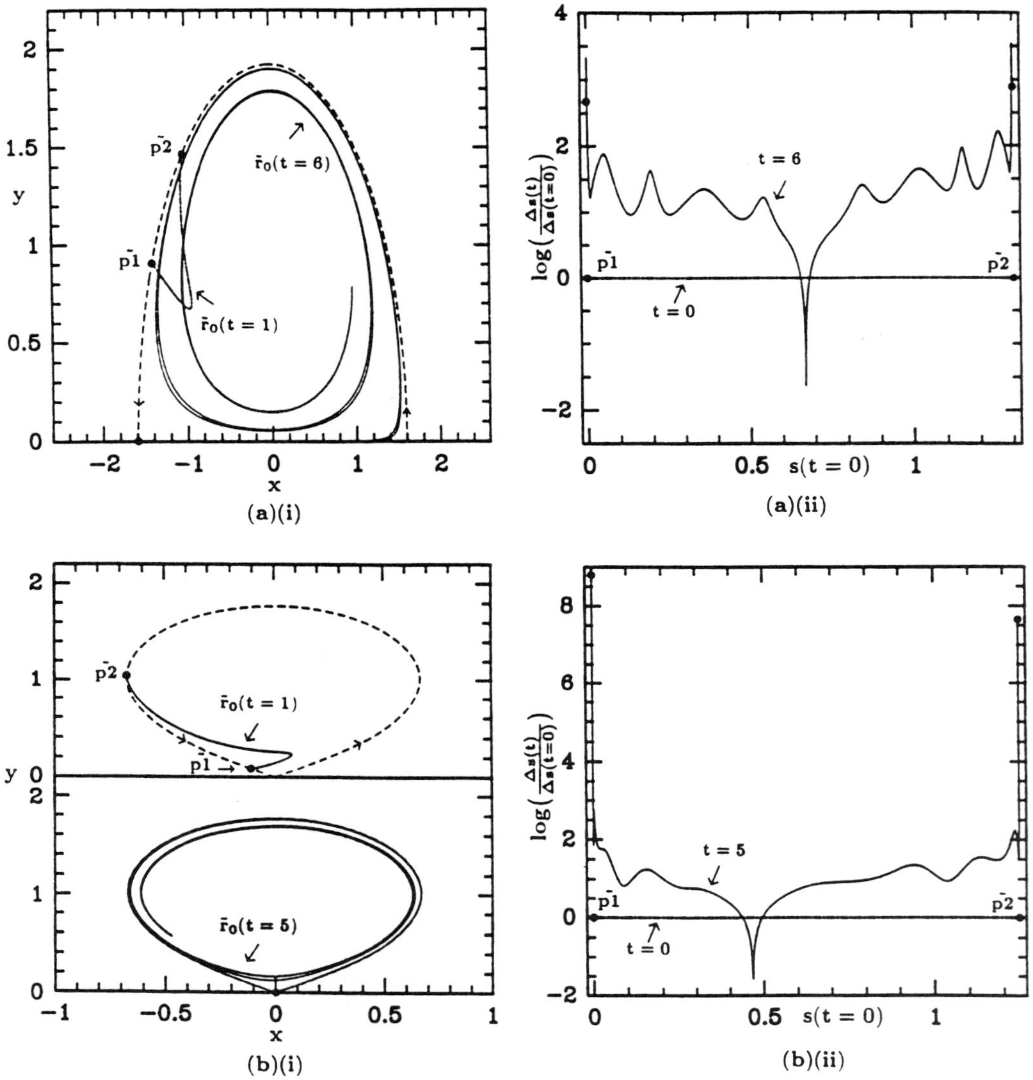

Fig. 37. Evolution, according to the unperturbed (a) open and (b) closed flows, of the material curve $\tilde{r}_0(t)$. The material curves are shown in (i) (the dashed line represents the unperturbed separatrix), and the associated stretch profiles are shown in (ii) (s denotes arclength along the material curve and log denotes the common logarithm). The magnitude of the circulation associated with each vortex is $\Gamma = 0.4(2\pi)^2$. For the stretch profiles, note that $\tilde{r}_0(t = 0) = P_\varepsilon^{-1}(\tilde{r}_0(t = 1))$, i.e., we evolve $\tilde{r}_0(t)$ from $t = 0$ to $t = 1$ according to the *perturbed* flow (see Fig. 43 for the parameters) and from then on according to the *unperturbed* flow. For the perturbed flows, the vortices are at $(0, \pm 1)$ at $t = 0$; for the unperturbed flows the vortices are set at the time average of the perturbed vortex coordinates, i.e., at $(0, \pm 0.92183)$ for the open flow and $(0, \pm 0.98039)$ for the closed flow.

overall $\tilde{r}_0$ stretches only asymptotically linearly in time. Hence as time progresses, the stretch profiles in Fig. 37 will be peaked more and more towards the endpoints, with an overall dip in the middle corresponding to poor stretching. In the perturbed case, r_0 is entrained into the bounded core at $t = 1$, and then for near-integrable flows starts to wind around the tangle not much unlike $\tilde{r}_0$ in the unperturbed case, except that an arbitrarily small time-periodic modulation of the velocity field suffices to cause r_0 to intersect the stable manifold segment of the partition (these intersection points are SIPs), as shown heuristically in Fig. 38 for the open flow (numerical simulations are shown presently). In

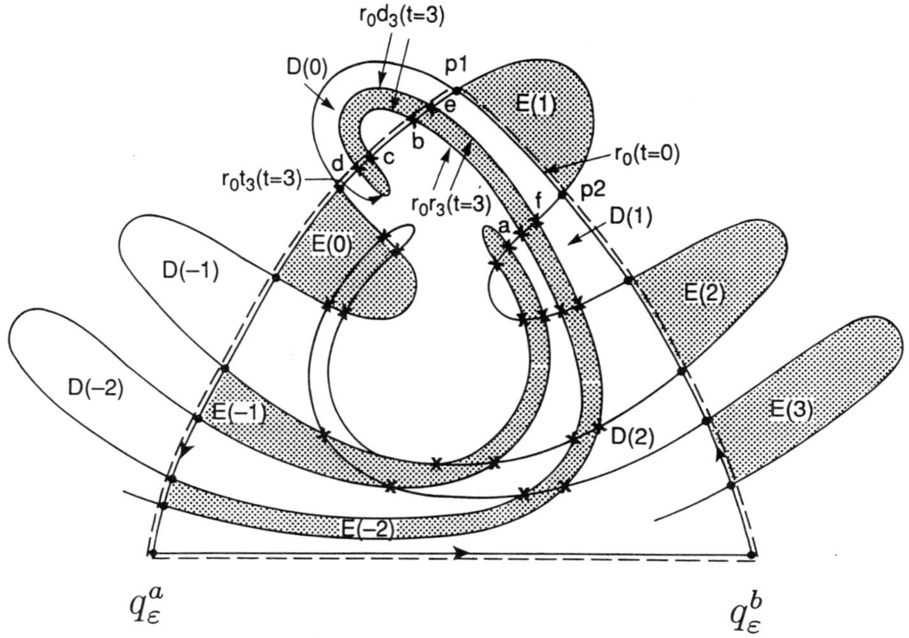

Fig. 38. The upper ($y \geqslant 0$) invariant lobe structure of the perturbed open flow (the system is symmetric about $y = 0$).

particular, it is a straightforward consequence of the invariance and non-self-intersecting property of the global stable and unstable manifolds that the first intersection of $r_0(t)$ with the stable manifold segment of the partition, $S[q_\varepsilon^a, p1]$, must be with the stable manifold segment of the first image of the turnstile lobe $D(0) \equiv P_\varepsilon(D(1))$, as shown in Fig. 38. The SIPs have a two-fold significance: first, like the PIPs they asymptotically approach a hyperbolic fixed point and thus the infinitesimal line elements of r_0 associated with the SIPs stretch asymptotically exponentially in time; second, they are the dividing points between segments of r_0 that behave in a qualitatively different fashion, as shown in Fig. 38.

(i) A 'returned' part r_0r_3, at $t = 3$ between SIPs a, b and e, f, and which at $t = 3$ is close to $r_0(t = 0)$, and hence will evolve from time $t = 3$ approximately the way r_0 does from time $t = 0$ (the SIPs playing the role that the PIPs did originally for r_0). The subscript on the second r specifies the number of time samples it takes for $r_0r_3(t = 0)$ to 'return' to $r_0(t = 0)$ (the subscript on future symbols should be clear from this remark).

(ii) A detrained part r_0d_3, at $t = 3$ between SIPs b, c and d, e, which will remain outside the partition and stretch asymptotically linearly in time.

(iii) A tail r_0t_3, at $t = 3$ between SIPs c, d, which will wind around the tangle as its SIPs asymptotically approach the left fixed point, but in a way that is qualitatively different from r_0. (Though a general scheme needs to include a tail, for some system parameters there will be no tail, as can occur for example if in Fig. 38 the tip of $E(-2)$ lies in $D(0)$.)

(iv) The remainder of r_0, to which we add no further labels for the moment.

In this manner we construct for the open flow a symbolic description that applies to the entire material curve, as shown in Fig. 39. This description is based upon a characterization of the outcome of successive revolutions around the chaotic tangle, and is constructed in the following manner.

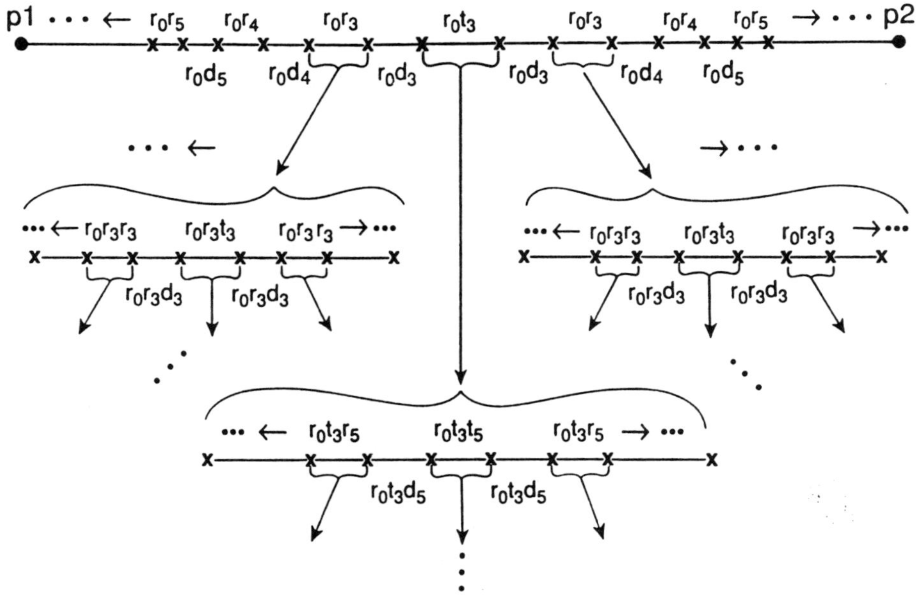

Fig. 39. The symbolic dynamics that applies to r_0 in the open flow. The upper line corresponds to r_0, and the lower lines to enlargements of pieces of r_0.

(i) With each additional application of P_ε, a portion of the remainder of r_0 will be 'returned' and another portion detrained. Hence, there will be an r_0r_4 and an r_0d_4 after the fourth iterate of P_ε, then an r_0r_5 and an r_0d_5 after the fifth iterate of P_ε, and so on, as shown in Fig. 40. Referring to Fig. 38, it is clear from the lobe dynamics that at $t = 3$ r_0r_n, $n \geqslant 3$, is a portion of the boundary of $E(-2)$ between $D(n-3)$ and $D(n-2)$, and r_0d_n is a portion of the boundary of $E(-2)$ inside $D(n-3)$. Hence the r_0r_ns and r_0d_ns join contiguously to cover r_0, as shown by the upper line in Fig. 39. These pairs of symbols on the upper line pertain to the first revolution around the tangle of successive pieces of r_0.

(ii) The tail and returned parts will further subdivide qualitatively the same way as does r_0 starting at $t = 0$, as shown in Fig. 41; the detrained parts do not keep subdividing. When we say that further subdivisions are qualitatively similar, we should note that successive revolutions around the tangle may take a different number of iterates of P_ε for the first return to occur, and that these numbers may in fact differ (by at most one) on either side of the tail (when this occurs we split the tail in two and have the t subscript on each side agree with that of the adjacent detrained part). Also, for large enough perturbations, though the tail subdivisions will for large enough n settle down to the alternating sequence of $r_0 \ldots t_m r_n$, $r_0 \ldots t_m d_n$, $r_0 \ldots t_m r_{n+1}$, $r_0 \ldots t_m d_{n+1}$, etc., for small n the ordering may be a bit erratic, due to the possibly complicated behavior of the tail. These minor complications do not detract from the overall usefulness of the symbolic dynamics in understanding the topology of enhanced stretching.

The result is a recursive partitioning of the interface, each partition represented by a string of symbols with three types of entries r, d, and t, and which terminates if and only if a 'd' is reached (since a 'd' signals detrainment, and hence the end of the stretch and fold process). Each entry denotes the outcome of a revolution around the tangle, and the subscript of the entry denotes the number of perturbation periods taken to obtain the outcome. For example, the string $r_0r_7r_{20}d_5$ denotes the segment of r_0 which takes 7 time samples to wind around the core and 'return' to r_0, then another 20 time samples to wind

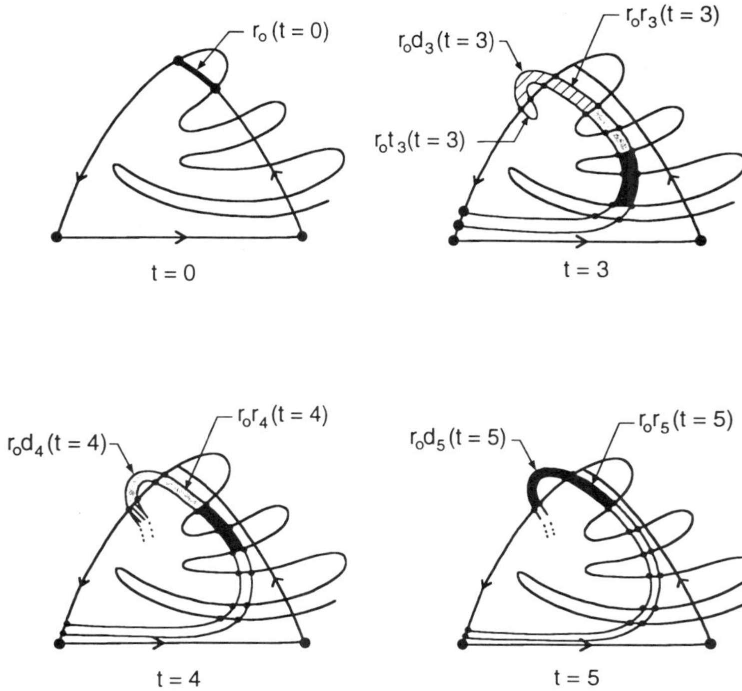

Fig. 40. A symbolic description of the open flow interface according to the first revolution around the tangle. The dashed, lightly shaded, and dark portions of the lobe indicate the portions of the lobe whose boundaries return, or are detrained, at $t = 3$, 4, and 5, respectively.

around the core once more and 'return', and then 5 time samples to wind around the core one last time to be detrained. The differences from conventional symbolic dynamics associated with the Smale horseshoe map are that the strings may be finite (if they contain a d), and that there is a subscript on each entry which specifies the time interval between successive entries. These differences in time scales, as we shall later discuss, are relevant to understanding the dynamics within the tangle regions. One can construct a similar symbolic dynamics for the closed flow, with two differences.

(i) Detrained segments can be re-entrained, so detrained segments will further subdivide, and the string of symbols will not terminate at a d.

(ii) The segment r_0 can pass through both the upper and lower tangle, which can be kept track of by using, say, small letters for the upper tangle and capital letters for the lower tangle.

The symbolic dynamics, coupled with a knowledge of the unperturbed stretch profile, provides a framework for studying the mechanism for, and topology of, enhanced stretching in chaotic tangles, as we shall describe more carefully momentarily. The basic picture to appreciate at the outset is that, since the 'returned' segments evolve approximately like $r_0(t = 0)$, one then understands the enhancement of the stretch profile, loosely speaking, in terms of the unperturbed profile approximately repeating itself on smaller and smaller scales, the topology of this repetition understood via the symbolic dynamics (refer back to Fig. 39, and see Fig. 42 for a heuristic—indeed rather cartoon-like—portrayal). For the open flow, the detrained segments stretch asymptotically linearly in time, so the stretch profile will contain gaps of poor stretching, the ordering of which is again understood by the symbolic dynamics.

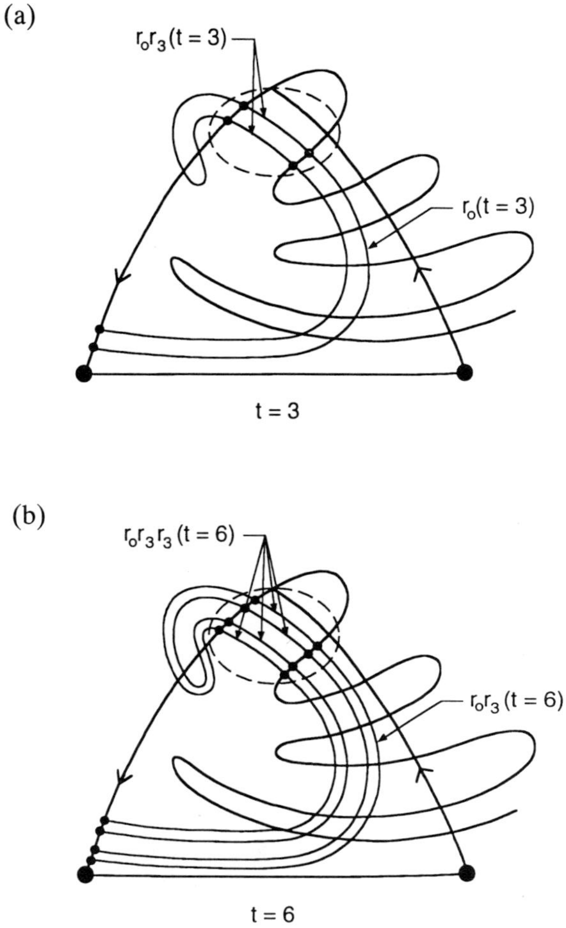

Fig. 41. Illustrating the approximate self-similar behavior of $r_0(t)$ as it evolves in time. (a) $r_0(t = 3)$ is shown, with the portion that has 'returned' at $t = 3$, $r_0 r_3(t = 3)$, encircled. This portion will evolve from $t = 3$ much as $r_0(t)$ evolves from $t = 0$. (b) This portion at $t = 6$, $r_0 r_3(t = 6)$, is portrayed, and yet another 'returned' portion $r_0 r_3 r_3(t = 6)$ is encircled.

In order to gain a better understanding of this symbolic description, let us consider in the context of a short-time simulation of the open and closed flow the evolution of r_0 and its associated stretch profiles with symbolic dynamics labeling, shown in Fig. 43. The closed flow clearly shows the highly non-uniform unperturbed stretch profile approximately repeating itself on smaller and smaller scales. The SIPs entering the region local to the hyperbolic fixed point play the role of the endpoints in the unperturbed case, and with increasing time there appear sharp peaks of extremely good stretching around these SIPs. One even sees the small dip in the unperturbed profile near the endpoints appear near the SIPs in the perturbed profile. Since the open-flow unperturbed profile is more uniform, it is more difficult to decipher the perturbed stretch profile without a somewhat belabored discussion, but with the help of SIP portrayal and symbolic dynamics labeling in Fig. 43 one can make some basic observations. The detrained segments clearly begin to form pockets of poor stretching, and the returned segments do show evidence of approximate self-similar behavior. An obvious example of this latter behavior is the sharp dip of poor stretching associated with the 'tip' of the lobe formed by $r_0(t)$ appearing in the returned segments when they perform another revolution around the heteroclinic tangle. More

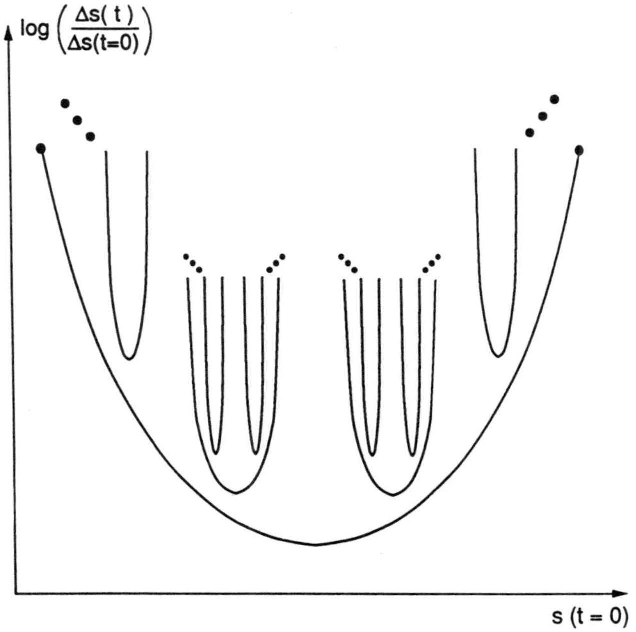

Fig. 42. The unperturbed stretch profile approximately repeating itself on smaller and smaller scales. This diagram gives only an extremely cartoon-like portrayal of the phenomenon: the unperturbed profile here is represented simply by a well-shaped curve (spatial oscillations and sharp dips are ignored) and we are indicating a sequence of events by a single diagram.

specifically, note in Fig. 43 the open-flow stretch profile of $r_0(t=3)$ approximately repeating itself on a smaller scale in the stretch profile of $r_0r_3(t=6)$: both show four significant peaks separated by significant dips, the middle dip being quite pronounced. Of course, the repetition is indeed only approximate. For example, the middle dip of $r_0r_3(t=6)$ associated with the 'tip' is much wider than the corresponding dip of $r_0(t=3)$, since the tail of the former is much larger as a consequence of $r_0r_3(t=3)$ lying further inside the heteroclinic tangle than $r_0(t=0)$. Additionally, since the stretch profile of $r_0r_3(t=3)$ has a mild non-uniformity, this non-uniformity carries through to give an additional mild non-uniformity in $r_0r_3(t=6)$ not found in $r_0(t=3)$.

It is with the above qualifying comments that we hope the spirit of our notion of approximate self-similar repetition is understood. Within this setting, let us now make a number of qualitative observations about the stretching.

(i) Notice the relationship between the non-uniformity of the unperturbed stretch profile and the non-uniformity of the perturbed profile. For example, the unperturbed closed-flow profile, for the parameters chosen, has good stretching highly localized around the endpoints, with a large dip of poor stretching between the endpoints; hence the perturbed profile has good stretching highly localized around the SIPs, with significant dips of poor stretching between SIPs. In contrast, the unperturbed open-flow profile is more uniform (due in part to the presence of the other hyperbolic fixed point); hence the perturbed profile has a more uniform distribution of good stretching. We see then that the unperturbed stretch profile provides qualitative and yet basic information about the non-uniformity of the perturbed profile, which is relevant to efficiency of mixing in the context of diffusion across interfaces, as we will describe in the next section, and to the statistics of stretching, as we will study later in this section. Primarily in this context is our interest in approximate repetition on smaller and smaller scales.

(ii) The mechanism for enhanced stretching can be understood partially in a variety of ways, and the different flows highlight these different ways. For example, consider *residence time near the hyperbolic fixed points*. The creation of SIPs by external forcing entails a creation of a countable infinity of points along the interface that eventually reside near the fixed point for all time, and hence entails an overall increase in the interface's residence time near the hyperbolic fixed points. This mechanism is dominant in the closed flow, where the stretching is highly localized around SIPs. Consider also *the repeated stretch, fold and return mechanism*, which need not be tied to residence near the hyperbolic fixed point. For example, referring back to Fig. 38, after three perturbation periods r_0 has been stretched and folded. The portion r_0r_3 will repeat the stretch and fold process, as will

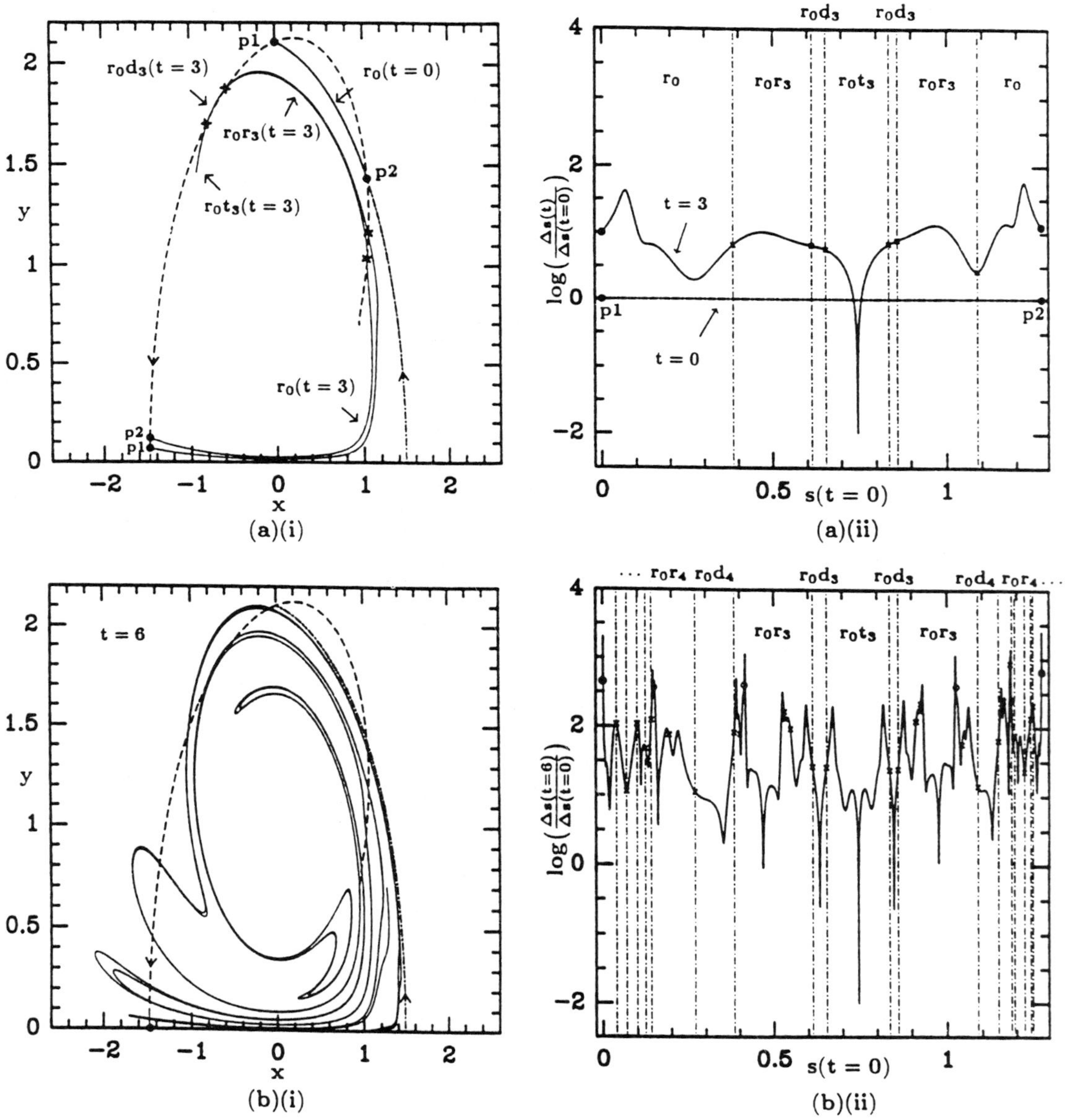

Fig. 43(a) and (b).

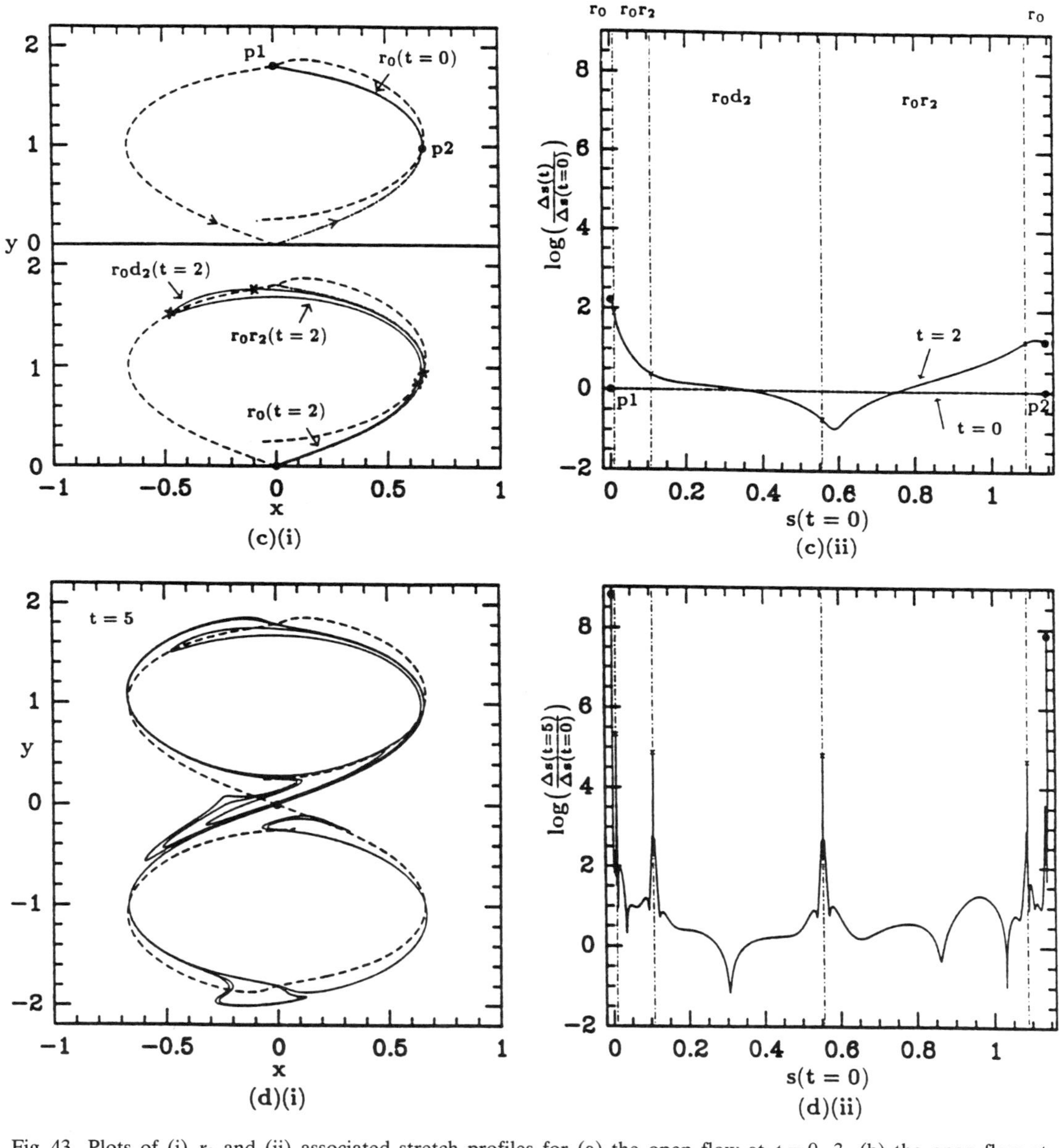

Fig. 43. Plots of (i) r_0 and (ii) associated stretch profiles for (a) the open flow at $t = 0$, 3, (b) the open flow at $t = 6$, (c) the closed flow at $t = 0$, 2, and (d) the closed flow at $t = 5$, with symbolic dynamics labeling. In (i) the stable manifolds are shown by dashed lines, and the unstable manifolds by dashed-dotted lines; in (ii) the dashed lines indicate SIPs. In plots (b) and (d) we do not explicitly label all the subdivisions of r_0, since the labeling would become much too crowded on the scale shown (plot (b) does indicate additional SIPs by crosses, and points which are 'almost SIPs' (i.e. almost intersect the stable manifold segment of the partition) by open dots). The flows are specified by the parameters $\Gamma = 0.4(2\pi)^2$ and (a, b) $\varepsilon = 0.085 \cdot 2\pi$, (c, d) $\varepsilon = 0.02 \cdot 2\pi$, with the vortices at $(0, \pm 1)$ at $t = n \in Z$.

then $r_0r_3r_3$, and so on. As long as the concerned segment stretches by a factor greater than one with each fold and return, it will be undergoing exponential stretching. Here, then, exponential stretching is not necessarily due to good local stretching near a hyperbolic fixed point, but to the repeated stretch and fold mechanism. This mechanism is highlighted by the open flow, where stretching is more evenly distributed between the SIPs. Note that the stretch profile of $r_0r_3(t = 3)$ need not be uniformly greater than one, so that one will have to compute the profile (and then further profiles) to determine which portions of r_0 are candidates for exponential stretching. In such a manner one determines 'where the horseshoe maps are', so to speak. A third explanation is that turnstile lobes act in a certain sense as *reorientation* mechanisms, which we will pursue in Section 6.2. All of the above-mentioned mechanisms for stretch enhancement are really part of the same overall mechanism we previously mentioned, that is the *approximate self-similar behavior of r_0 as it evolves*.

(iii) Notice the great variation of stretch histories (the vertical scale of the profiles in Fig. 43 is logarithmic) and the spatial complexity of the interface produced in a very short time by a very simple velocity field. As we shall discuss, mixing in the context of diffusion of passive scalars across an interface depends on the stretch *histories* of the interface and, as the simulations in Fig. 43 suggest, there may be little obvious connection between a velocity field's properties and the corresponding advection properties. In particular, simple velocity fields can of course produce complicated advection, and more generally the scaling properties of turbulent velocity fields may have little to say about the mixing properties. In addition, the presence of hyperbolicity in the chaotic tangles implies substantial temporal variation of the strain experienced by an infinitesimal line element, as well as great spatial variation of strain history along a material curve. In studies of turbulent mixing, sometimes an assumption of constant strain is made at some point (see Section 6.2); such an assumption certainly would not apply in chaotic tangles, and we see no particular reason for it to be valid for turbulent flows.

(iv) Though previous investigators have commonly referred to chaotic zones as regions of 'enhanced' stretching, notice how in both the open and closed flows there are gaps of poor stretching on finite time scales (we are not referring here to regular islands within the tangles, but to chaotic zones themselves). As previously mentioned, these gaps in the open flow are partially explained by detrainment. For both flows, however, the additional explanation is offered by the previously described dip in the unperturbed stretch profile, which carries through to the perturbed profile by the previously described approximate self-similar enhancement of the stretch profile. Note how the regions of poor stretching typically correspond to 'bulbous' parts of the lobe, and turning points such as the 'tip' of the interface. These bulbous parts can rotate throughout the chaotic zones with ineffectual stretching on finite time scales, not coming close to the fixed point, and not undergoing the repeated stretch and fold mechanism. We should stress here a basic difference between the open and closed flow: in the closed flow the chaotic zone associated with the homoclinic tangle is finite, re-entrainment occurs, and poor stretching eventually gives way to enhanced stretching; in the open flow the chaotic zone associated with the heteroclinic tangle is infinite, and detrainment leads to advection out to $x = -\infty$ in an asymptotically linear manner, so that poor stretching can remain so for all time.

(v) Note the variety of stretch scales, spatial scales, and temporal scales involved in the interface stretching. Basically r_0 is undergoing a repeated stretch and fold process, with the magnitude of stretching depending on the location relative to the fixed point, and this stretch, fold, and return process takes place on all time scales (above some minimum time), as monitored by the symbolic dynamics: for example, $r_0 r_n(t = 0)$, $n > n_{\min}$, takes n time samples for the first stretch, fold, and return. Hence horseshoe maps occur on all time scales above some minimum time. The variety of stretch scales and spatial scales is indicated by the non-uniformity of the stretch profiles. For each revolution around the homoclinic/heteroclinic tangle, one can have anything from *contraction* on sizable spatial scales to extremely good stretching on very small spatial scales. This range of scales will be relevant to the statistical analysis of stretching given momentarily.

(vi) Consider the total length of the interfaces, as shown in Fig. 44. Up until the first

(a)

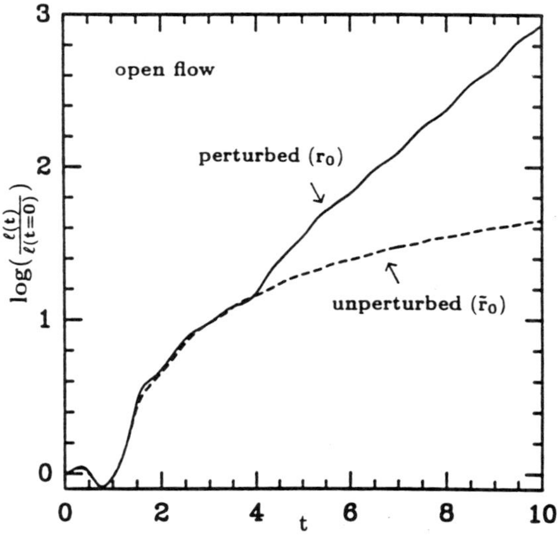

(b)

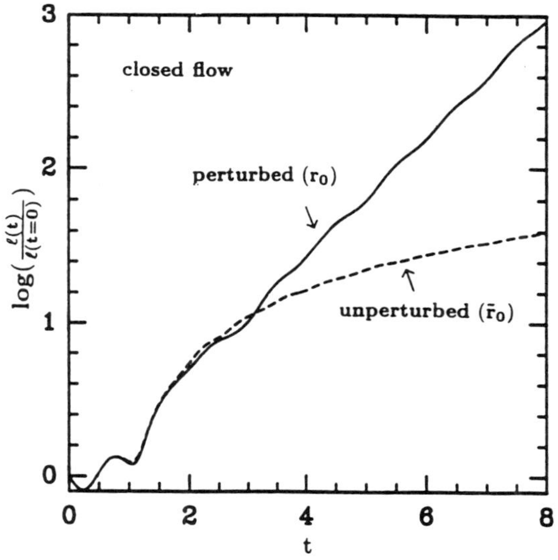

Fig. 44. Total length $l(t)$ of r_0 in the perturbed case and of $\tilde{r}_0$ in the unperturbed case for (a) the open flow and (b) the closed flow.

'returned' portion of r_0 wraps around $E(0)$ (i.e. a portion of r_0 begins its second wind around the tangle), the total length of r_0 does not differ significantly from that of $\tilde{r}_0$ in the unperturbed problem; however, thereafter (after $t \approx 4$ for the open flow and $t \approx 3$ for the closed flow) the length of r_0 quickly converges to mild oscillation about exponential stretching. In this context we see one indication of how the stretch process associated with a single revolution of an entrained lobe around a homoclinic/heteroclinic tangle is similar in the perturbed and unperturbed case (up until a re-orientation process in the perturbed flow at the end of a revolution, to be discussed in Section 6.2), and great regularity underlies the approximate self-similar behavior of r_0.

(vii) We lastly note that an alternative symbolic description of the evolution of segments of unstable manifolds is given by Rom-Kedar [84]. It is unclear to us how the variety of stretch histories exhibited by our stretch profiles is captured by this alternative scheme.

To summarize the geometric understanding thus far, it is a consequence of the lobe dynamics that SIPs naturally partition a lobe boundary into segments of qualitatively different behavior, based on outcomes of, and elapsed times for, revolution times around the chaotic tangle. A symbolic description allows one to uncover the global topology associated with the recursive partitioning of the lobe boundary by the SIPs. For any line element along the lobe boundary defined by SIP boundaries, one natural way to conceptualize the stretching process is in terms of products of stretches associated with successive revolutions around the chaotic tangle. For example, the stretch experienced by an open-flow line element $r_0r_8r_2r_8r_5d_{10}$ up until detrainment (which occurs at period $n = 8 + 2 + 8 + 5 + 10$) is

$$\mu(r_0r_8r_2r_8r_5d_{10}) = \mu_1(r_8)\mu_2(r_2)\mu_3(r_8)\mu_4(r_5)\mu_5(d_{10}), \tag{61}$$

where $\mu_1(r_8)$ denotes the stretch experienced by the line element under the first revolution around the tangle through eight periods to a 'return' outcome, $\mu_2(r_2)$ denotes the stretch experienced by the line element under the second revolution through two periods to a 'return' outcome, and so on. An exact determination of stretch experienced by line elements requires a numerical evaluation; however, such an evaluation for ensembles of line elements can quickly become numerically intensive and may not shed much light on the underlying stretch process. As a result, past studies of stretching under chaotic flows often resort to models for the stretch process. Such models are typically fairly elementary, and suffer from the lack of an explanation of whether such models actually capture the true stretch process under chaotic flows. For example, in their study of the kinematic dynamo problem under chaotic advection, Ott and Antonsen [72] assume simple enough models for the stretch processes, such as a generalized baker's map, that stretch distributions are described by standard multifractals from a binomial multiplicative process. However, we shall see in Section 6.3 how such a model can be inappropriate. We argue that the invariant manifold analysis of this section, exploited in the context of symbolic dynamics and stretch profiles, provides a framework for more realistic models, in which the topology of, and multiple scales involved in, chaotic flows are respected. To model the stretch process described in expression (61) for all line elements on the lobe boundary, we note that there are two levels of infinities to address, since in the term $\mu_i(r_n)$ both i (revolution number) and n (revolution time) can take on all positive integer values. We can characterize the stretch process by a finite number of values, however, with the following two observations.

(i) For a fixed revolution number i, we have

$$\lim_{n\to\infty} \frac{\mu_i(r_{n+1})}{\mu_i(r_n)} = \text{constant}. \tag{62}$$

As $n \to \infty$ the percentage of the revolution time that the line element spends in the vicinity of the hyperbolic fixed points goes to one, from which the above relation follows.

(ii) For a fixed revolution time n, invoking the approximate self-similar behavior of the stretching process, it is reasonable, at least for the sake of modeling, to put upper and lower bounds, $\pm\delta$, on the variation of $\mu_i(r_n)$ with each successive revolution around the tangle. Hence we write

$$\mu_i(r_n) \equiv S_n \pm \delta. \tag{63}$$

For the sake of modeling, one might wish to simply take $\mu_i(r_n) = S_n$ (i.e. model the approximately self-similar process by an exactly self-similar one).

With these two observations, one can make some reasonable models based on the symbolic dynamics. For example, one could choose a reasonable analytical model for the unperturbed stretch profile, one whose non-uniformity can be adjusted by a parameter, and have the profile repeat on smaller and smaller scales according to the symbolic dynamics (with the SIP locations chosen by some reasonable model). The result is a multifractal with a much richer construction, one that would be expected to be truer to the actual stretch process in chaotic tangles. In Section 6.3 we will consider some elementary models in the context of a study of distributions of finite-time Lyapunov exponents in chaotic tangles. There we will see how the non-uniformity in the stretch profiles plays a basic role in the stretch process, and the resulting stretch distributions. To study realistic flows rather than models can involve intensive computation; however, an appreciation of the underlying geometric template in chaotic tangles can be helpful in formulating and then interpreting these computations. For example, we have seen how, in tangles with highly non-uniform stretch processes (as in the closed flow), it is possible for the SIPs to be the seeds of greatly enhanced stretching in phase space. Identifying these intersection points would then be helpful in maximizing a stretching (and later mixing) process. A goal of this section is thus to encourage future studies to address the underlying geometry within chaotic tangles and the presence of great variation of finite-time dynamics. There are a number of practical examples where such a consideration would be helpful. For example, in their study of separation under flow reversal, Aref and Jones [3] present a computational example illustrating that separation of a material blob is better in chaotic zones than in regular zones, and argue that this enhancement is due to the enhanced stretching in chaotic zones. A consideration of the topology of stretching on finite timescales would be useful in maximizing this effect.

6.2. *Rates of strain*

Whereas the previous section used the geometric template of invariant manifolds to obtain some understanding of stretching in chaotic tangles, we now turn our attention to *rates* of stretching, i.e. rates of strain, under chaotic flows. This latter study is perhaps more directly related to conventional studies of stretching under fluid flows. It is interesting to note that, whereas studies of turbulent fluids sometimes assume a constant rate of strain (e.g. see Batchelor [5] and more recently Dimotakis [24, 25]), chaotic flows have been claimed to have rates of strain which 'oscillate wildly' (e.g. Dresselhaus and Tabor [27]). Given this backdrop, we now attempt to obtain a geometric understanding of the rate of strain experienced by infinitesimal line elements evolving in chaotic tangles. Our principal interest will be in the notion of *irreversible rate of strain*, the role of hyperbolic fixed points as engines for good irreversible straining, and the role of turnstiles as mechanisms for increasing straining efficiency via *re-orientation* of line elements and *transport* of line elements to regions of superior straining. In Section 6.2.1 we consider the unperturbed flows, and study numerically the rate of strain associated with sets of infinitesimal line elements distributed along the material interfaces $\tilde{r}_0(t = 1)$ of Section 6.1. Continuous-time strain-rate plots contain a large-amplitude oscillatory component associated with the component of the line element along the streamlines, which has no net contribution to line-element stretch or contraction. We thus subtract out this contribution to define the *irreversible rate of strain* directly responsible for net stretch. Continuous-time plots of this irreversible rate of strain make apparent the contributions to net stretching associated with the fact that the line element has a component normal to the streamlines. Distinct peaks are observed with each passage by a hyperbolic fixed point, and the peak amplitudes decay in time due to increasing line-element orientation along the streamlines as time progresses. In Section 6.2.2 we address the perturbed flows, and discuss the enhancement of straining efficiency via re-orientation of line elements by the turnstile lobes. The role of turnstiles in the study of *transport* in phase space, described in Section 5, has had fundamental and far-reaching consequences, and we offer here a complementary observation about turnstiles relevant to *stretching* in phase space. We study numerically the conventional rate of strain in the non-integrable flows, observe the enhancement of straining, and note the particular relevance of SIPs discussed in Section 6.1. As in the integrable case, there is typically a large-amplitude oscillatory straining component which again complicates the study. For non-integrable flows, lacking streamlines, there is no *exact* expression afforded that one can subtract out, but one can subtract out an *approximate* contribution which can give reasonable results for near-integrable flows. Of central importance will be that the re-orientation process does not apply (or apply well) to *all* line elements of the flow. In particular, those line elements near the tips of lobes, and successive folds of the lobes, are not re-oriented in such a way as to enhance stretching.

6.2.1. *Integrable flows*

(a) *Conventional rate of strain*. Again the unperturbed flow will provide the setting with which to understand the perturbed flows. Fig. 45 shows the initial location of the infinitesimal line elements whose straining we study, distributed along $\tilde{r}_0(t=1)$ from Fig. 37, with initial orientation defined by the local tangent to the material curve on which they lie. The rate of strain experienced by an infinitesimal line element $(\delta l_1, \delta l_2)$ located at $(x_1, x_2) \equiv (x, y)$ at time t in the presence of a two-dimensional velocity field $(u_1(x_1, x_2, t; \varepsilon), u_2(x_1, x_2, t; \varepsilon))$ is given by

$$\sigma \equiv \frac{1}{\delta l}\frac{\mathrm{d}(\delta l)}{\mathrm{d}t} = m_i m_j \frac{\partial u_i}{\partial x_j}(x_1, x_2, t; \varepsilon), \tag{64}$$

where

$$\delta l \equiv \{(\delta l_1)^2 + (\delta l_2)^2\}^{1/2}$$

is the line-element length,

$$m_i \equiv \frac{\delta l_i}{\delta l}, \qquad i = 1, 2,$$

defines the components of the unit vector oriented along the line element, and the Einstein summation convention is used. The spatial location of the line element is evolved by

$$\dot{x}_i = u_i(x_1, x_2, t; \varepsilon), \qquad i = 1, 2, \tag{65}$$

and its orientation by

$$\dot{m}_i = m_k \frac{\partial u_i}{\partial x_k} - m_k m_l \frac{\partial u_k}{\partial x_l} m_i, \qquad i = 1, 2. \tag{66}$$

Evolving the line elements of Fig. 45 according to equations (65) and (66), the continuous-time evolution of the rate of strain experienced by each line element via equation (64) is shown in Figs 46 and 47 for the open and closed flow, respectively. We make the following observations.

(i) There is a large-amplitude oscillatory component associated with the translation of the line elements around the closed streamlines. If the line elements were oriented along

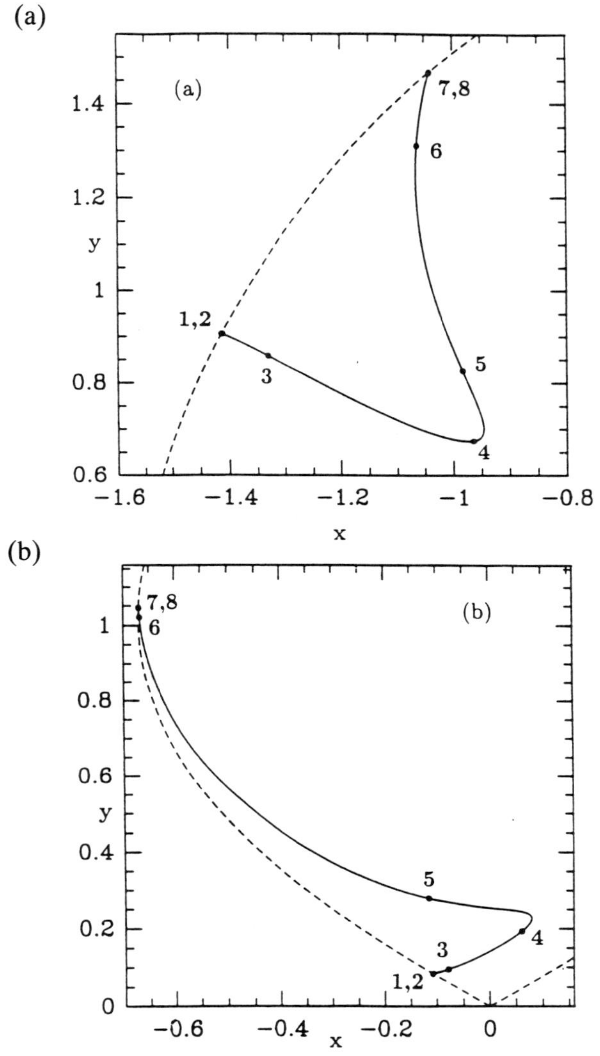

Fig. 45. The set of infinitesimal line elements distributed along the material interface $\tilde{r}_0(t=1)$ of Fig. 37 for (a) the unperturbed open flow and (b) the unperturbed closed flow ($x \equiv x_1$, $y \equiv x_2$). Points 1 and 8 are situated identically at the endpoints (i.e. they are points $\tilde{p}1$ and $\tilde{p}2$, respectively), while points 2 and 7 are very close by: $s_2(0) - s_1(0) = 8.39 \times 10^{-4} (2.47 \times 10^{-8})$ and $s_8(0) - s_7(0) = 1.29 \times 10^{-3} (5.86 \times 10^{-7})$ for the open (closed) flow, where $s_i(0)$ denotes the initial arclength along the curve for the ith point.

the streamlines, these oscillations would have zero mean, and hence no net contribution to stretch or contraction. *Note that there is significant instantaneous straining away from the hyperbolic fixed points*. Indeed, for the open flow, the most significant straining occurs away from the region local to the hyperbolic fixed point.

(ii) The oscillations are asymmetric about $\sigma = 0$; the area above $\sigma = 0$ bounded by the positive strain rates is greater than the area bounded below $\sigma = 0$ by the negative strain rate, entailing net positive stretch. The imbalance is manifested by larger positive strain amplitudes and longer time intervals over which the positive strain is defined. These asymmetries are due to the initial orientation of the line elements having a component normal to the streamlines.

(iii) The asymmetries decay in time due to the increasing orientation of line elements along the streamlines as time progresses (refer back to Fig. 37), entailing that the net positive stretch is poor, only asymptotically linear in time (refer back to Section 6.1). The decay is fairly rapid with successive revolutions, entailing that the obvious asymmetries of the first revolution quickly become difficult to decipher with the backdrop of large-amplitude oscillations.

(iv) The further away the line element is from the endpoints $\tilde{p}1$, $\tilde{p}2$, (i.e. the deeper the line element is into the integrable bounded core), the higher the frequency of the straining oscillations (since the line elements translate faster around the bounded core).

(v) Besides the difficulty associated with the large-amplitude oscillations that have no net contribution to stretch, note the limitations of the conventional rate of strain due to the lack of a normalization—for an individual orbit one cannot identify the overall efficiency of the straining process.

The difficulties in (i), (iii), and (v) motivate an improved definition of rate of strain.

(b) *Irreversible rate of strain*. We address the difficulties in (i), (iii), and (v) of the previous section by defining a new rate of strain. If a line element were oriented along a streamline, then due to the invariance of the streamline, the line element would be

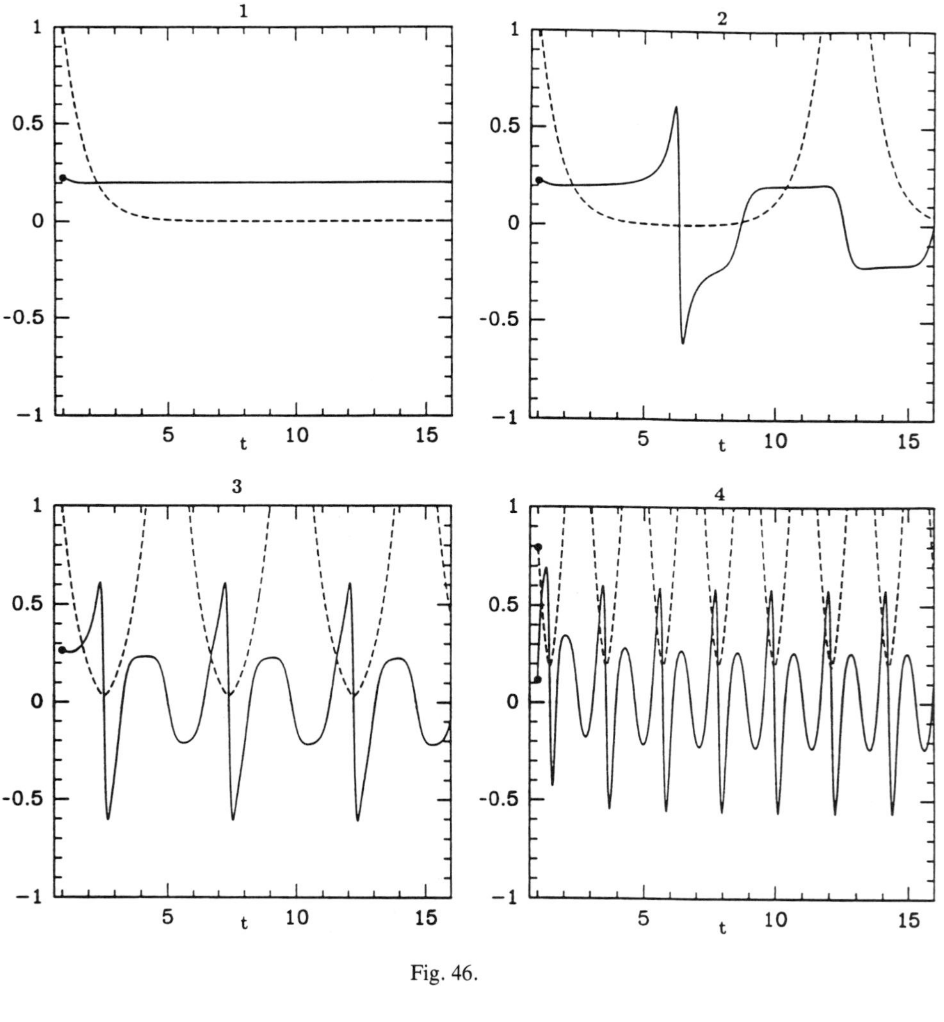

Fig. 46.

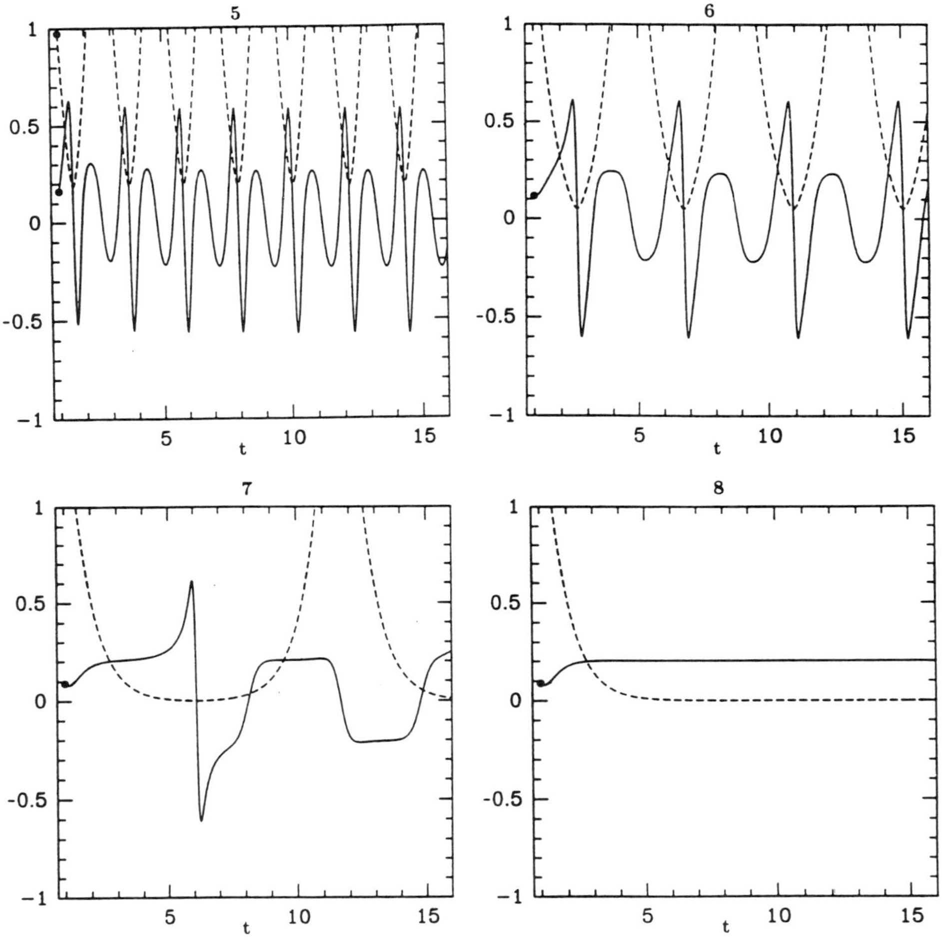

Fig. 46. Continuous-time open flow strain-rate plots for the points shown in Fig. 45(a). The solid line is the strain rate, and the dashed line is the vertical ($x_2 \equiv y$) location of the infinitesimal line element. *Note that the strain rates are given in units of* 2π. This convention is a consequence of the fact that the units originally used in the strain-rate study differed from those of Sections 6.1 and 6.3, and the above convention renders the present units consistent with the other sections. Further note that the unstable eigenvalue at the hyperbolic fixed points is found analytically to be $\lambda_u = 1.281 = 0.204 \times 2\pi$.

destined to remain oriented along the streamline, oscillating around the closed curve *ad infinitum* with no *net* stretching and contraction. The strain rate could of course oscillate from positive to negative values, but with zero mean. Hence, though the strain rate at any given moment might have what may seem an impressive contribution, the time integral of this contribution would offer no net stretch. For any line element, we thus subtract out this *reversible rate of strain* σ_{rev} to define the *irreversible rate of strain* σ_{irr} responsible for the net stretch

$$\sigma_{\text{irr}} \equiv \sigma - \sigma_{\text{rev}}, \tag{67}$$

where

$$\sigma_{\text{rev}} \equiv \widetilde{m}_i \widetilde{m}_j \frac{\partial u_i(x_1, x_2, t; \varepsilon = 0)}{\partial x_j}, \tag{68}$$

and

$$\widetilde{m}_1 = \frac{u_1(x_1, x_2, t; \varepsilon = 0)}{\{u_1(x_1, x_2, t; \varepsilon = 0)^2 + u_2(x_1, x_2, t; \varepsilon = 0)^2\}^{1/2}}$$
$$\widetilde{m}_2 = \frac{u_2(x_1, x_2, t; \varepsilon = 0)}{\{u_1(x_1, x_2, t; \varepsilon = 0)^2 + u_2(x_1, x_2, t; \varepsilon = 0)^2\}^{1/2}} \tag{69}$$

defines the orientation of an infinitesimal line element tangent to the streamline. This will allow us to ignore the large-amplitude oscillations of (i) and focus on the subtle asymmetries of (iii).

In addition to addressing (i) and (iii), one might like to address (v) by normalizing the strain rate. One choice of normalization (e.g. see [74]) is to define a *stretch efficiency* by dividing the conventional rate of strain by $\{D: D\}^{1/2}$, where D is the symmetric 2-tensor associated with spatial derivatives of the velocity field

$$D_{ij} \equiv \frac{1}{2}\left(\frac{\partial u_i}{\partial x_j} + \frac{\partial u_j}{\partial x_i}\right). \tag{70}$$

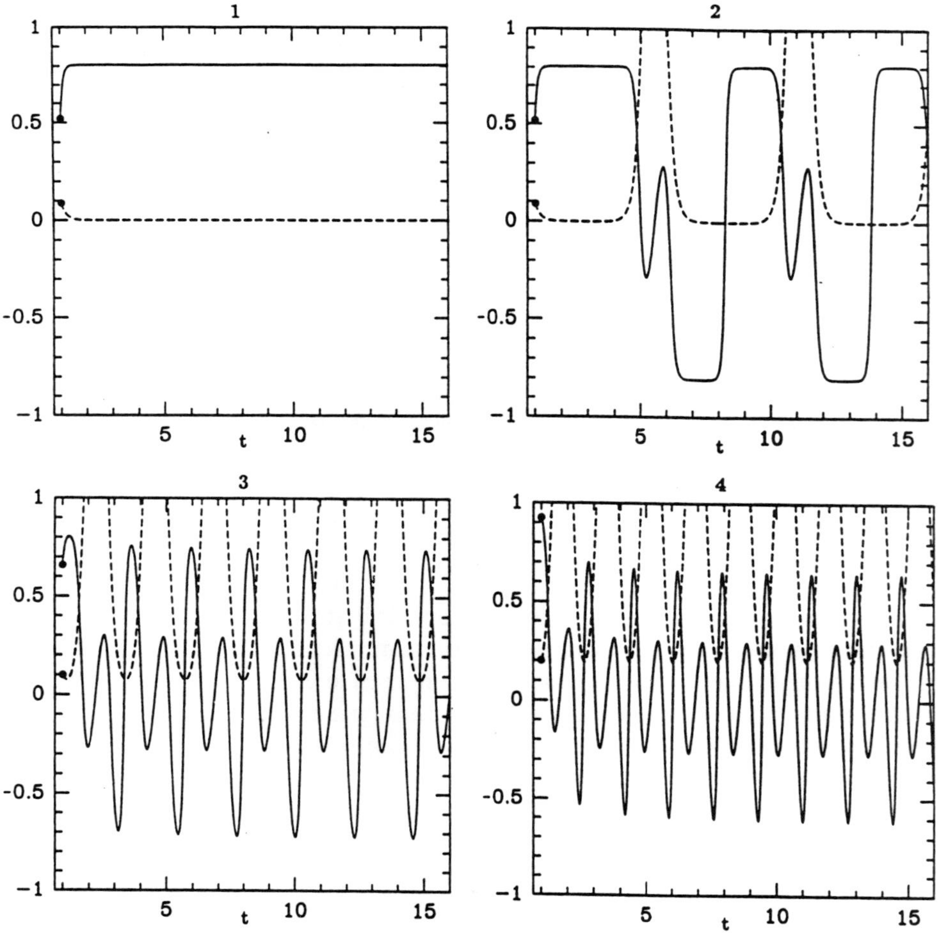

Fig. 47.

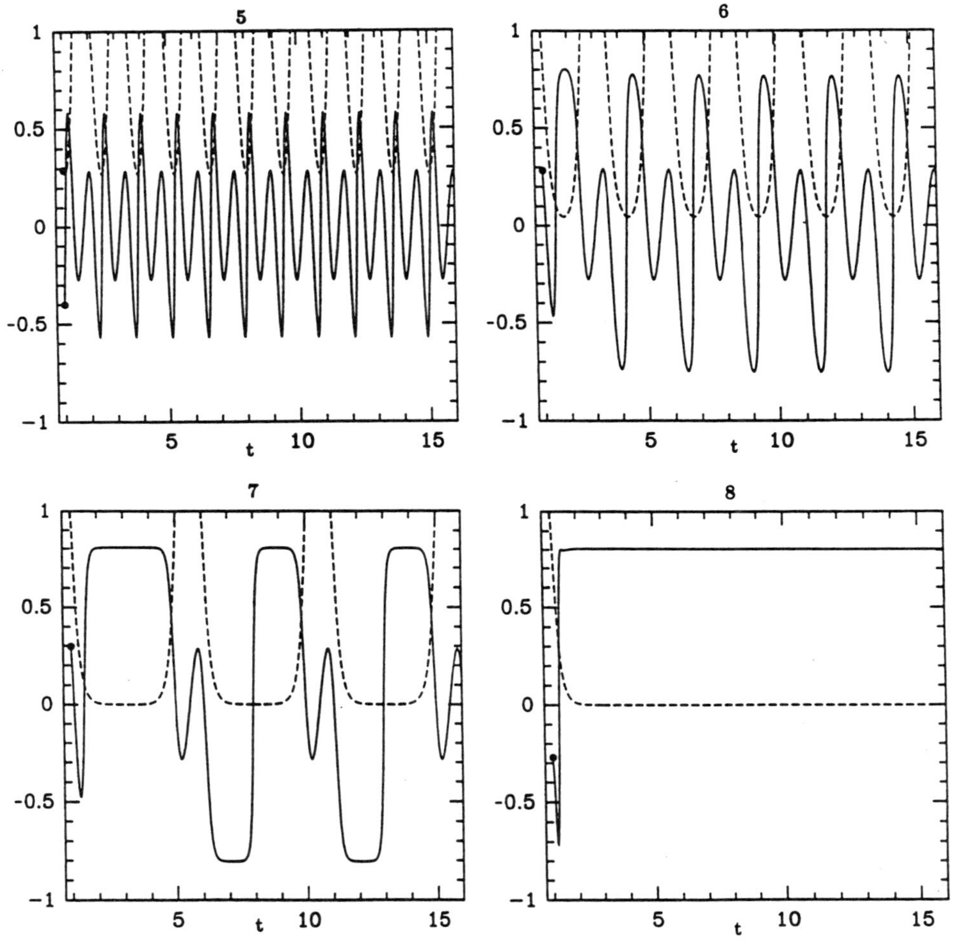

Fig. 47. Continuous-time closed flow strain rate plots for the points shown in Fig. 45(b). The solid line is the strain rate, and the dashed line is the vertical location of the infinitesimal line element. Note that the unstable eigenvalue of the hyperbolic fixed point located at the origin is found analytically to be $\lambda_u = 5.064 = 0.806 \times 2\pi$.

Such a definition of stretch efficiency has the property that it does not equal one in such instances where it would seem to be natural (as discussed in [11]). Instead we choose a normalization for positive (negative) strain rate with division by the absolute value of the maximum (minimum) rate of strain a line element can experience at its given spatial location (in practice one can rotate a hypothetical line element about its spatial location to determine the orientation corresponding to maximal (minimal) strain rate). More formally, we define a *straining efficiency* by

$$\sigma^{\mathrm{e}} \equiv \frac{\sigma}{\sigma_{\mathrm{m}}} \tag{71}$$

and the *irreversible straining efficiency* by

$$\sigma^{\mathrm{e}}_{\mathrm{irr}} \equiv \frac{\sigma}{\sigma_{\mathrm{m}}} - \frac{\sigma_{\mathrm{rev}}}{\sigma_{\mathrm{m}}}, \tag{72}$$

where σ_m denotes the maximum or minimum strain rate as discussed above. By definition $\sigma^e \in [-1, 1]$ and $\sigma^e_{irr} \in [-2, 2]$. The value $\sigma^e = -1$ corresponds to maximally efficient contraction, $\sigma^e = 1$ to maximally efficient stretching. The value $\sigma^e_{irr} = 0$ corresponds to rate of strain equal to the reversible rate of strain; $\sigma^e_{irr} = 2$ corresponds to maximally efficient stretching for the strain rate and maximally efficient contraction for the reversible strain rate (and conversely for $\sigma^e_{irr} = -2$).

Though a study of *efficiencies* has advantages, we should stress its limitations. Straining efficiency indicates the *percentage* of the maximum available strain rate being exploited, while ignoring what this maximum strain rate is. Hence continuous-time straining efficiency plots for an individual trajectory offer no accounting of the relative importance of a given efficiency at two different times. For example, a very efficient straining when the maximal strain rate is close to zero will appear significant in efficiency plots, but it may be negligible compared with a comparatively inefficient straining when the maximum available strain rate is quite large. A study of both strain rates and strain efficiencies will thus be useful, and we will consider the two interchangeably.

Figures 48 and 49 show continuous-time plots of the irreversible straining efficiency

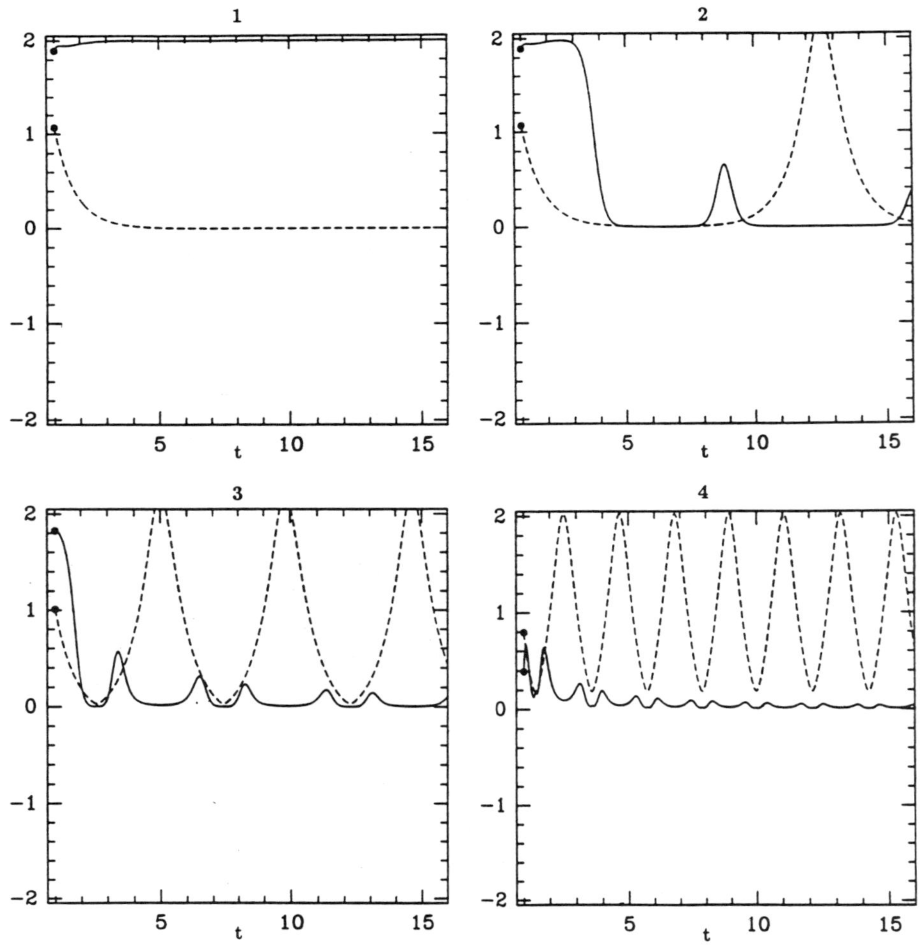

Fig. 48.

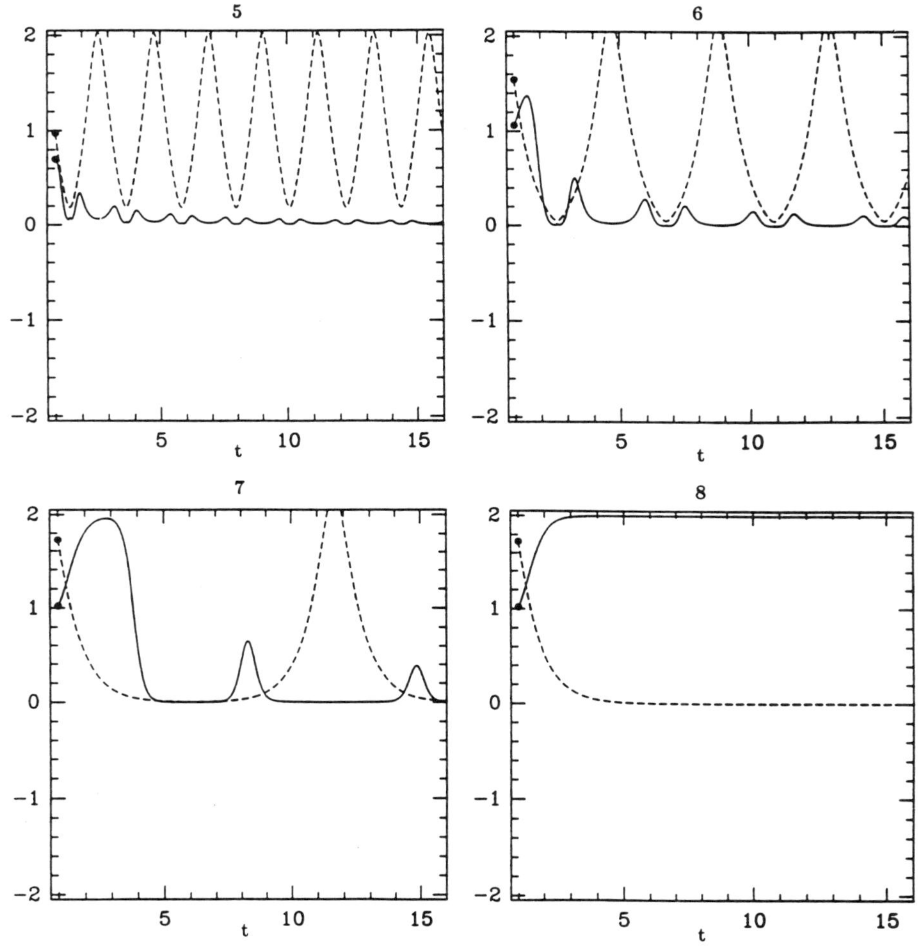

Fig. 48. Open-flow irreversible straining efficiency (solid line) for the points shown in Figure 45(a). The dashed line is the vertical location of the infinitesimal line element.

associated with Figs 46 and 47, respectively. Subtracting out the mean-zero oscillations and normalizing makes quite apparent the nature of the straining. We observe the following.

(i) The irreversible strain rate is always positive in these plots, and there are isolated peaks associated with passage by a hyperbolic fixed point (there are thus two peaks per revolution in the heteroclinic tangle and one peak per revolution in the homoclinic tangle). This effect is explained momentarily in the context of the linear flow local to the hyperbolic fixed point.

(ii) Peak amplitudes decay in time due to the increasing orientation of line elements along the streamlines as time progresses.

(iii) The further away the line element is from the endpoints $\tilde{p}1$, $\tilde{p}2$, (i.e. the deeper the line element is into the integrable core), the higher is the frequency of peak occurrence (since the line elements revolve faster around the core), and yet *the smaller the peaks* (since the line elements are further away from the hyperbolic fixed points).

The reversible straining can be interpreted in the context of streamline geometry, and the irreversible straining in terms of the linear flow local to hyperbolic fixed points. First, the reversible strain rate large-amplitude oscillations are easily interpreted in light of the incompressible nature of the flow. When neighboring streamlines squeeze together (expand apart), area elements are squeezed (expanded) normal to the adjacent streamlines, and hence stretched (contracted) parallel to the streamlines. The temporal reversible straining oscillations are thus understood from the spatial oscillations in neighboring streamline separation. We reiterate that substantial reversible strainings occur outside the region local to the hyperbolic fixed point. Second, the irreversible strain rate peaks associated with line elements having a component normal to the streamlines can be understood with the help of Fig. 50. Consider the linear flow local to a hyperbolic fixed point in the form

$$\begin{aligned} \dot{x} &= \lambda_u x \\ \dot{y} &= \lambda_s y, \end{aligned} \tag{73}$$

where $\lambda_u > 0$ and $\lambda_s < 0$ are the unstable and stable eigenvalues, respectively, of a hyperbolic fixed point located at the origin. Let α_d denote the angle between the line element and the tangent to the streamline, referred to as the *angle of deviation* (see Fig. 50(a)). We write α_d as the difference between the angles the line element and streamline

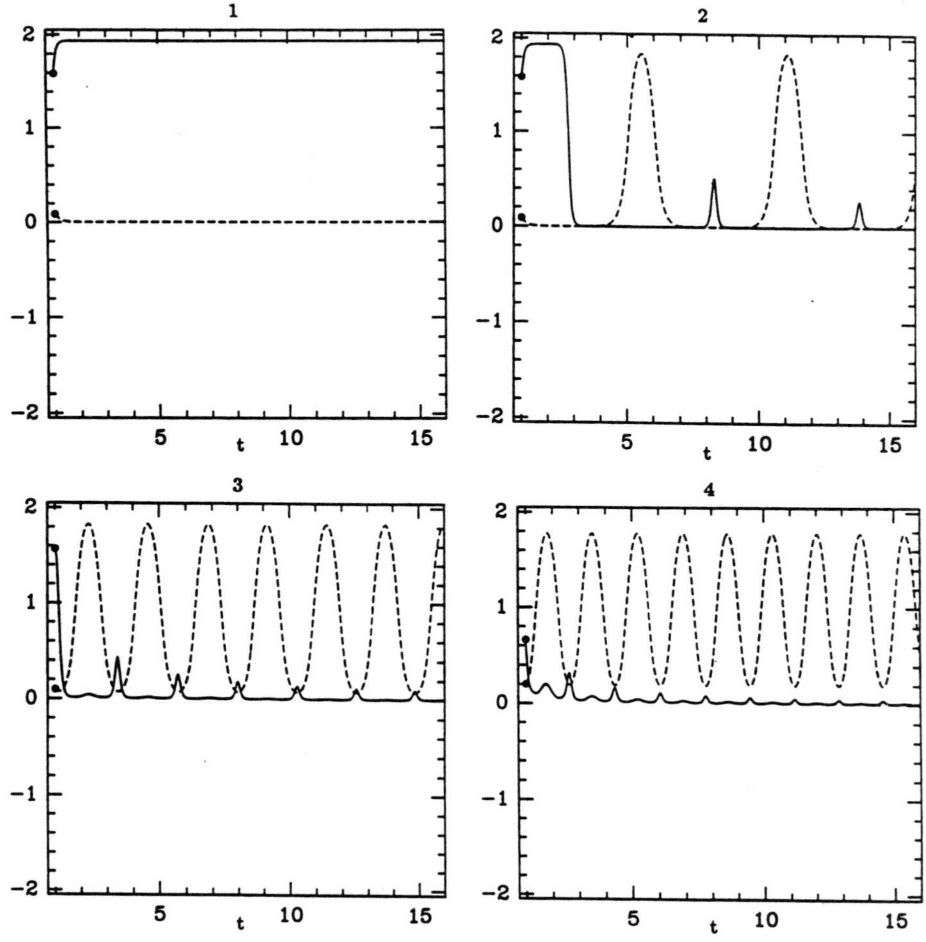

Fig. 49.

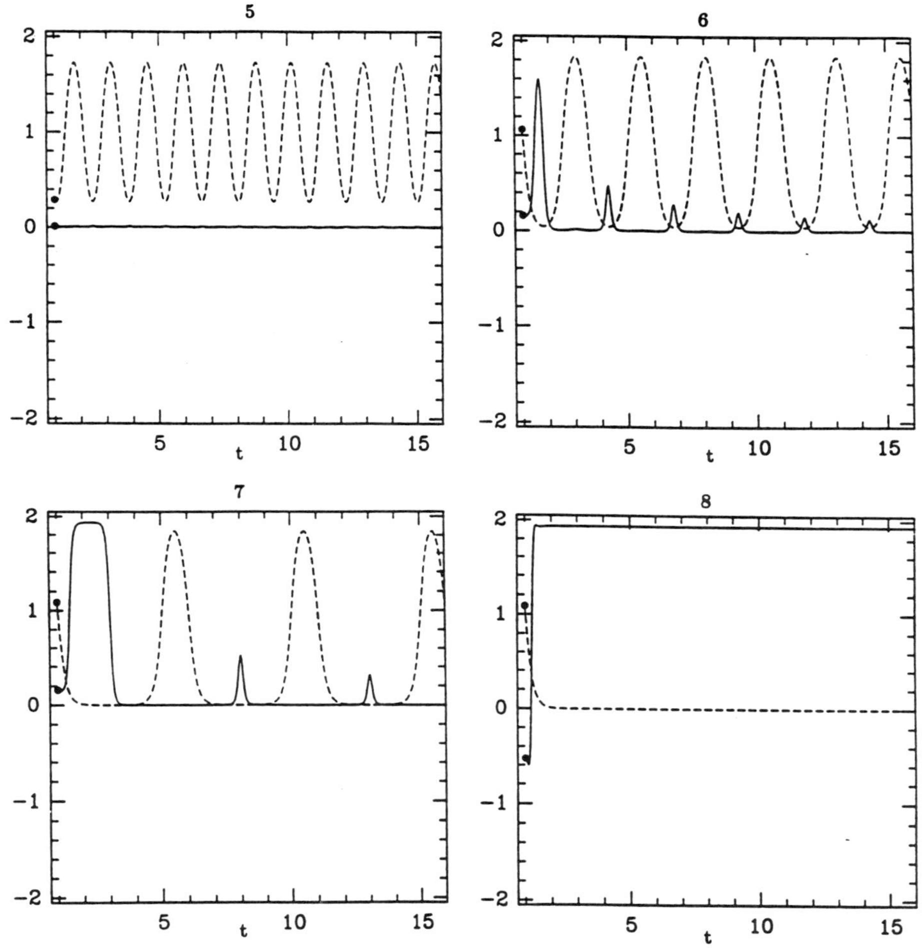

Fig. 49. Closed-flow irreversible straining efficiency (solid line) for the points shown in Fig. 45(b). The dashed line is the vertical location of the infinitesimal line element. Notice that line elements asymptotically approaching the hyperbolic fixed point oriented along the local unstable manifold correspond to $\sigma_{\mathrm{irr}}^{\epsilon}$ close to two, but not identically two, a consequence of the non-orthogonality of the local stable and unstable manifolds.

tangent make with the vertical axis, α_l and α_t, respectively:

$$\alpha_d = \alpha_l - \alpha_t, \tag{74}$$

where

$$\begin{aligned} \alpha_l &= \tan^{-1}(\Delta x_l/\Delta y_l) \\ \alpha_t &= \tan^{-1}(\Delta x_t/\Delta y_t) \end{aligned} \tag{75}$$

and $\alpha_l > \alpha_t$ (see Fig. 50(b)). For the moment, assume α_l and α_t (hence α_d) are small as they enter the region local to the hyperbolic fixed point. Equation (73) implies that, as the line element passes by the hyperbolic fixed point Δx_l, Δx_t increase exponentially in time, and Δy_l, Δy_t decrease exponentially in time; hence, using that $\tan^{-1}(\Delta x/\Delta y) \approx \Delta x/\Delta y$ for small $\Delta x/\Delta y$ one easily obtains that $\alpha_d = \alpha_l - \alpha_t$ grows exponentially in time from some small initial value. This growth in α_d is saturated and then reversed, due to two effects (as

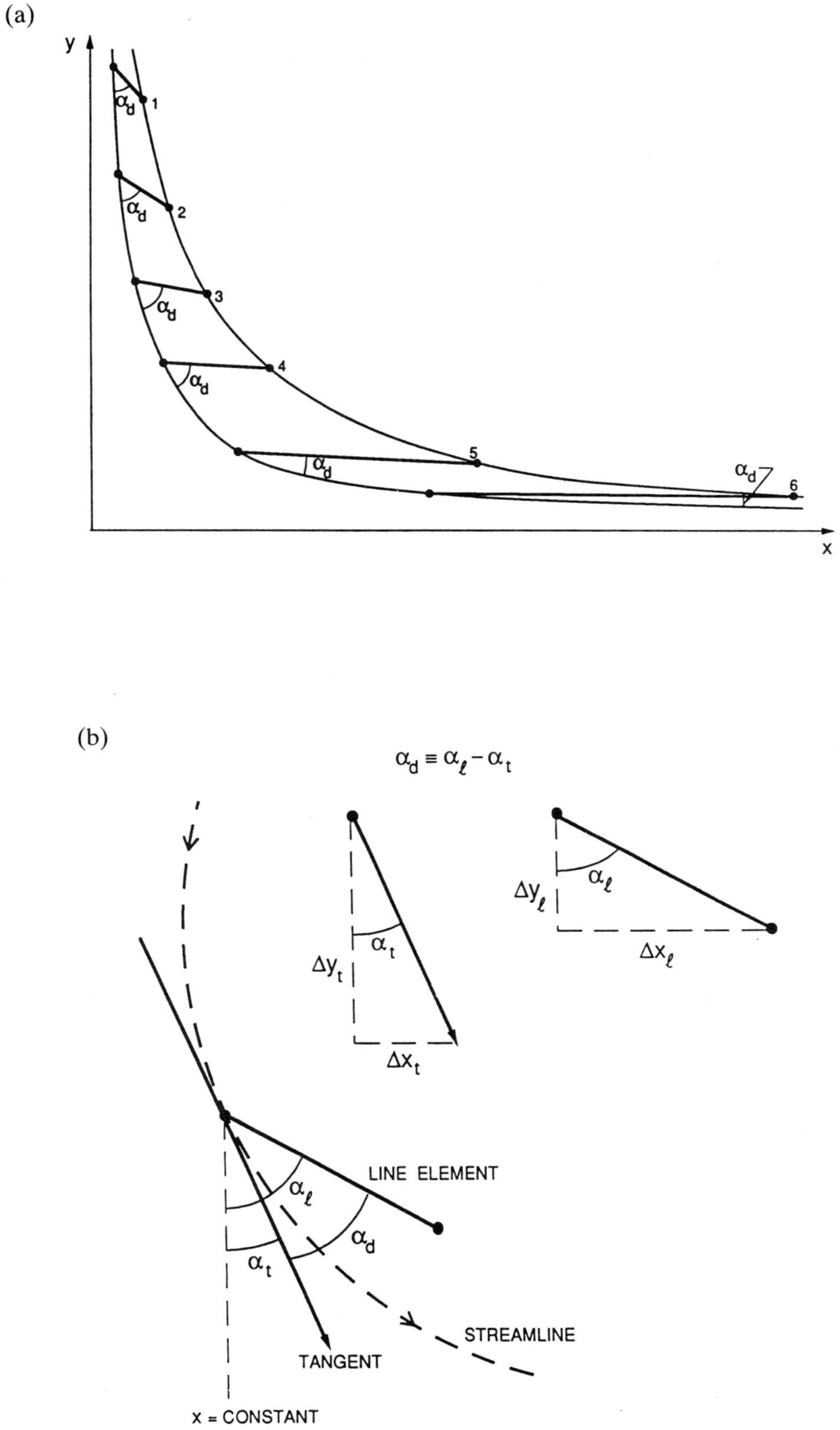
(a)
y
x
1
2
3
4
5
6
α_d
(b)
$\alpha_d \equiv \alpha_\ell - \alpha_t$
α_t
Δy_t
Δx_t
α_ℓ
Δy_ℓ
Δx_ℓ
LINE ELEMENT
STREAMLINE
TANGENT
x = CONSTANT

Fig. 50 (a) and (b).

(c)

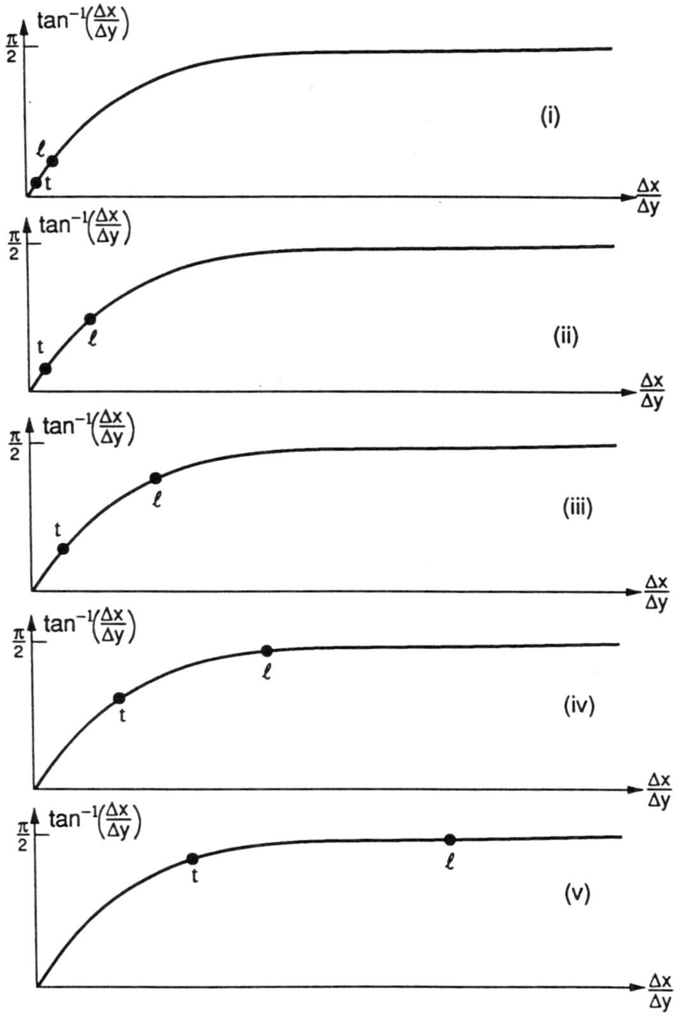

(d)

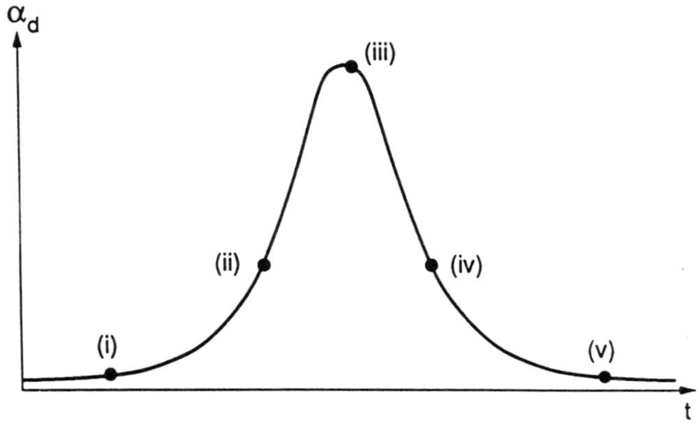

Fig. 50 (c) and (d).

(e)

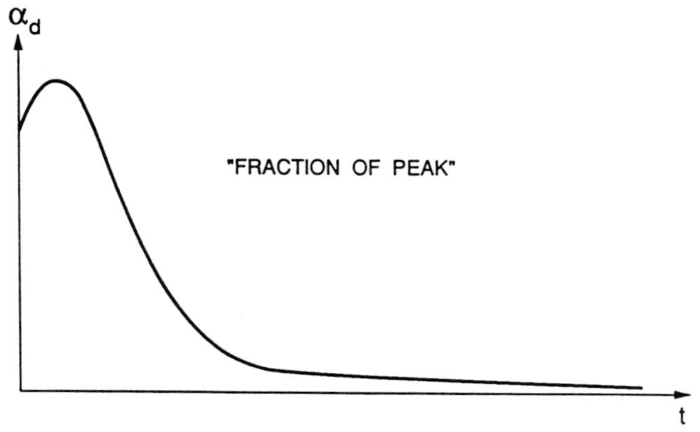

Fig. 50. Understanding the irreversible strain rate peaks associated with passage by a hyperbolic fixed point. (a) Portrayal of the peak in the deviation angle α_d as the line element passes through the region local to the hyperbolic fixed point. (b) Expressing the deviation angle in terms of α_t and α_l. (c) Converting slope to angle via $\tan^{-1}(\Delta x/\Delta y)$, and visualizing, in terms of the lag of α_t behind α_l, the growth in α_d followed by saturation and reversal. (d) The resulting isolated peak in α_d, and (e) a 'fraction' of a peak experienced in the region local to the hyperbolic fixed point when α_d is initially not small.

should be made clear in Fig. 50(c)):

(i) The growth in each of α_l and α_t is saturated at $\pi/2$ by the higher order terms in $\tan^{-1}(\Delta x/\Delta y)$.

(ii) The angle α_l 'saturates' at (comes close to) $\pi/2$ before α_t does, the latter continuing to grow until it then 'saturates' at $\pi/2$.

The net result of the pass by the fixed point is then an isolated peak in α_d, as portrayed in Fig. 50(d), entailing an isolated peak in the irreversible rate of strain. *Note then that the influence of hyperbolic fixed points on good stretching is not merely 'good straining', but rather good irreversible straining due to the lag of α_t behind α_l.* We have assumed in our explanation of the peaks that α_d and α_t are small as the line element enters the region local to the hyperbolic fixed point. For any pass by a hyperbolic fixed point but the first, this assumption on α_d is good. During the very first pass α_d may be initially large, however, which means only that one gets a 'fraction' of a peak in the region local to the hyperbolic fixed point, as seen in Figs 48 and 49 and portrayed in Fig. 50. The assumption on α_t is good as long as the line element enters the region local to the hyperbolic fixed point sufficiently close to the local stable manifold; the further a line element is from the stable manifold (i.e. the further into the core it is), the larger α_t will initially be, hence the smaller the lag of α_t behind α_l, and the smaller the straining peak, consistent with numerical observation. With successive passes the approaching value of α_d will be increasingly smaller, making the lag of α_t behind α_l progressively smaller, and hence the peaks less pronounced.

One needs to take care in interpreting stretching in the context of these observations before proceeding to the non-integrable case. At first glance, one might conclude that the hyperbolic fixed points are the engines for good irreversible straining, and hence a goal to maximize strain rates would be to be near these engines. Note, however, that the peaks can correspond to very sizable regions around these fixed points, as should be clear from the *y* component of the line elements in Figs 48 and 49. More significantly, as one moves deeper into the core, there is a balance between decreasing peak size (hinders stretching) and increasing peak frequency (helps stretching). It remains to be determined whether having a few large peaks is better or worse than having many small peaks, which necessitates integrating the strain rates, i.e. performing a study of stretch *history*. The stretch analysis given in the previous and forthcoming subsections is thus essential for a complete understanding of the problem.

6.2.2. *Chaotic flows.* In Section 5 we studied the role of *turnstile lobes* as mechanisms for transport across partial barriers in phase space. In a similar manner, turnstiles play a fundamental role in increasing straining efficiency via re-orientation of line elements and transport of line elements to regions in phase space of superior straining (as illustrated in Figs 51 and 52). Figure 51(a) shows again for the perturbed open flow the material

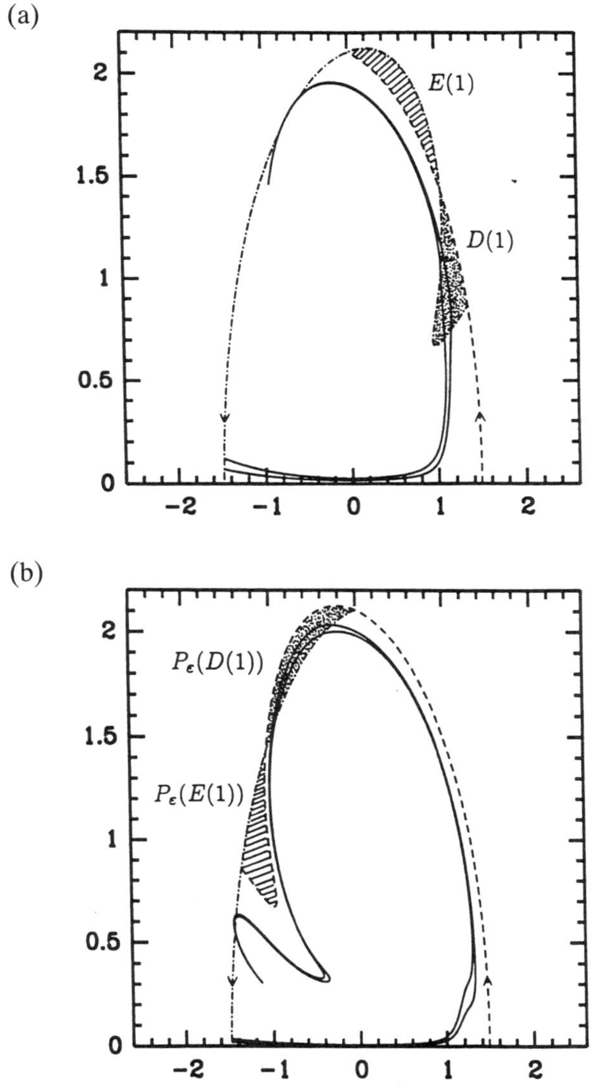

Fig. 51. The perturbed open flow material interface at successive times (a) $t = 3$ and (b) $t = 4$. Substantial portions of the interface are re-oriented and some portions are transported to regions of superior straining. In particular a portion of the lobe wraps around $r_0(t = 1)$ (i.e. around $P_\varepsilon(E(1))$), thus approximately regaining its orientation at $t = 1$.

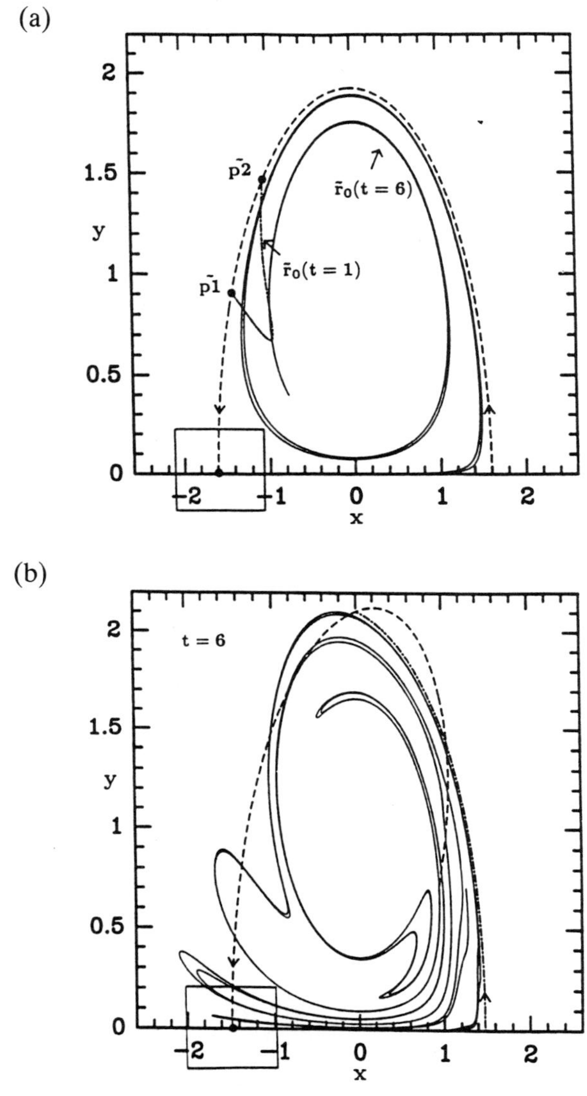

Fig. 52. Revisiting the (a) integrable and (b) near-integrable open flow at $t = 6$. The box highlights the results of re-orientation and transport via the turnstiles: a significant portion of the interface has been re-oriented and transported to the region local to the hyperbolic fixed point.

interface after one revolution around the tangle, a result not too dissimilar from the integrable case. Fig. 51(b) shows the dramatic change in the interface under the next iterate of the Poincaré map due to the turnstile lobes. In contrast to the integrable case, where line elements continue to wind around the integrable core with increasing orientation along the streamlines, significant portions of the interface have been dramatically *re-oriented* to recover a large component normal to the unperturbed streamlines (for near-integrable flows these streamlines are relevant, as we shall see). Indeed a significant portion of the interface (and successive portions with increasing time) wraps around $r_0(t = 1)$ (which recall is the portion of the unstable manifold bounding $P_\varepsilon(E(1))$), and *will thus approximately recover the orientation it had at* $t = 1$. Additionally, portions of the interface have been displaced by the turnstiles leftwards relative to the stable manifold, so that they will soon enter the region local to the hyperbolic fixed point with significant re-orientation. The SIPs seen in Fig. 51(b) that are a result of this leftward displacement are an extreme example of this effect—they will in short order enter the region local to the hyperbolic fixed point, and they indeed asymptotically approach the fixed point while oriented parallel to the unstable manifold, corresponding to optimal straining (or near-optimal straining when the local stable and unstable manifolds are not orthogonal). Two periods later, the consequence of this transport is apparent (see Fig. 52)—substantial portions of the interface find themselves near the hyperbolic fixed point, oriented essentially parallel to the local unstable manifold, ideal for irreversible straining efficiency. The net result of re-orientation and transport is portrayed heuristically in Fig. 53. In identifying these mechanisms for enhanced straining efficiency, we should stress that they, of course, will not necessarily enhance the straining of the *entire interface* over finite or even infinite time intervals, but rather *portions* of the interface. For example, in the open flow, portions of the interface are *detrained* with each new time sample, and this effect of *transport* via the turnstiles is ultimately detrimental to straining (since the detrained portion ends up infinitely far away from the hyperbolic fixed points). In a similar vein, some of the interface that wraps around $r_0(t=1)$ is not dramatically re-oriented (e.g. near the 'tip' of the lobe, and of successive folds of the lobes). These realities do not detract from our message, however, for the goal in explaining good stretching in chaotic tangles is not to argue its occurrence everywhere in the tangle on finite or even infinite time scales, but on a significant subset of the ensemble of interest. Further, as should be clear from Section 6.1, we view the portions of phase space in the tangles associated with poor stretching (indeed contraction) over finite times as a central part of the stretching process in chaotic tangles, one whose geometric interpretation is clear in the context of our template. We should also stress how the two mechanisms of re-orientation and transport to regions of superior straining of course need not go hand-in-hand; significant portions of the interface can be re-oriented without any obvious gainful transport.

Fig. 54 shows typical strain-rate plots for the closed flow (see Beigie [11] for similar open-flow plots). The basic observation to make is that, as in the integrable case, there are asymmetries (more positive strain rates than negative), but in contrast to the integrable case the asymmetries do not appear to be decaying. Though this non-decay is plain to see, the large-amplitude oscillations clearly render the conventional strain rate difficult to decipher, motivating some sort of irreversible strain rate. As mentioned earlier, lacking streamlines, there is no *exact* expression to subtract out. There are several possible expressions, however, that can work reasonably well in the near-integrable case, depending

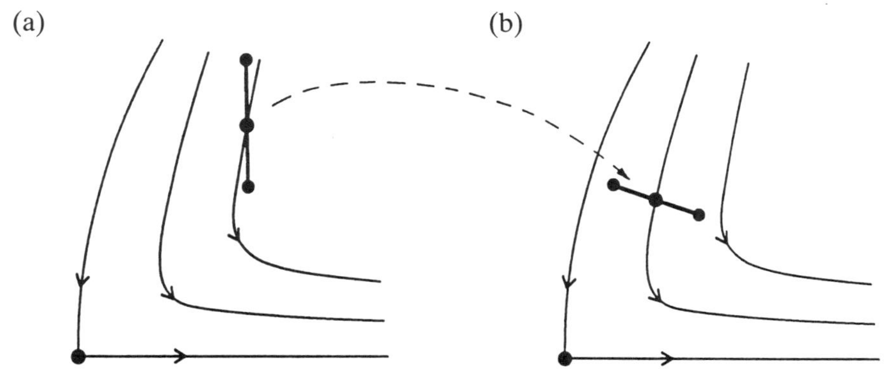

Fig. 53. Portrayal of the effects of re-orientation and transport: (a) a line element after several revolutions around the bounded core under the integrable flow, and (b) where that line element might end up after evolution under the near-integrable flow over the same time interval.

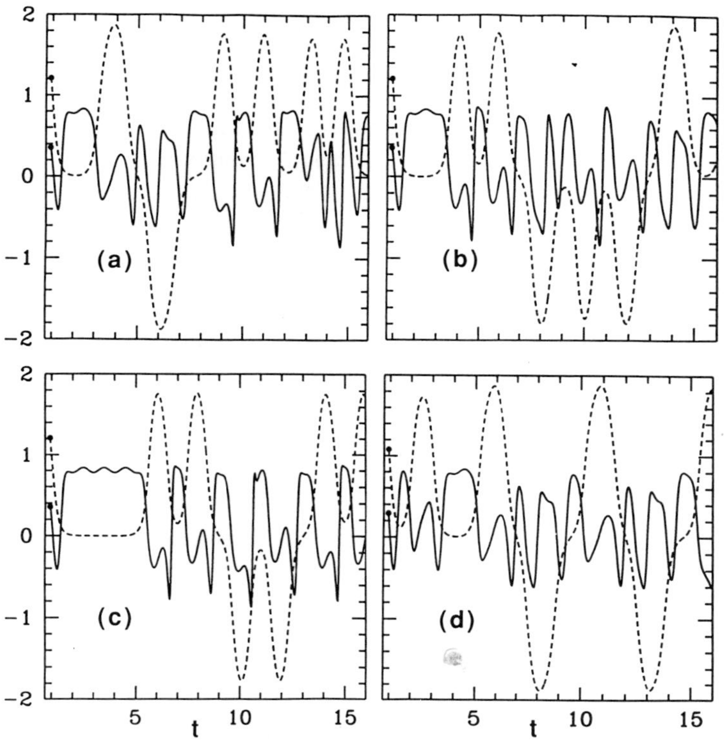

Fig. 54. Typical strain-rate plots (solid lines) associated with infinitesimal line elements distributed along r_0 for the perturbed closed flow. The negative values of the y component of the line element location (dashed lines) are due to the freedom of line elements to revolve around both upper and lower tangles.

on the situation at hand. We choose here to define an irreversible strain rate simply by subtracting out the *same term as in the integrable case*, to give the expressions identical to equations (67) to (69). We stress that, though we are in the non-integrable regime, we are subtracting out the straining term associated with the integrable flow. Beigie [11] examines the validity of such a subtraction term, including comparison with other reasonable choices for expression, as well as verification that, even though the line elements in the perturbed flow are not confined to a streamline, the expression subtracted out indeed does have a mean quite close to zero. The resulting irreversible strain rates are shown in Fig. 55, rendering the essence of the straining process relatively apparent. In particular we observe the following.

(i) For each sufficiently close pass by the hyperbolic fixed point in the closed flow, there is a sizable peak in the irreversible strain rate, and these peaks are not decaying with increasing number of revolutions.

(ii) Besides these isolated peaks, there are some additional secondary peaks, dips, or small oscillatory behavior. Though the absence of such behavior would have given a much more visually appealing plot (isolated non-decaying peaks separated by dead intervals), the reality of non-integrable flows is the existence of these relatively minor variations, attributable to the following. First, the turnstile can re-orient line elements well before they reach the region local to the hyperbolic fixed point, and this re-orientation can entail peaks and dips before they pass by the fixed point. The major peaks, however, are associated with the pass by the hyperbolic fixed point. Second, the overall breathing or pulsating of

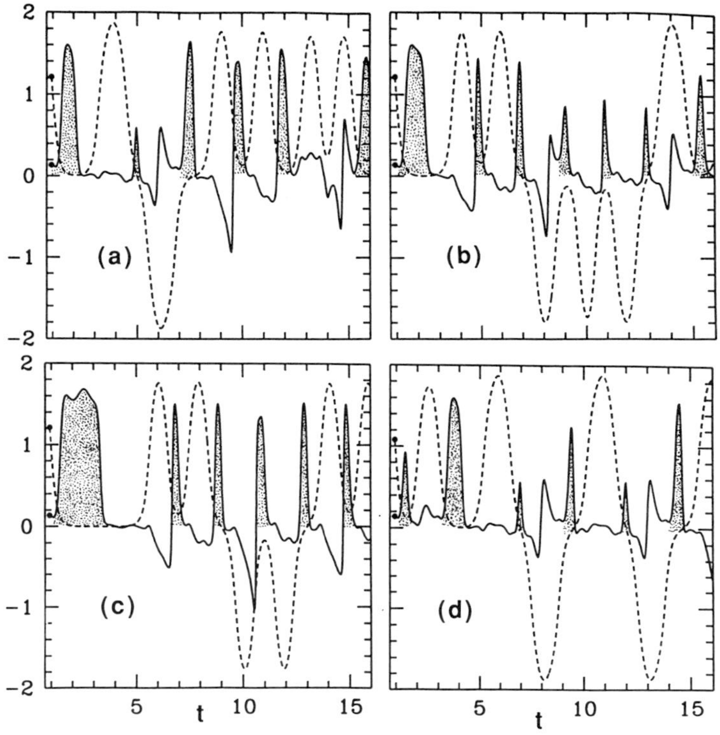

Fig. 55. Irreversible strain rates (solid lines) associated with the plots of Fig. 54. Peaks associated with passage by a hyperbolic fixed point are shaded. The dashed line is the vertical location y of the infinitesimal line element.

the lobe structure, and line elements within the structure, and the oscillating perturbing straining field can entail additional variations in the irreversible strain rates. As an example of the first cause, in the closed flow the unstable manifold bounding the image of the detraining turnstile, $P_\varepsilon(D(1))$, shows significant deviation angles at very large values of y (refer back to Fig. 43).

The basic message of non-diminishing peaks associated with passes by hyperbolic fixed points via re-orientation by turnstiles should thus be clear. The effect of transport of line elements is perhaps slightly less clear, demanding a qualification in conjunction with our previous remarks about whether passing closer to the hyperbolic fixed points is necessarily more beneficial to stretching. Referring back to the stretch-history analysis, the closed flow's highly non-uniform stretch profile shows extremely good stretching highly localized near the PIPs and SIPs. Clearly in this case, passing near the hyperbolic fixed point is most beneficial. The more uniform open flow profile, however, indicates that some of the best stretching can occur well between the PIPs and SIPs. In this case, remaining further away from the hyperbolic fixed point is not necessarily unhelpful to straining. The effect of transport thus depends on the spatial disparity of stretching in phase space, namely, whether or not the linearized flow about the hyperbolic fixed point corresponds to significantly better straining than elsewhere. Nonetheless, being transported near hyperbolic fixed points with proper re-orientation can of course never *hurt*, and under fairly common conditions it is unquestionably helpful.

So far we have only shown 'typical' trajectories. From the stretch analysis of Section 6.1, in particular the symbolic description of lobe-boundary evolution, it should be clear that the trajectories can exhibit a range of behavior, but also equally clear how to understand this range of behavior. Any trajectory will involve a succession of revolutions around the chaotic tangle, the periods of any revolution can vary from some minimum value all the way to infinity, and any combination of periods can occur under successive revolutions. From the symbolic dynamics, we know the topology associated with these various combinations. Since segments of the lobe boundary corresponding to longer revolution times around the tangles also correspond to vanishingly smaller initial arclengths (refer back to Section 6.1, and see also Section 6.3), trajectories with longer revolution times are less likely to occur in the tangle. As an extreme example, SIPs (whose final revolution time is by definition infinitely long) correspond to isolated points in the tangles, which will thus not be found unless explicitly searched for. We shall nevertheless see in Section 6.3 how these isolated points can have a significant impact on stretching in tangles. Figure 56 shows these extreme examples in the context of the open flow. Without resorting to an exhaustive numerical study, the basic point to appreciate is that the longer the time interval associated

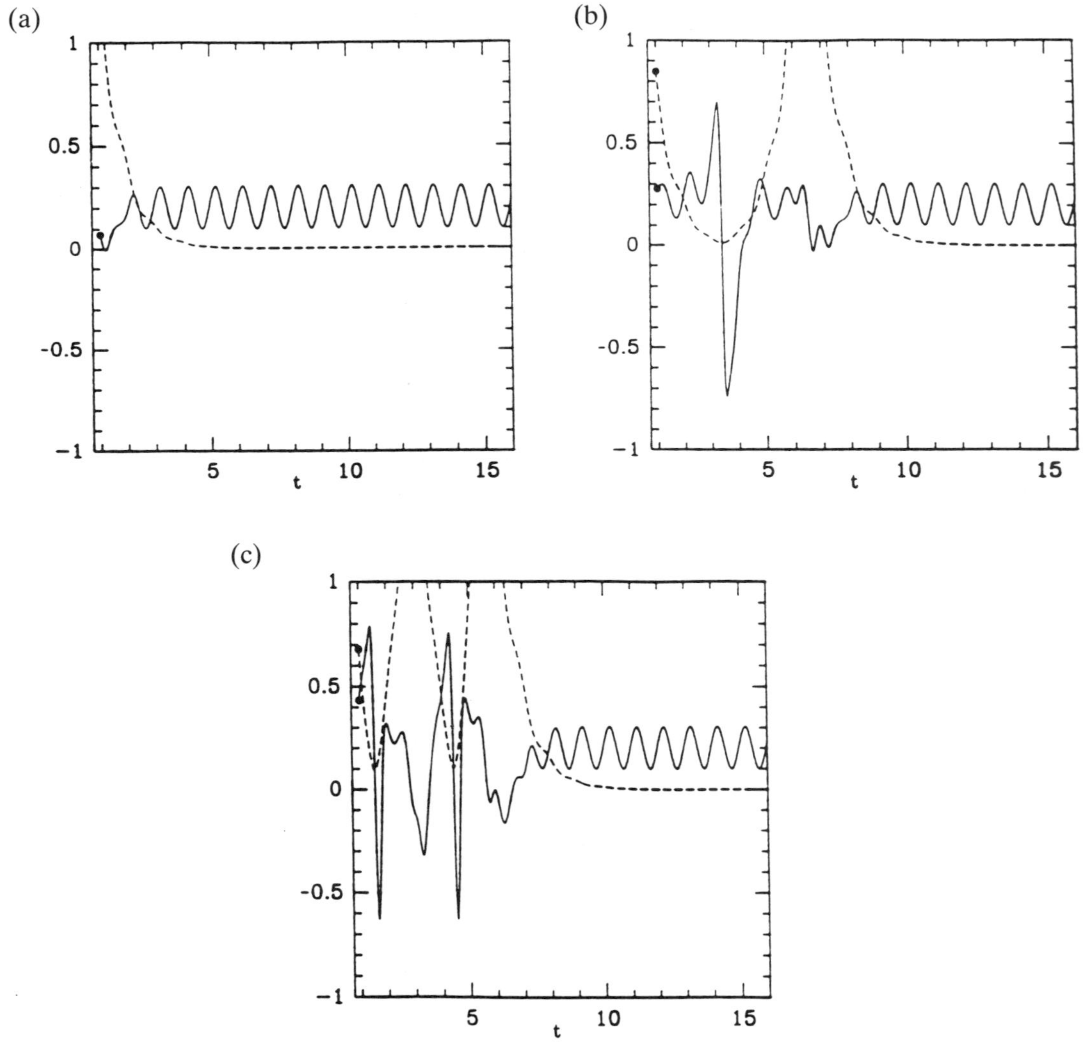

Fig. 56. Perturbed open-flow strain rate plots (solid lines) for infinitesimal line elements of r_0 associated with (a) a PIP, (b) a SIP that revolves around the tangle once, and (c) another SIP that revolves around the tangle twice. The dashed line is the vertical location y of the infinitesimal line element.

with a revolution around the tangle, the more spread apart will be the peaks in irreversible strain rate associated with successive passes by the hyperbolic fixed point, and the longer the 'dead interval' between peaks. Hence, for longer periods and smaller perturbation values the basic re-orientation picture will be more apparent.

6.3. *Stretch statistics*

We now turn our attention to the statistics associated with stretching under chaotic advection, one of the many features of stretching one might wish to study in the context of an invariant manifold template. In particular, we investigate probability distributions of finite-time Lyapunov exponents (or simply *stretch distributions*) associated with interfaces evolving within 2D chaotic tangles. Such statistics are relevant, for example, to the mixing process in fluid flows, where the stretching of an interface between two species affects the rate of mixing, and to kinematic dynamo phenomena, where the stretching of fluid elements affects magnetic field amplification. In contrast to previous studies of stretch statistics associated with chaotic flows [35, 38, 42, 71, 72, 77, 86, 99], we focus explicitly on the high-stretch tails of these distributions, which are relevant for several reasons. First, though the high-stretch tails correspond to small probability values, we find they can play a significant role in interfacial stretching. Second, for incompressible flows these tails correspond to the limit of small spatial scales (small striation width), and thus have direct impact on the multifractal characteristics of passive scalars in the small-striation-width regime. Third, the turbulence community has shown significant recent interest in probability distribution function (PDF) tails [4, 16, 22, 36, 45, 91, 100], and in a related context fractal and multifractal characterizations associated with distributions (e.g. see Sreenivasan [94] and references therein), and it should be of interest to understand distribution tails associated with nonturbulent chaotic flows. We thus perform a high-resolution numerical study of these high-stretch tails by implementing a dynamic point-insertion scheme to maintain good interfacial covering. Such a scheme allows one to determine tail behavior right up to the maximum stretch in the distribution. We observe that high-stretch tails can take on a range of behavior, depending on the system at hand, varying from essentially Gaussian to nearly exponential, and show that the non-Gaussian deviations can play a significant role in interfacial evolution. Such an observation indicates the need for sufficiently high-resolution experiments to capture tail behavior when studying stretch statistics associated with chaotic flows. The range of high-stretch tail statistics is understandable in the context of the symbolic dynamics construction of Section 6.1, which shows how, though one can indeed view stretching in chaotic tangles in terms of products of weakly correlated events (each 'event' can be thought of as a revolution around the chaotic tangle, although other interpretations may be chosen), there is a rich variety of stretch scales, spatial scales, and temporal scales, allowing for a range of tail statistics. A way to appreciate this variety is to reconsider the interfacial stretch profiles of Fig. 43. The open- and closed-flow profiles are markedly different as a simple consequence of different flow geometry and flow parameters (e.g. the hyperbolic fixed point's unstable eigenvalue is significantly higher in the closed-flow tangle, and there is an additional hyperbolic fixed point in the open-flow tangle), which influences the nonuniformity of the profiles. The closed flow exhibits a highly nonuniform stretch profile with isolated sharp peaks of very large stretch values localized over very small initial interfacial arclengths. Additionally, the vertical dashed lines of Fig. 43 denote points on the interface which intersect the stable manifold, and hence asymptotically approach a hyperbolic fixed point. Therefore as one gets closer to these points along the interface, the time interval associated with revolution around the chaotic tangle goes to infinity. Hence the closed-flow profile contains isolated sharp peaks of good stretching associated with very small spatial scales and large temporal scales. In contrast, the open-flow profile is substantially more uniform, and has, loosely speaking, evolved from a more even mix of scales. It seems plausible that the stretch statistics are qualitatively different for the two flows.

6.3.1. *Distributions of finite-time Lyapunov exponents.* Given an ensemble of infinitesimal line elements distributed along an interface, let $P(\lambda(n); n)$ denote the probability distribution at the nth cycle ($t = n \in Z$) of the finite-time Lyapunov exponent $\lambda(n) \equiv \ln(\delta s(n)/\delta s(0))/n$, where $\delta s(n)$ denotes the length at the nth cycle of an infinitesimal line segment that originates along our interface. Studies of finite-time Lyapunov exponent distributions that monitor a fixed number of particles would approximate $P(\lambda(n); n)$ as the percentage of particles per bin width $\delta\lambda$ at a given $\lambda(n)$ bin. In contrast, we employ a point-insertion scheme that maintains the covering of our interface to within a certain width (chosen to be about 1% of the mean vortex separation). When we insert a new point along the interface, its associated stretch history is interpolated from the two neighboring points (accurate for a dense enough interfacial grid), and adjust the initial arclengths of the line segments associated with the inserted and neighboring points. The result at the nth cycle is thus a partitioning of the interface, of total length $s_{tot}(n)$, into a set of *small* line elements $\{\delta s_i(n) | \sum_i \delta s_i(n) = s_{tot}(n)\}$, each of which has a specified *stretch* $\delta s_i(n)/\delta s_i(0)$ and initial arclength $\delta s_i(0)$. The stretch and initial arclength can vary greatly along the interface; hence, an improved approximation of $P(\lambda(n); n)$ is given by the *percentage of the interface's initial arclength per bin width* $\delta\lambda$ *in the* $\lambda(n)$ *bin*:

$$P(\lambda(n); n) = \frac{1}{\delta\lambda} \sum_{\substack{i \\ \lambda_i(n) \in \lambda(n)\text{bin}}} \frac{\delta s_i(0)}{s_{tot}(0)}, \tag{76}$$

where $\lambda_i(n)$ is the finite-time Lyapunov exponent associated with the ith line element. Our prescription is similar to a fixed particle-number scheme with appropriately weighted initial particle separations, and we checked our results against this alternative scheme to confirm negligible point insertion error.

We find for our class of chaotic tangles that high-stretch tails of finite-time Lyapunov exponent distributions can range from essentially Gaussian to nearly exponential. This range of behavior is illustrated by plotting the distributions on a logarithmic scale and then scaling the vertical axis with division by n, which exploits our high-resolution calculation at the high-stretch tail (see Figs 57 and 58). The open-flow tail, which corresponds to a more uniform stretch profile, is essentially Gaussian by $n = 10$. The closed-flow tail, which corresponds to a much more non-uniform stretch profile with isolated sharp peaks of good stretching highly localized over small spatial scales, appears to be nearly exponential by $n = 10$. With increasing n the closed-flow tail, in addition to smoothing out, appears to be flattening and extending downwards and to the right. This transient behavior is easily understood by the fact that it takes a few cycles for the first few portions of the interface to approach the region local to the hyperbolic fixed point; hence with increasing n the line segments in this local region will have spent a larger percentage of time there and their associated $\lambda(n)$ increases. The rightmost point in the distribution is thus asymptotically approaching $\lambda \simeq 5$ associated with the linearized flow local to the hyperbolic fixed point. By $n = 10$ the transient behavior of the scaled distributions has decayed considerably, so that, for example, the $n = 9$ and $n = 10$ distributions lie essentially on top of one another. One can thus confidently claim in the closed-flow example significant non-Gaussian behavior on short and medium time scales, and it seems plausible this behavior persists asymptotically under the present scaling. Though one might be tempted to draw a definitive conclusion

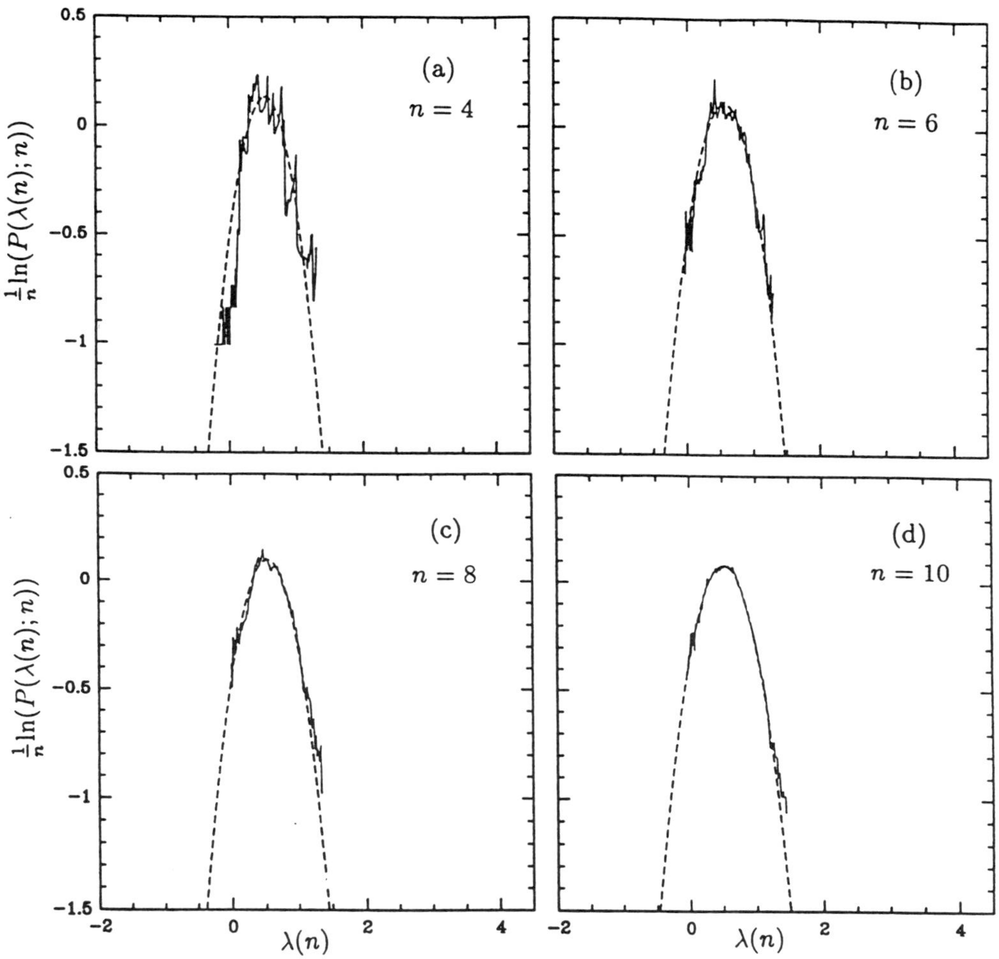

Fig. 57. Scaled open-flow stretch distributions (solid lines) associated with the material interface of Fig. 43 at sample times $n = 4, 6, 8, 10$. The dashed lines are Gaussian approximations, defined by having the same mean and standard deviation of the actual distributions.

about the asymptotic distribution, we avoid this temptation for two basic reasons. First, an appearance of convergence can be deceptive in the context of stretch distributions, for the distributions can vary slowly over long time scales; for example, because all particles on the open flow interface (except those which intersect the stable manifold) asymptotically approach infinitely far away from the hyperbolic fixed points, the maximum of the *open* flow distribution asymptotically approaches $\lambda = 0$ very slowly, and asymmetries in the distribution can slowly set in. Hence, comparing a few successive iterates for convergence can be meaningless in the context of distributions. Second, asymptotic results can depend on choice of scaling. Our scaling via contraction of the vertical scale (dividing by n) is consistent with that of previous investigators [38, 99] (we neglect a small transient term that some choose to keep in the scaling). There are, of course, other possible scalings, however, such as the one employed in the central limit theorem, where one expands the horizontal scale $\lambda(n) \rightarrow (\lambda(n) - \mu(n))\sqrt{n}/\sigma(n)$, with $\mu(n)$ the mean and $\sigma(n)$ the standard deviation of the distribution. Different scalings can give different asymptotic results; for example, a binomial distribution asymptotically approaches a Gaussian over any finite interval of its domain under the above horizontal scaling, but *not* under the previous vertical scaling.

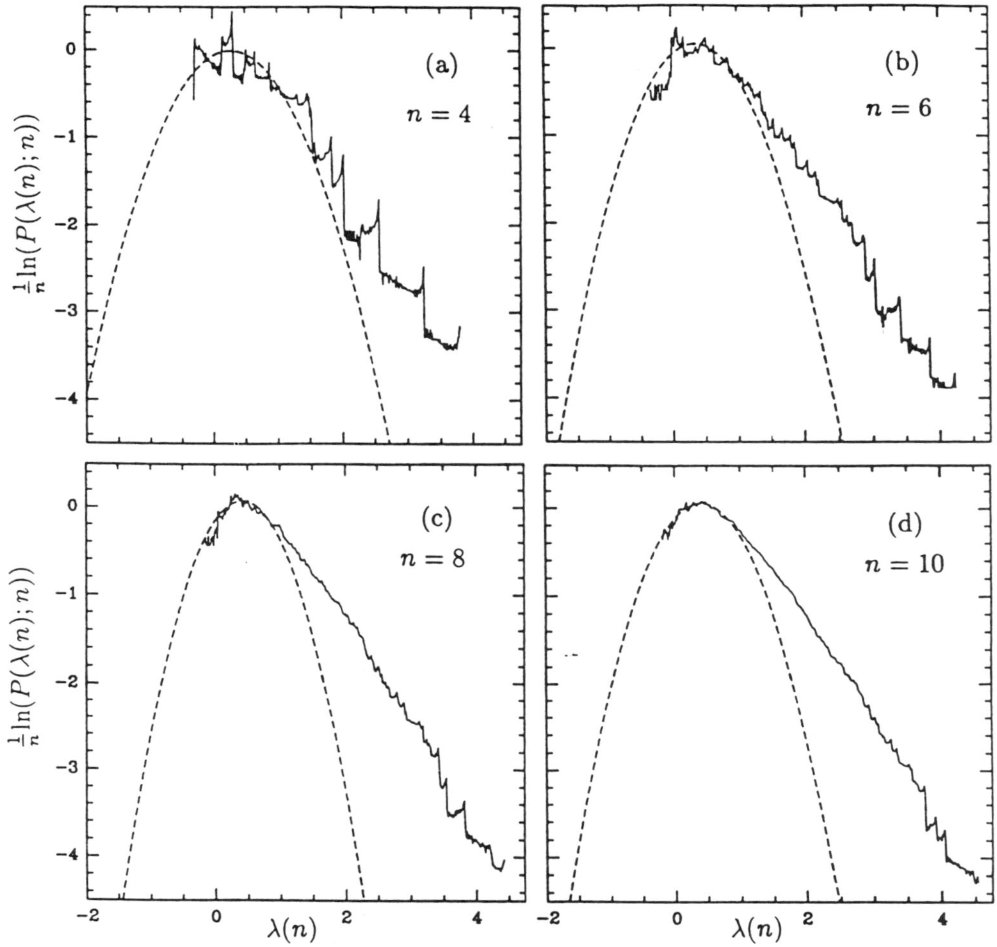

Fig. 58. Scaled closed-flow stretch distributions (solid lines) associated with the material interface of Fig. 43 at sample times $n = 4, 6, 8, 10$. The dashed lines are Gaussian approximations, defined by having the same mean and standard deviation of the actual distributions. The Gaussian approximation *fits the actual distributions well except for very small probability values*.

This horizontal scaling result (in keeping with the central limit theorem) is caused by pushing any tail deviations out to infinity, a wise scaling if one wishes to ignore vanishingly small probability values. However, in the context of stretching, it is not obvious that we wish to scale these deviations away, since small probability values are associated with high stretch values. Hence, the vertical scaling is useful, and we are additionally motivated to study the effects of tail deviations in a *scaling-independent* framework, which is the focus of the next section.

6.3.2. *Asymptotic result.* Since the non-Gaussian deviations associated with the high-stretch tail are also associated with small probability values (e.g. see Fig. 59), we need to ask whether they can have a significant effect on the evolution of our interface. This question can be studied in a simple setting that exploits rapid convergence of a single scalar quantity, the exponent associated with the growth rate of the total length of the interface. Let $l_a(n)$ denote the total length of the *actual interface*, and $l_g(n)$ denote the total length of the *interface whose stretch statistics are described by the corresponding Gaussian*

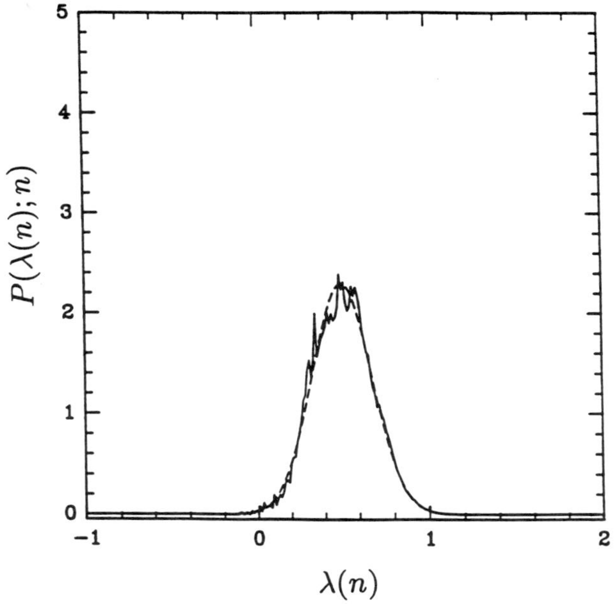

Fig. 59. The closed flow stretch distribution (solid line) at $n = 10$ agrees well with the Gaussian approximation (dashed line) for the range of probabilities shown. The non-Gaussian deviations at the high-stretch tail are indetectable at the present scale.

approximation. A measure of the relevance of the non-Gaussian deviations is thus the length ratio $l_g(n)/l_a(n)$. If this ratio is significantly less than one, then the non-Gaussian deviations have a significant effect on the interface. For the open-flow interface the values of $\ln(l_a(n))/n$ and $\ln(l_g(n))/n$ are essentially equal over the short-time calculation (which went to $n = 11$), as one would expect from Fig. 57; hence, one cannot easily conclude that any non-Gaussian deviations are significant. For the closed flow, however, Fig. 60 indicates for $l_a(n)$ and $l_g(n)$ fairly rapid convergence with increasing time to the relations

$$\begin{aligned} \frac{\ln(l_a(n))}{n} &\to 0.87 \\ \frac{\ln(l_g(n))}{n} &\to 0.65. \end{aligned} \tag{77}$$

This would indicate rapid convergence of the length ratio to a time dependence

$$\frac{l_g(n)}{l_a(n)} \sim \mathrm{e}^{-0.22n}. \tag{78}$$

Hence it appears that the length ratio asymptotically approaches zero, giving strong indication of the relevance of the non-Gaussian tail in the evolution of the closed-flow interface.

6.3.3. *Models.* Given a numerical observation of non-Gaussian high-stretch tails, and an indication of their significance, one would like to have a better appreciation of the range of high-stretch statistics. A key feature in determining the statistics is seen to be the non-uniformity of the perturbed stretch profiles, which in turn is related to the non-uniformity of the corresponding *unperturbed* stretch profiles, as described in Section 6.1

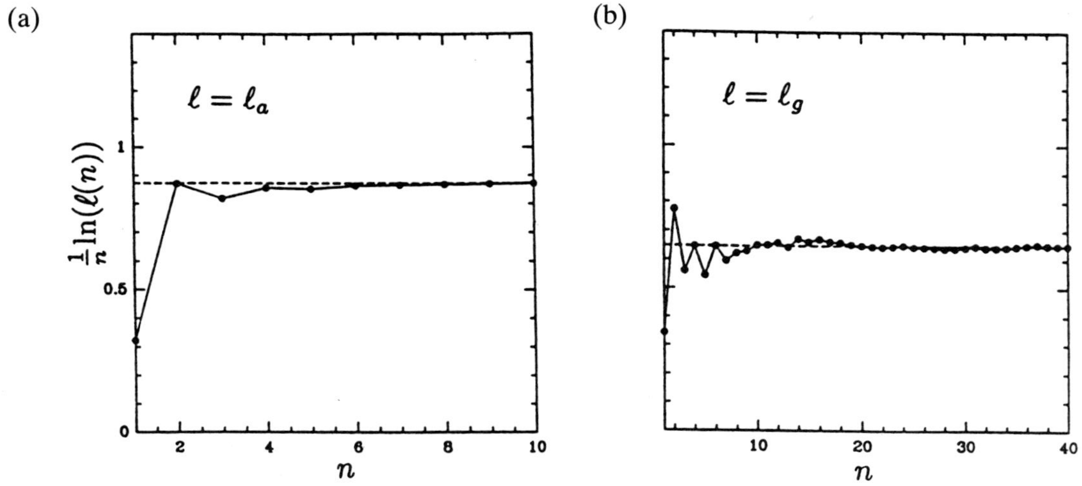

Fig. 60. Exponents associated with the growth rate of the closed flow interfacial length for (a) $l = l_a$ and (b) $l = l_g$. Note that the closed flow computation of $l_g(n)$ is performed over 40 cycles to ensure convergence. This computation is performed with a *fixed* number of points; we verified that it does not affect the statistics associated with the Gaussian approximation $l_g(n)$ (since ignoring the high-stretch tail does not significantly alter the Gaussian approximation).

(recall the cartoon-like portrayal of Fig. 42). Loosely speaking, as the unperturbed profiles become more non-uniform, good stretching is weighted more towards the *endpoints* of the profiles (see Fig. 61). As a result, the perturbed profiles become more non-uniform, with good stretching weighted more towards the *secondary intersection points* (SIPs) of the profiles (refer back to the closed flow example of Section 6.1), entailing progressively better stretching associated with progressively smaller spatial scales (smaller initial arclength of the interface), and progressively longer time intervals for revolution around the homoclinic/heteroclinic tangle.

One approach to understanding the stretch statistics exhibited by interfaces evolving in chaotic tangles is to search for elementary models for the stretch processes that capture the observed statistical behavior, the goal being to obtain the simplest model possible which respects the essential features of the actual stretch processes. We consider here some elementary models for stretch processes under chaotic flows consisting of iterative processes acting on the unit interval [0, 1]. The domain represents an interface, parameterized by initial arclength, and the value defined over this domain represents the stretch experienced by the interface, taken initially to be one throughout the domain for all the models. The simplest model, motivated by the horseshoe map, would have a *single* stretch p acting over the entire unit interval. The result would be a finite-time Lyapunov exponent equal to $\ln(p)$ defined over the entire domain, giving for a stretch distribution a delta function centered at $\lambda(n) = \ln(p)$. Such a model is obviously inadequate as a consequence of containing only a single stretch, so we proceed to a more complicated model that has *two* stretches, entailing a *binomial multiplicative process* (see Fig. 62). Partition the unit interval into two equal halves and multiply by p on one half, $q < p$ on the other half. Repeat this process to each half, then to each resulting quarter, and so on. This procedure is a canonical construction of a *multifractal*, which is heuristically described as a set of fractals, each with its own weighting and dimension (see, for example, Feder [32] for an introductory account). The result at the nth step is a set of stretches

$$\{p^i q^{n-i};\ i \in [0, n]\} \tag{79}$$

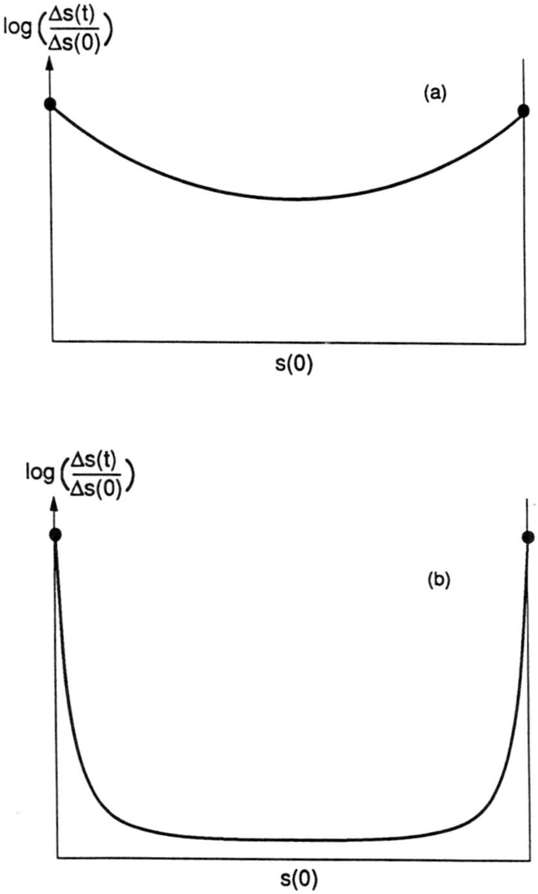

Fig. 61. As the unperturbed stretch profile becomes more non-uniform, good stretching is weighted more towards the endpoints, entailing that good stretching will be weighted more towards the SIPs in the perturbed case.

with corresponding domain widths given by the binomial distribution

$$\left\{\frac{n!}{i!(n-i)!}\frac{1}{2^n};\ i \in [0, n]\right\}. \tag{80}$$

These expressions give a set of finite-time Lyapunov exponents

$$\left\{\lambda_i(n) = \frac{i}{n}\ln(p) + \left(1 - \frac{i}{n}\right)\ln(q);\ i \in [0, n]\right\} \tag{81}$$

with corresponding asymptotically-invariant scaled probability distribution in the large n limit

$$\frac{\ln(P(\lambda(n);\ n))}{n} = -\{\xi\ln(\xi) + (1-\xi)\ln(1-\xi) + \ln(2)\}, \tag{82}$$

where

$$\xi \equiv \frac{\lambda(n) - \lambda(n)_{\min}}{\lambda(n)_{\max} - \lambda(n)_{\min}}, \tag{83}$$

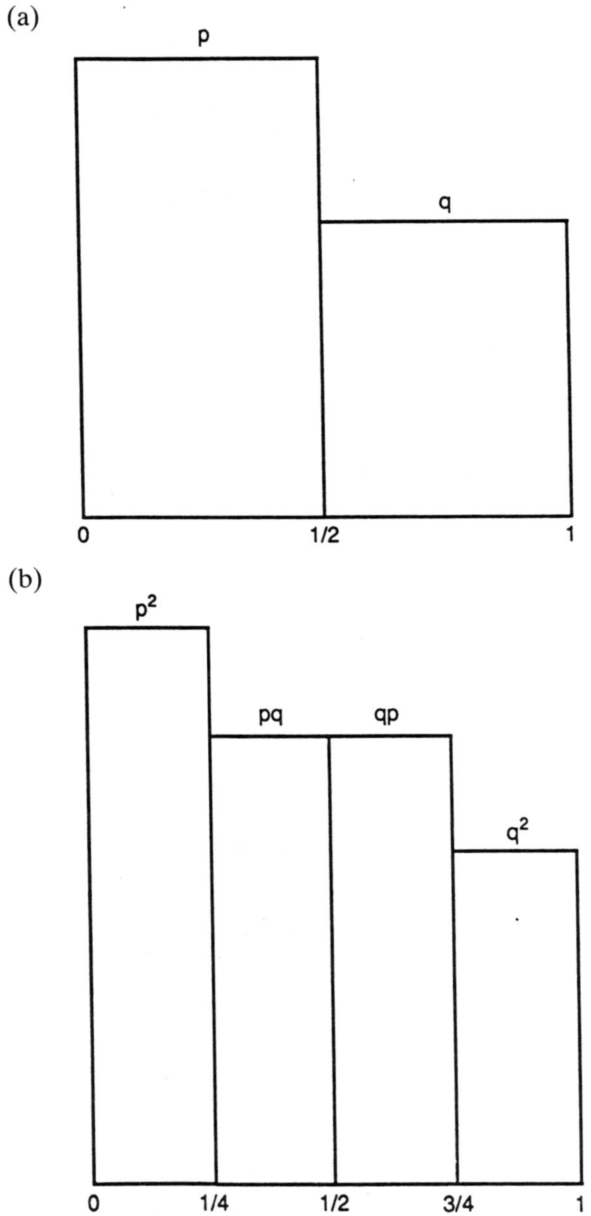

Fig. 62. The (a) first and (b) second iterate of the binomial multiplicative process.

as follows from Stirling's formula and sending the transients to zero. The Gaussian fit to this distribution, determined by requiring the same curvature at the maximum, is

$$\frac{\ln\left(P_{\mathrm{g}}(\lambda(n);\, n)\right)}{n} = -2(\xi - \tfrac{1}{2})^2. \tag{84}$$

The resulting stretch distribution for any $p \neq q$ is shown in Fig. 63. The model obviously lacks the freedom to capture the range of high-stretch tails found earlier. For example, there is no way to get a nearly exponential tail from this model in a limiting regime. This model appears inadequate since, although it contains two *stretch* scales, it contains only

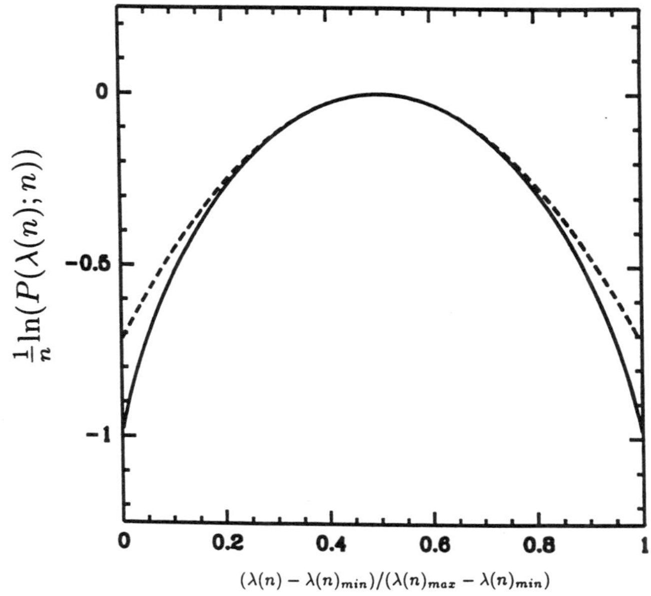

Fig. 63. The asymptotically invariant form of the scaled stretch distribution associated with the binomial multiplicative process for arbitrary $p \neq q$. The vertical scale is in units of $|\ln(P(\lambda(n);\, n)_{\min})|$, and the dashed line shows the Gaussian approximation.

one *spatial* scale, thus not respecting the possible occurrence of very good stretching distributed over very small spatial scales.

An improved model might be then to make the division of the domain and successive subdomains weighted by some ratio β: $(1-\beta)$, $0<\beta<1$, as shown in Fig. 64. The result at the nth step is then a set of stretches

$$\{p^i q^{n-i};\ i \in [0,\, n]\} \tag{85}$$

with corresponding domain widths

$$\left\{\frac{n!}{i!(n-i)!}\beta^i(1-\beta)^{n-i};\ i \in [0,\, n]\right\}. \tag{86}$$

For example, one might take $\beta/(1-\beta) \propto q/p$, so that as the disparity in the two stretches increases, the disparity in spatial scales increases, and in particular better stretching is associated with smaller spatial scales. Equation (85) again gives a set of finite-time Lyapunov exponents

$$\left\{\lambda_i(n) = \frac{i}{n}\ln(p) + \left(1-\frac{i}{n}\right)\ln(q);\ i \in [0,\, n]\right\}, \tag{87}$$

with corresponding asymptotically invariant scaled probability distributions in the large n limit given by

$$\frac{\ln(P(\lambda(n);\, n))}{n} = -\{\xi\ln(\xi) + (1-\xi)\ln(1-\xi) - \xi\ln(\beta) - (1-\xi)\ln(1-\beta)\}, \tag{88}$$

again from using Stirling's formula and sending transients to zero, and where ξ is again given by equation (83). The Gaussian approximation, using the same prescription as in the

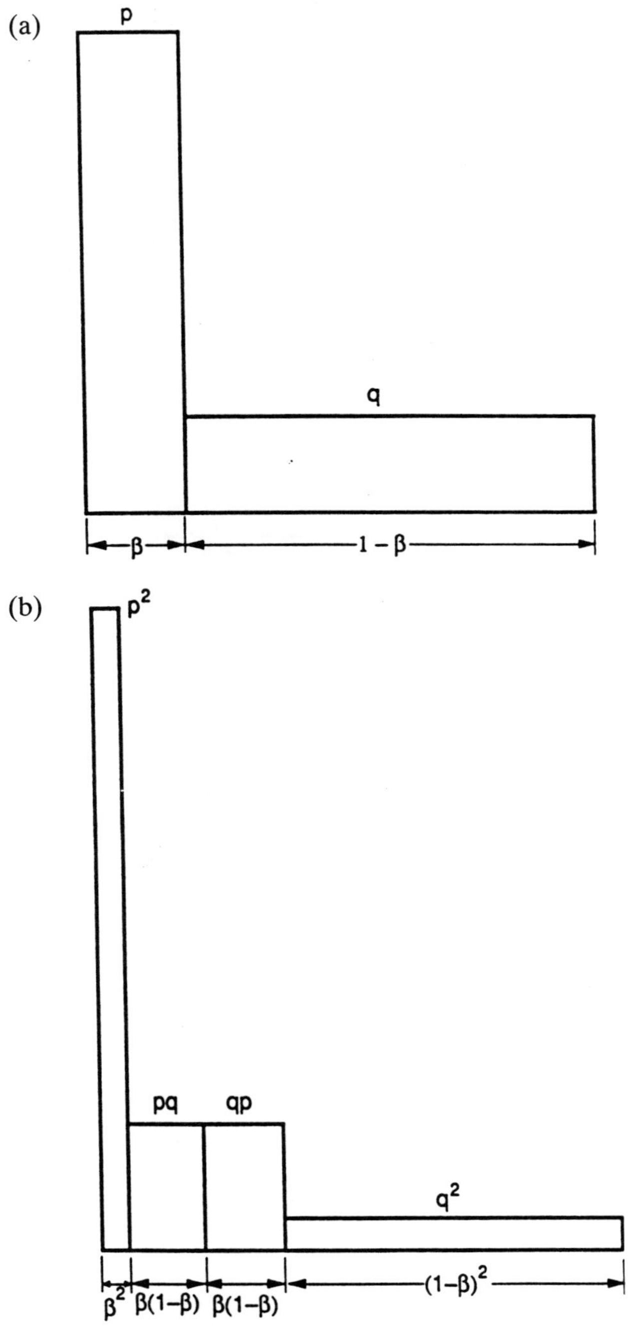

Fig. 64. The (a) first and (b) second iterate of the binomial multiplicative process with weighted subdomains.

previous binomial model, is

$$\frac{\ln\left(P_g(\lambda(n);\,n)\right)}{n} = -\frac{1}{2}\left\{\frac{1}{\beta} + \frac{1}{1-\beta}\right\}(\xi - \beta)^2. \tag{89}$$

The resulting stretch distributions are shown in Fig. 65 for decreasing values of β. On the one hand, the results look more promising, for in the limit $\beta \to 0$ one recovers an exponential tail. Hence by varying β from 1/2 to 0, one can vary the high-stretch tail from nearly Gaussian to nearly exponential. In the limit $\beta \ll 1$, equation (88) gives

$$\frac{\ln(P(\lambda(n);\, n))}{n} \approx \xi \ln(\beta), \tag{90}$$

valid for $\xi > \xi_{\min}$, where $\xi_{\min}$ is some minimum value that vanishes in the limit $\beta \to 0$. If we take $\beta \sim p^{-1}$ in the limit $\beta \ll 1$, a reasonable choice for a closed incompressible flow and consistent with the numerical observation of good stretching localized on small spatial scales, then $\lambda(n) \approx -\xi \ln(\beta)$ (again for $\xi > \xi_{\min}$, where $\xi_{\min} \to 0$ as $\beta \to 0$) and equation (90) gives

$$\frac{\ln(P(\lambda(n);\, n))}{n} \approx -\lambda(n), \tag{91}$$

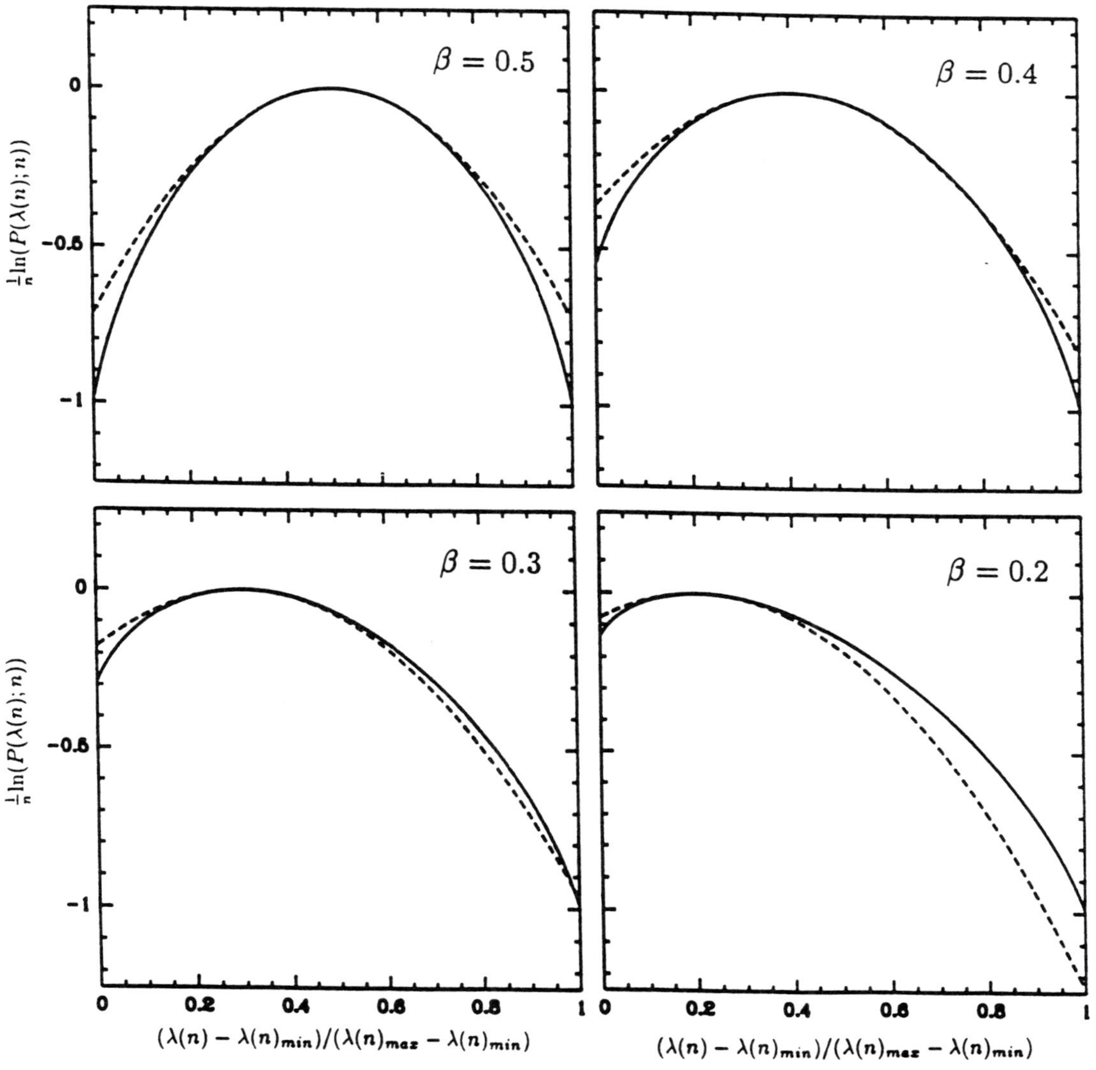

Fig. 65.

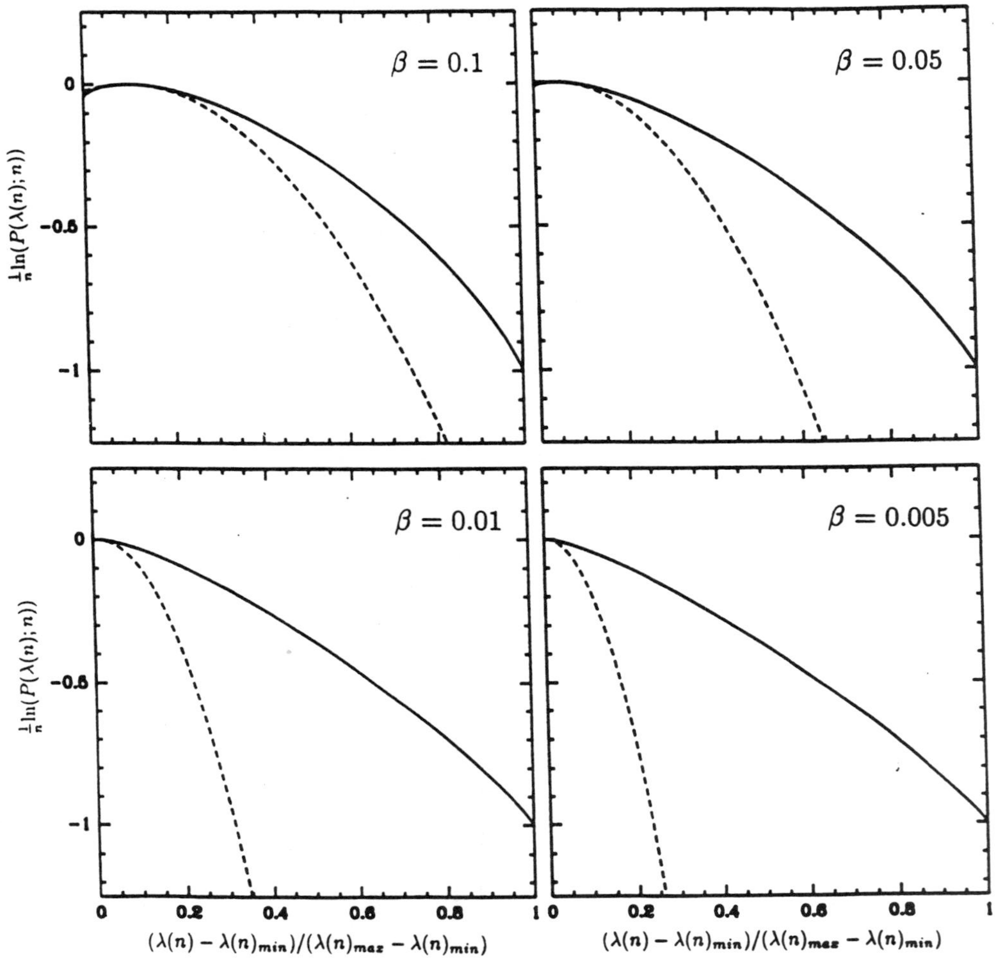

Fig. 65. Scaled probability distributions (solid lines) associated with the weighted binomial multiplicative process for decreasing values of β. The vertical scale is in units of $|\ln(P(\lambda(n); n)_{\min})|$, and the dashed lines show the Gaussian approximation.

an exponential tail with slope minus one. Clearly, however, this weighted binomial multiplicative process does not capture the numerically observed statistics, since it approaches an exponential high-stretch tail at the expense of the domain of the Gaussian hump, and for no parameter values do we find a nearly exponential tail with a Gaussian hump defined over a sizable domain, i.e. we can find nothing resembling the closed-flow distribution.

We thus again search for an improved model, and a *trinomial multiplicative process* (with weighted divisions of the domain) seems promising, the rationale being that as we let one of the three stretches become very large, we will have *two* other stretches to help us control the Gaussian hump. An example of this process is shown in Fig. 66. In an effort to capture the closed flow statistics, we will take p and q to be reasonably small ($O(1)$) with r much greater ($r \gg 1$). We take the width of the r strip to be r^{-1}, and the widths of the p, q strips to each be $(1 - r^{-1})/2$. This process then attempts to model a highly non-uniform stretch process, with good stretching highly localized on very small spatial

(a)

(b)

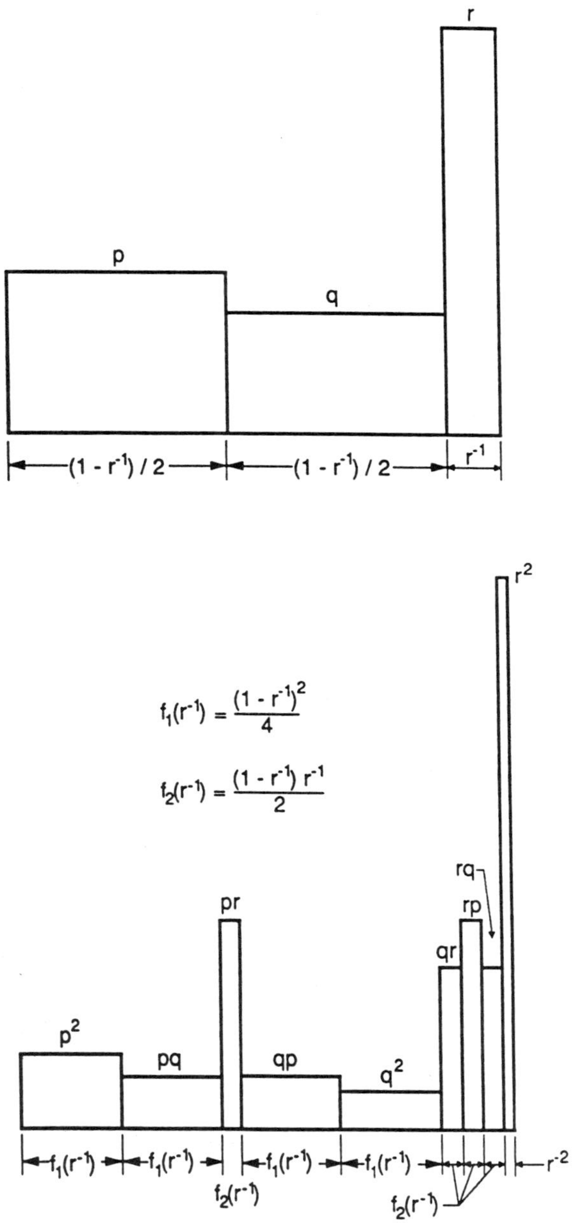

Fig. 66. The (a) first and (b) second iterate of the trinomial stretch process, with one stretch r much bigger than the other two stretches p, q.

scales. The result at the nth step is then a set of stretches

$$\{p^i q^j r^{n-i-j};\ i,\ j \in [0,\ n],\ i + j \leqslant n\} \tag{92}$$

with corresponding domain widths

$$\left\{\frac{n!}{i!j!(n-i-j)!}\ \frac{(1-r^{-1})^{i+j}}{2^{i+j}}\ \frac{1}{r^{n-i-j}};\ i,\ j \in [0,\ n],\ i + j \leqslant n\right\}. \tag{93}$$

The resulting set of finite-time Lyapunov exponents is

$$\left\{\lambda_{i,j}(n) = \frac{i}{n}\ln(p) + \frac{j}{n}\ln(q) + \left(1 - \frac{i}{n} - \frac{j}{n}\right)\ln(r);\ i,\ j \in [0,\ n],\ i + j \leq n\right\}. \tag{94}$$

Since this set depends on two independent variables i and j, we need an intermediate step to get the asymptotically invariant probability distribution in the large n limit. We first write

$$\begin{aligned}\frac{\ln(P(\lambda_{i,j}(n);\ n))}{n} = -\left\{\frac{i}{n}\ln\left(\frac{i}{n}\right) + \frac{j}{n}\ln\left(\frac{j}{n}\right) + \left(1 - \frac{i}{n} - \frac{j}{n}\right)\ln\left(1 - \frac{i}{n} - \frac{j}{n}\right)\right.\\ \left. + \frac{i}{n}\ln\left(\frac{2}{(1 - r^{-1})}\right) + \frac{j}{n}\ln\left(\frac{2}{(1 - r^{-1})}\right) + \left(1 - \frac{i}{n} - \frac{j}{n}\right)\ln(r)\right\},\end{aligned} \tag{95}$$

followed by

$$P(\lambda(n);\ n)\Delta\lambda(n) = \sum_{\substack{i,j \\ \lambda_{i,j}(n) \in \lambda(n)\text{bin}}} P(\lambda_{i,j}(n);\ n)\Delta\lambda_{i,j}(n). \tag{96}$$

Figure 67 shows the scaled probability distribution determined by the above prescription. Figure 67(a) is the result of an effort to obtain ballpark agreement with the closed-flow distribution. For example, we take $r \approx \exp(\lambda_u)$, where λ_u is the unstable eigenvalue of the closed-flow hyperbolic fixed point at the origin, and p, q are chosen with the aim of obtaining a similar Gaussian hump. For this still rather crude model we do not try to account for contraction, however, and simply deal with three stretches. At first glance the distribution compares well with that of the closed flow. The curvature of the high-stretch tails are nevertheless still significantly different, as highlighted in Fig. 68. As in the weighted binomial process, we can make the high-stretch tail of the trinomial model arbitrarily close to exponential, but at the expense of either

(i) making the width of the Gaussian hump smaller, by decreasing $(p - q)/r$, or
(ii) having the peak of the Gaussian hump occur at a smaller $\lambda(n)$, by decreasing $(p + q)/r$.

An example of (i) is shown in Fig. 67(b)—by making $(p - q)/r$ smaller, the high-stretch tail is slightly more exponential, but at the expense of a much skinnier Gaussian hump; indeed, non-Gaussian deviations exist for significant probability values, which is why we do not try to approximate the distribution with a Gaussian. For the weighted trinomial stretch process, then, one can choose parameter values to obtain approximate agreement with the closed-flow distribution, but it appears as though small but fundamental differences remain. There are several avenues one may pursue with more complicated models having multiple stretch, spatial, and even temporal scales. Such avenues are best pursued in the context of a more systematic numerical study of the stretch models (to obtain a precise idea of what needs to be modeled) and more serious effort at constructing models that adequately respect the fairly complicated underlying symbolic dynamics. We thus end our study of stretch models here, the primary purpose of this section being to point out the range of high-stretch statistics afforded by the fairly complicated stretch processes found in chaotic tangles, the significance of these high-stretch tails, and the need for high-resolution numerical studies of stretch statistics.

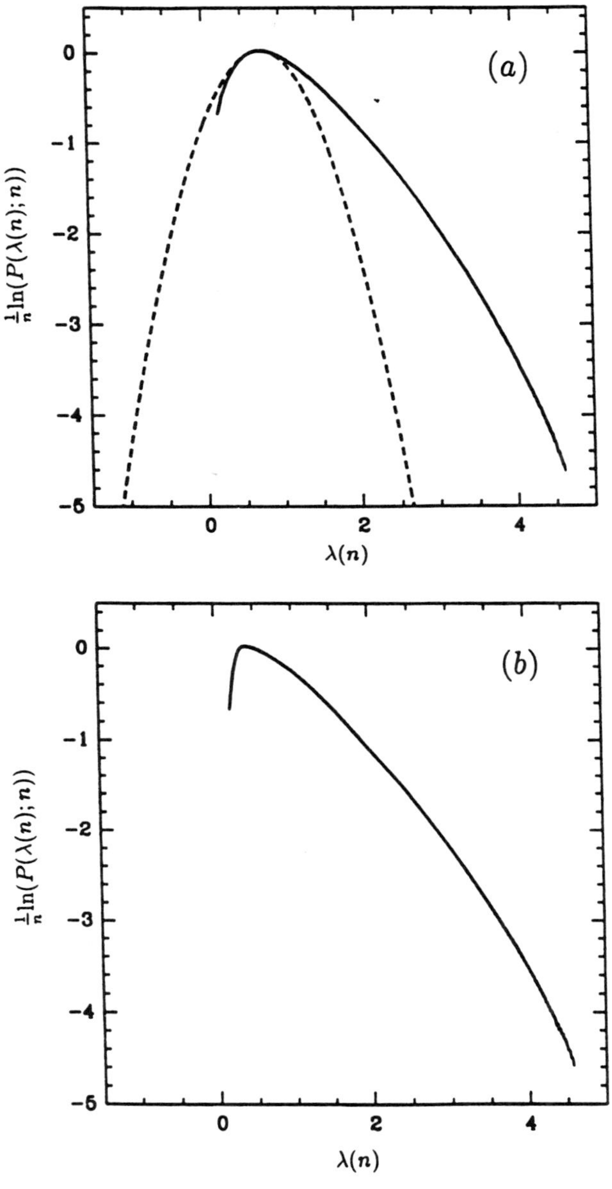

Fig. 67. Scaled probability distributions associated with the trinomial stretch process for (a) $p = 1.2$, $q = 3.4$, and $r = 100$, and (b) the same parameters as (a) except now $q = 1.7$. In (a) the Gaussian approximation is given by the dashed line; in (b) the approximation is omitted since the distribution is obviously non-Gaussian even at substantial probability values.

7. MIXING

7.1. *The interplay between advection and mixing*

Thus far we have considered chaotic advection, and the ensuing stirring, transport and stretching properties, in the absence of any mixing effects due to such stochastic phenomena as molecular diffusion or heat transfer. From a dynamical systems perspective, to incorporate these non-deterministic effects entails a fundamental departure from the

(a) (b)

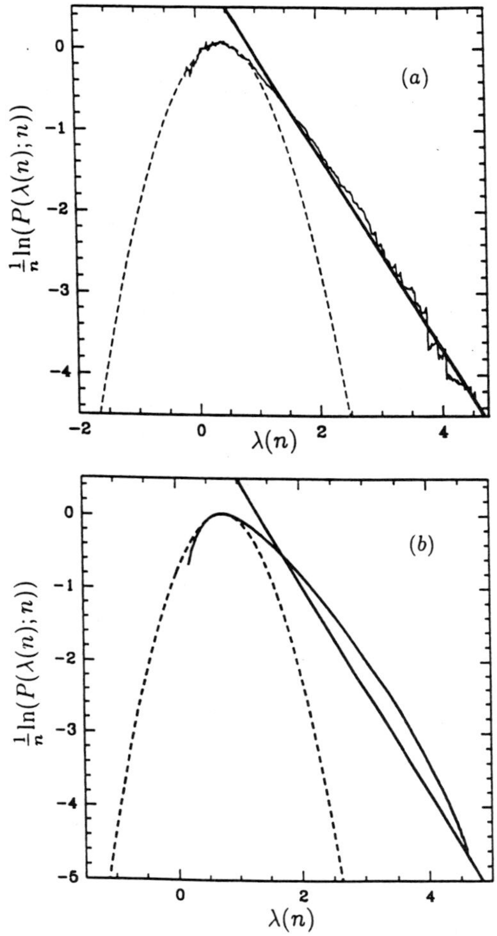

Fig. 68. (a) Attempting to draw a straight line through the nearly exponential tail of the closed flow at the $n = 10$ sample, and (b) a similar line for the trinomial distribution of Fig. 67(a).

analysis thus far. For example, to incorporate diffusive effects into the class of deterministic flows under consideration, one must add a stochastic term to the equations governing fluid particle motion. Rather than forsake the analysis applicable to deterministic flows, however, we would like to somehow use the analysis to understand the new role of stochastic effects. In particular, the aim is to understand the interplay between mixing and the advection properties we have studied, such as transport and stretching, with particular emphasis on efficiency of mixing and control of mixing. For concreteness of discussion, we henceforth consider mixing in the context of molecular diffusion.

In order to study such an interplay, let us first consider the scenarios of a 2D fluid with an interface separating two species of fluid, types A and B. For a stationary fluid the area of mixed fluid per unit initial arclength of the interface, $\delta\mathcal{A}$, increases as $\sqrt{(Dt)}$, where D is the diffusion coefficient. For a stirred fluid the length of the interface can increase dramatically in time, thereby enhancing the mixing. In particular, at a given point $\mathbf{p}$ on the interface the differential length has been increased by a factor defined as the interface stretch at $\mathbf{p}$, $S(\mathbf{p}, t) = \exp[\int_0^t \sigma(\mathbf{p}, \tilde{t})\,\mathrm{d}\tilde{t}]$, where $\sigma(\mathbf{p}, t)$ is the stretching rate at $\mathbf{p}$ (refer back to Section 6.2). On the other hand, interfacial stretching implies that diffusion normal to the interface is retarded because of the implied convective velocity normal and inwards towards the interface. If we assume for simplicity that σ is independent of time, the net result is that the length multiplying the differential length element to obtain the area of mixed fluid is replaced by a length that peaks out to a factor proportional to $\sqrt{[D/\sigma(\mathbf{p})]}$ as time increases. Thus the differential area element of mixed fluid increases as

$$\delta\mathcal{A} \sim \{D/\sigma(\mathbf{p})\}^{1/2} e^{\sigma(\mathbf{p})t} \tag{97}$$

for $t \geq 1/\sigma(\mathbf{p})$. In this manner, one obtains a connection between a mixing process and an underlying stretch process. Several additional realities must be considered, however, to address the relationship between mixing and chaotic advection, which we phrase in the context of a lobe of fluid evolving within a chaotic tangle, and hence conceptualize as two species of fluids diffusing across the invariant manifold boundary of a lobe.

(i) *The complexity of lobe boundary stretching*. The stretch analysis of Section 6 shows the presence of striking variation of finite-time stretch with initial condition, and significant variation of stretch rates with time. Indeed, well within the tangles stretching can be poor, certainly non-exponential, over finite time intervals. Phenomena such as these can invalidate expression (97).

(ii) *Lobe thickness and collision of mixed zones*. As a result of efficient stirring in the tangles, lobe boundaries will rapidly become close together, increasingly so with time, entailing an eventual 'collision' of the mixed zones associated with the lobe boundary (see Fig. 69). Such a collision will be seen to saturate the mixing process, thus altering expression (97) in a fundamental way. Hence, mixing efficiency will in a basic way depend on lobe thickness, and separation of neighboring segments of the lobe, as well as the stretching of the lobe.

(iii) *Lobe area and lobe content*. Of course, a basic influence on the mixing process is the amount of fluid available to be mixed. In our idealized setting this quantity depends on lobe area, and hence the splitting of the manifolds. Under more general settings, as we shall see, this quantity depends on both lobe area and lobe *content* (i.e. the percentage of a given species of passive scalar in a lobe), which will depend on lobe intersections. Such area and species content considerations will directly impact the saturation process referred to in (ii).

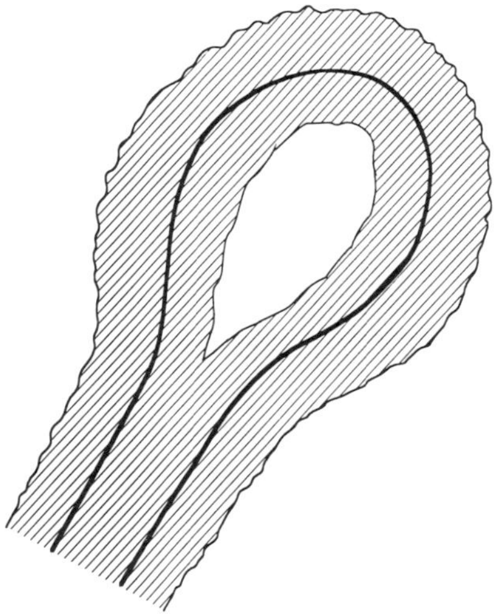

Fig. 69. Overlapping of neighboring mixed zones.

(iv) *Size of the mixing region—penetration of lobes into regions of physical space.* In addition to lobe area, the ability of the lobes to penetrate into regions of physical space affects the domain of the mixing region. For given lobe areas, the size of the mixing region affects the ability of different lobes to intersect one another, and the separation distances of neighboring segments of a given lobe, thus influencing lobe content and the above-mentioned saturation effect that we will soon analyze. More significantly, the global advection properties of the lobes can serve to efficiently spread the mixed fluid over different regions of physical space (e.g. different rolls in a Rayleigh–Bénard problem [18]).

Besides the above-mentioned influence of stretching, thickness, area, penetration, and transport properties of lobes on mixing, one is also interested in the converse problem, such as the effect of mixing on transport or statistical properties. In the context of transport, as a lobe of fluid evolves within the chaotic tangle, the effect of diffusion will be to smear out the lobe, i.e. to spread the contents of the lobe over the boundaries defined by the invariant manifolds of the corresponding deterministic flow, until eventually the passive scalar originally in the lobe bears little resemblance to the lobe defined by the above-mentioned invariant manifold boundaries. This smearing effect entails a fundamental limitation in the applicability of the transport theory of Section 5 in the presence of stochastics effects such as molecular diffusion. One would like to obtain a reasonable estimate of this smearing effect on transport. In the context of statistical properties, i.e. stretch statistics (Section 6.3) and in a related context the statistics associated with distributions of striation widths, the effect of diffusion is to smear out the thinnest striations (corresponding to highest stretch values), thus affecting statistics associated with thin striations and high stretch values.

Given an awareness of some of the realities associated with the interplay between chaotic advection and mixing, let us proceed with a more careful study of these issues.

7.2. *Lobe boundary evolution and mixing*

Let us consider mixing in the context of diffusion of passive scalars across an interface in a 2D fluid. Given that the analysis thus far has focused on the evolution of an individual lobe in a tangle, let us study the effects of mixing on lobe evolution by considering the scenario of a 2D fluid with passive scalar A originating in one of the turnstile lobes, and passive scalar B everywhere else. In particular, let us consider our time-periodic oscillating vortical flows, and focus on the entraining turnstile lobe. As the fluid advects according to the open or closed flow, let fluid A and B diffuse across the interface defined by the material curve that initially lies on the entraining turnstile lobe boundary. This curve thus consists of the segment r_0 from Section 6 and a segment of the stable manifold. The stable manifold part of the curve, as seen in Section 6, quickly shrinks to negligible length relative to r_0, and so with negligible error we ignore the diffusion across the stable manifold segment and focus only on r_0. For thin enough diffusion zones and sufficiently steep concentration gradients across the interface, which is valid for a given diffusion coefficient for small enough times, the diffusion process is essentially normal to the interface, and can be treated as one-dimensional. The amount diffused across the interface can then be related to the stretch history of the interface [19, 51, 59, 73]. The concentration of fluid A (and similarly fluid B) spreads with increasing time like an error function (see Fig. 70(a)):

$$\Phi_A(\mathbf{p}, y, t) = \frac{1}{2}\left\{1 + \operatorname{erf}\left[\frac{S(\mathbf{p}, t)y}{(4D\tau(\mathbf{p}, t))^{1/2}}\right]\right\}, \tag{98}$$

where $\mathbf{p}$ specifies a point on the interface, y is a relative coordinate normal to the interface at $\mathbf{p}$ and pointing into the lobe, D is the diffusion coefficient, $S(\mathbf{p}, t) = \delta s(\mathbf{p}, t)/\delta s(\mathbf{p}, t = 0)$

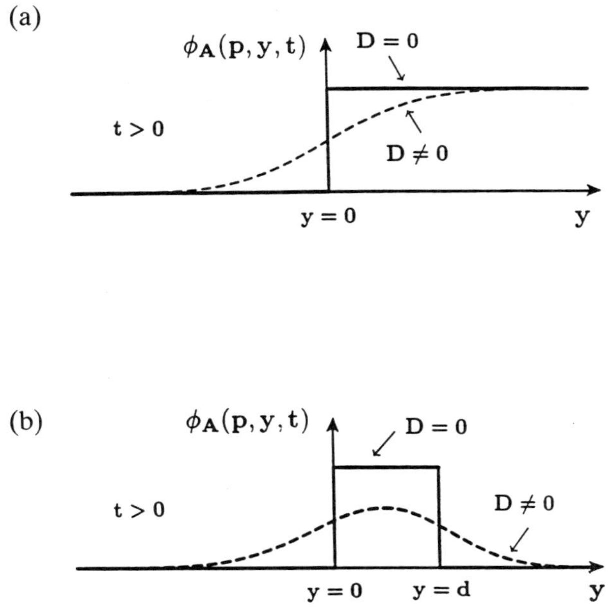

Fig. 70. The concentration profile of fluid A near the interface at **p** for (a) the non-overlapping thin diffusion zone approximation and (b) the overlapping thin difffusion zone approximation.

is the interface stretch at **p**, with $\delta s(\mathbf{p}, t)$ the arclength of the interfacial line element, and

$$\tau(\mathbf{p}, t) = \int_0^t S^2(\mathbf{p}, \tilde{t})\, d\tilde{t}. \tag{99}$$

The diffusion rate per unit initial arclength across the interface at **p**, $C(\mathbf{p}, t)$, is given by

$$C(\mathbf{p}, t) = D \frac{\partial \Phi_A}{\partial y}\bigg|_{y=0} S = \frac{1}{2}\left[\frac{D}{\pi\tau}\right]^{1/2} S^2. \tag{100}$$

The amount of fluid A diffused across the interface at **p** per unit initial arc length is then

$$\mathcal{A}(\mathbf{p}, t) = \int_0^t C(\mathbf{p}, \tilde{t})\, d\tilde{t} = \left[\frac{D\tau}{\pi}\right]^{1/2}. \tag{101}$$

Hence, the local time-dependent diffusion rate and the amount diffused across the interface is completely determined by the local stretch history of the interface. (Note that for σ constant in time, one recovers expression (97).) The total amount of fluid diffused out of the lobe is found by integrating equation (101) over all **p**. Enhanced interface stretching corresponds to enhanced diffusion across the interface, and the global understanding of stretching obtained in Section 6 directly translates into an understanding of diffusion. For example, the large variations in finite-time stretching seen in Section 6.1 translates to large variations in finite-time mixing, and the variations in rates of stretching studied in Section 6.2 translates to similar variations in mixing rates. However, the above analysis is valid only for sufficiently thin diffusion zones, and a notable exception to this case occurs when the interface folds back onto itself so that the diffusion zones overlap, as described earlier and portrayed in Fig. 69. As should be clear from the interface plots of Section 6, in chaotic zones interfaces quickly fold and wrap around themselves in a violent manner, and one would thus like to employ an analysis that addresses overlap of neighboring diffusion zones. A full analysis that takes into account the overlap of all neighboring diffusion zones would be quite laborious. Instead we employ an analysis that takes into account the overlap of a *nearest-neighbor* diffusion zone, which is sufficient to capture additional phenomena relevant to the notion of *efficiency of mixing*. The key to such an analysis is to notice, as seen in the interface plots of Section 6, that many of the neighboring sections of the interface are essentially parallel to one another and have small curvature (typically these portions of the interface have good stretching); in addition, the parts of high curvature tend to be bulbous and hence have less of a tendency for overlap of diffusion zones (typically these portions have poor stretching). Under the assumption that overlap of neighboring diffusion zones is not significant until the neighboring segments of the interface are essentially parallel, which is valid for a small enough diffusion coefficient, the diffusion process can still be treated as one-dimensional. We address overlap of diffusion zones when it is due to the lobe being sufficiently thin; we do not worry about overlap due to the lobe folding on top of itself. When there is overlap at $\mathbf{p}$, equation (98) is replaced by a superposition of two error functions

$$\Phi_A(\mathbf{p}, y, t) = \frac{1}{2}\left\{\operatorname{erf}\left[\frac{Sy}{(4D\tau)^{1/2}}\right] - \operatorname{erf}\left[\frac{S_d(y-d)}{(4D\tau_d)^{1/2}}\right]\right\}, \tag{102}$$

where $d = d(\mathbf{p}, t)$ is the normal distance at $\mathbf{p}$ between the two neighboring segments of the interface whose diffusion zones overlap, and S is understood to represent $S(\mathbf{p}, t)$, while $S_d \equiv S(\mathbf{p} + d\hat{y}, t)$ (and similarly for τ and τ_d). Note that in the limit $d \to \infty$ (with y fixed) the second error function goes to -1, and hence equation (102) recovers equation (98). The concentration spreads with increasing time like a sum of two error functions, one associated with each interface (see Fig. 70(b)). By a calculation similar to that in equation (100), except that one has to now take into account the presence of the other interface, the diffusion rate out of the lobe per unit initial length of neighboring segments of the lobe boundary is

$$C(\mathbf{p}, t) = \frac{1}{2}\left[\frac{D}{\pi}\right]^{1/2}\left\{\frac{S^2}{\tau^{1/2}}\left(1 - \exp\left[\frac{-(Sd)^2}{4D\tau}\right]\right) + \frac{S_d^{\,2}}{\tau_d^{\,1/2}}\left(1 - \exp\left[\frac{-(S_d d)^2}{4D\tau_d}\right]\right)\right\}. \tag{103}$$

The presence of a neighboring interface thus adds to the diffusion process a saturation term of the form $(1 - \exp[-(Sd)^2/4D\tau])$, so that the separation distance d is a factor in the diffusion process. To find the total amount of fluid diffused out of the lobe, a convenient trick, with physical meaning, is to separate the two terms in equation (103) and then integrate over *all* $\mathbf{p}$ in the interface, rather than integrate (103) over half the $\mathbf{p}$ of neighboring segments of the interface whose diffusion zones overlap. Hence for all $\mathbf{p}$ we write

$$C(\mathbf{p}, t) = \frac{1}{2}\left[\frac{D}{\pi\tau}\right]^{1/2} S^2\left(1 - \exp\left[\frac{-(Sd)^2}{4D\tau}\right]\right), \tag{104}$$

where it is understood that when there is negligible overlap at $\mathbf{p}$ we send $d \to \infty$. The amount of fluid A diffused across the interface at $\mathbf{p}$ per unit initial arc length, $\mathcal{A}(\mathbf{p}, t)$, is then given by integrating $C(\mathbf{p}, t)$ in time as done in equation (101), and then the total amount of fluid A that leaks out of the lobe is given by integrating $\mathcal{A}(\mathbf{p}, t)$ over *all* $\mathbf{p}$. Note that, to determine $\mathcal{A}(\mathbf{p}, t)$ at a particular time, one can of course also deal with a spatial integral of the concentration $\Phi_A(\mathbf{p}, y, t)$ rather than a time integral of $C(\mathbf{p}, t)$; the advantage of the former approach is that one need calculate $d(\mathbf{p}, t)$ only at that given time.

We present in Fig. 71 some plots of $\mathcal{A}(\mathbf{p}, t)$ when r_0 advects under the same velocity fields as considered in the examples of Section 6. Fig. 71 shows the total amounts of fluid A that diffuse out of the lobe. The saturation term is clearly relevant to the efficiency of the diffusion process. For example, in the closed-flow simulation, though the interface

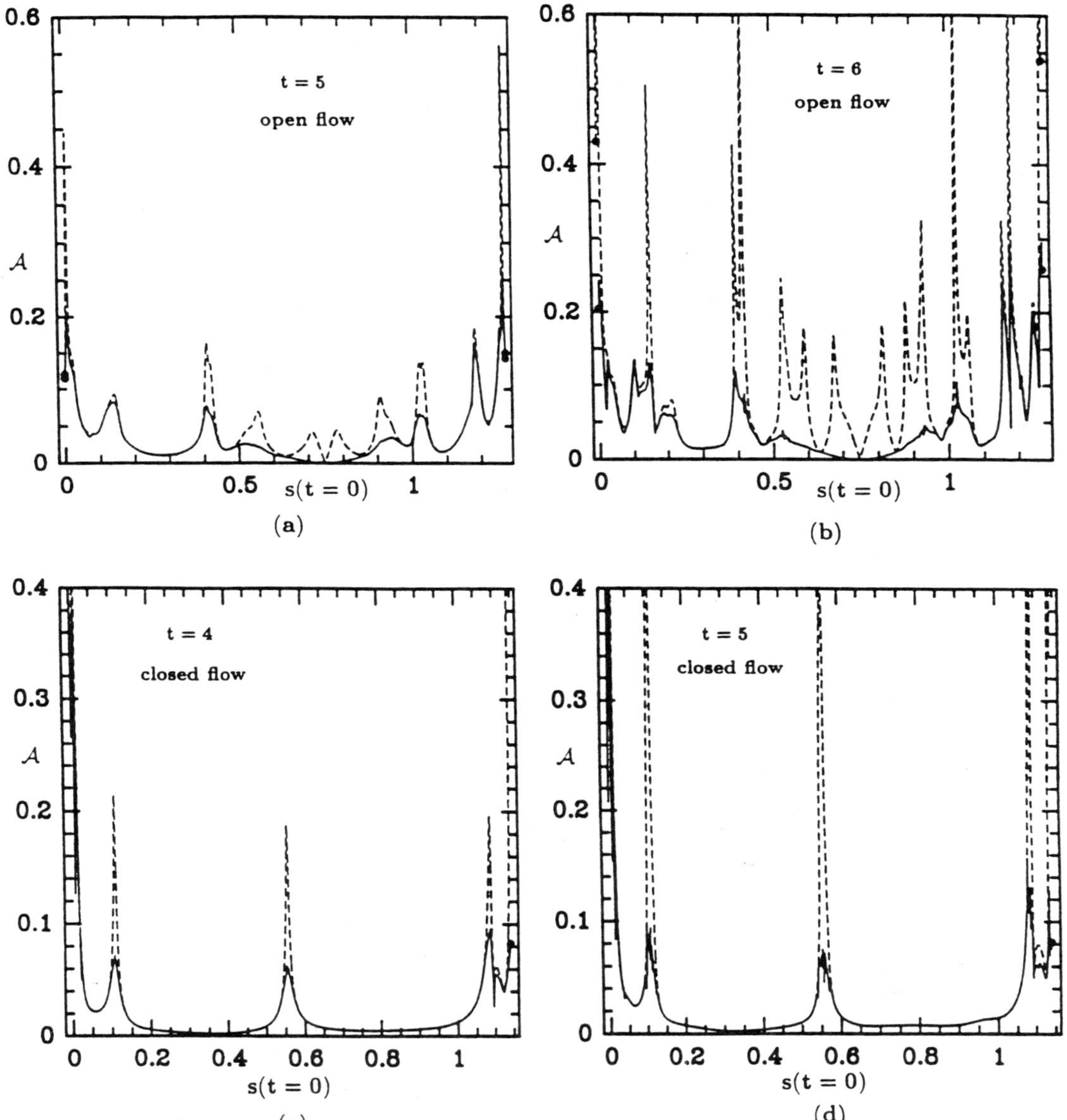

Fig. 71. $\mathscr{A}(\mathbf{p}, t)$ when r_0 advects under the same velocity fields as considered in the examples of Section 6 ($D = 2\pi \times 10^{-6}$ for both flows). The dashed line corresponds to the non-overlapping thin diffusion zone theory and the solid line to the overlapping thin diffusion zone theory.

stretches more than in the open flow, this stretching is highly localized in a few regions where the lobe is extremely thin, and the diffusion process quickly saturates since there is little fluid available to diffuse out of these parts of the lobe. Hence, we see from $t = 4$ to $t = 5$ there is little enhancement of diffusion since, though there is a great amount of stretching going on, it is localized in regions where the diffusion process has saturated (note that as more SIPs asymptotically approach the hyperbolic fixed point, the diffusion process will again be enhanced, so that the overall enhancement is staggered in time). This is in contrast to the open-flow example, where the stretching is more evenly distributed, and saturation is less of a factor. One sees then that, in the context of diffusion, the efficiency of mixing depends not only on the efficiency of stretching along the interface, but on how this stretching is distributed relative to the separation of neighboring segments of the interface. This effect is manifested by the fact that, even though r_0 stretches more in the closed flow, the open flow has a greater amount of diffusion on the time scale shown (the greater saturation in the closed flow is due in part to the smaller turnstile lobe area, but more importantly to the greater non-uniformity of the stretch distribution), as shown in Beigie [11]. On longer time scales the differences between a finite and infinite mixing region (the tangle) of the closed and open flow, respectively, will become more apparent. As mentioned earlier, regions of poor stretching in the closed flow, due to re-entrainment, eventually give way to enhanced stretching, which is not necessarily the case in the open flow. However, the finiteness of the closed-flow mixing region implies that diffusion zone overlap, and hence saturation, will become more of a factor than in the open flow (especially when one considers more than just nearest-neighbor overlap).

With Fig. 71 in hand, we emphasize the spirit of our investigation. Certainly the approach to studying diffusion is an approximate one, with several sources of error: we consider only nearest-neighbor overlap, we ignore diffusion across the stable manifold boundary, and the assumption of parallel neighboring segments of the interface and a one-dimensional diffusion process break down at sharp turning points of the interface, such as the tip of the lobe. However, the errors regarding the stable manifold and sharp turning points involve a truly negligible amount of fluid, and nearest-neighbor overlap is sufficient to capture approximately the effect of saturation. Our main goal is not to focus on a computational prescription for exact numerical results, but rather to take a first step past the non-overlapping thin diffusion zone theory, and to show how the efficiency of diffusion is affected by this extension. Until now our discussion of 'efficiency' has been informal; though there is more than one candidate for a formal definition of efficiency of diffusion in this context, none seems adequate for a *global* definition. For example, it seems natural to address the non-uniformity of stretching and interface separation, and hence to compare the amount of fluid diffused out of the lobe with an idealized case that has the same lobe area and lobe-boundary length, but spatially uniform stretching and manifold separation. Dividing the non-uniform result by the uniform result could define an 'efficiency' of diffusion, except that the result is not guaranteed to be less than one: though $\Delta A_o(t)/\Delta A_{no}(t)$ can be *decreased* by nonuniformity, where ΔA_o, ΔA_{no} denote the total amount of A diffused out of the lobe in the overlapping and nonoverlapping theory, respectively, $\Delta A_{no}(t)$ can be *increased* by nonuniformity, so that the above 'efficiency' can be greater than one or less than one, as shown in Beigie [11]. For a global consideration of diffusion, it is not clear that there is a 'best case' on which to base a definition of efficiency. Rather than focusing on a formal definition, and exact quantification, of efficiency, however, we would rather highlight the qualitative, practical aspects of the problem. For example, it is clear that the efficiency of diffusion across an interface is affected by the non-uniformity of stretching in the perturbed case, which, as discussed in Section 6, is related in near-integrable flows to the unperturbed stretch profile, which in turn is affected by the geometry of the unperturbed phase portrait (for example, the presence of an additional hyperbolic fixed point in the open flow tended to make the stretch distribution more even), and by the unperturbed system parameters (such as the strength of the vorticity). Thus one can alter the efficiency of the diffusion process by changing the geometry of the unperturbed phase portrait or the system parameters. In addition, we see from Section 6 that the local stretch and diffusion rates vary greatly with initial conditions, and by straightforward calculations one can identify the regions of good stretching and mixing; for example, in the closed-flow simulation good stretching and mixing is isolated near the SIPs, so identifying these points locates in phase space the 'seeds' of good stretching and mixing in chaotic tangles. Though exact results can only come from explicit numerical simulation, a knowledge of tangle geometry provides the framework for studying aspects of the relationship between stretching and mixing in chaotic tangles.

7.3. *Control of mixing and the interplay between transport, stretching and mixing at small spatial scales*

Having gained some understanding of the relation between lobe boundary evolution and mixing, let us explore further some of the relationships between transport, stretching, and mixing, in somewhat miscellaneous fashion (given the qualitative nature of the subject of mixing, a loosely ordered discussion seems not inappropriate). The goal here is merely to mention a few interesting points of what are a myriad of aspects of mixing one might wish to study and exploit.

7.3.1. *Control of mixing.* There are a variety of ways one might wish to enhance and control a mixing process. For example, in Section 7.2 we mentioned the possibility of enhancing mixing by identifying the seeds of good stretching in physical space and tuning system parameters to maximize the effects of these seeds. Another interesting and quite practical method for control involves the analytical technique of maximizing lobe area via Melnikov theory. As described in Sections 4 and 5, Melnikov theory allows one to measure analytically stable and unstable manifold separation, and hence lobe area, expressed as a function of system parameters, such as frequencies associated with the velocity field time dependence. With these convenient expressions one can efficiently sweep through parameter space to maximize lobe area and flux, and thus maximize one aspect of mixing.

7.3.2. *Quantifying the effect of diffusive spreading on fluid transport.* Section 5 studied the transport of fluid under classes of deterministic flows in the context of a lobe dynamics within invariant manifolds. Earlier in this section, however, we remarked that the introduction of stochastic processes such as molecular diffusion will entail a smearing effect manifested by the spreading of the initial contents of a lobe outside the invariant manifold boundaries defined by the original deterministic flow. Here we provide a rough analytical estimate of the significance of such a smearing effect, in particular, to determine the elapsed time before a significant amount of passive scalar has spread outside of the invariant manifold boundaries of a lobe. In their study of the relationship between transport and mixing, Camassa and Wiggins [18] estimate this time interval as the elapsed time before the area of mixed fluid in a stationary fluid is comparable to the lobe area for the tangle under consideration. Such an estimate, however, ignores stretching. A greatly improved estimate, though still quite approximate, is to use the concentration expression (102), which accounts for both interfacial stretching and lobe thickness. A reasonable criterion for significant smearing, associated with neighboring segments of the lobe boundary at point $\mathbf{p}$, is for the concentration function associated with one side of the lobe boundary to equal some small fraction, f, at the other side of the lobe boundary:

$$\frac{1}{2}\operatorname{erf}\left[\frac{Sd}{(4D\tau)^{1/2}}\right] = f. \tag{105}$$

For f sufficiently small we use $\operatorname{erf}(x) \approx (2/\sqrt{\pi})x$ to obtain the expression

$$\tau = \frac{1}{4\pi D}\left(\frac{Sd}{f}\right)^2. \tag{106}$$

Using that

$$\tau = \int_0^t \exp\,[2\lambda(\mathbf{p}, \tilde{t})\tilde{t}]\,\mathrm{d}\tilde{t}, \tag{107}$$

where $\lambda(\mathbf{p}, \tilde{t})$ denotes the finite-time Lyapunov exponent (e.g. refer back to Section 6.3), and assuming for simplicity that $\lambda(\mathbf{p}, \tilde{t})$ is independent of time $\tilde{t}$, we obtain for the critical time, t_{crit}, at which smearing is significant

$$t_{\text{crit}} = \frac{1}{2\lambda(\mathbf{p})} \ln \left\{ 1 + \frac{\lambda(\mathbf{p})}{2\pi D} \left[\frac{\{Sd\}(\mathbf{p})}{f} \right]^2 \right\} \tag{108}$$

where we have used that Sd asymptotically approaches a constant in time as a consequence of incompressibility, and thus write $Sd = \{Sd\}(\mathbf{p})$, meaning that to good approximation the quantity $\{Sd\}$ can be regarded as a function of only initial condition $\mathbf{p}$ along the interface. Equation (108) is of course to be interpreted as a rough estimate with a variety of qualifications: we consider spreading in the context of only two neighboring segments of the lobe, the finite-time Lyapunov exponent is assumed to be independent of time, and the desired fraction $f < 1$ is assumed to be small ($f \ll 1$). Nevertheless, basic dependences of mixing on stretching (through $\lambda(\mathbf{p})$), lobe area (through $\{Sd\}(\mathbf{p})$), and choice of smearing criterion (through f) are clear.

In this manner one obtains a measure of the time span over which the smearing effect on transport can be considered small. The significant variation of finite-time Lyapunov exponents along the lobe boundary will entail significant variation in the rate of diffusion along the lobe, and hence the rate at which the diffusive spreading becomes significant.

7.3.3. *Mixing at small spatial scales.* The subject of mixing at small spatial scales, and its relation to statistical, fractal, and multifractal properties of passive scalars, is a rich and active field, not without controversy (e.g. see Miller and Dimotakis [67] versus Sreenivasan and Meneveau [93] and Prasad and Sreenivasan [94]). We make only a few remarks here pertaining to our class of flows. Section 6.3 studies the high-stretch statistics associated with finite-time Lyapunov exponents, which for incompressible flows is related to fractal and multifractal properties associated with striations in the small spatial scale regime. There we saw how the range of stretch processes allow for a range of statistics at the high-stretch tail. In order to translate these results to flows in the presence of molecular diffusion, we must address the same diffusive smearing concerns affecting transport in the previous section. In particular, molecular diffusion will tend to smear out the smallest striations (the ones associated with the highest stretch). Using expression (105) in much the same way as for the t_{crit} estimate, one can obtain a rough estimate of the critical striation width d_{crit} (or finite-time Lyapunov exponent λ_{crit}) below which (above which) the striations are smeared out. The basic phenomenon to appreciate here is the possibility that the non-Gaussian deviations in the high-stretch tail for the stretch statistics under deterministic flows can be partially, or completely, wiped out by diffusive smearing effects. We believe this phenomenon may play a significant role in the fractal and multifractal properties associated with small spatial scale mixing.

8. CONCLUSION

We have given a detailed exposition on the use of invariant manifolds as a geometric template for the study of chaotic advection. Such a template is exploited in the context of lobe dynamics within tangle boundaries defined by global stable and unstable manifolds of invariant hyperbolic or normally hyperbolic sets. In conclusion we stress the following.

(i) Though invariant manifold templates have principally been exploited in the context of phase space transport, we have seen how they can be employed in the study of a variety of other features of the dynamics in chaotic tangles, such as those associated with stretching and mixing.

(ii) Since there is a tendency to fault an invariant manifold/phase space geometry approach to chaotic dynamics for its inability to provide easily obtained, comprehensive, and quantitative answers to the problem at hand, it is important to convey the spirit with which we exploit invariant manifolds. We feel a more realistic expectation is the use of invariant manifolds to obtain a qualitative understanding of the dynamics, formulate paradigms of behavior, initiate partial answers to certain features of the nonlinear dynamics, and provide a setting for intelligently chosen numerical studies.

(iii) Though chaotic advection has primarily been studied in the context of 2D maps derived from 2D time-periodic velocity fields, invariant manifold theory applies to a much wider class of flows and maps. The key to extension past 2D maps is to generalize from stable and unstable manifolds eminating from *hyperbolic fixed points* to stable and unstable manifolds eminating from *normally hyperbolic invariant manifolds*. Since the manifolds eminating from normally hyperbolic invariant manifolds are *codimension-one*, i.e. of one less dimension than the phase space in which they are embedded, the role they play as tangle boundaries in these more complicated settings is analogous to that of codimension-one tangle boundaries in the time-periodic setting. We would like to stress, then, the role of normally hyperbolic invariant manifolds as a major extension past the more standard geometric construct of hyperbolic fixed points. Further, we note that even the simplest dynamical issues associated with 2D maps derived from 2D time-periodic velocity fields can become quite challenging in the more complicated settings of higher-dimensional flows and velocity fields with quasiperiodic or aperiodic time dependences. Here, even the mere establishment and determination of invariant manifolds, tangles, partial barriers, turnstiles, lobe area/volume, flux, stretching and mixing properties, and so on, is of great value, and much remains to be pursued in exploiting a lobe dynamic formalism in these more complicated settings.

Acknowledgments—This material is based upon work supported by the National Science Foundation Postdoctoral Research Associateship Program, the Air Force Office of Scientific Research, Grant No. AFOSR-91-0329 (through Philip Holmes), the National Science Foundation Graduate Fellowship Program, the National Science Foundation Presidential Young Investigator Program, the Office of Naval Research Young Investigator Program, and the Air Force Office of Scientific Research, Grant No. AFOSR-91-0241. Some of the numerical computations were performed on Cornell Theory Center and Center for Applied Mathematics facilities.

REFERENCES

1. H. Aref, Stirring by chaotic advection, *J. Fluid Mech.* **143**, 1–21 (1984).
2. H. Aref and S. Balachandar, Chaotic advection in a Stokes flow, *Phys. Fluids* **29**, 3515–3521 (1986).
3. H. Aref and S. W. Jones, Enhanced separation of diffusing particles by chaotic advection, *Phys. Fluids A* **1**, 470–474 (1989).
4. S. Balachandar and L. Sirovich, Probability distribution functions in turbulent convection, *Phys. Fluids A* **3**, 919–927 (1991).
5. G. K. Batchelor, Small-scale variation of convected quantities like temperature in turbulent fluid, *J. Fluid Mech.* **5**, 113–133 (1958).
6. R. P. Behringer, S. D. Meyers and H. L. Swinney, Chaos and mixing in a geostrophic flow, *Phys. Fluids A* **3**, 1243–1249 (1991).
7. D. Beigie, A Leonard and S. Wiggins, A global study of enhanced stretching and diffusion in chaotic tangles, *Phys. Fluids A* **3**, 1039–1050 (1991).
8. D. Beigie, A Leonard and S. Wiggins, Chaotic transport in the homoclinic and heteroclinic tangle regions of quasiperiodically forced two-dimensional dynamical systems, *Nonlinearity* **4**, 775–819 (1991).
9. D. Beigie, A Leonard and S. Wiggins, The dynamics associated with the chaotic tangles of two-dimensional

quasiperiodic vector fields: theory and applications, in *Nonlinear Phenomena in Atmospheric and Oceanic Sciences, IMA Volumes in Mathematics and its Applications*, Vol. 40, edited by G. F. Carnevale and R. Pierrehumbert, pp. 47–138. Springer: New York (1992).
10. D. Beigie and S. Wiggins, Dynamics associated with a quasiperiodically forced Morse oscillator: Application to molecular dissociation, *Phys. Rev. A* **45**, 4803–4827 (1992).
11. D. Beigie, Transport, Stretching, and Mixing in Classes of Chaotic Tangles, Ph.D. Thesis, Department of Physics, California Institute of Technology (1992).
12. D. Beigie, A Leonard and S. Wiggins, Statistical relaxation under nonturbulent chaotic flows: Non-Gaussian high-stretch tails of finite-time Lyapunov exponent distributions, *Phys. Rev. Lett.* **70**, 275–278 (1993).
13. D. Beigie, Codimension-one partitioning and phase space transport in multi-degree-of-freedom Hamiltonian systems with non-toroidal invariant manifold intersections, Cornell University preprint (1994).
14. D. Beigie, Multiple separatrix crossing in multi-degree-of-freedom Hamiltonian flows: Global geometry and phase space transport associated with multiple partial barriers and turnstiles, Cornell University preprint (1994).
15. D. Bensimon and L. P. Kadanoff, Extended Chaos and Disappearance of KAM Trajectories, *Physica D* **13**, 82–89 (1984).
16. R. Benzi, L. Biferale, G. Paladin, A. Vulpiani and M. Vergassola, Multifractality in the statistics of the velocity gradients in turbulence. *Phys. Rev. Lett.* **67**, 2299–2302 (1991).
17. M. V. Berry, Regular and irregular motion, in Topics in nonlinear dynamics: a tribute to Sir Edward Bullard, edited by S. Jorna. *AIP Conference Proceedings* No. 46, pp. 16–120, AIP, New York (1978).
18. R. Camassa and S. Wiggins, Chaotic advection in a Rayleigh–Bénard flow, *Phys. Rev. A* **43**, 774–797 (1991).
19. G. F. Carrier, F. E. Fendell and F. E. Marble, The effect of strain rate on diffusion flames, *SIAM J. Appl. Math.* **28**, 463–500 (1975).
20. J. Chaiken, R. Chevray, M. Tabor and Q. M. Tan, Experimental study of Lagrangian turbulence in a Stokes flow, *Proc. Royal Soc. Lond. A* **408**, 165–174 (1986).
21. W. L. Chien, H. Rising and J. M. Ottino, Laminar and chaotic mixing in cavity flows, *J. Fluid Mech.* **170**, 355–377 (1986).
22. E. S. C. Ching, Probabilities for temperature differences in Rayleigh–Bénard convection, *Phys. Rev. A* **44**, 3622–3629 (1991).
23. W. Dahm, K. B. Southerland and K. A. Buch, Direct high resolution measurements of the fine scale structure of SC $\gg$ 1 molecular mixing in turbulent flows, *Phys. Fluids A* **3**, 1115–1127 (1991).
24. P. E. Dimotakis, Turbulent shear layer mixing with fast chemical reactions, in *Turbulent Reactive Flows*, pp. 417–485. Springer, New York (1989).
25. P. E. Dimotakis, Turbulent free shear layer mixing and combustion, *Prog. Astronaut Aeronaut* **137**, 265–340 (1991).
26. T. Dombre, U. Frisch, J. M. Greene, H. Hénon, A Mehr and A. M. Soward, Chaotic streamlines in the ABC flows, *J. Fluid Mech.* **167**, 353–391 (1986).
27. E. Dresselhaus and M. Tabor, The persistence of strain in dynamical systems, *J. Phys. A: Math. Gen.* **22**, 971–984 (1989).
28. E. Dresselhaus and M. Tabor, The kinematics of stretching and alignment of material elements in general flow fields, *J. Fluid Mech.* **236**, 415–444 (1991).
29. R. W. Easton, Trellises formed by stable and unstable manifolds in the plane, *Trans. Am. Math. Soc.* **294**, 719–732 (1986).
30. R. W. Easton, Transport through chaos, *Nonlinearity* **4**, 583–590 (1991).
31. M. P. Escudier, Observations of the flow produced in a cylindrical container by a rotating endwall, *Exp. Fluids* **2**, 189–196 (1984).
32. J. Feder, *Fractals*. Plenum Press, New York (1988).
33. M. Feingold, L. P. Kadanoff and O. Piro, Passive scalars, three-dimensional volume-preserving maps, and chaos, *J. Stat. Phys.* **50**, 529–565 (1988).
34. N. Fenichel, Persistence and Smoothness of Invariant Manifolds for Flows, *Ind. Univ. Math. J.* **21**, 193–226 (1971).
35. J. M. Finn and E. Ott, Chaotic flows and fast magnetic dynamos, *Phys. Fluids* **31**, 2992–3011 (1988).
36. J. P. Gollub and T. H. Solomon, Complex particle trajectories and transport in stationary and periodic convective flows, *Phys. Scripta* **40**, 430–435 (1989).
37. J. P. Gollub, J. Clarke, M. Gharib, B. Lane and O. N. Mesquita, Fluctuations and transport in a stirred fluid with a mean gradient, *Phys. Rev. Lett.* **67**, 3507–3510 (1991).
38. P. Grassberger, R. Badii and A. Politi, Scaling laws for invariant measures on hyperbolic and nonhyperbolic attractors, *J. Stat. Phys.* **51**, 135–178 (1988).
39. J. Guckenheimer and P. Holmes, *Nonlinear Oscillations, Dynamical Systems, and Bifurcations of Vector Fields*. Springer, New York (1983).
40. M. W. Hirsch, C. C. Pugh and M. Shub, *Invariant Manifolds. Springer Lecture Notes in Mathematics*, Vol. 583, Springer, New York (1977).
41. P. Holmes, Some remarks on chaotic particle paths in time-periodic, three-dimensional swirling flows, *Contemp. Math.* **28**, 393–404 (1984).
42. T. Horita, H. Hata, R. Ishizaki and H. Mori, Long-time correlations and expansion-rate spectra of chaos in Hamiltonian systems, *Prog. Theor. Phys.* **83**, 1065–1070 (1990).

43. K. Ide and S. Wiggins, The bifurcation of homoclinic tori in the quasiperiodically forced Duffing oscillator, *Physica D* **34**, 169–182 (1989).
44. S. W. Jones, O. M. Thomas and H. Aref, Chaotic advection by laminar flow in a twisted pipe, *J. Fluid Mech.* **209**, 335–357 (1989).
45. P. Kailasnath, K. R. Sreenivasan and G. Stolovitzky, Probability Density of Velocity Increments in Turbulent Flows, *Phys. Rev. Lett.* **68** 2766–2769 (1992).
46. I. S. Kang and L. G. Leal, Bubble dynamics in time-periodic straining flows, *J. Fluid Mech.* **218**, 41–69 (1990).
47. T. J. Kaper and S. Wiggins, Lobe area in adiabatic Hamiltonian systems, *Physica D* **51**, 205–217 (1991).
48. D. V. Khakhar, H. Rising and J. M. Ottino, Analysis of chaotic mixing in two model systems, *J. Fluid Mech.* **172**, 419–451 (1986).
49. Y. T. Lau and J. M. Finn, Dynamics of a three-dimensional incompressible flow with stagnation points, *Physica D* **57**, 283–310 (1992).
50. S. Leibovich, The form and dynamics of Langmuir circulations, *Ann. Rev. Fluid Mech.* **15**, 391–427 (1983).
51. A. Leonard, V. Rom-Kedar and S. Wiggins, Fluid Mixing and Dynamical Systems, *Nucl. Phys. (Proc. Suppl.) B* **2**, 179–190 (1987).
52. C. W. Leong and J. M. Ottino, Experiments on mixing due to chaotic advection in a cavity, *J. Fluid Mech.* **209**, 463–499 (1989).
53. J. M. Lopez and A. D. Perry, Axisymmetric vortex breakdown—Part 3: Onset of periodic flow and chaotic advection, *J. Fluid Mech.* **234**, 449–471 (1992).
54. R. S. MacKay, J. D. Meiss and I. C. Percival, Transport in Hamiltonian Systems. *Physica D* **13**, 55–81 (1984).
55. R. S. MacKay and J. D. Meiss, Flux and differences in action for continuous time Hamiltonian systems, *J. Phys. A: Math. Gen.* **19**, L225–L229 (1986).
56. R. S. MacKay, J. D. Meiss and I. C. Percival, Resonances in area-preserving maps, *Physica D* **27**, 1–20 (1987).
57. R. S. MacKay and J. D. Meiss, Relation between quantum and classical thresholds for multiphoton ionization of excited atoms, *Phys. Rev.* **A37**, 4702–4706 (1988).
58. R. S. MacKay, Transport in three dimensional volume-preserving flows, University of Warwick preprint.
59. F. E. Marble, Growth of a diffusion flame in the field of a vortex, in *Recent Advances in the Aerospace Sciences*, edited by C. Casci. pp. 395–413. Plenum, New York (1985).
60. E. Meiburg and P. K. Newton, Particle dynamics and mixing in a viscously decaying shear layer, *J. Fluid Mech.* **227** 211–244 (1991).
61. J. D. Meiss and E. Ott, Markov tree model of transport in area-preserving maps, *Physica D* **20**, 387–402 (1986).
62. J. D. Meiss, Symplectic maps, variational principles, and transport, *Rev. Mod. Phys.* **64**, 795–848 (1992).
63. V. V. Meleshko, M. Y. Konstantinov, A. A. Gurzhi and T. P. Konovaljuk, Advection of a vortex pair atmosphere in a velocity field of point vortices, *Phys. Fluids A* **4**, 2779–2797 (1992).
64. V. K. Melnikov, On the stability of the center for time periodic perturbations, *Trans. Mosk. Math. Soc.* **12**, 1–57 (1963).
65. K. R. Meyer and G. R. Sell, Melnikov transforms, Bernoulli bundles, and almost periodic perturbations, *Trans. Am. Math. Soc.* **314**, 63–105 (1991).
66. I. Mezić and S. Wiggins, On the integrability and perturbation of three-dimensional fluid flows with symmetry, Caltech preprint.
67. P. L. Miller and P. E. Dimotakis, Stochastic geometric properties of scalar interfaces in turbulent jets, *Phys. Fluids A* **3**, 168–177 (1991).
68. F. C. Moon and W. T. Holmes, Double Poincaré sections of a quasi-periodically forced, chaotic attractor, *Phys. Lett. A* **111**, 157–160 (1985).
69. F. J. Muzzio, P. D. Swanson and J. M. Ottino, The statistics of stretching and stirring in chaotic flows, *Phys. Fluids A* **3**, 822–834 (1991).
70. F. J. Muzzio, C. Meneveau, P. D. Swanson and J. M. Ottino, Scaling and multifractal properties of mixing in chaotic flows, *Phys. Fluids A* **4**, 1439–1456 (1992).
71. F. J. Muzzio, P. D. Swanson and J. M. Ottino, Mixing distributions produced by multiplicative stretching in chaotic flows, *Int. J. Bifurcation & Chaos* **2**, 37–50 (1992).
72. E. Ott and T. M. Antonsen, Jr., Fractal measures of passively convected vector fields and scalar gradients in chaotic fluid flows, *Phys. Rev. A* **39**, 3660–3671 (1989).
73. J. M. Ottino, Description of mixing with diffusion and reaction in terms of the concept of material surfaces, *J. Fluid Mech.* **114**, 83–103 (1982).
74. J. M. Ottino, *The Kinematics of Mixing: Stretching, Chaos and Transport*. Cambridge University Press, Cambridge (1989).
75. J. M. Ottino, Mixing, Chaotic Advection, and Turbulence, *Ann. Rev. Fluid Mech.* **22**, 207–254 (1991).
76. J. M. Ottino, Unity and diversity in mixing: Stretching, diffusion, breakup, and aggregation in chaotic flows, *Phys. Fluids A* **3**, 1417–1430 (1991).
77. R. T. Pierrehumbert, Large-scale horizontal mixing in planetary atmosphere, *Phys. Fluids A* **3**, 1250–1260 (1991).
78. H. Poincaré, *Les Méthodes Nouvelles de la Mécanique Céleste*. Gauthier-Villars, Paris (1899).
79. R. R. Prasad and K. R. Sreenivasan, The measurement and interpretation of fractal dimensions of the scalar

interface in turbulent flows, *Phys. Fluids A* **2**, 792–807 (1990).
80. R. Ramshankar, D. Berlin and J. P. Gollub, Transport by Capillary Waves—1: Particle Trajectories, *Phys. Fluids A* **2**, 1955–1965 (1990).
81. C. Robinson, Sustained resonance for a nonlinear system with slowly varying coefficients, *SIAM J. Math. Anal.* **14**, 847–860 (1983).
82. V. Rom-Kedar, A. Leonard and S. Wiggins, An analytical study of transport, mixing and chaos in an unsteady vortical flow, *J. Fluid Mech.* **214**, 347–394 (1990).
83. V. Rom-Kedar and S. Wiggins, Transport in two-dimensional maps. *Arch. Rat. Mech. Anal.* **109**, 239–298 (1990).
84. V. Rom-Kedar, Transport Rates of a Class of Two-Dimensional Maps and Flows, *Physica D* **43**, 229–268 (1990).
85. J. Scheurle, Chaotic solutions of systems with almost periodic forcing, *J. Appl. Math. Phys. (ZAMP)* **37**, 12–26 (1986).
86. M. A. Sepúlveda, R. Badii and E. Pollak, Spectral Analysis of Conservative Dynamical Systems, *Phys. Rev. Lett.* **63**, 1226–1229 (1989).
87. K. Shariff, A. Leonard, N. J. Zabusky and J. H. Ferziger, Acoustics and dynamics of coaxial interacting vortex rings, *Fluid Dyn. Res.* **3**, 337–343 (1988).
88. K. Shariff, A Leonard and J. H. Ferziger, Dynamics of a class of vortex rings, *NASA TM*-102257 (1989).
89. K. Shariff, T. H. Pulliam and J. M. Ottino, A dynamical systems analysis of kinematics in the time-periodic wake of a circular cylinder, in *Vortex Dynamics and Vortex Methods*, edited by C. R. Anderson and C. R. Greengard, pp. 613–646, *Lectures in Applied Mathematics*, Vol. 28. American Mathematical Society, New York (1991).
90. K. Shariff and A. Leonard, Vortex rings, *Ann. Rev. Fluid Mech.* **24**, 235–279 (1992).
91. Z. She and S. A. Orszag, Physical model of intermittency in turbulence—inertial-range non-Gaussian statistics, *Phys. Rev. Lett.* **66**, 1701–1704 (1991).
92. T. H. Solomon and J. P. Gollub, Chaotic particle transport in time-dependent Rayleigh–Bénard convection, *Phys. Rev. A* **38**, 6280–6286 (1988).
93. K. R. Sreenivasan and C. Meneveau, The fractal facets of turbulence, *J. Fluid Mech.* **173**, 357–386 (1986).
94. K. R. Sreenivasan, Fractals and multifractals in fluid turbulence, *Ann. Rev. Fluid Mech.* **23**, 539–600 (1991).
95. D. Stoffer, Transversal homoclinic points and hyperbolic sets for non-autonomous maps—I, *J. Appl. Math. Phys. (ZAMP)* **39**, 518–549 (1988).
96. D. Stoffer, Transversal homoclinic points and hyperbolic sets for non-autonomous maps—II. *J. Appl. Math. Phys. (ZAMP)* **39**, 783–812 (1988).
97. E. Stone and P. Holmes, Random perturbations of heteroclinic attractors, *SIAM J. Appl. Math.* **50**, 726–743 (1990).
98. E. Stone and P. Holmes, Unstable fixed points, heteroclinic cycles and exponential tails in turbulence production, *Phys. Lett. A* **155**, 29–42 (1991).
99. F. Városi, T. M. Antonsen, Jr., and E. Ott, Spectrum of fractal dimensions of passively convected scalar gradients in chaotic fluid flows, *Phys. Fluids A* **3**, 1017–1028 (1991).
100. A. Vincent and M. Meneguzzi, The spatial structure and statistical properties of homogeneous turbulence, *J. Fluid Mech.* **225**, 1–20 (1991).
101. J. B. Weiss and E. Knobloch, Mass transport and mixing by modulated traveling waves, *Phys. Rev. A* **40**, 2579–2589 (1989).
102. S. Wiggins, Chaos in the quasiperiodically forced Duffing oscillator, *Phys. Lett. A* **124**, 138–142 (1987).
103. S. Wiggins, *Global Bifurcations and Chaos – Analytical Methods*. Springer, New York (1988).
104. S. Wiggins, On the geometry of transport in phase space—I: Transport in k-degree-of-freedom Hamiltonian systems, $2 \leq k < \infty$, *Physica D* **44**, 471–501 (1990).
105. S. Wiggins, *Introduction to Applied Nonlinear Dynamical Systems and Chaos*. Springer, New York (1990).
106. S. Wiggins, *Chaotic Transport in Dynamical Systems*. Springer, New York (1992).
107. H. Yamada and T. Matsui, Preliminary study of mutual slip-through of a pair of vortices, *Phys. Fluids* **21**, 292–294 (1978).
108. M. Van Dyke, *An Album of Fluid Motion*. Parabolic, Stanford, CA (1982).

APPENDIX

Equations of motion for oscillating vortical flows

We consider the two-dimensional fluid flows induced by (i) a pair of equal and opposite point vortices and (ii) a pair of identical point vortices oscillating periodically in response to a time-periodic strain-rate field. We refer to these two flows as the open and closed flow, respectively, whose streamfunctions, in the comoving and corotating frames, respectively, can be written as

$$\psi_{\text{open}}(x, y, t) = \frac{-\Gamma}{4\pi}(\ln[(x - x_v(t))^2 + (y - y_v(t))^2] - \ln[(x - x_v(t))^2 + (y + y_v(t))^2]) + \varepsilon xy \sin(2\pi t) - \tilde{v}y, \tag{109}$$

for the open flow and

$$\psi_{\text{closed}}(x, y, t) = \frac{-\Gamma}{4\pi}(\ln[(x - x_v(t))^2 + (y - y_v(t))^2] - \ln[(x + x_v(t))^2 + (y + y_v(t))^2])$$
$$+ \varepsilon xy \sin(2\pi t) + \frac{\tilde{\omega}}{2}(x^2 + y^2) \tag{110}$$

for the closed flow, where $(x_v(t), y_v(t))$ represents the spatial coordinates of one of the vortices in the pair (whose evolution is easily determined from the above streamfunctions with the self-contribution term ignored), Γ is the magnitude of circulation associated with each vortex, ε is the strain rate amplitude, $\tilde{v}$ is the mean vortex pair velocity in the open-flow lab frame, and $\tilde{\omega}$ is the mean vortex-pair rotation frequency in the closed-flow lab frame.

We also consider for the open flow the multi-frequency generalization of the single-frequency problem, given by the streamfunction

$$\psi_{\text{open}}(x, y, t) = \frac{-\Gamma}{4\pi}\ln\left[\frac{(x - x_v(t))^2 + (y - y_v(t))^2}{(x - x_v(t))^2 + (y + y_v(t))^2}\right] - V_v y + \varepsilon xy\left(\sum_{i=1}^{l}\omega_i f_i \sin(\omega_i t + \theta_{i_0})\right), \tag{111}$$

where $(x_v(t), \pm y_v(t))$ are the vortex positions, V_v is the average velocity of the vortex pair in the lab frame, and $\varepsilon\omega_i f_i$ is the strain rate amplitude associated with the ith frequency (εf_i is non-dimensional). We note that in certain instances (explicitly mentioned in the text), the single-frequency computations are specified in terms of the single-frequency limit of this streamfunction; otherwise it is to be understood that the single-frequency streamfunction (109) is used. In the context of this multi-frequency velocity field, let us give a fuller discussion of the open flow. For $\varepsilon = 0$, $(x_v, y_v) = (0, d)$ and $V_v = \Gamma/4\pi d$. Of course $l = 1$ corresponds to the time-periodic case, and $l \geq 2$ corresponds to the time-quasiperiodic case. Let us specialize to the two-frequency case and use the dimensionless variables $x/d \to x$, $y/d \to y$, $\Gamma t/2\pi d^2 \to t$, $V_v 2\pi d/\Gamma \to v_v$, $2\pi\omega_i d^2/\Gamma \to \omega_i$ $(i = 1, 2)$. The flow is then specified by

$$\begin{aligned}\dot{x} &= -\left\{\frac{y - y_v(t)}{[x - x_v(t)]^2 + [y - y_v(t)]^2} - \frac{y + y_v(t)}{[x - x_v(t)]^2 + [y + y_v(t)]^2}\right\}\\ &\quad -v_v + \varepsilon x\{\omega_1 f_1 \sin(\omega_1 t + \theta_{1_0}) + \omega_2 f_2 \sin(\omega_2 t + \theta_{2_0})\}\\ \dot{y} &= [x - x_v(t)]\left\{\frac{1}{[x - x_v(t)]^2 + [y - y_v(t)]^2} - \frac{1}{[x - x_v(t)]^2 + [y + y_v(t)]^2}\right\}\\ &\quad - \varepsilon y\{\omega_1 f_1 \sin(\omega_1 t + \theta_{1_0}) + \omega_2 f_2 \sin(\omega_2 t + \theta_{2_0})\},\end{aligned} \tag{112}$$

where

$$\begin{aligned}x_v(t) &= \exp\{-\varepsilon[f_1 \cos(\omega_1 t + \theta_{1_0}) + f_2 \cos(\omega_2 t + \theta_{2_0}) - f]\}\\ &\quad \cdot \int(\tfrac{1}{2} - v_v \exp\{\varepsilon[f_1 \cos(\omega_1 t + \theta_{1_0}) + f_2 \cos(\omega_2 t + \theta_{2_0}) - f]\})\,\mathrm{d}t\\ y_v(t) &= \exp\{\varepsilon[f_1 \cos(\omega_1 t + \theta_{1_0}) + f_2 \cos(\omega_2 t + \theta_{2_0}) - f]\}\end{aligned} \tag{113}$$

and v_v is chosen so that

$$\lim_{T\to\infty}\int_0^T(\tfrac{1}{2} - v_v \exp\{\varepsilon[f_1 \cos(\omega_1 t + \theta_{1_0}) + f_2 \cos(\omega_2 t + \theta_{2_0}) - f]\})\,\mathrm{d}t = 0. \tag{114}$$

The vortex solutions $(x_v(t), \pm y_v(t))$ are found using that each vortex is advected by the flow due to the other vortex and the oscillating strain-rate field, so the vortices satisfy

$$\begin{aligned}\frac{\mathrm{d}x_v}{\mathrm{d}t} &= \frac{1}{2y_v} - v_v + \varepsilon x_v\{\omega_1 f_1 \sin(\omega_1 t + \theta_{1_0}) + \omega_2 f_2 \sin(\omega_2 t + \theta_{2_0})\}\\ \frac{\mathrm{d}y_v}{\mathrm{d}t} &= -\varepsilon y_v\{\omega_1 f_1 \sin(\omega_1 t + \theta_{1_0}) + \omega_2 f_2 \sin(\omega_2 t + \theta_{2_0})\}.\end{aligned} \tag{115}$$

Rom-Kedar *et al*. [83] choose, for the single-frequency analysis with $\theta_{1_0} = 0$, the initial conditions $(x_v(0), y_v(0)) = (0, 1)$, which guarantees a vortex response symmetric about $x = 0$ to first order in ε. A simple quasiperiodic generalization will need to keep this symmetry, which is done by choosing the constant of integration in the x_v expression to be zero. The y_v behavior is determined uniquely by the choice of f: a choice of $f = f_1 \cos\theta_{1_0} + f_2 \cos\theta_{2_0}$ or $f = f_1 + f_2$ follows the spirit of Rom-Kedar *et al*. [83]; alternatively, one could choose $f = 0$ to obtain a perturbation of mean zero in the y-direction (not necessary, as is the case for x, but nevertheless appealing). We choose $f = f_1 + f_2$ for our simulations. The periodic forcing simulation of Fig. 16 corresponds to equations (112) with $f_2 = 0$, $\varepsilon f_1 = 0.1$, $\omega_1 = 2.5$, $\theta_{1_0} = 0$, $\varepsilon f = 0.1$.

Computation of invariant manifolds

We describe here the numerical computation of invariant manifolds. Addressing periodic, quasiperiodic, and aperiodic time dependences involves successive generalizations of a similar procedure. We describe here the procedure for 2D multiple-frequency velocity fields, from which the single-frequency and aperiodic procedures should be clear (as we discuss at the end). For simplicity of discussion, let us consider a two-frequency homoclinic case (e.g. consider the upper homoclinic loop for the closed flow subjected to a two-frequency perturbation). Suppose we wish to portray the invariant manifolds in (x, y) at time $t = (2\pi/\omega_2)n$, or, equivalently, the invariant manifolds in the time slice $\chi[\theta_0 + 2\pi(\omega/\omega_2)n]$. Evolving the curve

$$W^{u}_{loc}(\tau_\varepsilon) \cap \chi\left(\theta_{1_0} + 2\pi\frac{\omega_1}{\omega_2}(n - i)\right) \tag{116}$$

forward in time according to (16) for i sample periods (where i is some positive integer) gives a curve which extends from

$$\tau_\varepsilon \cap \chi\left(\theta_{1_0} + 2\pi\frac{\omega_1}{\omega_2}n\right) \tag{117}$$

along a finite length of

$$W^{u}(\tau_\varepsilon) \cap \chi\left(\theta_{1_0} + 2\pi\frac{\omega_1}{\omega_2}n\right) \tag{118}$$

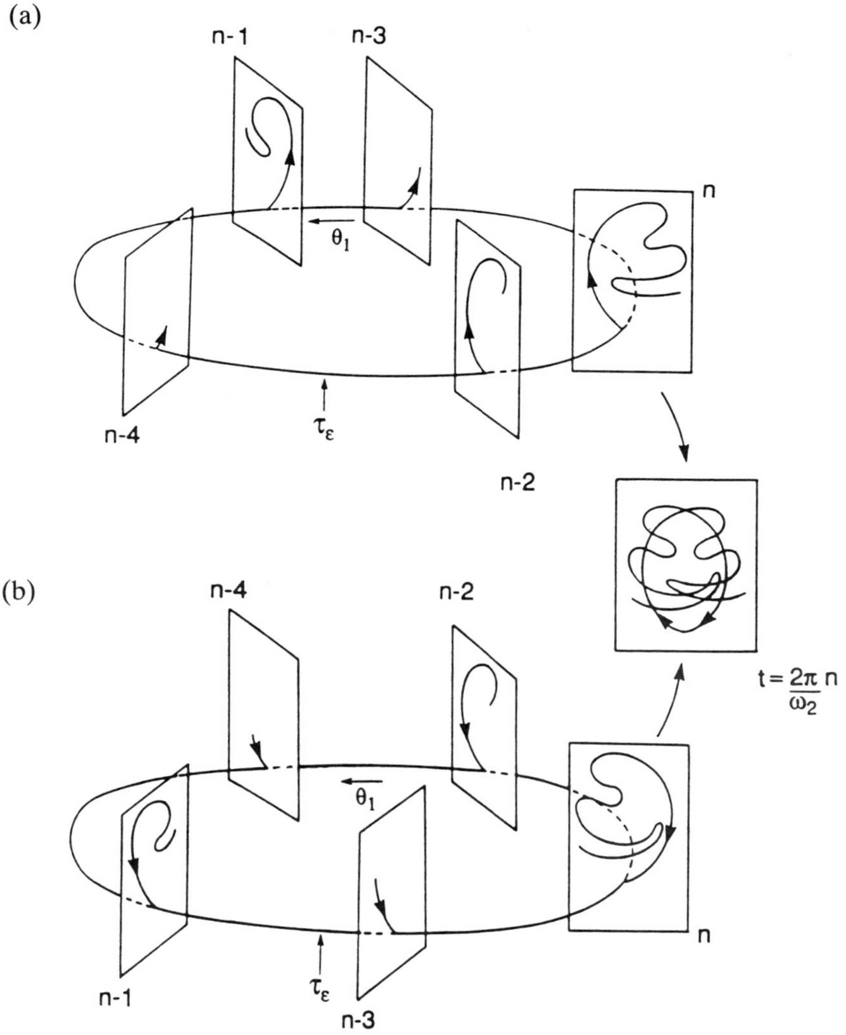

Fig. 72. Simulating a finite length of (a) $W^{u}(\tau_\varepsilon)$ and (b) $W^{s}(\tau_\varepsilon)$ in the nth time slice for a two-frequency homoclinic case.

(the greater i is, of course, the greater the length will be). Similarly, evolving the curve

$$W^{s}_{\text{loc}}(\tau_\varepsilon) \cap \chi\left(\theta_{1_0} + 2\pi\frac{\omega_1}{\omega_2}(n + i)\right) \tag{119}$$

backwards in time according to (16) for i sample periods gives a curve which extends from

$$\tau_\varepsilon \cap \chi\left(\theta_{1_0} + 2\pi\frac{\omega_1}{\omega_2}n\right) \tag{120}$$

along a finite length of

$$W^{s}(\tau_\varepsilon) \cap \chi\left(\theta_{1_0} + 2\pi\frac{\omega_1}{\omega_2}n\right) \tag{121}$$

(see Fig. 72). The resulting two curves form the boundary of a finite number of lobes of the two-dimensional lobe structure in the time slice $\chi[\theta_{1_0} + 2\pi(\omega_1/\omega_2)n]$.

We still need to be able to find $W^{u}_{\text{loc}}(\tau_\varepsilon)$ and $W^{s}_{\text{loc}}(\tau_\varepsilon)$ in the appropriate slices. The procedure for doing this is a straightforward generalization of the standard trial-and-error procedure used in the time-periodic case, so our discussion here will only be heuristic. Suppose for some arbitrary phase slice $\chi(\bar\theta_1)$ of $\Sigma^{\theta_{2_0}}$ we wish to find $\tau_\varepsilon \cap \chi(\bar\theta_1)$ and one or both of $W^{u}_{\text{loc}}(\tau_\varepsilon) \cap \chi(\bar\theta_1)$ and $W^{s}_{\text{loc}}(\tau_\varepsilon) \cap \chi(\bar\theta_1)$. Let $C_\pm$ be a closed curve in the phase slice $\chi[\bar\theta_1 \pm 2\pi(\omega_1/\omega_2)j]$ that contains the point $\tau_\varepsilon \cap \chi(\bar\theta_1 \pm 2\pi(\omega_1/\omega_2)j)$ and is pierced by $W^{u}_{\text{loc}}(\tau_\varepsilon)$ and $W^{s}_{\text{loc}}(\tau_\varepsilon)$, where j is some positive integer (see Fig. 73). Due to the normal hyperbolicity of τ_ε, $P^{j}_{\varepsilon}(C_-)$ will be stretched along $W^{u}(\tau_\varepsilon) \cap \chi(\bar\theta_1)$ and $P^{-j}_{\varepsilon}(C_+)$ will be stretched along $W^{s}(\tau_\varepsilon) \cap \chi(\bar\theta_1)$. The region in $\chi(\bar\theta_1)$ bounded by $P^{j}_{\varepsilon}(C_-)$ will intersect with the region in $\chi(\bar\theta_1)$ bounded by $P^{-j}_{\varepsilon}(C_+)$ in one or more disjoint regions, one of which, $\mathcal{Q}(\bar\theta_1)$, will contain $\tau_\varepsilon \cap \chi(\bar\theta_1)$ and shrink to zero area as $j \to \infty$ (again see Fig. 73). Thus, the trial-and-error procedure, similar to the time-periodic case, is to make reasonable tries at $C_\pm$ and evolve them to $\chi(\bar\theta_1)$ as described above to pinpoint $\tau_\varepsilon \cap \chi(\bar\theta_1)$ and its local stable and unstable manifolds by watching how the two curves stretch and intersect. By choosing $C_\pm$ sufficiently small, we can obtain an arbitrarily good approximation of $W^{u}_{\text{loc}}(\tau_\varepsilon) \cap \chi(\bar\theta_1)$ and $W^{s}_{\text{loc}}(\tau_\varepsilon) \cap \chi(\bar\theta_1)$. When the system is near-integrable, one can make a good initial estimate of $C_\pm$ due to the knowledge of the location of the fixed point of the unperturbed system; otherwise one must employ a more arduous trial-and-error effort, which would be the case under time-periodic vector fields as well.

The procedure for simulating global stable and unstable manifolds in the nth time slice of the autonomous system phase space is thus quite similar to the procedure for simulating the invariant lobe structure in the time-periodic case, except that one has to take into account that curves are, with each application of P_ε, mapped

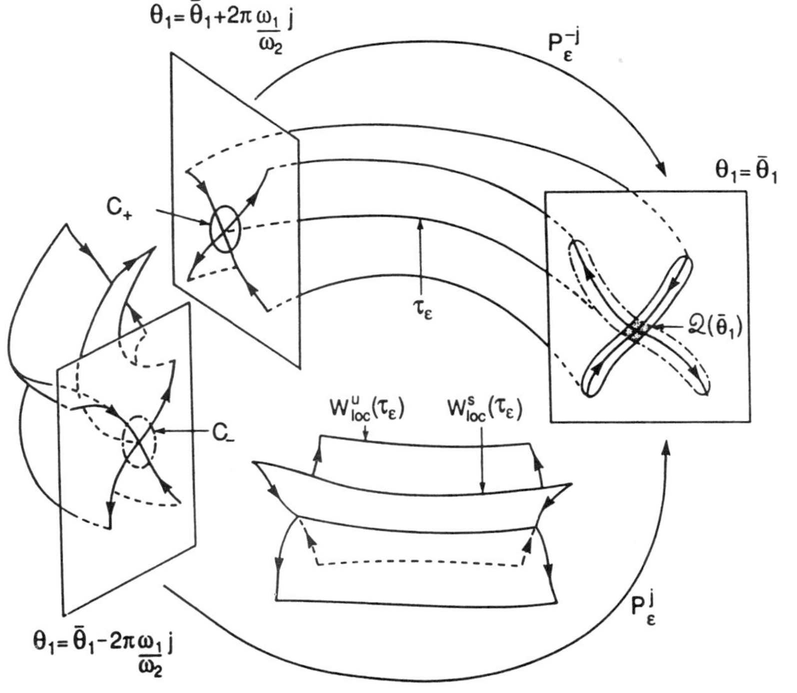

Fig. 73. Evolving $C_\pm$ to the phase slice $\chi(\bar\theta_1)$ in $\Sigma^{\theta_{2_0}}$ to obtain τ_ε and its local stable and unstable manifolds in that plane.

around T^1 in an enlarged phase space. Hence one finds the intersection of $W^{u}_{loc}(\tau_\varepsilon)$ and $W^{s}_{loc}(\tau_\varepsilon)$ with an appropriate pair of phase slices, which in turn are used to simulate a finite length of $W^u(\tau_\varepsilon)$ and $W^s(\tau_\varepsilon)$ in the desired phase slice. From our discussion of the two-frequency homoclinic case, the procedure for more frequencies and for the heteroclinic case should be clear. Note how the described procedure contrasts with a previous suggestion by Moon and Holmes [68] for analyzing the dynamics under quasiperiodic vector fields, a double Poincaré map method which essentially attempts to treat the system as periodic. Samples of an equation like (16) are taken only when both $\theta_2 = \bar{\theta}_2$ and $\theta_1 \in [\bar{\theta}_1 - \beta, \bar{\theta}_1 + \beta]$ (for some choice of $\bar{\theta}_1$, $\bar{\theta}_2$ and $\beta \ll 2\pi$), and the results are summed. The time between samples can be much longer with this approach, and there is a 'fuzziness' of the resulting structure due to the finite width of the sample window, 2β. In contrast to this method, we shall refer to our approach as a *double phase slice method*, since to simulate the lobe structure in any phase slice one evolves two curves, each originating in a different phase slice. For 2D time-aperiodic velocity fields, one conceptualizes the problem in the context of a 3D autonomous flow, in which the third dimension is the time coordinate, as described in Section 2.3. Here the same numerical procedure as in the quasiperiodic case applies, except that it is phrased in the context of 2D *time* slices, rather than 2D phase slices (hence referred to as a *double time slice method* in this instance). For the special case of 2D time-periodic velocity fields, one obtains the special circumstance where each time slice is identical.

Quantification of Mixing in Aperiodic Chaotic Flows

M. LIU

Department of Mechanical Engineering, Rutgers University, Piscataway, NJ 08855, USA

and

F. J. MUZZIO*

Department of Chemical and Biochemical Engineering, Rutgers University, Piscataway, NJ 08855, USA

and

R. L. PESKIN

Department of Mechanical Engineering, Rutgers University, Piscataway, NJ 08855, USA

Abstract—Most previous studies of mixing in deterministic flows have focused on time-periodic or spatially-periodic flows. In contrast, mixing processes in aperiodic flows have been considerably less studied. Four procedures are used in this paper to generate well-characterized aperiodic flows. These procedures are applied to the cavity flow and to a two-dimensional mapping with a sinusoidal velocity profile. Mixing in periodic and aperiodic flows is quantitatively compared. Since most of the available analytical tools developed in the context of periodic systems (Poincaré sections, periodic points and their associated manifolds) are poorly suited for the analysis of aperiodic systems, comparisons are based on measures, such as the structure and statistics of the stretching field and the rate of tracer spreading, that apply to both periodic and aperiodic systems. Aperiodicity enhances mixing enormously. Aperiodic perturbations generate widespread chaos under conditions where periodic flows generate minimal or no chaos. The average rate of stretching of material elements can be increased several orders of magnitude in brief intervals corresponding to just 10–20 periods of the periodic flow. The spatial distribution of stretching is much more uniform for aperiodic systems than for periodic ones, and tracers spread much more rapidly and uniformly.

1. INTRODUCTION

Mixing is perhaps one of the most widespread phenomena both in nature and in technological applications. However, although mixing is of interest to many scientific disciplines, mixing problems are still studied on a case-by-case basis; a general theory has yet to be developed. Recent advances in the study of mixing in chaotic flow systems have opened promising leads for the systematic understanding of mixing processes [1–16]. Chaotic motion of fluid has been studied in different types of flows, and it is well known that chaotic particle motions significantly improve mixing efficiency in many processes. The most frequently studied cases are two-dimensional time-periodic flows with small Reynolds numbers. Under such conditions, the velocity field is deterministic, greatly facilitating the

*To whom all correspondence should be sent

implementation and analysis of both experiments and computations. However, mixing in periodic flows is not guaranteed to be complete, because such systems often display coexisting chaotic and non-chaotic regions (regular islands). Fluid can neither penetrate nor leave islands of regular motion. Therefore, islands are barriers to mixing. Fluid inside islands can be homogenized only through diffusion, which is usually a very slow process. Moreover, several studies have shown that islands can greatly hinder mixing inside the chaotic region itself [12–14, 17–20]. Lack of homogeneity due to the presence of islands can have significant undesirable effects on other processes. Examples important to practical applications are systems with competing chemical reactions, in which incomplete mixing often results in acceleration of undesired reactions and excessive generation of byproducts.

Regular motions must be removed if the flow is to achieve complete mixing. Procedures that achieve global chaos (i.e. conditions that for practical purposes are free of islands) are therefore highly desirable and can be generated in several ways. The common starting point is to recognize that in a time-periodic flow, regions of regular motion are due to the periodicity. Periodic flows contain periodic points, some of which are elliptic and give rise to islands; in general, the lower the period, the larger the island. This observation suggests that mixing can be enhanced by searching for conditions that are free of low period elliptic points i.e. conditions for which all of the low period points are hyperbolic. Such conditions have been termed 'mixing windows' [21, 22]. However, this approach has some disadvantages. One is that the method assumes that islands generated by higher period points have negligible effects. Although this assumption agrees with a number of observations, there is no rigorous evidence supporting its general validity. Another disadvantage is that conditions determined for a given flow apply only to that flow, so that the search for mixing windows must be done on a case-by-case basis. Moreover, for some periodic flows, globally chaotic conditions may not exist at all.

Methods that are more generally applicable can be implemented by removing altogether the periodicity of the system. Aperiodic flows are devoid of periodic points, and hence they are free of islands. Such an approach is adopted in this paper. A price is paid for adopting aperiodic flow procedures; some of the procedures give up determinism, others result in somewhat complicated flow prescriptions. However, as is shown below, substantial benefits are achieved. The main purpose of this work is to compare mixing in periodic and aperiodic flows quantitatively. Three types of periodic systems are adopted as a basis for comparison: (1) mixed conditions with large regular and chaotic regions, (2) overwhelmingly regular conditions and (3) overwhelmingly chaotic conditions. Each type of condition is perturbed using several aperiodic procedures, which are then quantitatively compared with one another and with the periodic flow. Section 2 of this paper describes the flow systems: a rectangular cavity with movable top and bottom walls, and an idealized flow in the spatially periodic unit square generated by sinusoidal velocity profiles in the x- and y-directions. The procedures for generating aperiodic flows are also described in Section 2. The algorithms used to simulate mixing processes are presented in Section 3. In Section 4, the rate and the uniformity of mixing processes in the different periodic and aperiodic flows are quantified in terms of stretching field statistics and rate of spread of passive tracers. Summary and Conclusions are presented in Section 5.

2. FLOW SYSTEMS

Two flow systems are adopted as case studies. The first is a physically realizable flow in a two-dimensional rectangular cavity with an aspect ratio $H/L = 0.6$, where H is the height of the cavity and L is the length of the moving horizontal walls. A convenient dimensionless description is obtained by adopting as length and velocity scales the length L

and the velocity U of the moving walls. In this paper, the velocity field is solved numerically [16, 23], and particle trajectories are obtained by numerical integration of the velocity field, $d\mathbf{x}/dt = \mathbf{u}$. All of the results presented here correspond to small Reynolds numbers ($Re < 1.0$) so that, to a good approximation, the flow stays in the Stokes regime. The main advantage of this restriction is that transients occurring when one wall is stopped and the other is set in motion can be neglected. Periodic flows are generated by alternatively moving the top and bottom walls with constant velocity each for a time $T/2$, where T, the period of the flow, is defined in dimensionless terms as the combined displacement of both walls during one period divided by the length of the cavity. Mixing processes in periodic cavity flows have been studied both experimentally [4, 8] and computationally [16, 21–23]. Flows generated by imposing aperiodic sequences of wall motions have been studied considerably less often [22, 24] than periodic ones.

The second system investigated here is the highly idealized 'sine flow' defined in the unit box $0 \leqslant x \leqslant 1$, $0 \leqslant y \leqslant 1$ by the two motions

$$V_x = \sin(2\pi y), \qquad V_y = 0; \tag{1a}$$

$$V_x = 0, \qquad V_y = \sin(2\pi x). \tag{1b}$$

Similarly to the previous case, a periodic flow can be obtained by sequentially implementing each motion for a time $T/2$, where T is once again the period of the flow. The flow domain has periodic boundaries; whenever a particle exits the unit box, it re-enters the box through the opposite end. The sine flow is piecewise time-integrable; particle trajectories can be found in closed form, requiring no numerical integration and making it possible to obtain extensive results at moderate computational expense.

Both the cavity and the sine flows are composed of two independent steady motions, hereafter referred to as motions a and b, that are imposed in a sequential manner. For the cavity, motions a and b are respectively driven by the displacement of the upper and lower walls; for the sine flow, motion a is given by equation (1a) and motion b by equation (1b). Time-periodic flows are generated by the periodic sequence P,

$$ababababababababababababababab \ldots. \tag{2}$$

As mentioned above, aperiodic flows can be generated in several ways. For example, Khakhar and Ottino [25] used flows with randomly chosen strain rates, and Varosi *et al.* [26] investigated a standard map with a random phase angle. However, these studies did not perform an extensive comparison of mixing in periodic and aperiodic flows. Conducting such a comparison is the main objective of this study. Procedures that are simple to implement are emphasized in this paper. The simplest approach is to choose a random sequence of *a*s and *b*s, generated, for example, by tossing a coin. The result is either a head (a) or a tail (b), each with a probability of 0.5. A typical random sequence (RS) is

$$aabbbbaabbababbaaaaabbabbbbaabba \ldots. \tag{3a}$$

The disadvantage of such a completely random sequence is that it contains regular sub-sequences such as *aaaa* or *bbbb* which may decrease the efficiency of the mixing process. Restrictions can be applied to prevent such sub-sequences. A simple constraint is to forbid a and b from repeating more than twice. With this constraint, the random sequence becomes

$$aabbabaabbababbaabaabbabbabaabba \ldots. \tag{3b}$$

Sequences generated using this procedure are hereafter called restricted random sequences (RRS). Another possibility that also prevents more than two repetitions is the so-called symmetry-breaking prescription, proposed by Franjione *et al.* [24]. This prescription is based on the observation that Poincaré sections of periodic flows often display symmetries, and that periodic points are usually located either on the axes of those symmetries or in symmetric pairs with respect to the axes. Using both simulations and experiments, it was shown that islands were destroyed in a number of cases by using the recursive symmetry-breaking (SB) sequence

$$abbabaabbaababbabaababbaabbabaab \ldots \tag{3c}$$

All of the procedures introduced so far are constructed with identical building blocks, i.e. sequences of motions a and b of constant length. The fourth aperiodic procedure investigated in this paper imposes a random perturbation of restricted magnitude to the *duration* of a and b. Instead of $T/2$, the duration of each motion is given by $T(1+\varepsilon_i)/2$, where ε_i is a random number in the interval $(0, 1)$. The restricted random period (RRP) prescription generated by this procedure can be represented as

$$a(\varepsilon_1)b(\varepsilon_2)a(\varepsilon_3)b(\varepsilon_4)a(\varepsilon_5)b(\varepsilon_6)a(\varepsilon_7)b(\varepsilon_8)a(\varepsilon_9)b(\varepsilon_{10})a(\varepsilon_{11}) \ldots \tag{3d}$$

Ling [22] recently reported preliminary computational results for the RS and SB procedures in the cavity flow. However, comparisons were based on single realizations of each flow using statistical methods involving only a small number of particles. A considerably more extensive analysis, including a greater variety of systems, is presented in the next section. For the sake of brevity, the discussion is restricted to the four procedures described by equations (3a)–(3d), although many other aperiodic procedures can be designed easily.

3. SIMULATION OF TRACER STIRRING AND SEGMENT STRETCHING

Mixing experiments can be simulated by computing the trajectories of a large number of particles initially placed at properly chosen locations. Passive tracers move at the local flow velocity, i.e. $d\mathbf{x}/dt = \mathbf{u}$, where $\mathbf{x}$ is the particle position and $\mathbf{u}$ is the flow field. Particle positions are determined using a standard 4th order Runge–Kutta method. Consider Fig. 1(a), which shows the initial positions of two groups of particles inside the cavity flow domain. Figure 1(b) shows the final positions of the particles after they have been advected using the periodic flow, $T = 7.0$, for a total time of 8 periods. The figure shows a dramatic difference in the evolution of the two circles. While red particles are initially located inside a chaotic region and experience rapid mixing, blue particles are inside a regular region and undergo minimal mixing. All the blue particles remain in the regular region, and all the red particles remain in the chaotic region.

This type of behavior is well known and has been documented for a variety of systems using both experiments and computations. Mixing in the chaotic region can be very efficient, but if islands are present, complete mixing cannot be achieved. Islands are very common in periodic chaotic flows. As was reported by Franjione *et al.* [24], the islands in Fig. 1(b) can be destroyed using a flow generated by the SB prescription. Figure 1(c) shows the final positions of the particles when they are advected using the SB prescription (equation (3c)) for the same total time $t = 8T$. Particles of both colors are now more thoroughly mixed. It should also be pointed out here that the simulated mixing structures generated by both the P (Fig. 1(b)) and the SB (Fig. 1(c)) flows agree well with the experimental results reported by other authors [8, 24], demonstrating the validity of the computational algorithm.

However, aperiodic flows can accomplish substantially more than just destroying a few remaining islands. For example, they can generate widespread chaos under conditions for which the corresponding periodic flow is regular almost everywhere. This is illustrated in

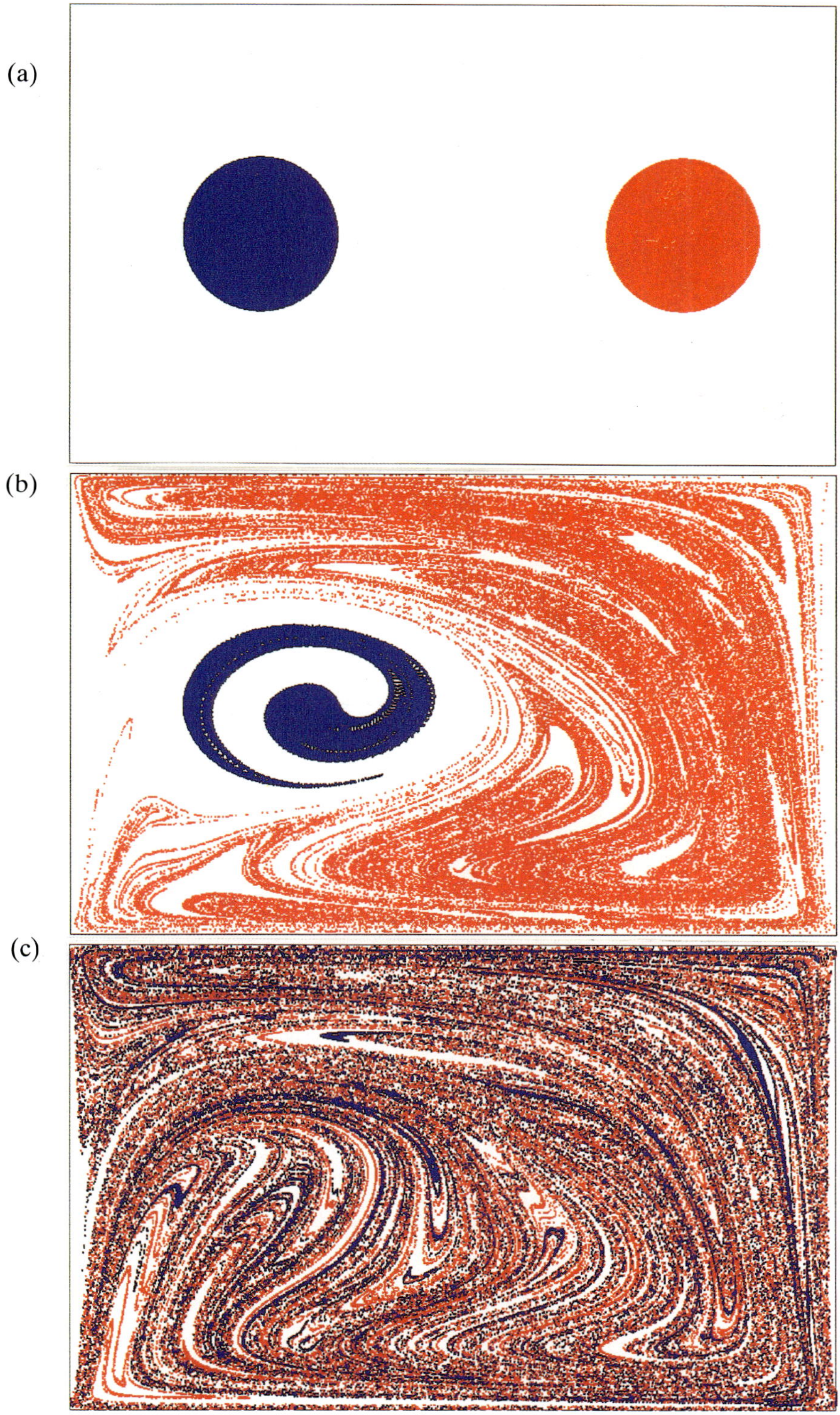

Fig. 1. (a) Initial location of particles in the cavity flow. (b) Positions of the particles after 8 periods of the periodic flow with $T = 7.0$. The red particles are located inside the chaotic regions and undergo rapid mixing. The blue particles are inside an island of regular motion and hardly mix at all. (c) Mixing is greatly enhanced by using the SB aperiodic flow (equation (3c)). After the same 8 periods, particles of both colors are thoroughly mixed throughout the flow domain.

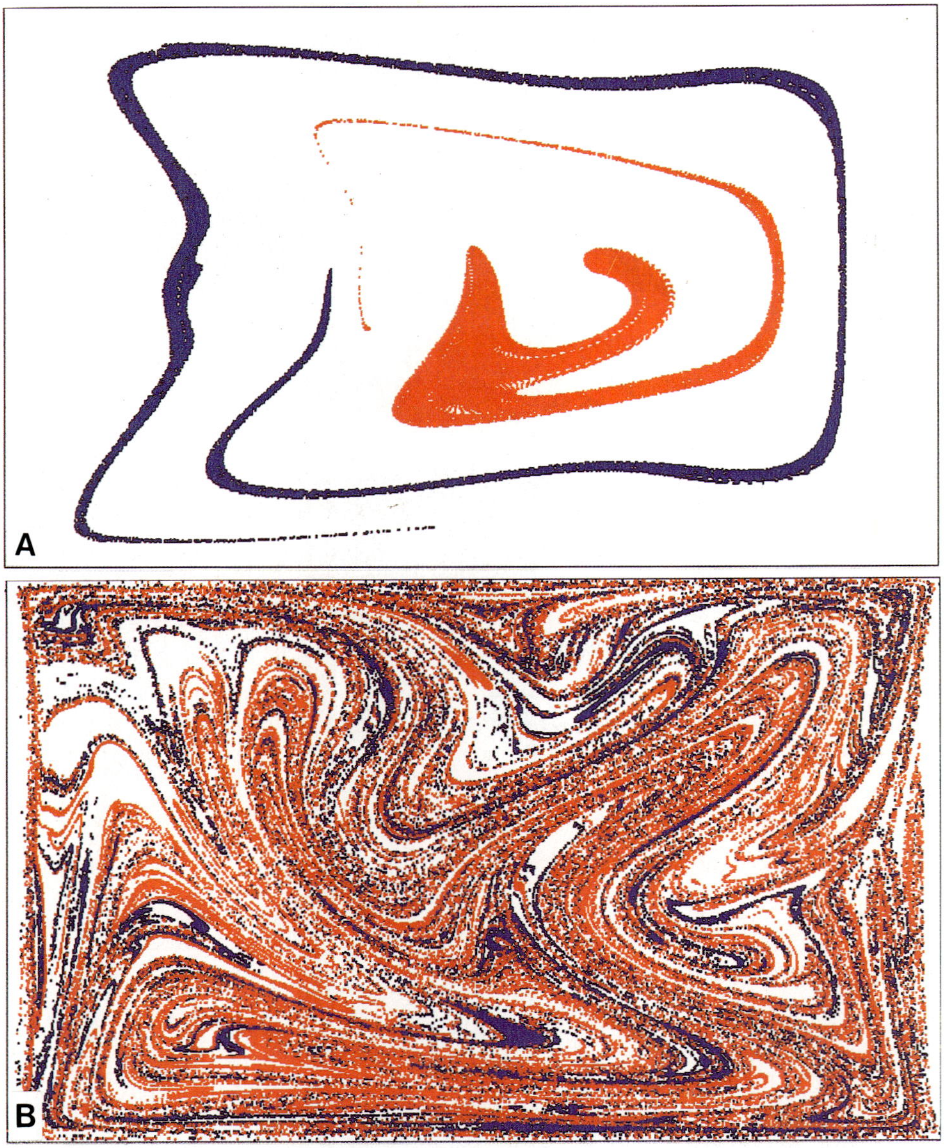

Fig. 2. (a) Mixing in the periodic cavity flow, $T = 2.0$. The particles were initially placed at the same position as in Fig. 1(a). This periodic flow is mostly regular, and after 50 periods of advection, minimal mixing has been achieved. (b) Mixing is greatly enhanced by using the RRS aperiodic flow (equation (3b)). Particles of both colors spread throughout the system. The remaining empty regions will disappear as mixing proceeds.

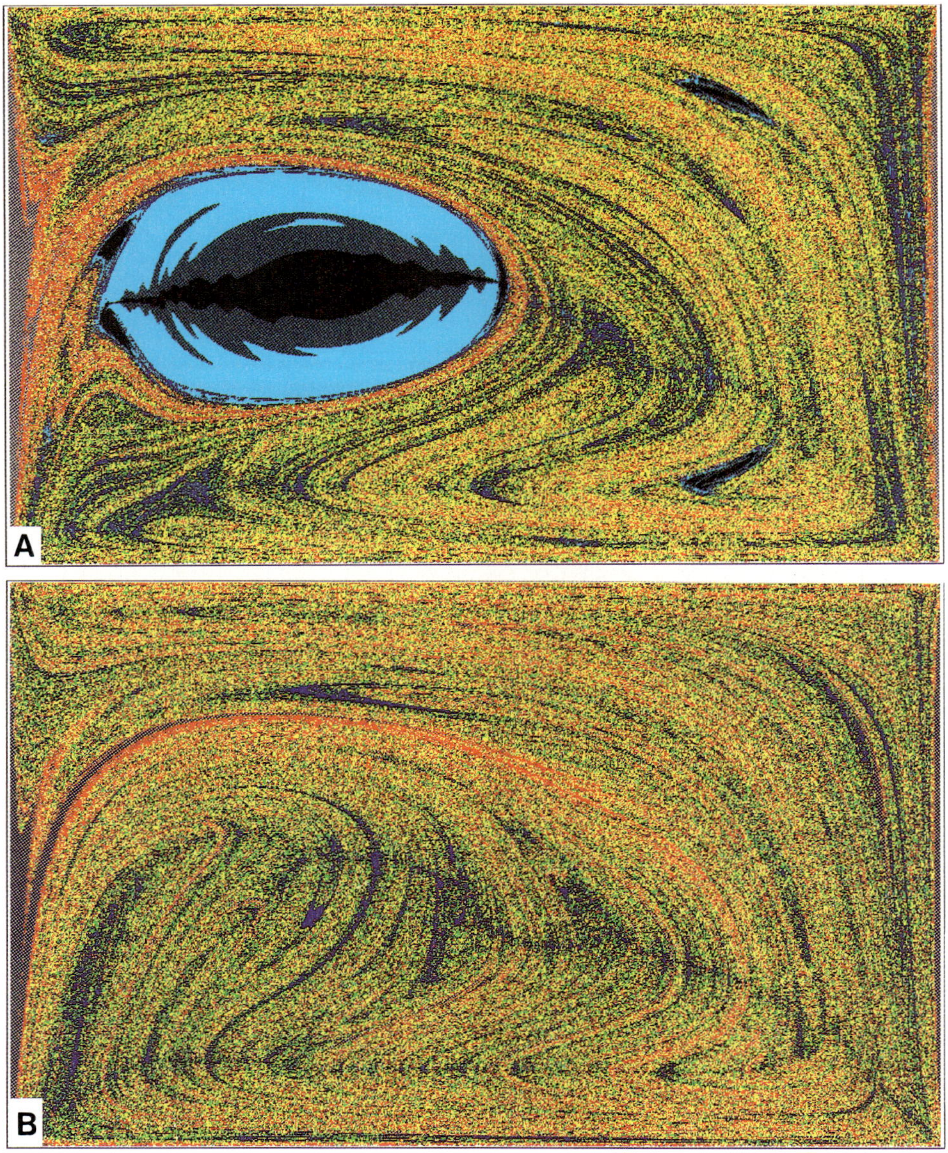

Fig. 3(a)–(c).

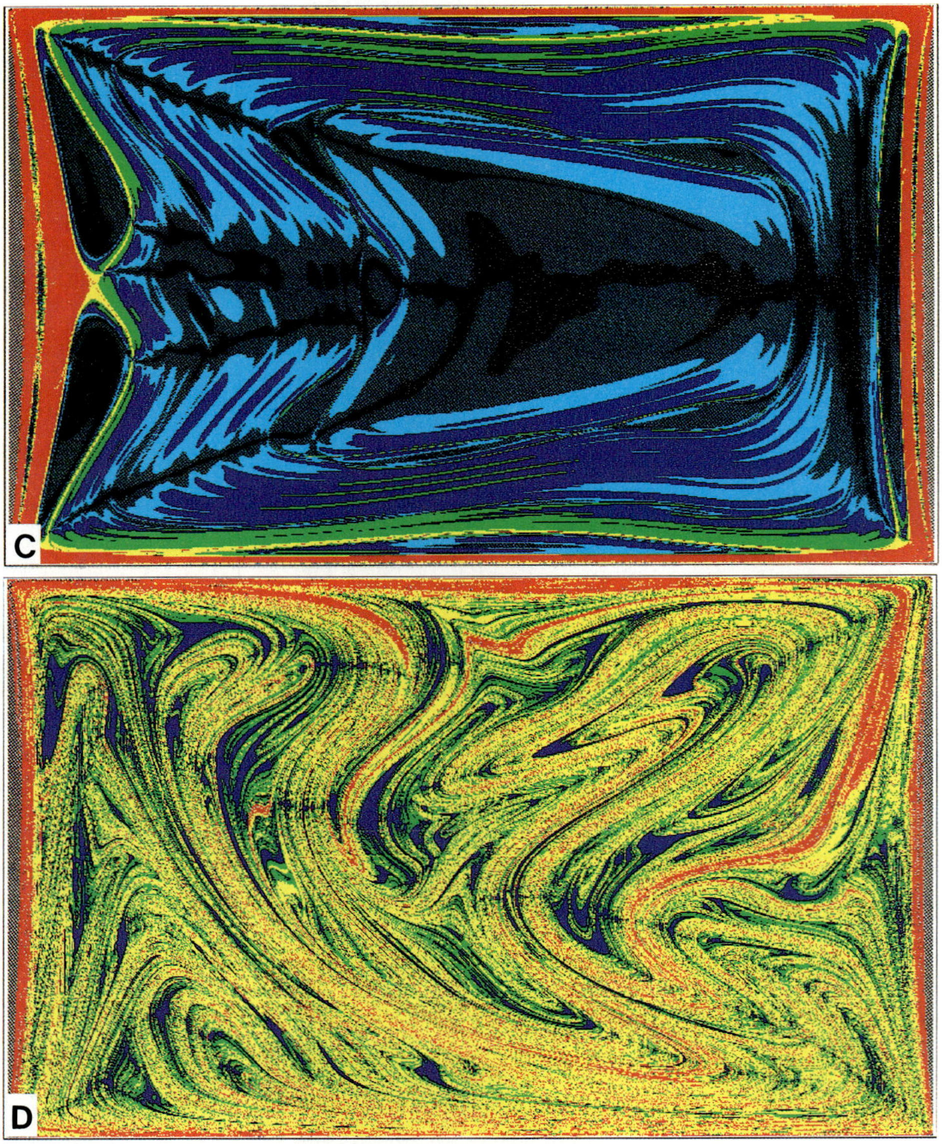

Fig. 3. (a) Stretching field in the periodic cavity flow for $T = 7.0$. The particles are color-coded according to the intensity of stretching experienced after 20 periods. The magnitude of stretching increases in the sequence of black, grey, light blue, dark blue, green, yellow and red. Particles in black, grey, light blue, dark blue and green have values of stretching intensities smaller than the mode, and those in yellow and red have stretching greater than the mode. The colors are assigned *ad hoc* to enhance the graphic display of (or the lack of) segregation effects. Black corresponds to $\log \lambda < 0.62$, grey to $0.62 < \log \lambda < 1.3$, light blue to $1.3 < \log \lambda < 3.5$, blue to $3.5 < \log \lambda < 6.7$, green to $6.7 < \log \lambda < 8.0$, yellow to $8.0 < \log \lambda < 10.0$, and red to $10.0 < \log \lambda$. (b) A considerably more uniform stretching field is generated by the SB aperiodic flow. Flow parameters are the same as in Fig. 3(a). As before, points are color-coded according to the intensity of stretching; blue corresponds to $\log \lambda < 6.9$, green to $6.9 < \log \lambda < 8.5$, yellow to $8.5 < \log \lambda < 10.0$, and red to $10.0 < \log \lambda$. (c) Stretching field generated by the periodic cavity flow, $T = 2.0$, 50 periods. Black corresponds to $\log \lambda < 0.07$, grey to $0.07 < \log \lambda < 0.89$, light blue to $0.89 < \log \lambda < 1.3$, blue to $1.3 < \log \lambda < 1.9$, green to $1.9 < \log \lambda < 3.0$, yellow to $3.0 < \log \lambda < 5.0$, and red to $5.0 < \log \lambda$. Regions of different stretching intensity are completely segregated for this mostly regular flow condition. Stretching is very low in most of the flow, except in red bands near the flow boundaries, where it is very high. (d) A very different situation is generated by using the RRS flow. The figure shows the stretching field for $T = 2.0$, $t = 50T = 100$. Blue corresponds to $\log \lambda < 1.9$, green to $1.9 < \log \lambda < 3.0$, yellow to $3.0 < \log \lambda < 5.0$, and red to $5.0 < \log \lambda$.

(a)

(b)

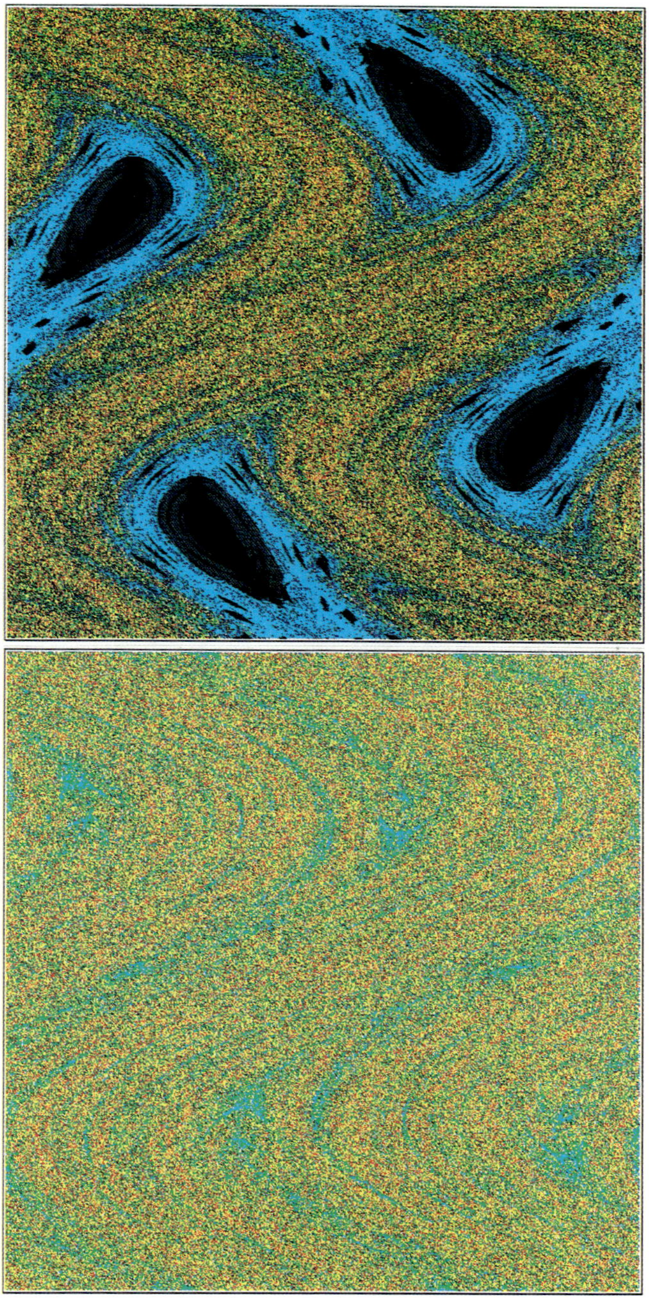

Fig. 4. (a) Stretching field generated by the periodic sine flow for $T = 0.8$, 50 periods. A number of islands of different sizes is revealed in the figure (black and grey). The colors represent stretching ranges of $\log\lambda < 1.0$ (black), $1.0 < \log\lambda < 2.8$ (grey), $2.8 < \log\lambda < 10.0$ (light blue), $10.0 < \log\lambda < 14.0$ (blue), $14.0 < \log\lambda < 17.0$ (green), $17.0 < \log\lambda < 19.0$ (yellow), and $19.0 < \log\lambda$ (red). Segregated low stretching regions (light blue) develop around the islands, occupying a considerable fraction of the chaotic region. (b) An almost completely uniform stretching field is generated for the sine flow, $T = 0.8$, by using the RRS aperiodic procedure. The ranges of stretching are: $\log\lambda < 17.0$ (blue), $17.0 < \log\lambda < 19.0$ (green), $19.0 < \log\lambda < 21.0$ (yellow), and $21.0 < \log\lambda$ (red).

Fig. 2(a) and (b) for the cavity flow, $T = 2.0$. At this low value of T, the periodic cavity is mostly regular, and its Poincaré section [23] is composed almost entirely of quasi-periodic orbits and chains of resonant islands. Not surprisingly, mixing is very poor. Figure 2(a) shows the final positions, after 50 periods, of particles initially placed in the same two circular regions as in Fig. 1(a). As is shown in the figure, even after this long time, only minimal mixing has taken place. However, considerably better mixing is achieved in the same amount of time by imposing the RRS aperiodic flow (equation (3b)). As is shown in Fig. 2(b), particles of both colors are significantly intermingled throughout the flow region.

However useful for developing intuition, qualitative descriptions of particle stirring such as those depicted in Figs 1 and 2 only provide a limited amount of information. A number of studies have used quantitative techniques based on computations of the deformation (stretching) experienced by populations of material segments [12–14, 16–19, 26–28]. Since regions populated by segments experiencing the highest and lowest stretching values respectively correspond to regions of best and worst local mixing, such computations can provide a considerable amount of quantitative information about the mixing process. Stretching can be computed in several ways. The approach adopted here monitors the evolution of an infinitesimal vector $\mathbf{I}$ as it is convected and stretched by the flow. This evolution is given by

$$d\mathbf{I}/dt = (\nabla\mathbf{u}).\mathbf{I}, \qquad \mathbf{I} = \mathbf{I}_0 \quad \text{at} \quad t = 0. \tag{4}$$

The stretching experienced by the vector is defined as $\lambda = |\mathbf{I}|/|\mathbf{I}_0|$. Figure 3(a) shows the stretching field for the periodic flow, $T = 7.0$, $t = nT = 140$ i.e. $n = 20$ periods. The figure is created by computing the stretching values of 365 820 (780 × 469) points initially placed in a lattice spanning the flow region. The points are color-coded according to intensity of stretching. In the order of increasing stretching, the color sequence is black, grey, light blue, dark blue, green, yellow and red. The structure of the field closely corresponds to the partially mixed structure shown in Fig. 1(b). As expected, stretching evolves very differently in chaotic and regular regions. Stretching in regular regions increases linearly with time, resulting in slow mixing, as was illustrated in Figs 1(b) and 2(a). Hence, points with very low stretching (black, grey, and light blue) populate a large, egg-shaped regular island to the left and the two smaller islands at symmetric positions to the right of the cavity. In contrast, stretching accumulates exponentially fast in the chaotic region (corresponding to positive Lyapunov exponents), which is populated by green, yellow and red high-stretching points.

However, even in the 'well mixed' chaotic region, stretching is considerably non-uniform. Segregated red bands are readily observed, indicating regions where stretching is extremely intense. Such regions are another manifestation of the flow periodicity. These high-stretching regions are created by unstable manifolds of hyperbolic periodic points that pass close to the cavity corners where the rate of stretching is unbounded. Points traveling along these manifolds accumulate extremely high stretching. Some branches of the manifolds tightly surround the island, creating a segregated high-stretching region around the island. This behavior is somewhat unusual; often, the points in the neighborhood of islands experience low stretching, creating poorly mixed areas within the chaotic region (for a detailed discussion of this and related effects, see Liu *et al.* [16]). In this case, however, the manifolds 'cut off' the island from the rest of the system, confining low-stretching points exclusively to the island.

For some mixing applications, segregated high-stretching regions can be as undesirable as islands; for example, shear-sensitive materials can be damaged in such regions. An ideal mixing process should generate a perfectly uniform stretching field, free both of islands and of high-stretching regions. Considerable progress towards achieving this goal can be observed in Fig. 3(b), which shows the stretching field generated by the SB flow (equation (3c)) for $t = 20T = 140$. As expected, the islands have disappeared; the black and light blue regions no longer exist. The colors are better mixed, indicating a more uniform stretching field. The red bands are still visible, revealing the persistence of segregated high-stretching regions generated by dynamics associated with cavity corners, but these regions are considerably smaller than for the periodic case.

The global structure of the periodic stretching field for $T = 2.0$, $t = 100$, shown in Fig. 3(c), closely resembles the corresponding Poincaré section [23]. Recall that for this condition the periodic cavity flow is almost entirely regular. The colors are highly segregated; while low-stretching points cover most of the flow region, high-stretching regions are restricted to sharp red and yellow bands near the flow boundaries. Figure 3(d) shows the stretching field obtained using the RRS flow (equation (3b)). The effects of aperiodicity are even larger than in the previous case. Stretching is not only much larger, but also much more uniformly distributed. High-stretching points are now present throughout the flow region, and the segregated red bands are much narrower than in Fig. 3(c). Low-stretching regions (blue) are also observed, but they are small and will disappear if the mixing process is allowed to continue for a longer time.

These effects are by no means unique to the cavity flow; corresponding calculations for the sine flow display similar types of behavior. Figure 4(a) shows the stretching field created by the periodic sine flow for $T = 0.8$, 50 periods. The stretching field is computed using 378 225 (615×615) points. Four large islands, as well as many small ones, can be readily observed (black and grey). In this case, segregated low-stretching regions (light blue) develop around the islands, occupying a considerable fraction of the chaotic region. These low-stretching regions occur as a result of the chaotic orbits that, upon approaching islands, remain in the neighborhood of islands for long times. 'Island stickiness' has been observed by other authors [12, 13, 16–19] and generates poorly mixed regions that are considerably larger than the islands themselves. The remainder of the chaotic region shows a fairly uniform mixture of green, yellow, and red points. Once again, as it is shown in Fig. 4(b), the structure of the stretching field is radically changed when the flow is made aperiodic. The figure shows the stretching field generated by the RRS flow (equation (3b)) for $t = 50T$. Black and grey points have disappeared. The figure is almost entirely devoid of structure, except for small light blue regions that will vanish if the process is allowed to continue.

4. QUANTITATIVE COMPARISON OF MIXING IN PERIODIC AND APERIODIC FLOWS

Stretching computations can be used to generate quantitative statistics of the stretching process, which provide a basis for comparing the mixing performance of different flows. Such statistics can describe the spatial distribution of mixing intensities observed in experiments. Another procedure useful for quantifying mixing, discussed later in this section, is based on the simple idea of discretizing the flow domain into small boxes, and finding the number of boxes containing tracer particles. In a mixing simulation, as particles spread, they occupy an increasing number of boxes. The rate of increase of box occupancy provides a gross estimate of the rate of mixing in the system.

4.1. *Stretching statistics*

A number of recent studies have focused on the statistics of the stretching process in chaotic flows. Often, the analysis is presented in terms of distributions of finite-time

Lyapunov exponents, which are immediately related to stretching (the finite-time Lyapunov exponent is simply the logarithm of stretching divided by time). Grassberger *et al.* [17] introduced the spectrum of finite-time Lyapunov exponents and computed results for the logistic map and for Henon's map. Sepulveda *et al.* [18] focused on the standard map for a condition with large non-chaotic regions. Horita *et al.* [19] also used the standard map but focused on conditions for which the map was chaotic everywhere or had only small islands. Spectra were also reported by Varosi *et al.* [26] for a randomized standard map. Pierrehumbert [27] calculated the structure and statistics of finite-time Lyapunov exponents for atmospheric flow in the northern hemisphere. Although a rigorous comparison was not performed, results clearly showed the same type of spectra as observed for the simple maps mentioned above. Muzzio *et al.* [12–14] computed spectra of stretching for the physically realizable flow between eccentric cylinders; and Beigie *et al.* [28] computed distributions of finite-time Lyapunov exponents along the unstable manifolds of perturbed heteroclinic tangles.

To keep the discussion more pertinent to fluid mixing, we focus on stretching, λ, rather than on Lyapunov exponents. Since in chaotic systems λ increases exponentially with time, the fluctuations of λ are best described by $H_n(\log\lambda)$, the probability density function of $\log\lambda$. Logarithms of base 10 are used throughout this paper. $H_n(\log\lambda) = \mathrm{d}N(\log\lambda)/\mathrm{d}(\log\lambda)$ is obtained from simulations similar to those used to generate Figs 3 and 4 by counting the number of points $\mathrm{d}N(\log\lambda)$ that have stretching values in the range of $[\log\lambda, \log\lambda + \mathrm{d}(\log\lambda)]$. Since in average the local striation thickness is inversely proportional to the stretching, $H_n(\log\lambda)$ can be conceptually interpreted as a spectrum of intensities of the fluid mechanical mixing process. In this paper $H_n(\log\lambda)$ is used to compare the stretching performance of periodic and aperiodic flows. Quantitative comparisons are performed using only two moments of $H_n(\log\lambda)$, Λ and σ, which are the mean and the standard deviation of $\log\lambda$, respectively. A detailed study of the statistics of stretching and its scaling properties in periodic cavity flow will be reported separately.

Figure 5(a) shows $H_n(\log\lambda)$ for the cavity flow, $T = 7.0$, $t = 20T$, computed for the periodic flow as well as for each of the four aperiodic protocols (equation (3a)–(3d)). As discussed above, for this value of T the periodic flow generates a mixed dynamical condition with large chaotic and regular regions. This is reflected by $H_n(\log\lambda)$, which displays a small peak to the left, generated by points inside the islands undergoing slow stretching, and a larger peak to the right, representing fast stretching in the chaotic domain. Another prominent feature of the periodic spectrum is the long and noisy tail for large values of $\log\lambda$, corresponding to the segregated red regions in Fig. 2(a). As mentioned above, such regions are produced by the combined effects of singularities at the cavity corners and of unstable manifolds associated with hyperbolic periodic points. The curves for the four aperiodic flows in Fig. 5(a) all look very similar to one another and show two major differences with respect to the curve for the periodic flow; the peak corresponding to the islands has disappeared, and the high stretching tail is considerably shorter. All four aperiodic flows enhance stretching; in quantitative terms, Λ, the mean of $\log\lambda$, increases from 7.20 (P) to 8.39 (RS), 8.76 (RRS), 8.70 (SB), and 9.06 (RRP), indicating an increased efficiency of the process. Moreover, σ, the standard deviation of $\log\lambda$, decreases from 3.55 (P) to 2.15 (RS), 2.13 (RRS), 2.15 (SB), and 2.41 (RRP), indicating a considerably more uniform process.

Figure 5(b) shows $H_n(\log\lambda)$ for the cavity flow, $T = 2.0$, $t = 50T$. Recall that for this value of T the periodic flow generates regular motion in most of the domain. $H_n(\log\lambda)$ shows a large peak to the left corresponding to the regular region, and a long, noisy tail generated by small chaotic regions. The four aperiodic procedures generate spectra that again look very similar to one another and also to the aperiodic curves in the previous

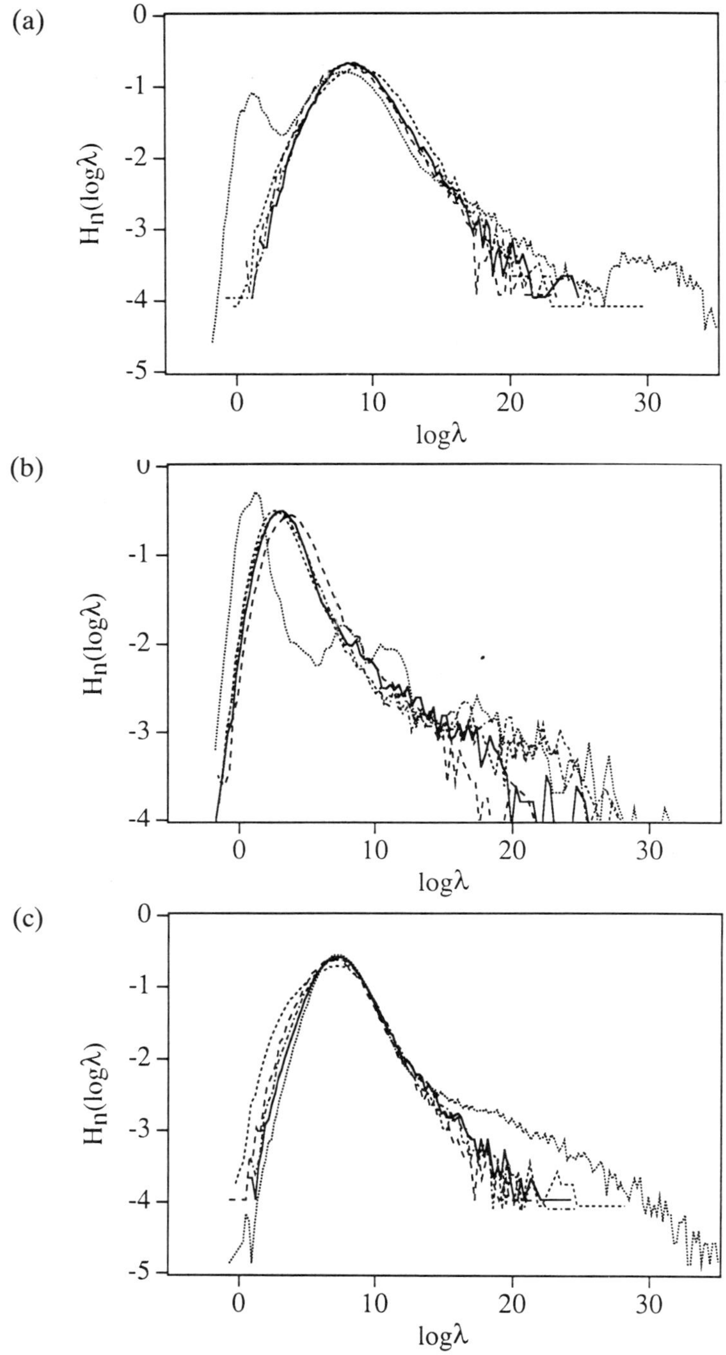

Fig. 5. The spectrum $H_n(\log\lambda) = dN(\log\lambda)/d\log\lambda$ for the cavity flows. The curves show $H_n(\log\lambda)$ for the periodic flow (P = dots) as well as for each of the four aperiodic flows (RRP = short dashes; RS = long dashes; RRS = solid curve; SB = dot–dash curve). (a) $T = 7.0$, $t = 20T$. The periodic spectrum shows an irregular peak to the left corresponding to stretching inside islands, a bell-shaped peak corresponding to the bulk of the chaotic region, and a long, noisy tail corresponding to segregated high-stretching regions. The aperiodic procedures generate spectra showing only the chaotic peak and a considerably shorter tail. (b) $T = 2.0$, $t = 50T$. The periodic spectrum is composed entirely of a regular peak and a high-stretching tail. The aperiodic procedures generate a bell-shaped, smooth peak indicative of global chaos. (c) $T = 5.6$, $t = 20T$. For this condition the periodic flow is globally chaotic, and all five flows generate similar looking spectra, except for the high-stretching tail, which is considerably larger for the periodic flow.

figure; in all cases, $H_n(\log\lambda)$ displays a smooth, bell-shaped peak that is typical of chaotic stretching processes, and a tail that is generated by segregated high-stretching regions. Once again, Λ increases from 1.89 (P) to 4.14 (RS), 3.60 (RRS), 3.57 (SB), and 3.33 (RRP) (i.e. the geometric mean stretching 10^Λ increases by about 10^2), and σ decreases from 2.98 (P) to 1.86 (RS), 2.08 (RRS), 2.38 (SB), and 2.31 (RRP).

The mixing enhancements observed in the cavity flow for $T = 7.0$ and $T = 2.0$ should have been expected, since in both cases aperiodicity considerably increases the extent of the chaotic region. The outcome is less predictable if aperiodic flows are compared with globally chaotic periodic flows. Such comparison is made in Fig. 5(c), which shows $H_n(\log\lambda)$ for the cavity flow, $T = 5.6$, $t = 20T = 112$. For this condition, both the periodic and the aperiodic flows generate spectra with a single, bell-shaped peak. Mixing enhancements are considerably less significant than in the previous two cases; in fact, in this case Λ *decreases* from 7.74 (P) to 7.10 (RS), 7.48 (RRS), 7.41 (SB), and 7.00 (RRP). However, it should be pointed out that the larger value of Λ for the P flow is due mostly to contributions from the high-stretching tail. As in previous cases, aperiodicity leads to more uniform stretching; the high-stretching tails in the aperiodic spectra are shorter than in the periodic case, and σ decreases from 2.25 (P) to 1.89 (RS), 1.85 (RRS), 1.95 (SB), and 2.24 (RRP).

The effects of aperiodic flow procedures on $H_n(\log\lambda)$ are further illustrated in Fig. 6(a–c) for the sine flow for three conditions: $T = 0.8$, for which the periodic sine flow corresponds to a mixture of chaos and large islands; $T = 0.4$, for which the periodic sine flow generates regular motion in most of the domain; and $T = 1.6$, for which the periodic sine flow is chaotic nearly everywhere. These conditions are dynamically similar to those discussed above for the periodic cavity flow. In all three cases, $H_n(\log\lambda)$ is computed for $t = 20T = 16$ using 40 000 particles initially placed uniformly in the entire domain. Figure 6(a) shows results for $T = 0.8$, $n = 20$. As discussed above, for this value of T, the periodic sine flow has four large islands and stretching is very slow not only inside the islands but also in large bands around them (see Fig. 4(a)). This effect is readily observed in Fig. 6(a) by the 'flat top' periodic curve, which shows a wide range of low-stretching high-frequency values. On the other hand, all four aperiodic procedures generate narrow spectra that consist of well-defined, bell-shaped peaks significantly displaced towards high-stretching values. These observations can be expressed in quantitative terms by noticing that when aperiodicity is imposed, Λ increases from 4.92 (P) to 6.66 (RS), 6.98 (RRS), 7.55 (SB), and 7.72 (RRP) (i.e. the geometric mean stretching 10^Λ increases 10^2–10^3 times). As expected for this case, the stretching process also becomes substantially more uniform, and σ decreases from 2.62 (P) to 1.30 (RS), 1.41 (RRS), 1.40 (SB), and 1.70 (RRP).

For $T = 0.4$, (Fig. 6(b)) the periodic sine flow shows only narrow strips of chaos in an otherwise regular flow, and therefore only a small fraction of the particles undergoes high stretching. Similarly to the case for the mostly regular periodic cavity (Fig. 5(b)), the periodic spectrum for this sine flow is composed almost entirely of a large regular peak to the left, followed by a broad, low-frequency chaotic tail. When aperiodic procedures are applied to this case, the spectrum changes dramatically. Single-peak bell-shaped curves, typical of globally chaotic systems, are observed for all the aperiodic procedures. In quantitative terms, Λ increases from 1.45 (P) to 5.39 (RS), 4.59 (RRS), 5.54 (SB), and 3.71 (RRP) i.e. the geometric mean stretching 10^Λ increases 10^3–10^4 times. In spite of this massive increase in the mean stretching, σ increases only moderately, from 1.29 (P) to 1.48 (RS), 1.49 (RRS), 1.69 (SB), and 1.73 (RRP) (i.e. the dispersion coefficient σ/λ decreases substantially). Finally, the results for $T = 1.6$, corresponding to a globally chaotic time-periodic sine flow, are presented in Fig. 6(c). As was also the case for the globally chaotic

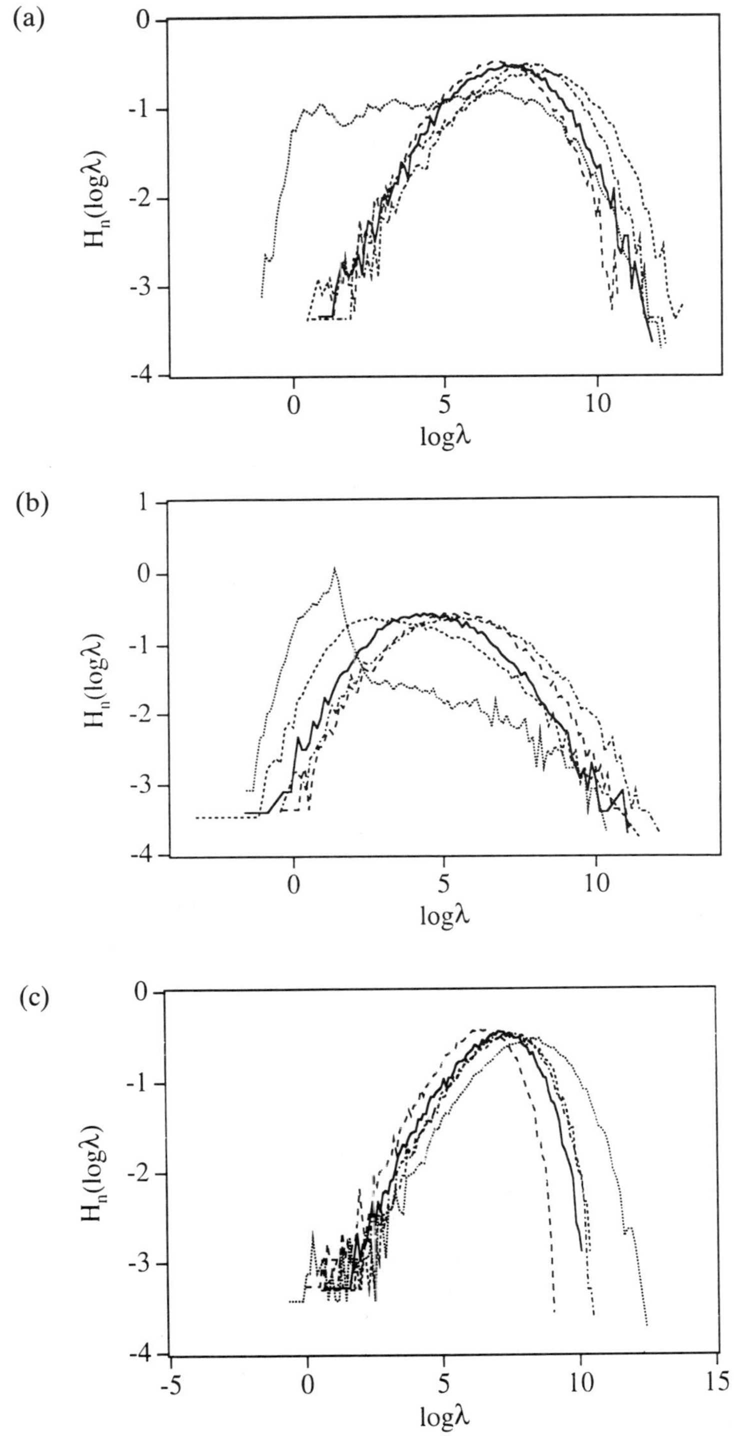

Fig. 6. $H_n(\log \lambda)$ for the sine flow; (P = dots; RRP = short dashes; RS = long dashes; RRS = solid curve; SB = dot–dash curve). (a) $T = 0.8$, $t = 20T$. The 'flat top' periodic spectrum reveals that low-stretching points occur with a high, nearly constant frequency. The aperiodic procedures again generate a single, broad peak indicative of global chaos (compare with Fig. 5(a)). (b) $T = 0.4$, $t = 40T$. For this condition the periodic flow is mostly regular (compare with Fig. 5(b)). (c) $T = 1.6$, $t = 10T$. All five flows generate very similar spectra. The periodic flow displays the fastest stretching.

periodic cavity, Λ decreases from 8.00 (P) to 6.28 (RS), 6.93 (RRS), 7.23 (SB), and 7.26 (RRP). Although σ also decreases from 1.46 (P) to 1.15 (RS), 1.24 (RRS), 1.25 (SB), and 1.26 (RRP), this decrease is roughly proportional to the decrease in Λ and therefore it is not statistically significant. These results indicate that, when evaluated in terms of stretching alone, aperiodicity does not seem to enhance mixing in a globally chaotic sine flow.

So far, all comparisons have been based on individual cases. However, due to the randomness incorporated by the RS, RRS, and RRP procedures, the performance observed from individual realizations is subject to some amount of fluctuation. Comparisons between individual realizations may be misleading, unless such fluctuations are characterized. These fluctuations must be quantified also for practical reasons. Due to the lack of time-periodicity, batches of material being mixed at different times would experience different extents of mixing. Arguably, this lack of uniformity is a disadvantage of aperiodic procedures. Therefore, the extent of random fluctuations deserves further examination.

In order to perform statistically meaningful comparisons of periodic and aperiodic stretching processes, a large number of realizations of each of these aperiodic procedures is conducted for the sine flow and ensemble averages are obtained. The symbols in Fig. 7(a) show the values of Λ for each of 100 realizations for $T = 0.8$, $t = 20T = 16$, using the RS, RRS and RRP procedures. The horizontal lines correspond to Λ_{av}, the overall average of Λ for all 100 realizations of each procedure; also shown are the corresponding values of Λ_{av} for the SB flow and the P flow. The order in average stretching performance in Fig. 7(a) happens to be the same as that in Fig. 6(a); Λ_{av} increases from 4.9224 (P) to 6.9007 (RS), 7.4392 (RRS), 7.5348 (SB), and 7.8120 (RRP). The amount of scattering generated by the procedures differs widely; the standard deviation of the 100 values of Λ is 0.3657, 0.2929, and 0.1861 for the RS, RRS, and RRP procedures, respectively. These results are statistically significant and indicate that for the sine flow, $T = 0.8$, the RRP flow generates both the largest stretching and the smallest fluctuations among realizations. All 100 RRP realizations have values of Λ higher than the value of Λ_{av} corresponding to the RS procedure. All aperiodic realizations generate much higher stretching than the periodic flow.

Figure 7(b) shows 100 realizations of sine flow for $T = 0.4$, $t = 40T = 16$. Symbols and lines have the same meaning as in the previous figure. A different order in stretching performance is found at this value of T; Λ_{av} now increases from 1.4620 (P) to 2.6141 (RRP), 4.5662 (RRS), 5.5313 (SB), and 5.8272 (RS). All the individual realizations of the RS procedure have higher values of Λ than the value of Λ_{av} corresponding to the RRS procedure. As in the previous case, fluctuations among realizations are largest for the RS flow (for which the standard deviation of the 100 values of Λ is 0.3967), and smallest for the RRP flow (the standard deviation is 0.1820). Thus, for this flow condition the RS flow generates the largest stretching, but also the largest scattering among realizations. Again, all aperiodic cases generate much higher stretching than the periodic flow. Finally, Fig. 7(c) shows 100 realizations of each procedure for $T = 1.6$, $t = 10T = 16$. In terms of Λ_{av}, the order is 6.3182 (RS), 7.2172 (SB), and 7.3387 (RRS), 7.5039 (RRP), and 8.0108 (P). The standard deviation of different values of Λ is 0.6915 (RS), 0.3178 (RRS), and 0.3152 (RRP). Unlike the previous two cases, the periodic flow now has the highest mean stretching; only 1% of the individual RRS realizations and 2% of RRP realizations are above the value of Λ_{av} corresponding to the periodic flow. The RRP procedure again generates the largest stretching of all four aperiodic systems and the smallest fluctuations between individual realizations.

The results presented in Fig. 7(a–c) clearly show that the stretching performance of

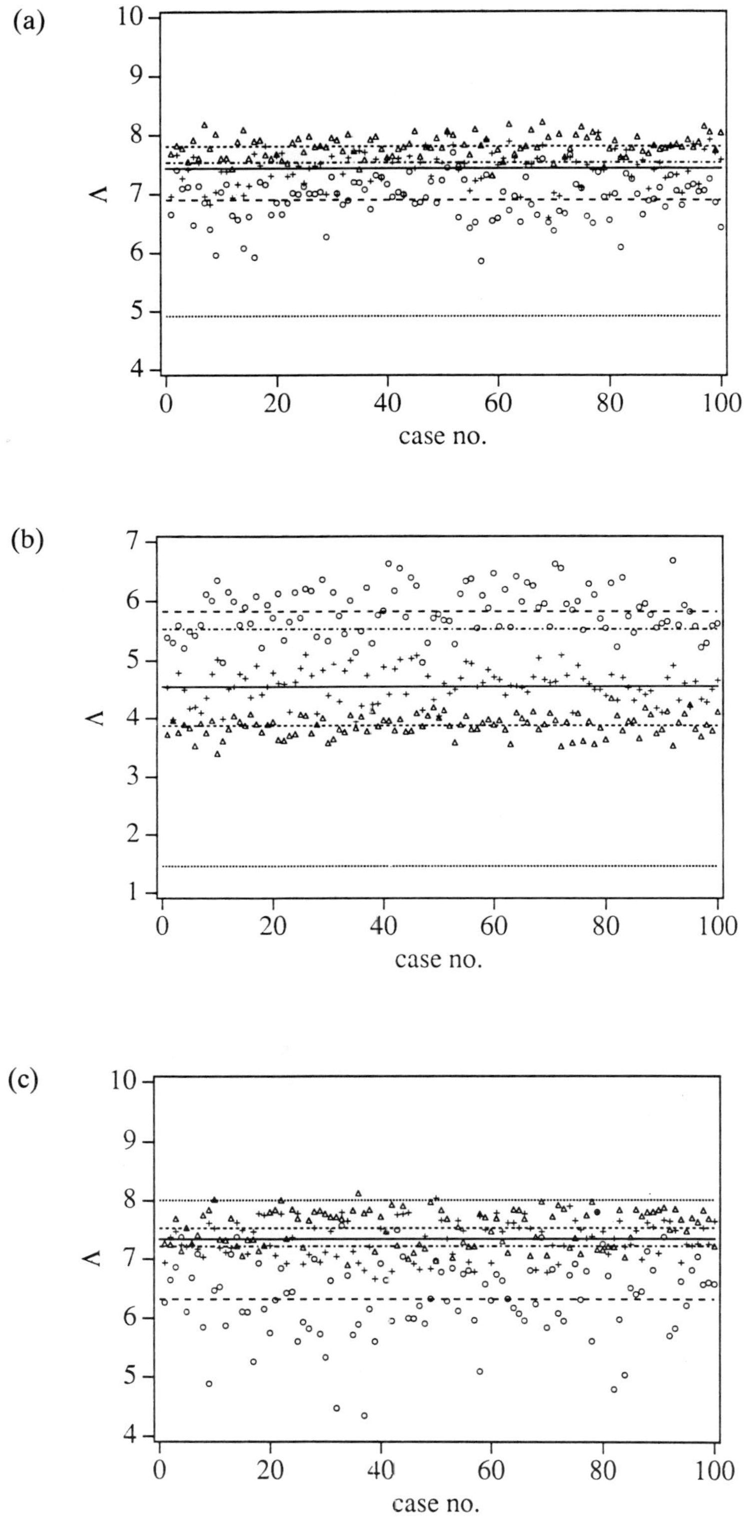

Fig. 7. Λ for each of 100 independent realizations of RS (circles), RRS (crosses), and RRP (triangles) sine flows. All realizations are for a total time of $t = 100$. The lines show Λ_{av} (P = dots; RRP = short dashes; RS = long dashes; RRS = solid curve; SB = dot–dash curve). (a) $T = 0.8$; (b) $T = 0.4$; (c) $T = 1.6$.

aperiodic flows is strongly affected by the value of T; the largest value of Λ_{av} is achieved by different procedures at different values of T. To analyze the dependence on T in more detail, Λ_{av} is calculated for 100 realizations of the sine flow for values of T from 0.4 to 2.0 at intervals of $\Delta T = 0.05$. Figure 8(a) shows Λ_{av} for all four aperiodic procedures as well as for the periodic flow. The results presented above in Fig. 7(a–c) correspond to three cuts at $T = 0.4$, 0.8 and 1.6 in Fig. 8(a). The change of stretching performance of different aperiodic flows can be seen clearly in this figure. At low value of $T < 0.6$, the order from the smallest to the largest stretching is P, RRP, RRS, SB and RS. The RRP flow generates the highest stretching among the aperiodic flows for all values of $T > 0.6$. At $T \approx 0.95$, the P flow catches up with the RS flow, and at $T \approx 1.2$, the P flow first surpasses the RRS and the SB flow, and then outperforms the RRP flow. For all values of T in the interval $1.25 < T < 2.0$, the order from the smallest to the largest stretching is RS, SB, RRS, RRP, and P. As expected, the RS procedure gives the worst stretching performance, except at low values of T for which the P flow is almost completely regular. For most flow conditions, the SB and RRS flows generate almost identical stretching, possibly due to the similarity between the sequences generated by these two procedures. Despite the random-

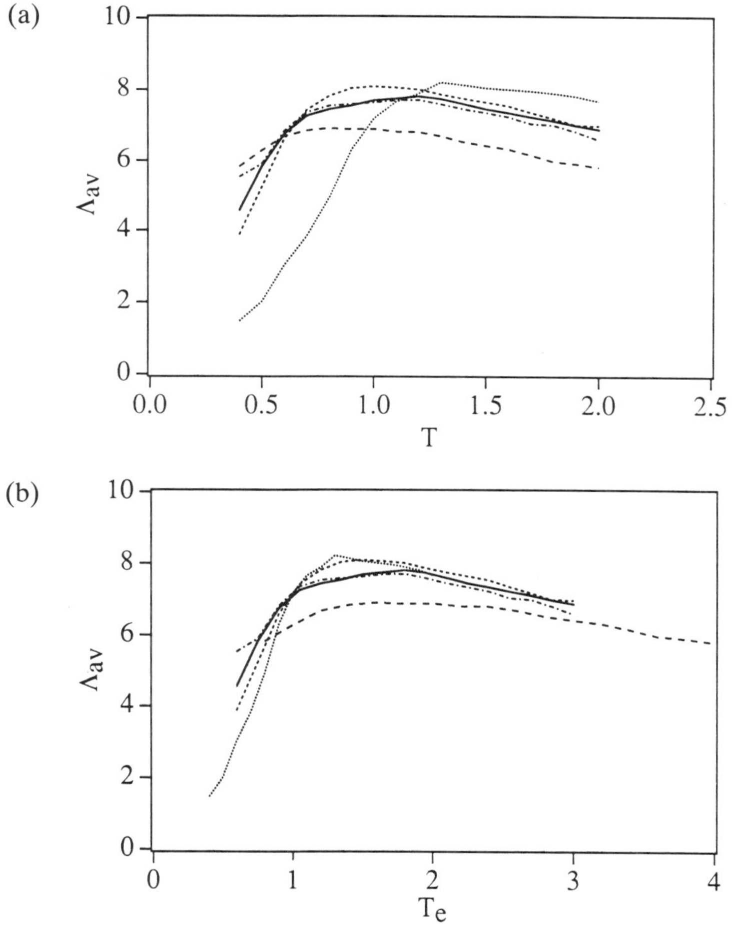

Fig. 8. (a) Comparisons of Λ_{av} as a function of T. (b) The same results as in Fig. 8(a) but plotted as a function of the effective period T_e. (P = dots; RRP = short dashes; RS = long dashes; RRS = solid curve; SB = dot–dash curve.)

ness in all the aperiodic flows, a maximum exists in each curve in Fig. 8(a); the RS procedure achieves its maximum the earliest, at $T \approx 0.8$; the RRP flow achieves its maximum at $T \approx 1.0$, and RRS and SB at $T \approx 1.2$. The maximum for the P flow is achieved at $T \approx 1.3$.

However, the comparison in Fig. 8(a) in terms of T is in some sense biased. For the P flow, continuous motion in each direction lasts exactly $T/2$. For the RS, RRP, and SB procedures, the average duration of continuous motions in each direction is longer than $T/2$, due to repetitions of motions such as *aa* and *bb*. A better comparison can be performed in terms of the effective period T_e, defined as $T_e = 2^t/(N+1)$, where t is the total time (in terms of T) and N is the total number of changes in motion direction i.e. the number of *ab* transitions plus the number of *ba* transitions. For example, for the P flow, $t = 10T$, we have $N = 19$ and $T_e = T$; for an SB flow, $t = 4T$, the sequence of motions is *abbabaab*, the number of changes is 5, and $T_e = 8/6T$. Flows with the same value of T_e have the same frequency of changes in flow direction regardless of the actual procedure employed to generate the flow. As the number of elements a and b increases, a limiting T_e is reached for each sequence. Sequences of 10 000 elements were used to approximate the limiting T_e, obtaining $T_e = 2T$ for the RS procedure and $T_e = 1.5T$ for the RRS, RRP and SB procedures. Figure 8(b) compares the flow in terms of T_e. All of the aperiodic flows achieve the maximum mean stretching in a narrow interval of $T_e \approx 1.5$. Curves for the aperiodic flows with $T = 1.5T_e$ cross the curve for the P flow at roughly the same value of $T_e \approx 1.0$. The RS procedure, on the other hand, crosses the periodic curve at $T_e \approx 0.9$ and generates much smaller stretching than all of the other flows for all values of $T_e > 0.9$. The P flow generates the largest stretching only in a narrow window $1.0 < T_e < 1.4$; the RRP flow generates larger stretching than the P flow for all other values of T_e. However, comparisons in terms of Λ_{av} alone do not take into account the fact that while the aperiodic flows achieve complete mixing for all values of T_e, mixing in the P flow is often hindered by the presence of islands and segregated low/high stretching regions. Although stretching computations provide straightforward avenues for quantifying a number of aspects of mixing, additional insight can be gained by considering alternative approaches.

4.2. *Particle spread*

The second method used in this paper quantifies the rate at which particles spread by dividing the flow domain into small uniform bins and counting the number of bins that contain particles. Similar 'box counting' techniques have been used by other authors to obtain gross quantifications of mixing processes [13, 29, 30]. Selecting bins of appropriate size is critical; if the bins are too large, the technique will quantify mixing only on a coarse scale. On the other hand, if the bins are too small, a very large number of particles will be required to achieve adequate computational resolution; otherwise, only a small fraction of the bins will be occupied at any given time. Several criteria can be adopted to select the bin size. Jones [30] recognized that a perfectly randomized population of particles has a Poisson bin-occupancy distribution, and used this information to determine a bin size so that on average there was one particle per bin. In this paper, the bin size is chosen so that for a perfectly random distribution of particles, 98% of the bins contain at least one particle. Numerical tests indicate that such a bin size is approximately given by $s \sim 2N^{-1/2}$, where s is the size of the bin and N the total number of particles. The simulations in this section use 40 000 particles and a bin size $s = 0.01$.

Figure 9(a) shows the time evolution of the particle spread in the cavity flow, $T = 7.0$, for the periodic and aperiodic flows described above. The particles are initially placed in a

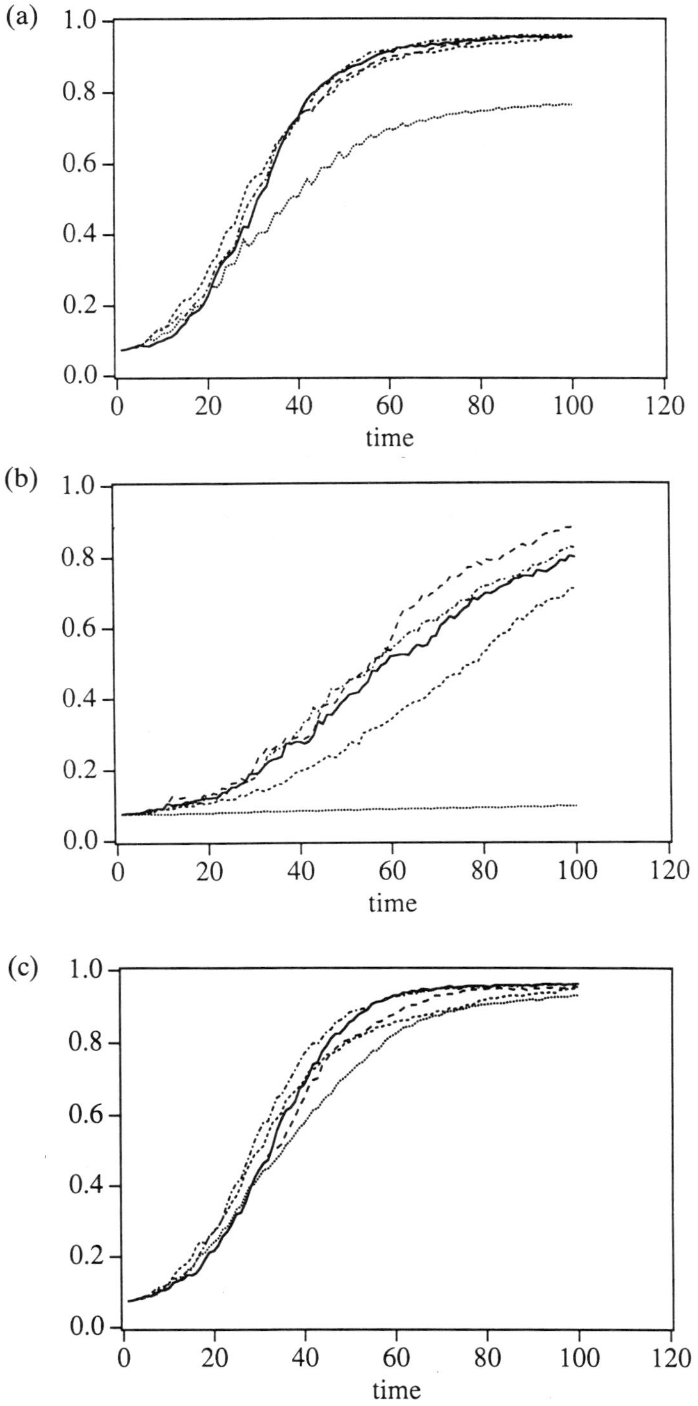

Fig. 9. Evolution of the particle spread in the cavity flow. The flow domain is divided into square boxes of size $s = 0.01$ and the percentage of boxes that contain particles is computed during a mixing simulation (P = dots; RRP = short dashes; RS = long dashes; RRS = solid curve; SB = dot–dash curve). (a) $T = 7.0$. The periodic flow achieves only partial box coverage. All of the aperiodic flows reach high coverage at essentially the same rate. (b) $T = 2.0$. Minimal spread is achieved by the (mostly regular) periodic flow. (c) $T = 5.6$. Although the periodic flow is globally chaotic and achieves high box coverage, it does so considerably more slowly than the aperiodic flows.

small square centered at $(x, y) = (0.75, 0.3)$, from where they are convected for a maximum time $t = 100 \simeq 14T$ i.e. about 14 periods of the P flow. For all five flows, particles initially spread at an exponential rate, but the rate decreases as the boxes fill up; at $t = 100$, the curves have leveled off. It should be pointed out that the curves shown in Fig. 9(a) are similar to those obtained from experiments [31]. As expected, due to the presence of large islands, the P flow achieves only incomplete spread with just 75% of the boxes occupied by particles. On the other hand, all four aperiodic sequences yield almost equivalent results, with particles covering approximately 95% of the boxes at $t = 100$. The RRP flow seems to spread particles slightly faster for short times, and the SB flow seems to achieve higher spread at later times, but the differences are small and are probably due to statistical noise.

Substantially different behavior is displayed in Fig. 9(b) for $T = 2.0$. None of the flows achieve box saturation in the timescale of the simulation. For the P flow, particle motion is regular in almost the entire flow, thus it is not surprising that only a small fraction of the boxes are covered even after $t = 100$ (50 periods). Differences in performance among the different aperiodic flows are considerably more pronounced than for the previous case. At $t = 100$, 88% box coverage is achieved by the RS flow, 80% by the RRS flow, 82% by the SB flow, and about 72% by the RRP flow. This order in spread performance by the different flows is exactly the same as the order in stretching performance observed in Fig. 5(b) (the same sequences of motions were used to compute both figures); for $T = 2.0$, the RS flow generates both the highest mean stretching and fastest particle spread, and the RRP flow exhibits the lowest scores in both measures. This correspondence indicates that the two quantification techniques used here, stretching and particle spread, are consistent.

Figure 9(c) shows the time evolution of the particle spread for $T = 5.6$. For this condition, the P flow is globally chaotic, and particles cover a large fraction of the total number of boxes for all five flows. However, all four aperiodic procedures generate faster particle spread than the P flow. The poorer performance of the periodic flow is caused by the segregated high-stretching regions, because particles penetrate and leave such regions slowly. As a result, the P flow achieves a total area coverage that is comparable to that of the aperiodic flows, but only after a considerably longer time. As discussed before, the size of such segregated regions decreases when aperiodic procedures are used, leading both to more uniform stretching and faster particle spread.

The final set of results presented in this paper focuses on particle spread in the sine flow. In this case, the particles are initially placed in a small square of size 0.1×0.1 located at the center of the flow domain. The particles are then advected for a total time of $t = 16$. Figure 10(a) shows the particle spread for all five flows for $T = 0.8$. The periodic sine flow for this condition is only partially chaotic, and as a result, the maximum number of boxes covered by particles is about 70%; the remaining boxes are located inside the islands. Particle spread is slightly faster for the RRP flow; the curves corresponding to the other three procedures are essentially identical to one another. For all four aperiodic flows, box coverage exceeds 95% after $t \approx 10$. The figure is qualitatively identical to the corresponding case for the cavity flow ($T = 7.0$, Fig. 9(a)).

Figure 10(b) shows the particle spread for the sine flow, $T = 0.4$. Once again, the observed behaviors are qualitatively very similar to those described in Fig. 9(b) for the corresponding case of the cavity flow. Particles spread very slowly for the P flow, which is almost entirely regular. Significant differences can be observed in the spread generated by the different aperiodic procedures. As it is shown in the figure, in terms of particle spread, the RRP flow performs the worst and the RS flow performs marginally better than the RRS and SB flows. This order in particle-spread performance is the same as for the cavity flow, $T = 2.0$ (Fig. 9(b)), and is also consistent with the stretching performances in both

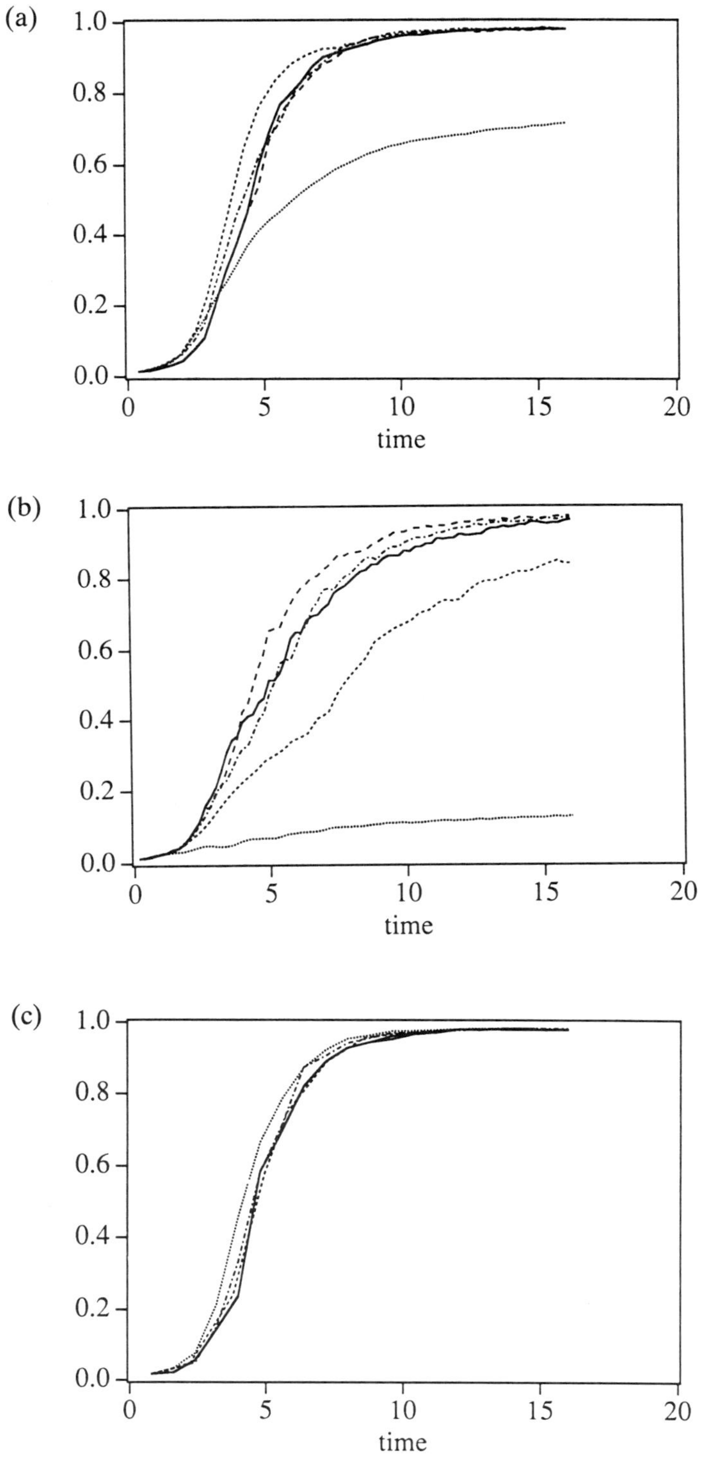

Fig. 10. Particle spread in the sine flow. Curves as in Fig. 9. (a) $T = 0.8$. The observed behaviors are very similar to those depicted for the corresponding cavity flow case (Fig. 9(a)). (b) $T = 0.4$. Again, the figure looks qualitatively very similar to the corresponding cavity flow case. (c) $T = 1.6$. Since mixing in the globally chaotic sine flow is not hindered by segregated high-stretching regions, particles spread at essentially the same rate for all five flows.

flows, since for both the cavity, $T = 2.0$ (Fig. 5(b)) and the sine flow, $T = 0.4$ (Fig. 7(d)), the RRP procedure generates considerably less stretching than the other aperiodic procedures. Finally, Fig. 10(c) shows the particle spread for the sine flow, $T = 1.6$. For this condition, the P flow is globally chaotic. In spite of the differences in stretching described in Fig. 7(c) and (d), from the point of view of particle spread, all five flows achieve almost identical performances.

5. SUMMARY AND CONCLUSIONS

This paper focuses on aperiodic chaotic mixing processes in the physically realizable cavity flow and in the highly idealized sine flow. The effects of aperiodicity were investigated using four aperiodic flow procedures: a completely random sequence of motions (RS), a restricted random sequence (RRS), a symmetry-breaking deterministic procedure (SB), and a procedure in which a restricted amount of randomness was applied to the duration of each motion (RRP). Many of the conceptual tools commonly used to analyze chaotic mixing (Poincaré sections, periodic points and their associated manifolds) require the flow to be periodic. Hence, other approaches need to be developed for aperiodic flows. The two techniques used in this paper, stretching statistics and particle spread, are simple to apply, consistent with each other, and provide considerable quantitative information about different aspects of the mixing process.

Three periodic flow conditions were used as a basis for comparison: (1) a mixture of chaos and large islands, (2) almost global regularity, and (3) almost global chaos. For conditions such that the periodic flow had a mixture of chaos and large islands, all four aperiodic procedures destroyed the islands and induced fast mixing, but the RRP flow performed better than any of the other flows, generating larger and more uniform stretching and (marginally) faster particle spread. Perhaps more significant, however, are the results presented here for conditions such that the periodic flow was almost completely regular. When aperiodic procedures were applied to such cases, global chaos was achieved and particles spread throughout the flow domain. Mixing is enhanced enormously, both in terms of stretching and particle spread, at the sole expense of a slightly more complicated flow procedure. Surprisingly, comparisons between aperiodic procedures indicate that for this case the RRP flow is the poorest mixer. Finally, mixing enhancements are also achieved for conditions such that the periodic flow is globally chaotic. For the cavity flow, aperiodicity reduces the size of segregated high-stretching regions, generating more uniform stretching and faster particle spread. A properly chosen aperiodic sine flow generates slightly more stretching than a globally chaotic periodic one for many conditions (Fig. 8(b)).

In summary, mixing in periodic flows is usually worse than in aperiodic flows, regardless of the procedures used to generate aperiodicity. The exception are globally chaotic periodic flows, which achieve roughly the same mixing efficiency than aperiodic procedures. However, global chaos in most periodic flows can be achieved only for specific parameter values; moreover, for some periodic flows, globally chaotic conditions might be nonexistent or impractical. On the other hand, aperiodic procedures achieve complete and efficient mixing for broad ranges of parameter values. The results presented here are not conclusive regarding which of the four aperiodic procedures, if any, is superior, but it is clear that much would be gained by achieving a better understanding of aperiodic mixing. Conceivably, aperiodic perturbations can be easily imposed on real mixers, both laminar and turbulent, and are likely to generate mixing enhancements in a wide variety of practical applications.

Acknowledgements—This work was supported by grants from the Exxon Education Foundation and the Merck Foundation to FJM, by NSF grant ECS 9110424 to RLP, and by a grant from the Electric Power Research Institute to RLP. Dr. Sandra Walther at CAIP developed the scientific data management and visualization tools used in the research.

REFERENCES

1. H. Aref, Stirring by chaotic advection, *J. Fluid Mech.* **143**, 1–21 (1984).
2. D. V. Khakhar, H. Rising and J. M. Ottino, Analysis of chaotic mixing in two model systems, *J. Fluid Mech.* **172**, 419–451 (1986).
3. H. Aref and S. Balachandar, Chaotic advection in a Stokes flow, *Phys. Fluids* **29**, 3515–3521 (1986).
4. W. L. Chien, H. Rising, and J. M. Ottino, Laminar and chaotic mixing in cavity flows, *J. Fluid Mech.* **170**, 355–377 (1986).
5. J. Chaiken, R. Chevray, M. Tabor and Q. M. Tan, Experimental study of Lagrangian turbulence in a Stokes flow, *Proc. R. Soc. Lond.* **A408**, 165–174 (1986).
6. H. Aref and S. W. Jones, Enhanced separation of diffusing particles by chaotic advection, *Phys. Fluids* **A1**, 470–474 (1988).
7. E. Ott and T. M. Antonsen, Chaotic fluid convection and the fractal nature of passive scalar gradients, *Phys. Rev. Lett.* **61**, 2839–2842 (1988).
8. C. W. Leong and J. M. Ottino, Experiments on mixing due to chaotic advection in a cavity, *J. Fluid Mech.* **209**, 463–499 (1989).
9. P. D. Swanson and J. M. Ottino, A comparative computational and experimental study of chaotic mixing of viscous fluids, *J. Fluid Mech.* **213**, 227–249 (1990).
10. V. Rom-Kedar, A. Leonard and S. Wiggins, An analytical study of transport, mixing and chaos in an unsteady vortical flow, *J. Fluid Mech.* **214**, 347–394 (1991).
11. D. Beigie, A. Leonard and S. Wiggins, A global study of enhanced stretching and diffusion in chaotic tangles, *Phys. Fluids* **A3**, 1039–1050 (1991).
12. F. J. Muzzio, P. D. Swanson and J. M. Ottino, The statistics of stretching and stirring in chaotic flows, *Phys. Fluids* **A3**, 822–834 (1991).
13. F. J. Muzzio, C. Meneveau, P. D. Swanson and J. M. Ottino, Scaling and multifractal techniques for analysis of mixing in chaotic flows, *Phys. Fluids* **A4**, 1439–1456 (1992).
14. F. J. Muzzio, P. D. Swanson and J. M. Ottino, Mixing distributions produced by multiplicative stretching in chaotic flows, *Int. J. Bifurcation Chaos* **2**, 37–50 (1992).
15. H. A. Stone, A. Nadim and S. H. Strogatz, Chaotic streamlines inside drops immersed in steady linear strokes flows, *J. Fluid Mech.* **232**, 629–646 (1991).
16. M. Liu, F. J. Muzzio and R. L. Peskin, *Am. Inst. Chem. Engs. J.* (in press).
17. P. Grassberger, R. Badii and A. Politi, Scaling laws for invariant measures on hyperbolic and nonhyperbolic attractors, *J. Statist. Phys.* **51**, 135–178 (1988).
18. M. A. Sepulveda, R. Badii and E. Pollak, Spectral analysis of conservative dynamical systems. *Phys. Rev. Lett.* **63**, 1226–1229 (1989).
19. T. Horita, H. Hata and H. Mori, Long-time correlations and expansion-rate spectra of chaos in Hamiltonian systems, *Prog. in Theor. Phys.* **83**, 1065–1070 (1990).
20. M. F. Shlesinger, G. M. Zaslavsky and J. Klafter, Strange kinetics, *Nature Lond.* **363** (6 May), 31–37 (1993).
21. F. Ling and G. Schmidt, Mixing windows in discontinuous cavity flows, *Phys. Lett.* **A165**, 221–230 (1992).
22. F. Ling, The effect of mixing protocol on mixing in discontinuous cavity flow, *Phys. Lett.* **A177**, 331–337 (1993).
23. M. Liu, Numerical study of nonlinear dynamics of particles and chaotic mixing in two dimensional periodic driven cavity flows, Ph.D. dissertation, Rutgers, The State University of New Jersey (1992).
24. J. G. Franjione, C. W. Leong and J. M. Ottino, Symmetries within chaos: a route to effective mixing, *Phys. Fluids* **A1**, 1172–1783 (1989).
25. D. V. Khakhar and J. M Ottino, Fluid mixing (stretching) by time periodic sequences for weak flows, *Phys. Fluids* **29**, 3503–3505 (1986).
26. F. Városi, T. M. Antonsen and E. Ott, The spectrum of fractal dimensions of passively convected scalar gradients in chaotic fluid flows. *Phys. Fluids* **A3**, 1017–1028 (1991).
27. R. T. Pierrehumbert, Large scale horizontal mixing in planetary atmosphere, *Phys. Fluids* **A3**, 1250–1260 (1991).
28. D. Beigie, A. Leonard and S. Wiggins, Statistical relaxation under non-turbulent chaotic flows: non-Gaussian high stretch tails of finite-time Lyapunov exponent distributions, *Phys. Rev. Lett.* **70**, 275–278 (1993).
29. J. C. H. Fung and J. C. Vassilicos, Fractal dimensions of lines in chaotic advection, *Phys. Fluids* **3**, 2725–2733 (1991).
30. S. W. Jones, The enhancement of mixing by chaotic advection, *Phys. Fluids* **A3**, 1081–1086 (1991).
31. C. W. Leong, Chaotic mixing of viscous fluids in time-periodic cavity flows, Ph.D. dissertation, University of Massachusetts (1990).

Hamiltonian Formulation of the Equations of Streamlines in Three-dimensional Steady Flows

K. BAJER*

University of Warsaw, Institute of Geophysics, ul. Pasteura 7, 02-093, Warsaw, Poland

Abstract—It is well known that the equations of motion of a fluid particle in a two-dimensional flow can be written in the canonical form with the stream function playing the role of the Hamiltonian. Here we extend this canonical formalism to three-dimensional steady flows. We show how the methods of the theory of Hamiltonian systems can be successfully used to investigate the advection of a passive particle. As an example of the perturbed closed streamline flow we consider a rigid rotation with added small quadratic velocity field and explain the structure of streamlines by averaging the corresponding Hamiltonian. We also show how the Hamiltonian formulation can be used to find the invariants of the fluid particle motion which are then used for non-canonical averaging.

1. INTRODUCTION

Many authors studying the structure of the magnetic fields have tried to explore the formal analogy between the equations of the field lines and the equations of a one degree-of-freedom (d.o.f.) Hamiltonian systems. The aim was to design an equilibrium configuration ensuring plasma confinement. In the most favoured toroidal geometry of tokamaks and spheromaks the problem consists of finding the magnetostatic equilibria in which the field-lines lie on a family of nested toroidal surfaces with the winding number continuously varying with radius. Such fields correspond to one d.o.f. non-autonomous, non-degenerate, time-dependent integrable Hamiltonian systems. The fields realized in practice differ from the ideal designs and thus correspond to perturbed integrable Hamiltonians.

It has been known for a long time that symmetric fields, i.e. fields with one ignorable coordinate, can be written in a Hamiltonian form. This was done, for example, for the two-dimensional fields [1] and fields with helical symmetry [2]. The Hamiltonians of the symmetric fields are one-dimensional time-independent and hence are always integrable.

More recently it was realized that any magnetic field, even lacking symmetry, can also be written as a one d.o.f. system with time-dependent Hamiltonian [3–5]. This result enables one to treat the symmetry-breaking distortions of the ideal symmetric designs with methods from the perturbation theory for Hamiltonian systems [4].

The Hamiltonian formalism for the field-line flow was derived for magnetic fields, but may well be applied to any solenoidal vector field like, for example, steady flows of an incompressible fluid. In fact, as we show in Section 2, it can be easily generalized to compressible steady flows satisfying the continuity equation. When applied to the equations of streamlines the formalism can be even more useful than in the magnetic case, where the interest is usually restricted to the non-degenerate integrable fields in the toroidal geometry.

*Present address: University of Cambridge, DAMTP, Silver Street, Cambridge, CB3 9EW, UK.

In this paper we formulate the canonical Hamiltonian formalism for the kinematics of fluid flows and demonstrate some interesting applications. This is a natural extension of the well-known Hamiltonian approach to two-dimensional flows based on a stream function (Section 2.2). One immediate benefit from the canonical formulation is the conclusion that any flow which has an autonomous Hamiltonian is integrable. In Section 2.1 we obtain a family of flows, parameterized with two arbitrary functions, corresponding to the Hamiltonian of a particle in a potential field of force. The time-independence of the potential guarantees the integrability of all flows in the family.

In Section 3 we outline the theory of small perturbations of the flows with closed streamlines. The canonical, as well as non-canonical averaging procedures, are tools for calculating the level surfaces of an adiabatic invariant which organize the streamlines of the perturbed flow. In the canonical case the adiabatic invariant can be obtained directly by averaging the Hamiltonian, and this is when the formalism proves particularly useful. A general theory of 'nearly periodic' systems was advanced by Kruskal [6]. His study was motivated by the problem of a charged particle gyrating in the magnetic field of a fusion device. The nearly periodic trajectory of such a particle stays on a level surface of an adiabatic invariant, which must be calculated in order to determine whether the magnetic trap is effective.

In Section 4 we consider a fluid occupying a spherical domain. The fluid is in motion which is a rigid rotation perturbed by a general quadratic velocity field. The adiabatic invariant is calculated by canonical averaging and its level surfaces compared with the Poincaré sections of streamlines. The shape of the surfaces is the same for all quadratic perturbations.

In Section 5 we consider perturbations of a particular quadratic flow in a sphere, called STF flow, and compute the adiabatic surfaces by means of non-canonical averaging. This has been done before [7] but here we show it as an application of the canonical formalism which makes finding the two required invariants straightforward. Finally the results are summarized in Section 6.

2. CANONICAL FORMALISM FOR THE FLOW KINEMATICS

Let us consider a steady fluid flow $\mathbf{u}'(\mathbf{x})$ in a three-dimensional (3D) domain D. The fluid is, in general, compressible and its density $\rho(\mathbf{x})$ satisfies the continuity equation $\nabla \cdot (\rho \mathbf{u}') = 0$. Hence the vector field $\mathbf{u} = \rho \mathbf{u}'$ is solenoidal and it defines a conservative dynamical system [19]

$$\frac{\mathrm{d}x}{\mathrm{d}t} = u_1(\mathbf{x}) = \rho(\mathbf{x}) u_1'(\mathbf{x}), \tag{1a}$$

$$\frac{\mathrm{d}y}{\mathrm{d}t} = u_2(\mathbf{x}) = \rho(\mathbf{x}) u_2'(\mathbf{x}), \tag{1b}$$

$$\frac{\mathrm{d}z}{\mathrm{d}t} = u_3(\mathbf{x}) = \rho(\mathbf{x}) u_3'(\mathbf{x}), \tag{1c}$$

$$\nabla \cdot \mathbf{u}(\mathbf{x}) = 0, \tag{1d}$$

whose orbits are identical with the streamlines of $\mathbf{u}'$. Let $\mathbf{A}(\mathbf{x})$ be the vector potential of $\mathbf{u}(\mathbf{x})$:

$$\mathbf{u}(\mathbf{x}) = \nabla \wedge \mathbf{A}(\mathbf{x}). \tag{2}$$

When $\mathbf{u}(\mathbf{x})$ is two-dimensional (2D), e.g. $u_z \equiv 0$, the gauge can be chosen such that $\mathbf{A}$ has only one nonzero component called the stream function. Then the equations (1) take the canonical form of a one d.o.f. time-independent Hamiltonian system:

$$\mathbf{A} = (0, 0, \Psi\,(x, y)),$$
$$\frac{dq}{dt} = \frac{\partial H}{\partial p}, \qquad \frac{dp}{dt} = -\frac{\partial H}{\partial q}, \tag{3}$$
$$q = x, \qquad p = y, \qquad H(q, p) = \Psi(q, p).$$

Here we demonstrate that the streamlines of a steady 3D flow can also be described by Hamiltonian equations. Dividing (1a), (1b) by (1c) we obtain an even-dimensional system

$$\frac{dx}{dz} = \frac{u_1}{u_3}, \qquad \frac{dy}{dz} = \frac{u_2}{u_3}, \tag{4}$$

which can be made explicitly Hamiltonian by a suitable choice of variables. Clearly, the orbits of (4) coincide with the streamlines of $\mathbf{u}'(\mathbf{x})$ and $\mathbf{u}(\mathbf{x})$. Hence the Hamiltonian for the system (4) will describe the geometry of streamlines. However, all information about the magnitude of the fluid particle velocity will be lost.

Let us consider the vector potential $\mathbf{A}(x, y, z)$ of $\mathbf{u}(x, y, z)$ in the following gauge:

$$A_y(x, y, z) \equiv 0, \tag{5a}$$

$$u_x = \frac{\partial A_z(x, y, z)}{\partial y}, \quad u_y = \frac{\partial A_x(x, y, z)}{\partial z} - \frac{\partial A_z(x, y, z)}{\partial x}, \quad u_z = -\frac{\partial A_x(x, y, z)}{\partial y} \tag{5b}$$

which is called 'axial gauge' in quantum field theory (ref. [8], p. 566) and will be called 'canonical gauge' in this paper.

We define the phase space variables in the following way:

$$t = z, \qquad q = x, \tag{6a}$$

$$p = -A_x(x, y, z), \tag{6b}$$

$$H(q, p\ t) = A_z(x, y, z) = A_z(q, y(q, p, t), t), \tag{6c}$$

where the function $y(q, p, t)$ is defined implicitly by (6b). Now we have:

$$\frac{\partial y}{\partial q} = -\frac{\partial A_x}{\partial x}\left(\frac{\partial A_x}{\partial y}\right)^{-1}, \qquad \frac{\partial y}{\partial p} = -\left(\frac{\partial A_x}{\partial y}\right)^{-1}, \tag{7a}$$

$$\frac{\partial H}{\partial q} = \frac{\partial A_z}{\partial x} + \frac{\partial A_z}{\partial y}\frac{\partial y}{\partial q} = \frac{\partial A_z}{\partial x} - \frac{\partial A_z}{\partial y}\frac{\partial A_x}{\partial x}\left(\frac{\partial A_x}{\partial y}\right)^{-1} = \frac{\partial A_z}{\partial x} + \frac{\partial A_x}{\partial x}\frac{u_x}{u_z}, \tag{7b}$$

$$\frac{\partial H}{\partial p} = \frac{\partial A_z}{\partial y}\frac{\partial y}{\partial p} = -\frac{\partial A_z}{\partial y}\left(\frac{\partial A_x}{\partial y}\right)^{-1} = \frac{u_x}{u_z}, \tag{7c}$$

$$\frac{dp}{dt} = -\frac{\partial A_x}{\partial x}\frac{dx}{dz} - \frac{\partial A_x}{\partial y}\frac{dy}{dz} - \frac{\partial A_x}{\partial z} = -\frac{\partial A_x}{\partial x}\frac{dx}{dz} + u_z\frac{dy}{dz} - \frac{\partial A_x}{\partial z}, \tag{7d}$$

$$\frac{dq}{dt} = \frac{dx}{dz} \tag{7e}$$

and we obtain

$$\frac{dp}{dt} + \frac{\partial H}{\partial q} = -\frac{\partial A_x}{\partial x}\left(\frac{dx}{dz} - \frac{u_x}{u_z}\right) + u_z\left(\frac{dy}{dz} - \frac{u_y}{u_z}\right), \tag{8a}$$

$$\frac{dq}{dt} - \frac{\partial H}{\partial p} = \frac{dx}{dz} - \frac{u_x}{u_z}. \tag{8b}$$

If **u** vanishes only at isolated points or lines then the equations (8) imply a canonical form of the equations of streamlines of $\mathbf{u}(\mathbf{x})$*;

$$\begin{aligned}\frac{dq}{dt} &= \frac{\partial H}{\partial p} \\ \frac{dp}{dt} &= -\frac{\partial H}{\partial q}\end{aligned} \quad \text{if and only if} \quad \begin{aligned}\frac{dx}{dz} &= \frac{u_x}{u_z} \\ \frac{dy}{dz} &= \frac{u_y}{u_z}.\end{aligned} \tag{9}$$

The Hamiltonian $H(q, p, t)$ is, in general, explicitly dependent on the canonical time t. Hence the streamlines of a 3D steady flow correspond to what is sometimes called a $1\frac{1}{2}$ d.o.f. Hamiltonian.†

Formal equivalence between the equations of fluid kinematics and those of Hamiltonian mechanics of a material point is of great conceptual importance. Just as in classical mechanics the theory can well be developed without the canonical formalism, but the Hamiltonian formulation provides a framework and enables the interpretation of the analytical results in terms of symplectic geometry. We may expect similar benefits in flow kinematics. There are, however, some important complications.

Firstly, the relation (6b) between p and y may not be uniquely invertible. The necessary condition for invertibility is

$$\frac{\partial p}{\partial y} = -\frac{\partial A_x}{\partial y} = u_z \neq 0. \tag{10}$$

This condition is never satisfied at the stagnation points of the flow, and the Hamiltonian is often singular there. Hence the 3D flows correspond to the systems which are locally Hamiltonian, but due to the singularities may, perhaps, have a much richer global structure than those studied in classical mechanics.

Secondly, even where the relation (6b) is invertible, it cannot always be explicitly inverted. This is not a fundamental difficulty, but it limits the use of the canonical formalism for practical calculations. Fortunately this is not always the case.

2.1. *Flows corresponding to given Hamiltonians*

Specifying a time-dependent Hamiltonian $H(q, p, t)$ uniquely determines the phase-space dynamics of a one-dimensional system. Such a system corresponds to the entire family of three-dimensional fluid flows. In order to select one particular flow from the family we have to choose, in an arbitrary way, the function $A_x(x, y, z)$ which, through the relations (6) and (5), defines all velocity components.

Let us consider the Hamiltonian of a partice of unit mass in a potential well:

$$H = \tfrac{1}{2}p^2 + V(q). \tag{11}$$

The corresponding family of flows has the form

$$u_x = A_x \frac{\partial A_x}{\partial y}, \tag{12a}$$

*We shall see later that the above analysis can be generalized to any orthonormal co-ordinates (x_1, x_2, x_3). We can choose them in such a way that $u_{x_3} \neq 0$ except at the stagnation points of $\mathbf{u}(\mathbf{x})$.

†A one d.o.f. time-dependent Hamiltonian $H(q, p, t)$ in the phase space (q, p) can be represented as a two d.o.f. system in the extended phase space (q, p, t, H). Such special kind of systems are sometimes called $1\frac{1}{2}$ d.o.f. systems. Time always increases along their trajectories, so they cannot, for example, have closed orbits in the extended phase-space.

$$u_y = \frac{\partial A_x}{\partial z} - A_x\frac{\partial A_x}{\partial x} - \frac{\partial V(x)}{\partial x}, \tag{12b}$$

$$u_z = -\frac{\partial A_x}{\partial y}, \tag{12c}$$

where $A_x = A_x(x, y, z)$ is an arbitrary function. The Hamiltonian is autonomous, integrable and hence the flow (12) has integrable streamlines for every function $A_x(x, y, z)$. In general we have

$$\frac{dA_z}{dt} = \frac{\partial H}{\partial t}u_z. \tag{13}$$

If the Hamiltonian is autonomous then the corresponding 3D flows are integrable as they have an invariant $A_z(x, y, z)$ defined by (6c). For example, the streamlines of the flow (12) lie on the surfaces $\frac{1}{2}A_x^2 + V(x) = \text{const}$.

When the Hamiltonian is non-integrable some of the flows in the corresponding family are always integrable. The choice

$$A_x(x, y, z) = f(x, z) \tag{14}$$

gives, for example, a one-dimensional flow

$$u_x = 0, \qquad u_y = u_y(x, z), \qquad u_z = 0. \tag{15}$$

An interesting question arises: what are the necessary conditions on $A_x(x, y, z)$ which guarantee 'faithful' mapping between the orbits in the phase-space and the streamlines of the corresponding three-dimensional flow?

2.2. *Two-dimensional flows*

Let us consider a time-dependent 2D flow defined by stream function $\Psi(x, y, t)$:

$$u_x = \frac{\partial \Psi(x, y, t)}{\partial y}, \qquad u_y = -\frac{\partial \Psi(x, y, t)}{\partial x}. \tag{16}$$

Identifying time with the third spatial coordinate we obtain an equivalent steady 3D flow

$$u_x = \frac{\partial \Psi(x, y, z)}{\partial y}, \qquad u_y = -\frac{\partial \Psi(x, y, z)}{\partial x}, \qquad u_z = 1. \tag{17}$$

The vector potential in the canonical gauge has the form

$$\mathbf{A} = (-y, 0, \Psi(x, y, z)). \tag{18}$$

Following the procedure for 3D flows we obtain from (6):

$$q = x, \qquad p = y, \qquad t = z, \qquad H(q, p, t) = \Psi(x, y, z). \tag{19}$$

This means that for 2D flows the Hamiltonian formalism given by (6) reduces to a standard one.

The canonical formalism for 3D flows enables us, in principle, to associate a 2D time-dependent flow with every 3D steady flow by calculating the Hamiltonian of the latter and regarding it as a stream function of the former. Potentially this could be used to simplify numerical computations. However, we need to know to what extent the structure of a 3D flow is represented in the corresponding 2D flow. This remains an open problem.

2.3. *Curvilinear co-ordinates*

The formalism can be generalized to arbitrary orthonormal co-ordinates

$$\mathrm{d}\mathbf{x} = \hat{\mathbf{i}}_1 h_1\,\mathrm{d}x_1 + \hat{\mathbf{i}}_2 h_2\,\mathrm{d}x_2 + \hat{\mathbf{i}}_3 h_3\,\mathrm{d}x_3. \tag{20}$$

The equations of streamlines now take the form

$$\frac{\mathrm{d}x_i}{\mathrm{d}t} = \frac{1}{h_i} u_i(x_1, x_2, x_3), \qquad i = 1, 2, 3, \tag{21}$$

where

$$u_1 = \frac{1}{h_2 h_3}\frac{\partial(h_3 A_3)}{\partial x_2}, \qquad u_2 = \frac{1}{h_1 h_3}\left[\frac{\partial(h_1 A_1)}{\partial x_3} - \frac{\partial(h_3 A_3)}{\partial x_1}\right],$$
$$u_3 = -\frac{1}{h_1 h_2}\frac{\partial(h_1 A_1)}{\partial x_2}, \qquad \mathbf{A} = (A_1, 0, A_3). \tag{22}$$

The analogue of (6) is:

$$t = x_3, \qquad q = x_1, \tag{23a}$$

$$p = -h_1 A_1(x_1, x_2, x_3), \tag{23b}$$

$$H(q, p, t) = h_3 A_3(x_1, x_2(q, p, t), x_3); \tag{23c}$$

and we obtain

$$\frac{\mathrm{d}p}{\mathrm{d}t} + \frac{\partial H}{\partial q} = h_1 h_2 h_3\left(\frac{\mathrm{d}x_2}{\mathrm{d}x_3} - \frac{h_3}{h_2}\frac{u_2}{u_3}\right) - \frac{\partial(h_1 A_1)}{\partial x_1}\left(\frac{\mathrm{d}x_1}{\mathrm{d}x_3} - \frac{h_3 u_1}{h_1 u_3}\right), \tag{24a}$$

$$\frac{\mathrm{d}q}{\mathrm{d}t} - \frac{\partial H}{\partial p} = \frac{\mathrm{d}x_1}{\mathrm{d}x_3} - \frac{h_3 u_1}{h_1 u_3}. \tag{24b}$$

The formula (24) proves that the equations of streamlines written in curvilinear coordinates can be cast in a canonical form.

We shall now demonstrate how the canonical formulation can help the analysis of perturbations of the flows with closed streamlines.

3. SMALL PERTURBATIONS OF FLOWS WITH CLOSED STREAMLINES

The Hamiltonian formalism outlined in the previous section can be successfully used for practical calculations. The most tractable are the systems which are, in some sense, close to being integrable. The canonical formulation of flow kinematics enables us to study the structure of streamlines of the 3D steady flow in terms of the perturbation theory of integrable Hamiltonian systems with two d.o.f.

The steady 3D flows with integrable streamlines belong to two categories.* Typically the streamlines lie on a family of nested closed surfaces which are topologically equivalent to tori. We shall call them stream surfaces. The equations of streamlines have, in this case, one invariant, say $I(\mathbf{x})$, and the stream surfaces are the level surfaces of this invariant: $I(\mathbf{x}) = \text{const}$. A winding number $W(I)$ can be ascribed to each stream surface. The surfaces with irrational W are densely covered by quasi-periodic streamlines while the streamlines on the rational surfaces are periodic.

*Clearly the same flow may belong to different categories in different regions of physical space but here we disregard such possibilities.

If $W(I)$ is a continuous function of I then small perturbations* of such a velocity field can be described by the KAM theory. In such generic cases the streamlines are labelled by two numbers. One is the value of $I(\mathbf{x})$ which identifies a stream surface, and the other is an angle which identifies the streamline on the stream surface. Let $\mathbf{x}(t, \mathbf{x}_0)$ be the position of a fluid particle which started from $\mathbf{x}_0$ at $t = 0$. According to the KAM theorem there is a dense set of closed stream surfaces in the perturbed flow which means that $I(\mathbf{x}(t, \mathbf{x}_0))$ remains, for all times, close to its original value $I(\mathbf{x}_0)$ even if $\mathbf{x}_0$ is on a chaotic streamline. If there are any particles of a passive scalar carried by the fluid then they will also stay roughly where they were in the absence of perturbation.

Here we are more interested in a non-generic case when the equations (1) have two invariants $I(\mathbf{x})$ and $J(\mathbf{x})$ which are independent, i.e. their level surfaces intersect transversally. The streamlines are the intersections of the surfaces $I(\mathbf{x}) = \text{const}$ and $J(\mathbf{x}) = \text{const}$. Hence most of them are closed loops.† There are no uniquely defined stream surfaces which would preserve their identity under small perturbations. Every surface $F(I(\mathbf{x}), J(\mathbf{x})) = \text{const}$, where F is an arbitrary continuous function of its arguments, is made of streamlines.

The effect of small perturbations on a closed streamline flow is best explained in terms of Poincaré maps. The Poincaré sections of the individual streamlines of an unperturbed flow consist of isolated points. Every point p in the plane of section $\mathcal{P}$ is uniquely labelled by the values of I and J and is a fixed point of the map M_0: $p \to M_0(p)$ which is trival: $M_0(p) = p$. Introducing a perturbation is equivalent to *adding* a map of 'small magnitude'‡, that is,

$$M: p \to M(p) = M_0(p) + M_\varepsilon(p) = p + M_\varepsilon(p). \tag{25}$$

When ε is small a typical point in the Poincare section drifts slowly under the action of the map M_ε outlining a curve C in $\mathcal{P}$ which, in general, depends on the perturbation. The curve C is an intersection of $\mathcal{P}$ with one of the surfaces $G(\mathbf{x}) = F(I(\mathbf{x}), J(\mathbf{x})) = \text{const}$ for some function $F(\cdot, \cdot)$ which is thereby defined by the perturbation $\varepsilon\mathbf{u}(\mathbf{x})$ (see Fig. 1).

The level surfaces of $G(\mathbf{x})$ are the stream surfaces of the perturbed flow and therefore $G(\mathbf{x})$ is invariant along the perturbed streamline. The perturbation destroys two invariants

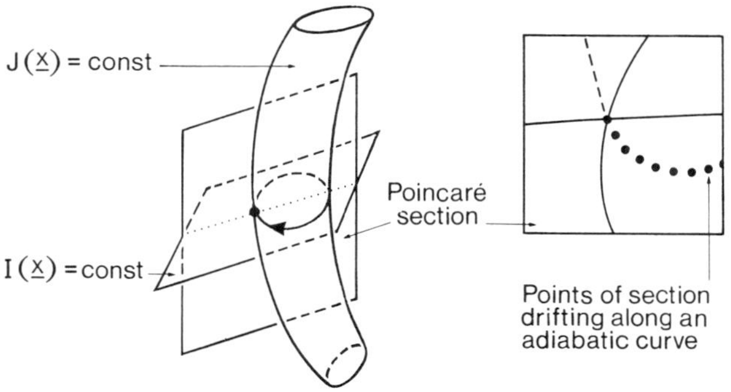

Fig. 1. The schematic view of the Poincaré section of a perturbed closed streamline flow.

*By small perturbation we mean adding an arbitrary non-singular solenoidal field of small magnitude.

†We are concerned with flows in a bounded domain. In such confined flows the streamlines which are not closed loops must end in stagnation points.

‡We identify points in the plane of section $\mathcal{P}$ with their position vectors relative to an arbitrary origin. We would say the map has 'small magnitude' when $|M_\varepsilon(p) - p|/L \ll 1$ where L is the total arc-length of the closed streamline to which p belongs.

$I(\mathbf{x})$ and $J(\mathbf{x})$, now they slowly change along streamlines, and creates a new one. This new constant of the fluid particle motion is called adiabatic invariant of the perturbed flow. Thus, the degeneracy is removed, the stream surfaces are introduced and the flow falls into the category mentioned earlier in this section.

The scenario for the perturbation of a closed streamline flow which was outlined above hinges on the assumption of 'small magnitude' of the map M_ε which is equivalent to saying that $\varepsilon\mathbf{u}_1(\mathbf{x})$ is small compared with $\mathbf{u}_0(\mathbf{x})$. However, in the neighbourhood $O(\mathbf{x}_s)$ of a stagnation point $\mathbf{x}_s$ of $\mathbf{u}_0$ this assumption may be false, no matter how small ε is. This will be the case when

$$\frac{|\mathbf{u}_1(\mathbf{x})|}{|\mathbf{u}_0(\mathbf{x})|} \underset{\mathbf{x}\to\mathbf{x}_s}{\longrightarrow} \infty. \tag{26}$$

Inside such neighbourhoods the level surfaces of $G(\mathbf{x})$ *are not* the stream surfaces of $\mathbf{u}_0 + \varepsilon\mathbf{u}_1$ and the passage of a fluid particle through $O(\mathbf{x}_s)$ may result in a jump in the value of $G(\mathbf{x})$. The existence of such 'holes' in physical space where fluid particles may jump between different surfaces $G(\mathbf{x}) = \text{const}$ has important topological implications. In consequence the global topology of the adiabatic surfaces, in the limit of $\varepsilon \to 0$, may be very different from that of ordinary stream surfaces. In Section 5 we shall see, for example, how all surfaces may have a common point of tangential contact.

3.1. *Averaging*

The mathematical tool for calculating the stream surfaces of a perturbed degenerate flow is called the 'principle of averaging'. The history of the problem is long (refs [9], Section 52; [10], p. 167) but it still inspires both mathematicians and physicists (see, for example, ref. [11]).

A general description of the principle of averaging was given by V. I. Arnold (ref. [9], Section 52.2) who writes:

> We note that this principle is neither a theorem, an axiom, nor a definition, but rather a physical proposition, i.e. a vaguely formulated and, strictly speaking, untrue assertion. Such assertions are often fruitful sources of mathematical theorems.

The closed streamline flows are, however, single-frequency systems for which the averaging principle can be rigorously proved (ref. [9], Section 52.3). The relevant conclusion is the following.

Consider the following system of equations:

$$\dot{\phi} = \omega(I, J) + \varepsilon f(I, J, \phi), \tag{27a}$$

$$\dot{I} = \varepsilon g_1(I, J, \phi), \tag{27b}$$

$$\dot{J} = \varepsilon g_2(I, J, \phi), \tag{27c}$$

where f, g_1, g_2 are 2π-periodic in ϕ and $(I, J) \in G \subset \Re^2$. Consider also the averaged system

$$\dot{\mathcal{I}} = \varepsilon\overline{g_1}(\mathcal{I}, \mathcal{J}), \tag{28a}$$

$$\dot{\mathcal{J}} = \varepsilon\overline{g_2}(\mathcal{I}, \mathcal{J}), \tag{28b}$$

where $\overline{g_i}(\mathcal{I}, \mathcal{J}) = (1/2\pi)\int_0^{2\pi} g_i(\mathcal{I}, \mathcal{J}, \phi)\,d\phi$. Suppose the functions ω, g_1, g_2 together with their first and second derivatives are bounded in G and ω is bounded away from zero.

Then, for $0 \leq t \leq \varepsilon^{-1}$, the solution of (28), provided it stays in G, well approximates the solution of (27) with the same intial condition, i.e.

$$[(I(t) - \mathcal{I}(t))^2 + (J(t) - \mathcal{J}(t))^2]^{1/2} \leq C\varepsilon, \tag{29}$$

where C is a constant independent of ε. Applying the method of averaging to a perturbed closed streamline flow $\mathbf{u}_0 + \varepsilon\mathbf{u}_1$ we identify I, J with the two invariants labelling the closed streamlines of $\mathbf{u}_0(\mathbf{x})$ and ϕ some angle coordinate along the loops. The functions f, g_1, g_2 can, in principle, be calculated for any given perturbation $\varepsilon\mathbf{u}_1(\mathbf{x})$. We can solve the averaged system (28) and write its integral curves in the form

$$\mathcal{G}(\mathcal{I}, \mathcal{J}) = \text{const.} \tag{30}$$

Then the function $G(\mathbf{x}) = \mathcal{G}(I(\mathbf{x}), J(\mathbf{x}))$ is an adiabatic invariant of the perturbed flow.

The non-canonical averaging procedure described above can be relatively easily applied to any system in which we can identify 'slow variables' (here I and J) and a 'fast variable' (here ϕ). For an explicitly Hamiltonian system such distinction is achieved by writing the Hamiltonian in action–angle variables (A, θ). The Hamiltonians corresponding to weakly perturbed closed streamline flows have the general form

$$H = H_0(A) + \varepsilon H_1(A, \theta, t). \tag{31}$$

For $\varepsilon = 0$ the unperturbed Hamiltonian H_0 is, obviously, a constant of motion. If H_1 is a periodic function of time, then the perturbed system has a new constant of motion which, to first order in ε, is equal to

$$\bar{H} = H_0 + \varepsilon\langle H_1 \rangle, \tag{32}$$

where the average $\langle \cdot \rangle$ is taken over both θ and t oscillations. (For a guide to the impressive theory behind this deceptively simple statement see, for example, ref. [12], especially Section 2.2b.)

A concise review of the averaging methods for both Hamiltonian and non-Hamiltonian systems is presented in ref. [11].

4. GENERAL QUADRATIC FLOW IN A SPHERE IN THE LIMIT OF RIGID ROTATION

The method of averaging is, perhaps, best demonstrated when we consider small perturbations of flows with circular streamlines. The angle co-ordinate parameterizing the circles is a natural 'fast' variable and hence a good candidate for canonical time.

We consider a fluid occupying a spherical domain of unit radius. The fluid is in motion which is predominantly a rigid rotation $\mathbf{u}_0$ around the y-axis passing through the centre. The rotation is slightly perturbed by an additional velocity field $\varepsilon\mathbf{u}_1(\mathbf{x})$, where $\mathbf{u}_1(\mathbf{x})$ is the general quadratic, incompressible flow in a sphere, i.e. $u_i(x, y, z)$, $i = 1, 2, 3$ are second-order polynomials in Cartesian coordinates and the conditions:

$$\nabla \cdot \mathbf{u}_1(\mathbf{x}) = 0, \qquad \mathbf{n} \cdot \mathbf{u}_1(\mathbf{x})|_{\mathbf{x}\cdot\mathbf{x}=1} = 0 \tag{33}$$

are satisfied. The rigid rotation $\mathbf{u}_0$ is, obviously, linear. Hence, for small ε, the flow $\mathbf{u}(\mathbf{x}) = \mathbf{u}_0(\mathbf{x}) + \varepsilon\mathbf{u}_1(\mathbf{x})$ is the most general quadratic, incompressible flow in a sphere in the limit of rigid rotation. Clearly the unit sphere is a material surface for all times and all values of ε.

Such a flow is, in general, the sum of rigid rotation with constant angular velocity $\mathbf{\Omega} + \varepsilon\boldsymbol{\omega}$, the Hill's vortex with the axis of symmetry in the direction of an arbitrary vector $\varepsilon\mathbf{a}$, and a combination of twists defined by a symmetric matrix of constant coefficients T_{ij} [7]:

$$\mathbf{u}^Q(\mathbf{x}) = \mathbf{u}_0(\mathbf{x}) + \varepsilon\mathbf{u}_1(\mathbf{x}), \tag{34a}$$

$$\mathbf{u}_0(\mathbf{x}) = \mathbf{\Omega} \wedge \mathbf{x}, \tag{34b}$$

$$\mathbf{u}_1(\mathbf{x}) = \boldsymbol{\omega} \wedge \mathbf{x} + [(1 - 2r^2)\mathbf{a} + (\mathbf{a}\cdot\mathbf{x})\mathbf{x}] - \mathbf{x} \wedge \nabla(\mathbf{x}\hat{T}\mathbf{x}), \tag{34c}$$

where $r^2 = \mathbf{x}\cdot\mathbf{x}$, $(\mathbf{x}\hat{T}\mathbf{x}) = \sum_{i,j=1}^{3} x_i T_{ij} x_j$ and $\mathbf{\Omega}$, $\boldsymbol{\omega}$, $\mathbf{a}$ are arbitrary constant vectors.

The flow (34) is a Stokes flow and, as such, is the unique flow of a highly viscous fluid determined by the velocity field at the boundary. For $\varepsilon > 0$ fluid particles on the unit sphere move relative to each other; hence (34) is a flow in a spherical droplet with free surface. The real low Reynolds number flows in droplets are more likely to be cubic functions of Cartesian co-ordinates [13, 14]. The relative simplicity of the general quadratic flow (34) makes it a particularly good example of an application of the canonical formalism, but the analysis can, in principle, be extended to other types of perturbations of the rigid rotation.

We choose cylindrical polar co-ordinates (ρ, y, ϕ) with the y-axis in the direction of $\mathbf{\Omega}$:

$$x = \rho \sin\phi, \qquad z = \rho\cos\phi, \tag{35a}$$

$$\hat{\mathbf{i}}_\rho = \hat{\mathbf{i}}_x \sin\phi + \hat{\mathbf{i}}_z \cos\phi, \; \hat{\mathbf{i}}_\phi = \hat{\mathbf{i}}_x \cos\phi - \hat{\mathbf{i}}_z \sin\phi, \tag{35b}$$

and find the Hamiltonian of the flow (34) following the procedure described in Section 2.

The vector potential for the flow (34) has the form

$$\mathbf{A}^Q(\mathbf{x}) = (\mathbf{\Omega}\cdot\mathbf{x})\mathbf{x} + \varepsilon(\boldsymbol{\omega}\cdot\mathbf{x})\mathbf{x} + \varepsilon\tfrac{1}{2}(1 - r^2)(\mathbf{a} \wedge \mathbf{x}) + \varepsilon(\mathbf{x}\hat{T}\mathbf{x})\mathbf{x} + \nabla(\Phi_0 + \varepsilon\Phi_1). \tag{36}$$

We want $\mathbf{A}^Q(\mathbf{x})$ in the canonical gauge, $\mathbf{A}^Q\cdot\mathbf{\Omega} = 0$, and this ensured by the following choice of Φ:

$$\begin{aligned}\Phi_0 &= -\tfrac{1}{3}\Omega^{-2}(\mathbf{\Omega}\cdot\mathbf{x})^3 = -\tfrac{1}{3}\Omega y,\\ \Phi_1 &= -\tfrac{1}{3}\omega^{-2}((\boldsymbol{\omega}\cdot\mathbf{x})^3 - \tfrac{1}{2}[(\hat{\mathbf{\Omega}} \wedge \mathbf{a})\cdot\mathbf{x}](\hat{\mathbf{\Omega}}\cdot\mathbf{x})[1 + \tfrac{2}{3}(\hat{\mathbf{\Omega}}\cdot\mathbf{x})^2 - r^2]\\ &\quad -(\hat{\mathbf{\Omega}}\cdot\mathbf{x})^2)[\tfrac{1}{2}(\mathbf{x}\hat{T}\mathbf{x}) - \tfrac{1}{3}(\hat{\mathbf{\Omega}}\cdot\mathbf{x})(\hat{\mathbf{\Omega}}\hat{T}\mathbf{x}) + \tfrac{1}{12}(\hat{\mathbf{\Omega}}\cdot\mathbf{x})^2(\hat{\mathbf{\Omega}}\hat{T}\hat{\mathbf{\Omega}})],\end{aligned} \tag{37}$$

where $\hat{\mathbf{\Omega}}$ is the unit vector in the direction of $\mathbf{\Omega}$. From (23) we obtain:

$$t = \phi, \qquad q = \rho, \tag{38a}$$

$$\begin{aligned}p &= -A^Q_\rho \\ &= -\rho(\mathbf{\Omega}\cdot\mathbf{x}) - \varepsilon[\rho(\boldsymbol{\omega}\cdot\mathbf{x}) + \tfrac{1}{2}(1 - r^2)y(a_x\cos\phi - a_z\sin\phi) + \rho(\mathbf{x}\hat{T}\mathbf{x})] + \varepsilon\frac{\partial\Phi_1}{\partial\rho},\end{aligned} \tag{38b}$$

$$H = \rho A^Q_\phi = \varepsilon\tfrac{1}{2}(1 - r^2)\rho[\rho a_y - y(a_x\sin\phi + a_z\cos\phi)] + \varepsilon\frac{\partial\Phi_1}{\partial\phi}, \tag{38c}$$

where a_x, a_y, a_z are Cartesian components of $\mathbf{a}$.

To the lowest order in ε the relation (38c) is invertible:

$$y = -\frac{p}{q\Omega} + O(\varepsilon) \tag{39}$$

$$\begin{aligned}H(q, p, t) &= \varepsilon H_1(q, p, t) + O(\varepsilon^2)\\ &= \varepsilon\tfrac{1}{2}\left(1 - q^2 - \frac{p^2}{q^2\Omega^2}\right)\left[q^2 a_y + \frac{p}{\Omega}(a_x\sin t + a_z\cos t)\right] + \varepsilon\frac{\partial\Phi_1}{\partial\phi} + O(\varepsilon^2).\end{aligned} \tag{40}$$

As was explained in the previous section, an adiabatic invariant of the perturbed flow is obtained directly by averaging the Hamiltonian. Here the situation is somewhat special because $H_0 \equiv 0$ (cf. equation (31)), i.e. both q and p are slow variables. It follows that we can take a temporal average of the perturbation Hamiltonian $H_1(q, p, t)$ directly as a function of canonical co-ordinates and no action–angle variables are needed:

$$\bar{H} = \varepsilon\langle H_1\rangle + O(\varepsilon^2) = \varepsilon\tfrac{1}{2}\left(1 - q^2 - \frac{p^2}{q^2\Omega^2}\right)qa_y + \varepsilon\frac{1}{q}\left\langle\frac{\partial\Phi_1}{\partial\phi}\right\rangle + O(\varepsilon^2). \tag{41}$$

In order to calculate $\partial\Phi_1/\partial\phi$ we use the following simple rules:

$$\frac{\partial \mathbf{x}}{\partial\phi} = \hat{\mathbf{\Omega}} \wedge \mathbf{x}, \qquad \frac{\partial}{\partial\phi}(\hat{\mathbf{\Omega}}\cdot\mathbf{x}) = 0, \qquad \frac{\partial}{\partial\phi}(\boldsymbol{\omega}\cdot\mathbf{x}) = \mathbf{x}\cdot(\mathbf{\Omega}\wedge\mathbf{x}),$$
$$\frac{\partial}{\partial\phi}[(\hat{\mathbf{\Omega}}\wedge\mathbf{a})\cdot\mathbf{x}] = \mathbf{a}\cdot\mathbf{x} - (\hat{\mathbf{\Omega}}\cdot\mathbf{x})(\hat{\mathbf{\Omega}}\cdot\mathbf{a}), \qquad \frac{\partial}{\partial\phi}(\mathbf{x}(\hat{T}\mathbf{x}) = 2[(\hat{\mathbf{\Omega}}\wedge\mathbf{x})\hat{T}\mathbf{x}], \tag{42}$$

and obtain

$$\frac{\partial\Phi_1}{\partial\phi} = -\tfrac{1}{2}(\hat{\mathbf{\Omega}}\cdot\mathbf{x})[1 + \tfrac{2}{3}(\hat{\mathbf{\Omega}}\cdot\mathbf{x})^2 - r^2][\mathbf{a}\cdot\mathbf{x} - (\hat{\mathbf{\Omega}}\cdot\mathbf{x})(\hat{\mathbf{\Omega}}\cdot\mathbf{a})]$$
$$- (\hat{\boldsymbol{\omega}}\cdot\mathbf{x})^2[\boldsymbol{\omega}\cdot(\hat{\Omega}\wedge\mathbf{x})] - (\hat{\Omega}\cdot\mathbf{x})^2[(\hat{\mathbf{\Omega}}\wedge\mathbf{x})\hat{T}\mathbf{x}] = \tfrac{1}{3}(\hat{\mathbf{\Omega}}\cdot\mathbf{x})^3[\hat{\mathbf{\Omega}}\hat{T}(\hat{\mathbf{\Omega}}\wedge\mathbf{x})], \tag{43}$$

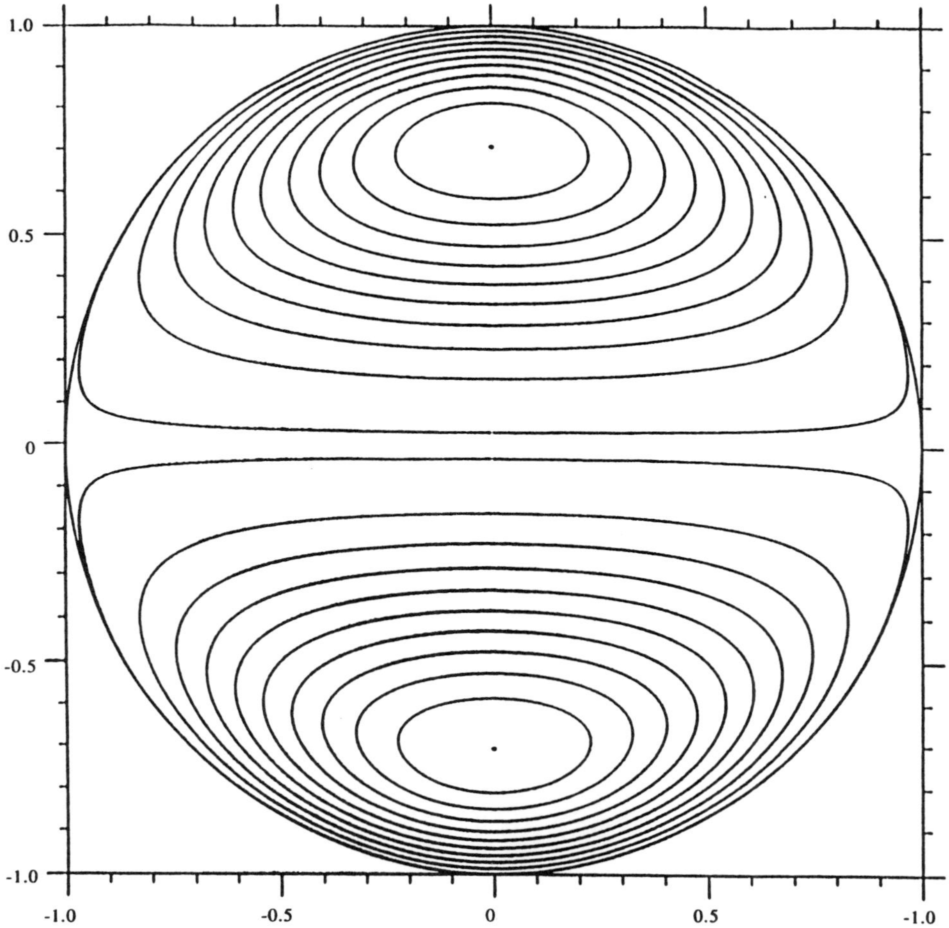

Fig. 2. The level surfaces of the adiabatic invariant $G(\mathbf{x}) = \rho^2(1 - \rho^2 - y^2)$ (equation 46).

where $\hat{\boldsymbol{\omega}}$ is the unit vector along $\boldsymbol{\omega}$. By a straightforward calculation we find that the time average, which is the same as the average over ϕ, vanishes for each term of (43) separately. Hence we obtain:

$$\left\langle \frac{\partial \Phi_1}{\partial \phi} \right\rangle = 0, \tag{44}$$

and the adiabatic invariant for the flow (34) has the form

$$\bar{H} = \tfrac{1}{2}\varepsilon a_y q^2\left(1 - q^2 - \left(\frac{p}{q\Omega}\right)^2\right) + O(\varepsilon^2). \tag{45}$$

Hence, for small ε the streamlines of the quadratic flow (34) lie on a family of nested toroidal surfaces (Fig. 2) given by

$$G(\mathbf{x}) = \rho^2(1 - \rho^2 - y^2) = \text{const.} \tag{46}$$

This is confirmed by computing the Poincaré sections of the flow (34) (Fig. 3).

When the axis of the Hill's vortex component of the general quadratic perturbation is perpendicular to the rotation axis, i.e. $\mathbf{a} \cdot \boldsymbol{\Omega} = 0$, the $O(\varepsilon)$ term in $\bar{H}$ vanishes identically. Then the adiabatic drift is 'super-slow'. In such special cases it is necessary to calculate the

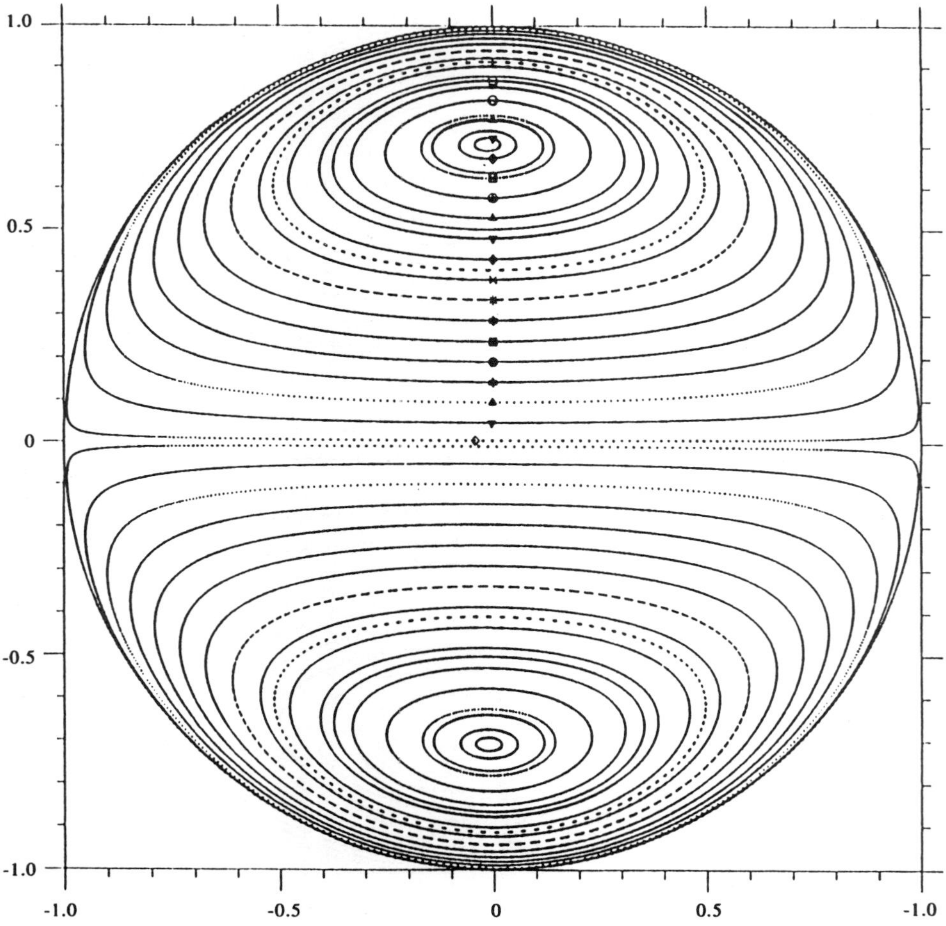

Fig. 3. The Poincaré section of the flow (34) with $\varepsilon = 1/700$ (from ref. [7]).

$O(\varepsilon^2)$ term in $\bar{H}$ in order to find the shape of the stream surfaces created by the perturbation.

5. NON-CANONICAL AVERAGING

In the previous section we have shown how useful the canonical formalism can be for analysing small perturbations of rigid rotation. The cylindrical polar co-ordinates are, in this case, natural slow (ρ, y) and fast (ϕ) variables. In a general case of an arbitrary closed streamline flow we need to write the Hamiltonian in the action–angle variables (cf. equations 31, 32), which may be difficult in practice. In such a case we can employ a non-canonical averaging procedure (equations 27, 28).

We shall now consider one particular quadratic closed streamline flow $\mathbf{u}^C(\mathbf{x})$ in a sphere, which is perturbed by adding a slow rigid rotation $\varepsilon\mathbf{u}^R(\mathbf{x})$:

$$\mathbf{u}^{STF}(\mathbf{x}) = \mathbf{u}^C(\mathbf{x}) + \varepsilon\mathbf{u}^R(\mathbf{x}), \tag{47a}$$

$$\mathbf{u}^C(\mathbf{x}) = (-8xy,\ 11x^2 + 3y^2 + z^2 - 3,\ 2yz), \tag{47b}$$

$$\mathbf{u}^R(\mathbf{x}) = (z,\ 0,\ -x). \tag{47c}$$

The flow (47) is a member of a one-parameter family of flows proposed and studied by Bajer and Moffatt as prototypes of the stretch-twist-fold (STF) mechanism of the magnetic field generation [15, 16, 18].

The structure of the STF flow with small ε is best obtained by non-canonical averaging, but the Hamiltonian form of the equations of streamlines still proves very useful. It turns out that the general solution of the canonical equations for the unperturbed flow can be found analytically. In doing so we obtain two constants of integration which make the two invariants I and J required for non-canonical averaging.

The flow $\mathbf{u}^C(\mathbf{x})$ has a vector potential which, in the canonical gauge $A_y \equiv 0$, takes the form:

$$\mathbf{A}^C(\mathbf{x}) = (-y^2z,\ 0,\ -\tfrac{11}{3}x^3 - 4xy^2 - xz^2 + 3x). \tag{48}$$

The relation (6b) is easily invertible and we obtain an explicit expression for the Hamiltonian:

$$\begin{gathered} t = z, \qquad q = x, \qquad p = y^2z, \\ H(q, p, t) = -4pqt^{-1} - \tfrac{11}{3}q^3 - qt^2 + 3q. \end{gathered} \tag{49}$$

The equations of streamlines of $\mathbf{u}^C(\mathbf{x})$ are:

$$\frac{dq}{dt} = -4qt^{-1}, \tag{50a}$$

$$\frac{dp}{dt} = 4pt^{-1} + 11q^2 + t^2 - 3. \tag{50b}$$

They can be easily solved:

$$q = It^{-4}, \qquad p = -I^2t^{-7} - t^3 + t + Jt^4, \tag{51}$$

where I, J are the constants of integration. Inverting the relations (51) we obtain:

$$I = qt^4, \qquad J = pt^{-4} + q^2t^{-3} + t^{-1} - t^{-3}. \tag{52}$$

Writing I and J in Cartesian co-ordinates we obtain the invariants of the unperturbed flow:

$$\mathbf{I}(\mathbf{x}) = xz^4, \tag{53a}$$

$$\mathbf{J}(\mathbf{x}) = z^{-3}(x^2 + y^2 + z^2 - 1). \tag{53b}$$

These two invariants are constants of motion of a fluid particle in the flow $\mathbf{u}^C(\mathbf{x})$. In the perturbed flow $\mathbf{u}^{STF}(\mathbf{x})$ they slowly change with the rate of change proportional to ε:

$$\dot{\mathbf{I}} = \mathbf{u}^{STF} \cdot \nabla I = \varepsilon I(w - 4w^{-1}), \tag{54a}$$

$$\dot{\mathbf{J}} = \mathbf{u}^{STF} \cdot \nabla J = 3\varepsilon w^{-1}, \tag{54b}$$

where $w = x^{-1}z$. The corresponding averaged system (equation 28) has the form

$$\dot{\mathcal{I}} = \varepsilon\mathcal{I}(\langle w^{-1}\rangle - 4\langle w\rangle), \tag{55a}$$

$$\dot{\mathcal{J}} = 3\varepsilon\langle w^{-1}\rangle, \tag{55b}$$

where $\langle \cdot \rangle$ is the temporal average over the streamlines of $\mathbf{u}^C(\mathbf{x})$ expressed in terms of I and J.

Every solution of the single equation

$$\frac{d\mathcal{I}}{d\mathcal{J}} = \tfrac{1}{3}\mathcal{I}\left(1 - 4\frac{\langle w\rangle}{\langle w^{-1}\rangle}\right) \tag{56}$$

yields a relation between $\mathcal{I}$ and $\mathcal{J}$ defining one level surface of the adiabatic invariant of $\mathbf{u}^{STF}(\mathbf{x})$. Hence the general solution of (56) describes a one-parameter family of adiabatic surfaces foliating the domain occupied by the fluid. Now we need to write the r.h.s. of (56) as a function of $\mathcal{I}$ and $\mathcal{J}$.

The unperturbed streamlines are symmetric with respect to $y \to -y$, so the average can be taken over the $y > 0$ part of every orbit. Changing the variable of integration we can write, for any function $f(\mathbf{x})$,

$$\langle f\rangle = \frac{\oint f\,dt}{\oint dt} = \frac{\displaystyle\int_{x_{min}}^{x_{max}} f(dx/dt)^{-1}\,dt}{\displaystyle\int_{x_{min}}^{x_{max}} (dx/dt)^{-1}\,dx}. \tag{57}$$

where $x_{min,max}$ are the minimal and maximal values of x on the orbit. Because of the $y \to -y$ symmetry the points where $x = x_{min,max}$ are those with $y = 0$, i.e. x_{min} and x_{max} are the roots of the equation:

$$y^2 = JI^{3/4}x^{-3/4} - x^2 - I^{1/2}x^{-1/2} + 1 = 0, \tag{58}$$

which is an 11th order polynomial in $s = x^{1/4}$. Changing the variable of integration again we obtain:

$$\frac{\langle w\rangle}{\langle w^{-1}\rangle} = I^{1/2}\frac{\displaystyle\int_{s_{min}}^{s_{max}} s^{-9/2}[P(I, J, s)]^{-1/2}\,ds}{\displaystyle\int_{s_{min}}^{s_{max}} s^{11/2}[P(I, J, s)]^{-1/2}\,ds}, \tag{59}$$

$$P(I, J, s) = s^{11} - s^3 + I^{1/2}s - JI^{3/4},$$

where $s_{min,max} = x_{min,max}^{1/4}$ are the roots of $P(I, J, s)$ satisfying $0 < s_{min} < s_{max} < 1$. It is now possible to integrate (56) numerically but the r.h.s. given by (59) has to be computed at every time-step which involves finding $s_{min,max}$ and evaluating the integrals.

In Fig. 4 we plot the adiabatic surfaces for the flow $\mathbf{u}^{STF}(\mathbf{x})$ and in Fig. 5 the Poincaré section of a single streamline with $\varepsilon = 0.01$. The flow $\mathbf{u}^R(\mathbf{x})$ is linear in (x, y, z) while $\mathbf{u}^C(\mathbf{x})$ is quadratic. Hence, near the stagnation points $\mathbf{x}_s^{\pm} = (0, \pm 1, 0)$ of $\mathbf{u}^{STF}(\mathbf{x})$ the

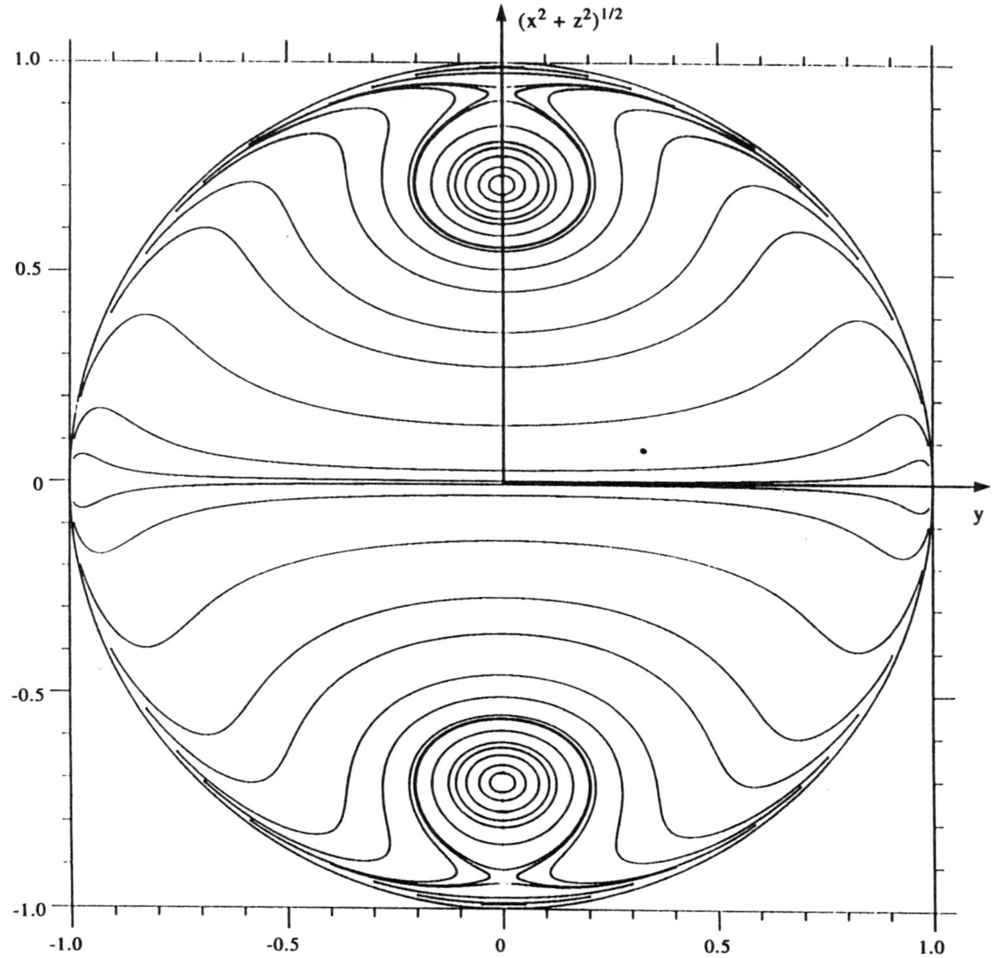

Fig. 4. The adiabatic surfaces of the flow $\mathbf{u}^{STF}(\mathbf{x})$, equation (47).

condition (26) is satisfied. Different adiabatic surfaces have common points of tangential contact at $x_s^{\pm}$. In Fig. 5 we see that a streamline jumps between different adiabatic surfaces as explained in Section 3. These jumps are called 'trans-adiabatic' drift [7].

The non-canonical averaging was recently applied by Stone and his colleagues to investigate the streamlines of a cubic flow inside a neutrally buoyant spherical droplet immersed in a general linear flow [14]. The method proved useful even for the flow which is far from being nearly periodic and has large chaotic regions. They used averaging to estimate numerically the size of the central regular island in the chaotic 'sea'. The same method was used to explain an interesting pattern of streamlines in a low Reynolds number flow between two concentric spheres steadily rotating around different axes [17].

6. CONCLUSIONS

We have shown how the equations of streamlines of a steady three-dimensional flow can be written in the form of a one degree-of-freedom non-autonomous Hamiltonian system. The correspondence is not unique. The entire family of flows parameterized by one scalar

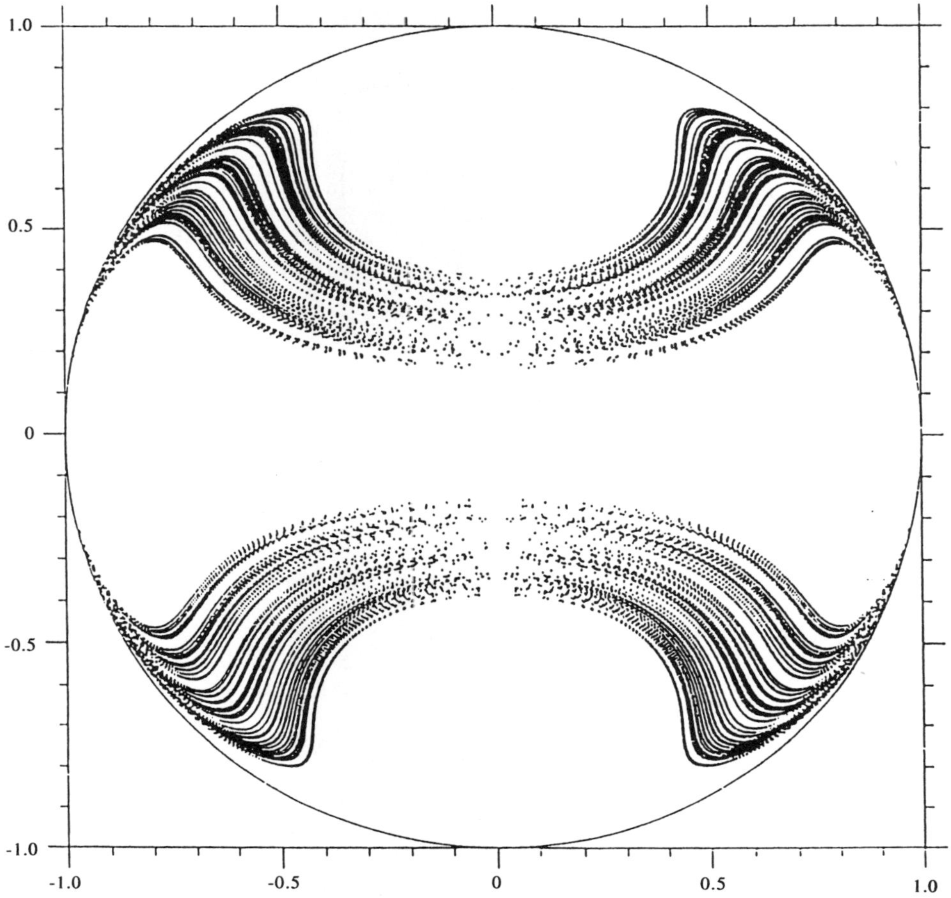

Fig. 5. The Poincaré section of the flow $\mathbf{u}^{STF}(\mathbf{x})$, equation (47) with $\varepsilon = 0.01$ (ref. [7]).

function of space corresponds to any given Hamiltonian. Also many different Hamiltonians can be assigned to the same flow. First, we have the freedom to choose the co-ordinates. Then, the canonical gauge is not unique. We can add to the vector potential a gradient of an arbitrary function $\Psi(x_1, x_3)$ without affecting the condition $A_2(x_1, x_2, x_3) = 0$. An interesting question arises: what is the mathematical structure of the family of Hamiltonians corresponding to a given steady flow?

A complementary problem is also of interest. The same Hamiltonian system has many different phase-space representations related by the canonical transformations. It would be interesting to know what these transformations mean in terms of the corresponding three-dimensional flows.

The canonical formalism enables us to associate a time-dependent 2D flow with every steady 3D flow. Sometimes this mapping is 'faithful', i.e. the structure of the latter is well represented in the former. In such case the computations of, say, transport of a passive scalar are greatly simplified [20].

The Hamiltonian formalism can also be applied to the instantaneous vorticity field $\boldsymbol{\omega}(\mathbf{x})$ of any fluid flow whether compressible or not. Now the velocity field $\mathbf{u}(\mathbf{x})$ is a vector potential of $\boldsymbol{\omega}(\mathbf{x})$ and taking the canonical gauge means adding a suitable potential flow to $\mathbf{u}(\mathbf{x})$, such that, $u_2(\mathbf{x}) = 0$. Then, the velocity component $u_3(\mathbf{x})$ is the Hamiltonian for the vorticity field.

The canonical formalism for the vortex lines helps to analyse their structure in the same way it does for streamlines. From the physical point of view there is, however, an important difference. The topological structure of vortex lines, unlike that of the streamlines, is conserved during the evolution governed by the Euler equation. Suppose at time $t = 0$ the vortex lines are 'nearly closed', i.e. $\boldsymbol{\omega}(\mathbf{x}, t = 0)$ is a slightly perturbed solenoidal field with all field-lines being closed loops. Then the 'flow' along vortex lines has an adiabatic invariant $G_0(\mathbf{x})$. If we are lucky then this invariant is, simply, equal to the averaged third component of the velocity in the canonical gauge. The structure of the vorticity field is invariant under the inviscid dynamics which, presumably, means that $\boldsymbol{\omega}(\mathbf{x}, t)$ must have an adiabatic invariant $G(\mathbf{x}, t)$ for every $t > 0$. Thus the Euler equations determine the time evolution of $G(\mathbf{x}, t)$ which is a single scalar function and yet contains most information about the topological structure of vortex lines.

Acknowledgements—I would like to thank Ben Mestel for drawing my attention to the paper by Cary and Littlejohn. This work was supported by the Committee for Scientific Research (KBN), grant No. 6-6317-91-02.

REFERENCES

1. D. W. Kerst, The influence of errors on plasma-confining magnetic fields, *Plasma Phys.* **4**, 253–262 (1962).
2. N. N. Filonienko, R. Z. Sagdeev and G. M. Zaslavsky, Destruction of magnetic surfaces by magnetic field irregularities, *Nucl. Fusion* **7**, 253–266 (1967).
3. K. J. Whiteman, Invariants and stability in classical mechanics, *Rep. Prog. Phys.* **40**, 1033–1069 (1977).
4. J. Cary and R. G. Littlejohn, Non-canonical Hamiltonian mechanics and its applications to magnetic field line flow, *Ann. Phys.* **151**, 1–34 (1983).
5. M. P. Bernardin and J. A. Tataronis, Hamiltonian approach to the existence of magnetic surfaces, *J. Math. Phys.* **26**, 2370–2380 (1985).
6. M. Kruskal, Asymptotic theory of Hamiltonian and other systems with all solutions nearly periodic, *J. Math. Phys.* **3**, 806–828 (1962).
7. K. Bajer and H. K. Moffatt, On a class of steady confined Stokes flows with chaotic streamlines, *J. Fluid Mech.* **212**, 337–363 (1990).
8. C. Itzykson and J.-B. Zuber, *Quantum Field Theory*. McGraw-Hill, New York (1980).
9. V. I. Arnold, *Mathematical Methods of Classical Mechanics*. Springer, Berlin (1978).
10. J. Guckenheimer and P. Holmes, *Nonlinear Oscillations, Dynamical Systems and Bifurcations of Vector Fields*. Springer, Berlin (1983).
11. P. Lochak and C. Meunier, *Multiphase Averaging for Classical Systems*. Springer, Berlin (1988).
12. A. J. Lichenberg and M. A. Lieberman, *Regular and Stochastic Motion*. Springer, Berlin (1983).
13. H. K. Moffatt, Electromagnetic stirring, *Phys. Fluids* **A2**, 1336–1343 (1991).
14. H. A. Stone, A. Nadim and S. H. Strogatz, Chaotic streamlines inside drops immersed in steady Stokes flows, *J. Fluid Mech.* **232**, 629–646 (1991).
15. H. K. Moffatt and M. R. E. Proctor, Topological constraints associated with fast dynamo action, *J. Fluid Mech.* **154**, 493–507 (1985).
16. K. Bajer, Flow kinematics and magnetic equilibria, Ph.D. thesis, University of Cambridge (1989).
17. K. Bajer and H. K. Moffatt, Chaos associated with fluid inertia, in *Topological Aspects of the Dynamics of Fluid and Plasmas*, edited by H. K. Moffatt, G. M. Zaslavsky, P. Comte and M. Tabor, pp. 517–534. Kluwer, Amsterdam (1992).
18. K. Bajer, H. K. Moffatt and F. H. Nex, Steady confined Stokes flows with chaotic streamlines, in *Topological Fluid Mechanics*, edited by H. K. Moffatt and A. Tsinober, pp. 459–466. Cambridge University Press, Cambridge (1990).
19. G. K. Batchelor, *An Introduction to Fluid Dynamics*. Cambridge University Press, Cambridge (1967).
20. S. Wiggins, *Chaotic Transport in Dynamical Systems*. Springer, Berlin (1992).

Orientation Dynamics and Stretching of Particles in Unsteady, Three-dimensional Fluid Flows: Unsteady Attractors

ANDREW J. SZERI

Department of Mechanical & Aerospace Engineering, University of California, Irvine, CA 92717-3975, USA

and

L. GARY LEAL

Department of Chemical & Nuclear Engineering, University of California, Santa Barbara, CA 93106, USA

Abstract—Recent studies of the nonlinear dynamics of deformable particles suspended in fluid flows have revealed a rich geometric structure. However, an analytic expression of the flow characteristic of principal interest in the dynamics of polymer solutions, namely the stretching power of the flow over deformable particles, has remained elusive when the history of particle orientations is complicated. This is the case, for example, when the orientation dynamics of particles is characterized by an unsteady attractor. This dynamical behavior is considered in some detail, both in steady and in time-periodic external flows. Strong flow criteria are developed by consideration of the statistical distribution of orientations in time on the relevant unsteady attractor.

1. INTRODUCTION

We are interested, in this paper, in a class of problems that relate to the interactions between a complex fluid motion, and the dynamics of particles that are suspended in the fluid. To date, there have been two general problems of this type investigated.

The first class of particle dynamics problems is concerned with the trajectories of single, mainly spherical non-Brownian particles, primarily to understand the conditions in which heavy particles may remain suspended indefinitely against gravity (see refs [1–6]). The flow itself is not modified by the particles. This class of problems can be seen as a generalization of the kinematical study of the advection of passive markers as material points in a specified flow. The key generalization is that the particles are assumed to be of finite size and mass, and thus move relative to the fluid in a manner described by a set of nonlinear dynamical equations. Hence, a steady flow that is three-dimensional generates a six-dimensional dynamical system describing the motion of a particle, while a two-dimensional flow yields a four-dimensional dynamical system.

The second class of particle dynamics problems is concerned with the orientation and stretch (or deformation) of neutrally buoyant, non-spherical particles in steady or unsteady three-dimensional flows. The particles may be thought of as very simple models for polymer molecules suspended in a flow, or they may represent drops or other deformable bodies, as in Tjahjadi and Ottino [7]. The objective in these problems is often a priori flow classification based upon whether, or to what degree, the configuration of the suspended phase is driven from its equilibrium state by the flow. This class of problems can be viewed as a generalization of the kinematics of the orientation and stretch of passive material filaments (or surfaces) within a fluid as studied by Ottino [8], Wiggins [9] and other researchers in the context of advective mixing phenomena. The principle distinctions between particles and fluid filaments are (a) that the particles have a finite aspect ratio and thus do not orient affinely as a passive material filament, and (b) the particles have at least one internal mechanism to resist deformation by the flow.

Recent studies [10–12] have revealed a very rich geometric structure in the nonlinear dynamics of deformable particles, and have led to the development of a general framework for classifying flows in terms of their potential for orientation and stretch of the suspended phase. However, to date these flow classification criteria have led to analytic results only for relatively simple steady or time-periodic flows, where there is either a steady attractor for the particle orientation or a steady attractor for orientation in the Poincaré map, respectively. On the other hand, for many of the most important applications we need to understand the stretching and/or alignment of particles in chaotic (or at least much more complicated) flows. The present paper represents a significant step toward the development of flow classification schemes for such cases, although this may not appear obvious at first.

In terms of organization, we adopt the following plan. First, we formulate a specific problem which, as we shall see, is a direct extension of recently published work on microstructural dynamics in a three-dimensional flow [10]. Then, we show that a strong flow criterion for particle orientation and stretch can be developed for the previously unresolved case of a *steady* flow in which the orientation dynamics is characterized by an *unsteady* attractor. This is a accomplished by using a new exact solution for particle orientation on the attractor, from which one can obtain an exact expression for the probability of finding the particle in a specific orientation. For *time-periodic* flows [10] the particle rotation is either periodic (representing a fixed point of a Poincaré map) or aperiodic (representing an unsteady attractor of the Poincaré map). In the former case, we were able to develop flow classification criteria, as indicated above. For aperiodic motions, the history of particle orientations on the attractor becomes sufficiently complicated that it is *impossible* to use predicted particle orientation trajectories to develop a well-defined strong flow criterion, except for finite times, where we may use direct numerical simulation of the full particle orientation trajectory for the specific flow and initial conditions. The main result of this paper, described in the last section, is a new approach to the derivation of strong flow criteria for this case, which adopts the key idea from the steady flow problem of using the orientation evolution equations *themselves* to estimate the probability of specific orientations, rather than the solutions of these equations. Finally, in the conclusions we speculate on a connection between this resolution of the problem for aperiodic rotation in a periodic flow and the development of strong flow criteria for much more complicated flows.

2. PROBLEM FORMULATION

We consider the dynamics of elastic, axisymmetric orientable particles in an unsteady, three-dimensional fluid flow. The particles are neutrally buoyant, and sufficiently small with respect to the length-scale of the flow that the orientation and stretch dynamics are forced by the velocity gradient of the surrounding flow field.

Of particular interest is whether or not the particles are stretched by the flow. Of course, this depends on the details of the flow field seen by the particles over some interval of time, and on the magnitude of the elastic restoring force. In addition, there is dependence on the orientation dynamics of particles, as a given flow will, at any instant, be stretching in some directions and contracting in others.

These assertions are more easily understood if one examines the equations that govern the orientation and stretch of a particle in a flow field [13, 14]:

$$\frac{d\mathbf{u}}{dt} = \boldsymbol{\kappa}\mathbf{u} - (\mathbf{u} \cdot \boldsymbol{\kappa}\mathbf{u})\mathbf{u} \tag{1}$$

$$\frac{1}{\rho}\frac{d\rho}{dt} = \frac{1}{F+1}(\mathbf{u} \cdot \boldsymbol{\kappa}\mathbf{u} - \alpha). \tag{2}$$

Here the vector $\mathbf{u}$ is the orientation of the particle, which has length ρ in the direction $\mathbf{u}$. The tensor $\boldsymbol{\kappa}$ is closely related to the velocity gradient tensor, but is modified by the parameter G which models the inefficiency of rotation of particles of finite aspect ratio in straining flows: $\boldsymbol{\kappa} = \frac{1}{2}(\nabla\mathbf{v} - \nabla\mathbf{v}) + G\frac{1}{2}(\nabla\mathbf{v} + \nabla\mathbf{v})$. Finally, the parameters α and F are the elastic modulus and internal viscosity of the particle that resist deformation by the flow. We refer the reader to [12, 14] for a discussion of various parameter combinations in the model.

One feature of the model represented by (1) and (2) that deserves special mention is the equilibrium length, which is predicted by (2) to be zero in the absence of flow. Of course, if $\rho = 0$ initially, then $\rho = 0$ for all times according to (2). These features are not in accord with the expected behavior of real macromolecules, or other elastic bodies, but are a consequence of adopting the simplest possible model. One could avoid the problem entirely by simply modifying (2) slightly to yield $\rho = \rho_0 \neq 0$ for $\boldsymbol{\kappa} \equiv 0$. In the case of a macromolecule, Brownian motion prevents the chain from collapsing, producing instead an equilibrium radius of gyration. Although the deterministic effect of Brownian motion is included via the elastic *entropic* spring in (2), which drives the extended chain back toward its equilibrium structure, the randomizing influence of Brownian motion has been neglected in the deterministic model (1) and (2). In the work that follows, it is not our goal in any case to integrate (2) from some initial condition. Rather, we attempt to establish criteria for the onset of stretching by the flow, assuming that the initial state is any arbitrarily small but nonzero value for ρ.

Formally, one may solve the orientation equation as follows [15]: put

$$\mathbf{u}(t;\, \mathbf{u}_0) = \frac{\mathbf{Q}(t)\mathbf{u}_0}{\|\mathbf{Q}(t)\mathbf{u}_0\|}, \tag{3}$$

then $\mathbf{Q}$ is the solution of the equation

$$\frac{d\mathbf{Q}}{dt} = \boldsymbol{\kappa}(t)\mathbf{Q} \tag{4}$$

with initial condition $\mathbf{Q}(0) = \mathbf{I}$. This procedure works whether $\boldsymbol{\kappa}$ depends on time or not. In [10], we establish that in incompressible flow where $\mathrm{tr}\boldsymbol{\kappa} = 0$, $\det \mathbf{Q} = 1$ for all time.

The flow has sufficient power to stretch particles over the time interval from 0 to t if

$$\frac{1}{t}\int_0^t \mathbf{u}(\tau;\, \mathbf{u}_0) \cdot \boldsymbol{\kappa}(\tau)\mathbf{u}(\tau;\, \mathbf{u}_0)\, d\tau > \alpha \tag{5}$$

for some initial orientation $\mathbf{u}_0$. This statement is readily understood from an integration of (2); see [10] for the details.

Our earlier work in this area has focused on: the dynamics of particles in two-dimensional steady and time-periodic flows [12], experimental verification of predictions of the orientation dynamics of rigid particles in unsteady two-dimensional flows [11], and a study of orientation dynamics and stretching of particles in three-dimensional flows [10]. In the last paper, in the case of a steady attractor for the orientation dynamics in a steady flow, or a steady attractor for the Poincaré map of the orientation dynamics in a time-periodic flow, we were able to use (5) to develop criteria to distinguish flows in which a particle is stretched from those in which it is not. However, when the orientation dynamics was associated with an unsteady attractor (i.e. limit cycle in a steady flow, or quasi-periodic dynamics in a time-period flow) we were unable to develop analytic criteria for particle stretch.

In the present paper, we determine the stretching power of the flow for particles that follow an unsteady attractor. Along the way, we develop analytic solutions for the orientation dynamics on the attractors.

3. UNSTEADY ATTRACTORS FOR THE ORIENTATION DYNAMICS IN STEADY FLOW

As described in [10], in a steady flow the orientation dynamics are characterized by either of two global attractors: a fixed orientation, or a limit cycle that corresponds to a great circle in orientation space (a sphere). These cases are distinguished by the spectrum of $\boldsymbol{\kappa}$, which has three real distinct eigenvalues in the first case, and a single real eigenvalue and a complex-conjugate pair in the second case.

In the second case, we have shown that there is an invariant great circle for the orientation dynamics. If one works in Cartesian coordinates such that $\boldsymbol{\kappa}$ takes the form

$$\boldsymbol{\kappa}(t) = \begin{bmatrix} \lambda_1 & \kappa_{12} & \kappa_{13} \\ 0 & \kappa_{22} & \kappa_{23} \\ 0 & \kappa_{32} & \kappa_{33} \end{bmatrix},$$

then the invariant great circle lies in the plane normal to the vector

$$\tilde{\mathbf{n}} = \begin{pmatrix} (\lambda_1 - \kappa_{22})(\lambda_1 - \kappa_{33}) - \kappa_{23}\kappa_{32} \\ \kappa_{12}(\lambda_1 - \kappa_{33}) + \kappa_{13}\kappa_{32} \\ \kappa_{13}(\lambda_1 - \kappa_{22}) + \kappa_{12}\kappa_{23} \end{pmatrix}.$$

In the situation where the real eigenvalue $\lambda_1 < 0$, the invariant great circle corresponds to an attracting limit cycle, and one can restrict the differential equation for the orientation of a particle to the invariant great circle as follows. Normal to the vector $\tilde{\mathbf{n}}$ we define the vector $\tilde{\mathbf{a}} = (-\tilde{n}_2, \tilde{n}_1 + \tilde{n}_3, -\tilde{n}_2)$. Normal to both $\tilde{\mathbf{n}}$ and $\tilde{\mathbf{a}}$, we define $\tilde{\mathbf{b}} = \tilde{\mathbf{n}} \times \tilde{\mathbf{a}}$. Finally, we obtain an orthonormal basis $(\mathbf{n}, \mathbf{a}, \mathbf{b})$ by simply normalizing.

Next, we write an orientation on the invariant great circle as $\mathbf{u}_{\text{lc}} = \cos\theta\mathbf{a} + \sin\theta\mathbf{b}$, where the subscript lc denotes limit cycle. On the invariant great circle,

$$\frac{\mathrm{d}\theta}{\mathrm{d}t} = (\mathbf{n} \times \mathbf{u}_{\text{lc}}) \cdot \boldsymbol{\kappa}\mathbf{u}_{\text{lc}} = c_1 + c_2 \sin 2\theta + c_3 \cos 2\theta,$$

where $c_1 = \frac{1}{2}(\mathbf{b} \cdot \boldsymbol{\kappa}\mathbf{a} - \mathbf{a} \cdot \boldsymbol{\kappa}\mathbf{b})$, $c_2 = \frac{1}{2}(\mathbf{b} \cdot \boldsymbol{\kappa}\mathbf{b} - \mathbf{a} \cdot \boldsymbol{\kappa}\mathbf{a})$, $c_3 = \frac{1}{2}(\mathbf{b} \cdot \boldsymbol{\kappa}\mathbf{a} + \mathbf{a} \cdot \boldsymbol{\kappa}\mathbf{b})$. Hence, the c_i are functions of the tensor components of $\boldsymbol{\kappa}$. Because the invariant circle is a limit cycle, the differential equation should have no fixed points; therefore, $c_2^2 + c_3^2 < c_1^2$. With this information, the orientation dynamics may be found by simple quadrature. However, we shall find a sufficient condition for stretch of particles on the attractor in another way, that is enlightening when we come to consider quasi-periodic attractors in the next section.

3.1. *Strong flow criterion on the limit cycle*

In general, a sufficient condition for stretch of particles on the attractor, or *strong flow criterion*, depends on the history of orientations of the particle as well as the history of flows seen by particle. In this section, we consider steady flows. Therefore, only the proportion of time spent by the particle in different directions with differing stretching power is of importance to the strong flow criterion. This is the significance of the change of variables in the integral that we make in what follows. To begin, the strong flow criterion is

$$\frac{1}{T}\int_0^T \mathbf{u}_{\mathrm{lc}}(\tau;\, \mathbf{u}_0)\cdot \boldsymbol{\kappa}\mathbf{u}_{\mathrm{lc}}(\tau;\, \mathbf{u}_0)\,\mathrm{d}\tau > \alpha.$$

We write this in terms of the orientation on the limit cycle and obtain

$$\frac{1}{T}\int_0^T [s_1 + s_2\cos 2\theta(\tau) + s_3\sin 2\theta(\tau)]\,\mathrm{d}\tau > \alpha,$$

where $s_1 = \frac{1}{2}(\mathbf{b}\cdot\boldsymbol{\kappa}\mathbf{b} + \mathbf{a}\cdot\boldsymbol{\kappa}\mathbf{a})$, $s_2 = -\frac{1}{2}(\mathbf{b}\cdot\boldsymbol{\kappa}\mathbf{b} - \mathbf{a}\cdot\boldsymbol{\kappa}\mathbf{a}) = -c_2$, $s_3 = \frac{1}{2}(\mathbf{b}\cdot\boldsymbol{\kappa}\mathbf{a} + \mathbf{a}\cdot\boldsymbol{\kappa}\mathbf{b}) = c_3$. Next, we use the knowledge that there are no fixed points on the invariant circle, and change variables in the integral:

$$\frac{1}{T}\int_0^{2\pi} \frac{s_1 + s_2\cos 2\theta + s_3\sin 2\theta}{c_1 + c_2\sin 2\theta + c_3\cos 2\theta}\,\mathrm{d}\theta > \alpha.$$

Upon integrating, we obtain the analytic strong flow criterion

$$\frac{2\pi s_1}{T(c_1^2 - c_1^2 - c_3^2)^{\frac{1}{2}}} > \alpha. \tag{6}$$

We may interpret the derivation leading to this result as follows. First, from the definition we see that the strong flow criterion is obtained by multiplying the stretching power of the flow at a given orientation (on the invariant circle) by the amount of time spent in that orientation, and integrating over the entire limit cycle. The variable transformation from τ to θ is therefore equivalent to evaluating the relative amount of time spent at orientation θ as $\mathrm{d}\theta/(\mathrm{d}\theta/\mathrm{d}t)$. We pursue a similar course in time-periodic flows, below.

3.2. *Stretching in the presence of a saddle-node bifurcation*

In effect, we have defined a distribution function for particle orientation on the limit cycle of the form $\psi(\theta) = 1/\dot{\theta}$. With this definition, $\psi(\theta)\,\mathrm{d}\theta (= \mathrm{d}t)$ is 2π times the fraction of a period spent by a particle in the range of orientations $\theta - \mathrm{d}\theta/2 \leqslant \theta \leqslant \theta + \mathrm{d}\theta/2$ on the attractor. Then the strong flow criterion may be written in the form

$$\frac{1}{T}\int_0^{2\pi} \mathbf{u}_{\mathrm{lc}}(\theta)\cdot\boldsymbol{\kappa}\mathbf{u}_{\mathrm{lc}}(\theta)\psi(\theta)\,\mathrm{d}\theta > \alpha. \tag{7}$$

This simple development will help motivate the theory for quasi-periodic dynamics.

As discussed and shown, for example, in ref. [10], it is possible to have a saddle-node bifurcation on the limit cycle. How then should one interpret the distribution function and the strong flow criterion?

When there is a saddle-node bifurcation, giving birth to a new fixed point at θ^*, then $\psi(\theta) = \delta(\theta - \theta^*)T$, where δ is a delta function. Then the strong flow criterion becomes

$$\int_0^{2\pi} \mathbf{u}_{\mathrm{lc}}(\theta)\cdot\boldsymbol{\kappa}\mathbf{u}_{\mathrm{lc}}(\theta)\delta(\theta - \theta^*)\,\mathrm{d}\theta = \mathbf{u}_{\mathrm{lc}}(\theta^*)\cdot\boldsymbol{\kappa}\mathbf{u}_{\mathrm{lc}}(\theta^*) = \lambda^* > \alpha. \tag{8}$$

Here λ^* is the (new) real eigenvalue of $\boldsymbol{\kappa}$ associated with the eigenvector in the direction θ^*. This is the strong flow criterion obtained in ref. [10] for the case of a steady attractor in a steady flow.

4. UNSTEADY ATTRACTORS FOR THE POINCARÉ MAP OF ORIENTATIONS IN TIME-PERIODIC FLOW

When the surrounding flow seen by the particles is time-periodic, as occurs in the important practical case of recirculating flow [16], it is convenient to work it terms of the Poincaré map

$$\mathbf{u}_1 \equiv \mathbf{u}(T;\, \mathbf{u}_0) = P(\mathbf{u}_0) = \frac{\mathbf{Q}(T)\mathbf{u}_0}{\|\mathbf{Q}(T)\mathbf{u}_0\|}. \tag{9}$$

The map may be concatenated, thus

$$\mathbf{u}_2 \equiv \mathbf{u}(2T;\, \mathbf{u}_0) = P^2(\mathbf{u}_0) = P(\mathbf{u}_1) = \frac{\mathbf{Q}^2(T)\mathbf{u}_0}{\|\mathbf{Q}^2(T)\mathbf{u}_0\|}.$$

The orientation dynamics one deduces from an analysis of the Poincaré map is of two varieties: either there is a globally attracting fixed point of the Poincaré map corresponding to a periodic solution of the underlying O.D.E., or there is a globally attracting invariant great circle of the Poincaré map which corresponds to (possibly) quasi-periodic motions in the underlying O.D.E. These motions have both been observed in experiments [11].

4.1. *Reduction of the Poincaré map to the invariant great circle*

In this section, we are interested in the case of possibly quasi-periodic dynamics, in which the tensor $\mathbf{Q}$ has a single real eigenvalue $\xi < 1$, and two complex conjugate eigenvalues. In the Cartesian basis in which $\mathbf{Q}$ takes the form

$$\mathbf{Q}(T) = \begin{bmatrix} \xi_1 & Q_{12} & Q_{13} \\ 0 & Q_{22} & Q_{23} \\ 0 & Q_{32} & Q_{33} \end{bmatrix},$$

there is an invariant great circle of the Poincaré map, which lies in the plane normal to

$$\tilde{\mathbf{n}} = \begin{pmatrix} (\xi_1 - Q_{22})(\xi_1 - Q_{33}) - Q_{23}Q_{32} \\ Q_{12}(\xi_1 - Q_{33}) + Q_{13}Q_{32} \\ Q_{13}(\xi_1 - Q_{22}) + Q_{12}Q_{23} \end{pmatrix}.$$

As in the case of steady flow, we can define an orthonormal basis $(\mathbf{n},\, \mathbf{a},\, \mathbf{b})$ in which a point on the invariant great circle $\mathbf{u}_{\mathrm{ic}} = \mathbf{a}\cos\theta + \mathbf{b}\sin\theta$. The subscript ic denotes an orientation on the invariant great circle. Then the Poincaré map specializes to

$$\mathbf{u}_{n+1} = \mathbf{a}\cos\theta_{n+1} + \mathbf{b}\sin\theta_{n+1} = \frac{\mathbf{Q}(T)(\mathbf{a}\cos\theta_n + \mathbf{b}\sin\theta_n)}{\|\mathbf{Q}(T)(\mathbf{a}\cos\theta_n + \mathbf{b}\sin\theta_n)\|}.$$

This may be rearranged to reveal the circle map

$$\tan\theta_{n+1} = \frac{1 + d_1\tan\theta_n}{d_2 + d_3\tan\theta_n}, \tag{10}$$

where $d_1 = (\mathbf{b}\cdot\mathbf{Qb})/(\mathbf{b}\cdot\mathbf{Qa})$, $d_2 = (\mathbf{a}\cdot\mathbf{Qa})/(\mathbf{b}\cdot\mathbf{Qa})$, and $d_3 = (\mathbf{a}\cdot\mathbf{Qb})/(\mathbf{b}\cdot\mathbf{Qa})$. Dynamics on the quasi-periodic attractor are thus described by a smooth, invertible nonlinear circle map, which, quite remarkably, has an analytic solution that we report next.

4.2. *Analytic solution of the nonlinear circle map*

The circle map just derived has a very special structure: it is a linear fractional transformation in $\tan\theta$. Because this property is preserved at *every* iterate of the map, e.g.

$$\tan\theta_{n+2} = \frac{1 + d_1\tan\theta_{n+1}}{d_2 + d_3\tan\theta_{n+1}} = \frac{1 + d_1(1 + d_1\tan\theta_n)/(d_2 + d_3\tan\theta_n)}{d_2 + d_3(1 + d_1\tan\theta_n)/(d_2 + d_3\tan\theta_n)},$$

which may be rearranged to yield

$$\tan\theta_{n+2} = \frac{(d_1 + d_2) + (d_1^2 + d_3)\tan\theta_n}{(d_2^2 + d_3) + d_3(d_1 + d_2)\tan\theta_n},$$

one can solve the map analytically, to obtain θ_n as a function of θ_0, say. This is a rather painful task, as it involves very lengthy (but not necessarily illuminating) counting arguments and searching for patterns in the formation of very long polynomials in $\tan\theta_0$. The result is

$$\tan\theta_n = \frac{f_1(n) + f_2(n)\tan\theta_0}{f_3(n) + f_4(n)\tan\theta_0}, \tag{11}$$

where

$$f_1(n) = \sum_{i=1}^{n} d_1^{(n-i)}d_2^{(i-1)} + \sum_{m=1}^{[(n-1)/2]} \frac{d_3^m}{(m!)^2}$$
$$\times \sum_{i=1}^{(n-2m)} \prod_{j=0}^{(m-1)} (n - i - m - j)(i + j)d_1^{(n-2m-i)}d_2^{(i-1)},$$

$$f_2(n) = d_1^n + \sum_{m=1}^{[n/2]} \frac{d_3^m}{m!(m-1)!}$$
$$\times \sum_{i=1}^{(n-2m+1)} \prod_{j=1}^{(m-1)} (i + j - 1) \prod_{j=m-1}^{(2m-2)} (n - i - j)d_1^{(n-2m+1-i)}d_2^{(i-1)},$$

$f_3(n)$ is the same as $f_2(n)$, but with d_1 and d_2 interchanged, and $f_4(n) = d_3 f_1(n)$. In these expressions, $[x]$ is the largest integer $\leqslant x$.

Of course, the expressions for the $f_i(n)$ are extremely complicated, for n even moderately large. The primary utility of this analytic solution is not to allow one to compute an orbit. This can be accomplished much more easily simply by iterating the Poincaré map. Instead, we use the analytic solution of the circle map to point out some very unusual features of the dynamical system of tumbling particles, restricted to the invariant great circle.

Before launching into a detailed study of this particular circle map, it is useful to recall the general theory of circle maps that depend on a parameter, see [17]. Typical circle maps are characterized by open sets of positive measure in parameter space for which there is a periodic attractor that attracts all orbits. The size of the open set decreases rapidly as the period of the attractor increases. Separating such open sets are other sets of parameter values for which the dynamics are truly quasi-periodic.

4.2.1. *Saddle-node bifurcation to a period-*1 *attractor*. In view of the linear-fractional nature of the map in $\tan\theta_n$, we work hereafter in the variable $\zeta_n \equiv \tan\theta_n$. The Poincaré map restricted to the circle is then

$$\zeta_{n+1} = P_{\text{ic}}(\zeta_n) = \frac{1 + d_1\zeta_n}{d_2 + d_3\zeta_n}.$$

A fixed point of this map would correspond to a globally attracting T-periodic orbit of the underlying, periodically forced O.D.E. A fixed point is a solution of the equation $P_{ic}(\zeta) = \zeta$; this equation has two solutions whenever $d_3 \geqslant -\frac{1}{4}(d_2 - d_1)^2$. This is a sufficient condition for a saddle-node bifurcation on the invariant closed curve, which leads to a globally attracting fixed point (the node) and a saddle, both on the invariant closed curve. If we put $d_3 = -\frac{1}{4}(d_2 - d_1)^2 + \epsilon$, where ϵ is a positive (but not necessarily small) number, then the saddle and node are at

$$\zeta = \frac{2}{(d_2 - d_1) \pm \epsilon^{\frac{1}{2}}}.$$

Before moving on to period $2T$ attractors, we note that $d_3 > -\frac{1}{4}(d_2 - d_1)^2$ is a rather large open set in (d_1, d_2, d_3) parameter space, for which the circle map has a period T attractor.

4.2.2. *An apocryphal saddle-node bifurcation to a period-*2 *attractor*. Following the development in the previous section, it is possible to look for a saddle-node bifurcation of P^2_{ic}, i.e. for the birth of a period-$2T$ attractor on the invariant circle of the Poincaré map P. To do so, we simply set $P^2_{ic}(\zeta) = \zeta$. One easily obtains the sufficient condition for bifurcation as

$$d_3 \geqslant \frac{(d_2^2 - d_1^2)^2}{4(d_1^2 + d_2^2)} = -\frac{1}{4}(d_2 - d_1)^2,$$

with no other solution. Hence, the only period-$2T$ orbits are period-T orbits of the underlying O.D.E. This is the first clue to the strange dynamics of the circle map. Typically, one would expect there to be an open set of positive measure in parameter space for which the circle map has a period-$2T$ attractor, but this appears not to be the case.

4.2.3. *Virtual saddle-node bifurcations to period-*n *attractors* (n > 2). Proceeding similarly, one can search for sufficient conditions for a saddle-node bifurcation to a period nT attractor ($n > 2$) by setting $P^n_{ic}(\zeta) = \zeta$ and searching for sufficient conditions for a solution, i.e. for *some* ζ to exist which solves the equation. This occurs when

$$(f_3(n) - f_2(n))^2 + 4f_1(n)f_4(n) \geqslant 0, \tag{12}$$

for then the new fixed points of P^n_{ic} are

$$\zeta = \frac{-(f_3(n) - f_2(n)) \pm ((f_3(n) - f_2(n))^2 + 4f_1(n)f_4(n))^{\frac{1}{2}}}{2f_4(n)}. \tag{13}$$

Typically, one would expect to find open sets of positive measure in (d_1, d_2, d_3) parameter space, of a size decreasing with n, for which the circle map has an attractor of period nT. As we show below, this is not the case here.

For $n = 3$ one obtains from (12) the conditions $d_3 = -\frac{1}{4}(d_2 - d_1)^2$ and $d_3 = -d_1^2 - d_1 d_2 - d_2^2$. The first corresponds to a period-T attractor, the second, new criterion, is for a period-$3T$ attractor. In what follows, we use the notation $p[3] = d_3 + d_1^2 + d_1 d_2 + d_2^2$. Now, if we set $p[3] = 0$ and evaluate the Poincaré map, we find $P^3_{ic}(\zeta) = \zeta$! In other words, all points are period-3 points when $p[3] = 0$; there is no period-$3T$ *attractor*. The analogous situation in a typical system would be very different. In a typical system, when the criterion for the existence of a period-3 point is satisfied, one would expect the emergence of a single period-3 orbit of the map, to which all other orbits are asymptotically attracted. In our system, the unusual feature is that when the criterion for the existence of *some* period-3 point is satisfied, every point on the invariant great circle is a period-3 point; there is no attracting period-3 orbit.

Even more bizarre is the following. The saddle-node bifurcation is incomplete, as can be seen by substituting $d_3 = -d_1^2 - d_1 d_2 - d_2^2 + \epsilon$ into the condition $(f_3(n) - f_2(n))^2 + 4f_1(n)f_4(n)$. One obtains, for small ϵ, $-\gamma\epsilon^2$ to leading order, with $\gamma > 0$. Hence the sufficient condition for a period-$3T$ orbit to exist is satisfied only *exactly* at $p[3] = 0$. We call this situation a *virtual bifurcation*. In a typical system, the sufficient condition for the existence of a period-3 orbit is satisfied for a range of parameters, i.e. for a set in parameter space of nonzero measure. In our case, parameter space is three-dimensional, so one would have expected to find a set of volume greater than zero within which the associated dynamical system has a period-3 attractor. Instead, we find that the parameter set is a *surface* with zero volume.

Similarly, strange things happen for $n = 4$. The sufficient condition for a period-$4T$ orbit to exist is $p[4] = d_3 + \frac{1}{2}(d_1^2 + d_2^2)$. Again, one finds $P_{\text{ic}}^4(\zeta) = \zeta$ when $p[4] = 0$. In addition, if one puts $d_3 = -\frac{1}{2}(d_1^2 + d_2^2) + \epsilon$, one finds $(f_3(n) - f_2(n))^2 + 4f_1(n)f_4(n) = -\gamma\epsilon^2$ to leading order, with $\gamma > 0$. Hence the sufficient condition for a period-$4T$ orbit to exist is satisfied only *exactly* at $p[4] = 0$.

The same is true for $n = 5$ and 6, where $p[5] = d_3 + \frac{1}{2}[(3d_1^2 + 4d_1 d_2 + 3d_2^2) \pm 5^{\frac{1}{2}}(d_1 + d_2)^2]$ and $p[6] = d_3 + \frac{1}{3}(d_1^2 - d_1 d_2 + d_2^2)$. Thereafter, the calculations become unwieldy. However, in view of the recurring structure of the polynomials $f_i(n)$, we speculate that the roots of $(f_3(n) - f_2(n))^2 + 4f_1(n)f_4(n) = 0$ are such that this behavior continues. Assuming this is the case, then the only periodic attractor one can find arising on the invariant great circle of the Poincaré map P is a period-T attractor. In addition, on codimension-1 submanifolds of (d_1, d_2, d_3) parameter space, one sees a virtual bifurcation with the peculiar dynamics of every point being a fixed point in the map P_{ic}^n.

4.3. *Dynamics of orientations on the invariant great circle of the Poincaré map*

The Poincaré map, together with the analytic solution developed above for dynamics on the invariant great circle, leads to the following picture of the dynamics. When $d_3 > -\frac{1}{4}(d_2 - d_1)^2$ there is a global attractor of period T for the orientation dynamics. When $d_3 < -\frac{1}{4}(d_2 - d_1)^2$, the dynamics are truly quasi-periodic, except on codimension-1 submanifolds of (d_1, d_2, d_3) parameter space, where every orientation is periodic with period nT, $n > 2$. If one were to plot a rotation number for dynamics on the invariant great circle versus a parameter that changes the flow, one would, therefore, expect to see something rather unusual. There would be no *devil's staircase* of parameter values associated with rational rotation numbers, which correspond to period-nT attractors. Instead, one would see a smooth, continuous curve for the rotation number, without steps. This was indeed the case in an example in ref. [11], see Fig. 7 of that paper.

Obviously, this behavior could be changed quite radically by a small perturbation to the differential equations. For example, if one changes the criterion $p[n] = 0$ for $n > 2$ to a nearby criterion of the form $p[n] = -\epsilon$, for ϵ small and positive, then the virtual saddle-node bifurcations at $p[n] = 0$ become true saddle-node bifurcations nearby, resulting in more typical behavior where there is a finite open set of parameter values for which there is a period-nT attractor. However, if we imagine perturbating the criterion $p[n] = 0$ for $n > 2$ to a nearby criterion of the form $p[n] = +\epsilon$, for ϵ small and positive, then the virtual saddle-node bifurcations at $p[n] = 0$ are annihilated! This nonlinear circle map would appear to be an interesting object for further study. However, in the present paper, our main interest is in stretching of particles that follow unsteady attractors. Therefore, we proceed with the analytic solution of the orientation dynamics.

4.4. *Stretching of particles that follow quasi-periodic orientation dynamics*

When the flow is time-periodic, and there is a globally attracting period-T trajectory for the orientation dynamics, then a strong flow criterion is easily derived, as in [10]. Here, we attempt to develop a strong flow criterion for the case where $d_3 < -\frac{1}{4}(d_2 - d_1)^2$, assuming that we avoid the subsets of zero measure in parameter space on which $p[n] = 0$ for $n > 2$. The problem is difficult, owing to the fact that trajectories (subject to the assumptions just stated) do not follow a repeating orientation time trace. Therefore, we will eventually be forced to *estimate* the amount of time spent by a given particle in a given orientation on the Poincaré section (the sphere of orientations), rather than calculating directly as we did earlier for steady flows. Note that this is equivalent to estimating the number density of particles in a suspension that lie in a given orientation on the attractor at a given instant of time. Therefore, our strong flow criterion will apply to a suspension of a great many particles which each evolve according to equations (1) and (2).

First, we investigate the relationship between the tensor $\mathbf{Q}$ and the sufficient condition for stretch of a particle. To begin, we compute

$$\frac{\mathrm{d}}{\mathrm{d}t}\|\mathbf{Q}(t)\mathbf{u}_0\| = \frac{1}{2}\,\frac{1}{\|\mathbf{Q}(t)\mathbf{u}_0\|}\,\frac{\mathrm{d}}{\mathrm{d}t}\|\mathbf{Q}(t)\mathbf{u}_0\|^2.$$

This is, using (3) and (4),

$$\frac{\mathrm{d}}{\mathrm{d}t}\|\mathbf{Q}(t)\mathbf{u}_0\| = \frac{\mathbf{Q}(t)\mathbf{u}_0}{\|\mathbf{Q}(t)\mathbf{u}_0\|}\cdot\dot{\mathbf{Q}}(t)\mathbf{u}_0 = \mathbf{u}(t;\,\mathbf{u}_0)\cdot\boldsymbol{\kappa}(t)\mathbf{Q}(t)\mathbf{u}_0 = \mathbf{u}(t;\,\mathbf{u}_0)\cdot\boldsymbol{\kappa}(t)\mathbf{u}(t;\,\mathbf{u}_0)\|\mathbf{Q}(t)\mathbf{u}_0\|.$$

Next, we divide through by $\|\mathbf{Q}(t)\mathbf{u}_0\|$ and integrate to obtain the result

$$\log\|\mathbf{Q}(t)\mathbf{u}_0\| = \int_0^t \mathbf{u}(\tau;\,\mathbf{u}_0)\cdot\boldsymbol{\kappa}(\tau)\mathbf{u}(\tau;\,\mathbf{u}_0)\,\mathrm{d}\tau.$$

Hence, the strong flow criterion may be written

$$\frac{1}{t}\log\|\mathbf{Q}(t)\mathbf{u}_0\| > \alpha. \tag{14}$$

In order to average over the entire attractor, we set $t = nT$ and will later take the limit as $n \to \infty$; this yields

$$\frac{1}{nT}\log\|\mathbf{Q}(nT)\mathbf{u}_0\| > \alpha.$$

Now, because the Poincaré map can be concatenated as described earlier, this may be simplified to read

$$\frac{1}{nT}\log\|\mathbf{Q}^n(T)\mathbf{u}_0\| > \alpha.$$

Next, we use the fact that

$$\begin{aligned}\mathbf{Q}^n(T)\mathbf{u}_0 &= \left[\prod_{i=1}^{n}\mathbf{Q}(T)\right]\mathbf{u}_0\\ &= \left[\prod_{i=2}^{n}\mathbf{Q}(T)\right]\mathbf{Q}(T)\mathbf{u}_0 = \left[\prod_{i=2}^{n}\mathbf{Q}(T)\right]\mathbf{u}_1\|\mathbf{Q}(T)\mathbf{u}_0\|.\end{aligned}$$

Continuing in this way, a pattern emerges

$$\mathbf{Q}^n(T)\mathbf{u}_0 = \left[\prod_{i=3}^{n}\mathbf{Q}(T)\right]\mathbf{u}_2\|\mathbf{Q}(T)\mathbf{u}_1\|\,\|\mathbf{Q}(T)\mathbf{u}_0\|.$$

Finally, one obtains

$$\mathbf{Q}^n(T)\mathbf{u}_0 = \mathbf{u}_n \prod_{i=0}^{n-1} \|\mathbf{Q}(T)\mathbf{u}_i\|.$$

Hence, upon taking the magnitude and using $\|\mathbf{u}\| \equiv 1$,

$$\|\mathbf{Q}^n(T)\mathbf{u}_0\| = \prod_{i=0}^{n-1} \|\mathbf{Q}(T)\mathbf{u}_i\|.$$

We use this result in the strong flow criterion to obtain

$$\frac{1}{nT}\log \prod_{i=0}^{n-1} \|\mathbf{Q}(T)\mathbf{u}_i\| > \alpha,$$

or,

$$\frac{1}{nT}\sum_{i=0}^{n-1} \log \|\mathbf{Q}(T)\mathbf{u}_i\| > \alpha.$$

Therefore, the stretching power of the flow, integrated over the orientation time trace $\mathbf{u}(t;\ \mathbf{u}_0)$ for $iT \leq t < (i+1)T$ is $\log \|\mathbf{Q}(T)\mathbf{u}_i\|$. For a general initial condition $\mathbf{u}$, the stretching power of the flow integrated over one period is $\log \|\mathbf{Q}(T)\mathbf{u}\|$. Now, assuming the dynamics are quasi-periodic, we next need to estimate the asymptotic proportion of iterates that lie in every possible orientation on the attractor; we want to do this in a way which does not involve determining the entire orbit $(\mathbf{u}_0, \mathbf{u}_1, \ldots)$.

We reason in a way that is roughly parallel to the development of the strong flow criterion for particles tumbling on a limit cycle in a steady flow: we use the circle map itself to deduce the asymptotic proportion of iteratives that lie in every possible orientation on the attractor. The analog of $\mathrm{d}\theta/\mathrm{d}t$ for a map is $\theta_{n+1} - \theta_n$. Another possibility is $\theta_n - \theta_{n-1}$. Rather than skew the results either way, we shall use a center difference $\frac{1}{2}(\theta_{n+1} - \theta_{n-1})$. The inverse of the map is easily computed as

$$\theta_{n-1} = \arctan \frac{1 - d_2 \tan \theta_n}{-d_1 + d_3 \tan \theta_n}.$$

Therefore we define an approximate distribution function $\psi(\theta)$ as follows:

$$\psi(\theta) \equiv \tag{15}$$

$$\frac{2c}{\pi - \mathrm{mod}\,[\arctan(1 + d_1 \tan\theta)/(d_2 + d_3 \tan\theta) - \arctan(1 - d_2 \tan\theta)/(-d_1 + d_3 \tan\theta),\ \pi]}.$$

The mod is a necessary technicality, in view of the repeating nature of the θ coordinate. In addition, we have written $\psi(\theta)$ for the case of retrograde motion, i.e. $\theta_{n+1} < \theta_n < \theta_{n-1}$.

The constant c allows for the normalization of the distribution function,

$$\int_0^\pi \psi(\theta)\,\mathrm{d}\theta = 1.$$

With this definition, $\psi(\theta)\,\mathrm{d}\theta$ is to be interpreted as the asymptotic proportion of iterates θ_i that fall into the range of orientations $\theta - \mathrm{d}\theta/2 \leq \theta_i \leq \theta + \mathrm{d}\theta/2$. Then the strong flow criterion may be written

$$\frac{1}{T}\int_0^\pi \log \|\mathbf{Q}(T)\mathbf{u}_{\mathrm{ic}}(\theta)\|\psi(\theta)\,\mathrm{d}\theta > \alpha. \tag{16}$$

4.5. *Example*

Finally, we turn to an example to demonstrate the utility and accuracy of this strong flow criterion and of the analytic estimate of the statistical distribution of orientations on the quasi-periodic attractor. The example is taken from [10] (example IV of that paper); the flow corresponds to an approximately two-dimensional flow with a *wobbling vorticity axis*: when $nT \leqslant t < (n+\frac{1}{2})T$,

$$\boldsymbol{\kappa}(t) = \boldsymbol{\kappa}^{(1)} = \begin{bmatrix} 1 & 2 & 0 \\ -2 & -1 & 0 \\ 0 & 0 & 0 \end{bmatrix},$$

and when $(n+\frac{1}{2})T \leqslant t < (n+1)T$,

$$\boldsymbol{\kappa}(t) = \boldsymbol{\kappa}^{(2)} = \boldsymbol{\Theta}\boldsymbol{\kappa}^{(1)}\boldsymbol{\Theta}^{T},$$

where

$$\boldsymbol{\Theta} = \begin{bmatrix} \cos \pi/30 & 0 & \sin \pi/30 \\ 0 & 1 & 0 \\ -\sin \pi/30 & 0 & \cos \pi/30 \end{bmatrix}.$$

There is an analytic solution for $\mathbf{Q}$, for any T, given in [10]. We evaluate the flow for the period $T = 3.7$, when the dynamics are quasi-periodic. Then

$$\mathbf{Q} = \begin{bmatrix} 1.0414 & 0.142434 & 0.211468 \\ -0.144013 & 0.919964 & -0.00756304 \\ -0.218156 & -0.0225751 & 0.977774 \end{bmatrix}$$

has an invariant great circle, which lies in the plane normal to the vector $\mathbf{n} = (-0.0980251, -0.848882, 0.519413)$. The orthonormal pair of vectors which span the invariant subspace are $\mathbf{a} = (0.667198, 0.331199, 0.667198)$ and $\mathbf{b} = (-0.738402, 0.411954, 0.533907)$. We set $\mathbf{u}_{\text{ic}} = \cos\theta\mathbf{a} + \sin\theta\mathbf{b}$, and $\theta_0 = 0$.

First, we test the distribution function (15) for the distribution of orientations in time on the attracting invariant great circle. Beginning at $\theta_0 = 0$, the nonlinear map is iterated 100,000 times for a single initial orientation. Next, we form the distribution function directly by counting the number of iterates that land in each of 100 bins spanning the variable $0 \leqslant \theta < \pi$. Upon normalizing the bin counts, we obtain an approximation to the distribution function. This is plotted in Fig. 1 as a bar chart, along with a smooth curve that gives our analytic estimate of the distribution function. One observes that the analytic estimate does very well indeed at capturing the statistical distribution of states on the attractor for a single particle. In view of the dense orbit of the particle and the (discrete) translation invariance in time of equations (1) and (2), the statistical distribution of the number density of (many) particles is also represented by the solid curve in Fig. 1.

Now, in order to test the strong flow criterion (16), we first compute

$$\frac{1}{T}\int_0^{\pi} \log \|\mathbf{Q}(T)\mathbf{u}_{\text{ic}}(\theta)\| \psi(\theta)\, d\theta = 0.00691058 > \alpha.$$

This result must be compared to the result one obtains by integration of the differential equations in time. Therefore, for the purposes of comparison, we define $\alpha_c(n;\, \mathbf{u}_0)$ to be the critical value of the elastic modulus of the particle below which stretch occurs for the orientation time trace chosen:

$$\alpha_c(n;\, \mathbf{u}_0) = \frac{1}{nT}\sum_{i=0}^{n-1} \log \|\mathbf{Q}(T)\mathbf{u}_i\|.$$

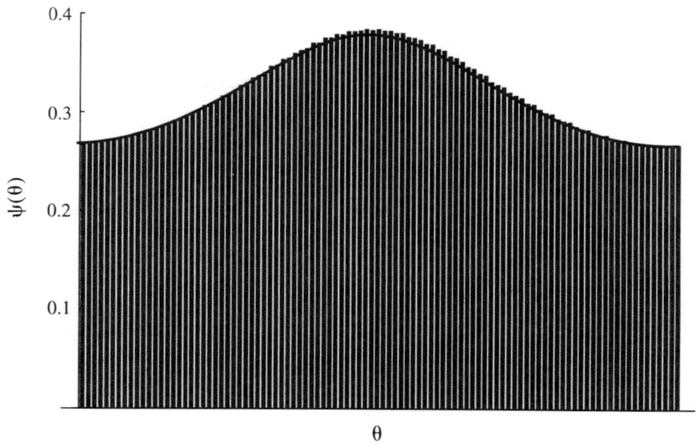

Fig. 1. Statistical distribution of orientations (in angle θ) on the invariant great circle of the Poincaré map for the example of Section 4.5. The bar chart indicates the distribution function computed for an orbit of 100,000 iterates, divided into 100 bins. The solid curve is the analytic estimate developed in the text.

In Fig. 2, we plot $\alpha_c(n;\ \mathbf{u}_0 = \mathbf{u}_{ic}(\theta = 0))$ versus n, the number of iterates of the flow over which the stretching power is measured. The solid line on Fig. 2 is the result of the analytic strong flow criterion, which does extremely well in capturing the asymptotic value of α_c for a single particle in a very few periods of the flow.

5. CONCLUSIONS

In this paper, we have described how to determine the stretching power of a fluid flow from the standpoint of a submerged, deformable particle which tumbles freely. Strong flow criteria, or sufficient conditions for stretching of particles, are developed in a way that

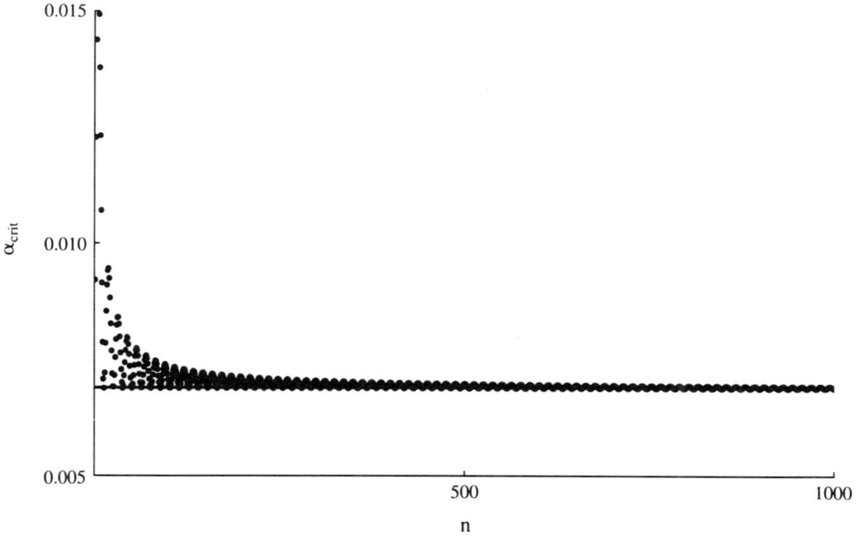

Fig. 2. Critical elastic modulus of the suspended particle, as a function of the number of iterates of the flow experienced by the particle. The solid curve is the analytic strong flow criterion developed without determining the orientation time trace of the particle. These results are for the example of Section 4.5.

accounts for the history of flows and orientation dynamics experienced by the particle. We focused on the case of orientation dynamics associated with an unsteady attractor. Strong flow criteria were developed by consideration of an analytical estimate of the statistical distribution of orientations (in time) on the relevant unsteady attractor.

Along the way, we specialized the orientation dynamical systems to the unsteady attractor, and solved analytically for the orientation as a function of time.

We have noted in the introduction that a most important problem is the ability to predict the behavior of orientable and deformable (stretchable) particles in complicated or even in chaotic flows. Clearly, if there is to be any hope of understanding this problem, it will be necessary to develop a technique to derive strong flow criteria in situations where we know the orientation evolution equations but not the detailed orientation history of a given particle. In other words, we will need to develop a statistical distribution of states of the suspended microstructure, and relate this to the stretching problem. That is precisely the step made in this paper in Section 4.4. There, we knew the orientation evolution equation on the attractor, we directly obtained the statistical distribution of states on the attractor, and we derived a strong flow criterion on that basis alone, i.e. without solving equations of motion.

This entire paper might be viewed as a paradigm for how to quantify stretching in complicated flows, where one cannot know the orientation dynamics. For example, consider taking the sphere of orientations where there is no attractor, and defining a distribution function of the form

$$\psi(\mathbf{u}) = \left\langle \frac{c}{|\mathrm{d}\mathbf{u}/\mathrm{d}t|} \right\rangle,$$

where $\mathbf{u}$ is a vector that denotes orientation, c is a constant, and the right-hand-side is averaged over some time interval $(\langle \cdot \rangle)$. Would this convey the statistical distribution of orientations of particles in the associated material point? Can this distribution function, suitably normalized, be used to develop a strong flow criterion? This is something worth considering, because $\mathrm{d}\mathbf{u}/\mathrm{d}t$ involves $\nabla\mathbf{v}(t)$ which may be a chaotic flow field.

REFERENCES

1. A. M. Gañán-Calvo and J. C. Lasheras, The dynamics and mixing of small spherical particles in a plane free shear layer, *Phys. Fluids* **A3**, 1207–1217 (1991).
2. K. K. Tio, A. M. Gañán-Calvo and J. C. Lasheras, The dynamics of small, heavy, rigid spherical particles in a periodic Stuart vortex flow, *Phys. Fluids* **A5**, 1679–1693 (1993).
3. R. Mallier and M. R. Maxey, The settling of nonspherical particles in a cellular flow field, *Phys. Fluids* **A3**, 1481–1494 (1991).
4. M. R. Maxey, The motion of small spherical particles in a cellular flow field, *Phys. Fluids* **30**, 1915–1928 (1987).
5. M. R. Maxey and S. Corrsin, Gravitational settling of aerosol particles in randomly oriented cellular flow fields, *J. Atmos. Sci.* **43**, 1112–1134 (1986).
6. J. B. McLaughlin, Particle size effects on Lagrangian turbulence, *Phys. Fluids* **31**, 2544–2553 (1988).
7. M. Tjahjadi and J. M. Ottino, Stretching and breakup of droplets in chaotic flows, *J. Fluid Mech.* **232**, 191–220 (1991).
8. J. M. Ottino, The kinematics of mixing: stretching, chaos and transport, Cambridge University Press, Cambridge (1989).
9. S. Wiggins, *Chaotic Transport in Dynamical Systems*. Springer, New York (1992).
10. A. J. Szeri and L. G. Leal, Microstructure suspended in three-dimensional flows, *J. Fluid Mech.* **250**, 143–167 (1993).
11. A. J. Szeri, W. J. Milliken and L. G. Leal, Rigid particles suspended in time-perdiodic flows: regular versus irregular motion, disorder versus order, *J. Fluid Mech.* **237**, 33–56 (1992).
12. A. J. Szeri, S. Wiggins and L. G. Leal, On the dynamics of microstructure in unsteady, spatially inhomogeneous, two-dimensional fluid flows, *J. Fluid Mech.* **228**, 207–241 (1991).

13. G. B. Jeffery, The motion of ellipsoidal particles immersed in a fluid, *Phil. Trans. R. Soc. Lond.* **A102**, 161–179 (1922).
14. W. L. Olbricht, J. M. Rallison and L. G. Leal, Strong flow criterion based on microstructure deformation, *J. Non-Newtonian Fluid Mech.* **10**, 291–318 (1982).
15. F. P. Bretherton, The motion of rigid particles in a shear flow at low Reynolds number, *J. Fluid Mech.* **14**, 284–304 (1962).
16. A. J. Szeri, Pattern formation in recirculating flows of suspensions of orientable particles, *Phil. Trans. R. Soc. Lond. A* **345**, 477–506 (1993).
17. J. Guckenheimer and P. Holmes, *Nonlinear Oscillations, Dynamical Systems and Bifurcations of Vector Fields*. Springer, New York (1983).

Interaction of Chaotic Advection and Diffusion

SCOTT W. JONES

Department of Theoretical and Applied Mechanics, University of Illinois, Urbana, IL 61801, USA

Abstract—The interaction of molecular diffusion and chaotic advection is examined in both unbounded and bounded flow domains. In unbounded flow domains molecular diffusion enables tracer particles to explore regions of regular particle motion and chaotic particle motion. Consequently the rate of particle dispersion in these mixed regimes by diffusive particles differs from that realized by perfect tracer particles. In bounded flow domains the effect of chaotic advection on the homogenization and dispersion of diffusive tracers is compared with some classical results on the integrable advection of diffusive tracers. Because the process of molecular diffusion is independent of the type of advection the $t \to \infty$ asymptotic behavior is the same in both chaotic advection and integrable advection. However, the rate at which these asymptotic states are approached is more rapid when the advecting flow generates chaotic particle trajectories. Simple scaling arguments are presented to determine the relevant timescale to achieve uniform mixing within a bounded domain. This timescale is also important in the process of shear dispersion.

1. INTRODUCTION

The phenomenon of chaotic advection highlights the distinction between the Lagrangian description of fluid motion by following marked fluid elements and the Eulerian description of motion at fixed points in space. It is now well known that a simple velocity field can yield chaotic trajectories of 'perfect' fluid particles [1]. Here the designation 'perfect' indicates that the particles move identically with the advecting velocity field. That is, the motion of particles by the velocity field $u_i(\mathbf{x}, t)$ is given by a set of ordinary differential equations of the form

$$\dot{x}_i = u_i(\mathbf{x}, t). \tag{1}$$

For two-dimensional unsteady flow or, three-dimensional steady flow, it is not possible, in general, to integrate equation (1). The subject of chaotic advection is based on this non-integrability and many studies of stirring and mixing use equations of this type to describe particle motion. Because in chaotic advection one is concerned with the fate of marked particles, the primary focus of these studies has been the issues of transport, dispersion and stirring of particles. Furthermore, because interesting behavior is observed at the deterministic and kinematic level of (1) relatively little attention has been paid to situations in which (1) might not be a valid representation of the true motion of the particle. However, (1) rarely provides a realistic portrait of the 'real' particle motion because real particles will either be large enough for inertial effects to be important or small enough to diffuse. For large particles the advection equations, (1), are augmented by additional equations of the form

$$\begin{aligned} \dot{v}_i &= A(u_i(\mathbf{x}, t) - v_i) \\ \dot{x}_i &= v_i(\mathbf{x}, t), \end{aligned} \tag{2}$$

and for small particles there are additional terms

$$\dot{x}_i = u_i(\mathbf{x}, t) + \zeta_i(t). \tag{3}$$

In (2) u_i is the ith component of the fluid velocity, v_i is the ith component of the particle velocity, and A is an inverse Stokes number. In (3) $\zeta_i(t)$ is a stochastic term with zero mean and covariance $\langle \zeta_i(t)\zeta_j(t+\tau)\rangle = 2D_{\text{mol}}\delta_{ij}\delta(\tau)$ with D_{mol} the molecular diffusivity. Several authors have studied systems described by (2) (cf. refs [2, 3]) and these will not be dealt with further here.

Aref [4] has coined the term 'kinematic template' to describe the motion given by the advection equations, which "... provide a base state of motion from which the motion of interest may deviate to a greater or lesser degree" (p. 275). In this paper I will focus my attention on the mixing and dispersion of diffusive particles given by (3). In particular I will examine how some classical results involving the advection and diffusion of passive scalars are modified by chaotic advection and attempt to describe the magnitude of deviation between the kinematic template (1) and the more realistic description given by (3). The relative importance of advection and diffusion in the transport of tracer is measured by the Péclet number, $Pe = UL/D_{\text{mol}}$, which is the ratio between a diffusive timescale (L^2/D_{mol}) and an advection timescale (L/U). The primary focus in this paper will be on the high Péclet number regime. However, it is important to note that in general the $Pe \to \infty$ limit is singular. This limit will be discussed in more detail below.

The effects of diffusion on the dispersion of tracers depend on the nature of the flow domain. In unbounded domains diffusion can either enhance or diminish the rate of dispersion with respect to non-diffusive tracers. This subject will be explored in Section 2, where some results from integrable velocity fields will also be discussed.

In bounded domains the general feature of the interaction between chaotic advection and diffusion is to diminish the duration of the transient phase leading to some asymptotic steady state. The scaling for the transient period in the homogenization process is examined in Section 3. The subject of Section 4 is shear dispersion and the scaling for the pre-asymptotic stage. In both of these situations the asymptotic scaling laws observed in the classical cases of integrable advection of diffusive tracers are recovered when the advection is chaotic.

2. DISPERSION IN UNBOUNDED DOMAINS

In flows of practical interest the Péclet number is typically large, and one might infer that the effect of molecular diffusion is negligible. However, the interaction of even weakly diffusive tracers with a flow field can lead to a significant enhancement (or, as we shall see below, diminution) of the rate at which the tracer is dispersed. I will begin by reviewing some results for the dispersion of diffusive tracers advected by integrable velocity fields. For example, it is well known that a typical lengthscale associated with a patch of diffusive tracer in a quiescent fluid grows like $t^{1/2}$ [5], while a patch of non-diffusive tracer in a linear shear flow grows like t in the direction of the flow. When both the shearing flow and diffusion are deforming the patch it is observed that the length of the patch grows like $t^{3/2}$ in the direction of the flow and like $t^{1/2}$ normal to the flow. That is, the two effects combine multiplicatively for the streamwise dispersion. This result appears in the review paper by Wang and Uhlenbeck [6] and was obtained by solving the advection–diffusion equation by the method of characteristics. Note that the same scalings are obtained in semi-infinite domains [7].

For this simple shear flow it is possible to understand the above result by explicitly writing the appropriate generalized Langevin equation (3)

$$\begin{aligned} \dot{x}(t) &= \alpha y \\ \dot{y}(t) &= \zeta(t). \end{aligned} \tag{4}$$

The mean square displacement of a particle in the y-direction is proportional to $t^{1/2}$ and the length associated with an ensemble of these particles grows at the same rate. This result then implies $\dot{x}(t) \sim t^{1/2}$ and performing the trivial integration yields the $t^{3/2}$ scaling noted above. The advantage of this Lagrangian 'analysis' is that it highlights the physical explanation for this rapid dispersion: diffusion across streamlines enables the tracer to consistently sample ever faster-moving regions of the flow.

What happens to the rate of tracer dispersion when the advection equations are non-integrable? In this case the flow domain is segregated into regions of regular particle motion and regions of chaotic particle motion. Figure 1 contrasts these two types of motion for the so-called ABC flow (cf. ref. [8])

$$\begin{aligned} \dot{x}(t) &= A \sin z + C \cos y \\ \dot{y}(t) &= B \sin x + A \cos z \\ \dot{z}(t) &= C \sin y + B \cos x, \end{aligned} \tag{5}$$

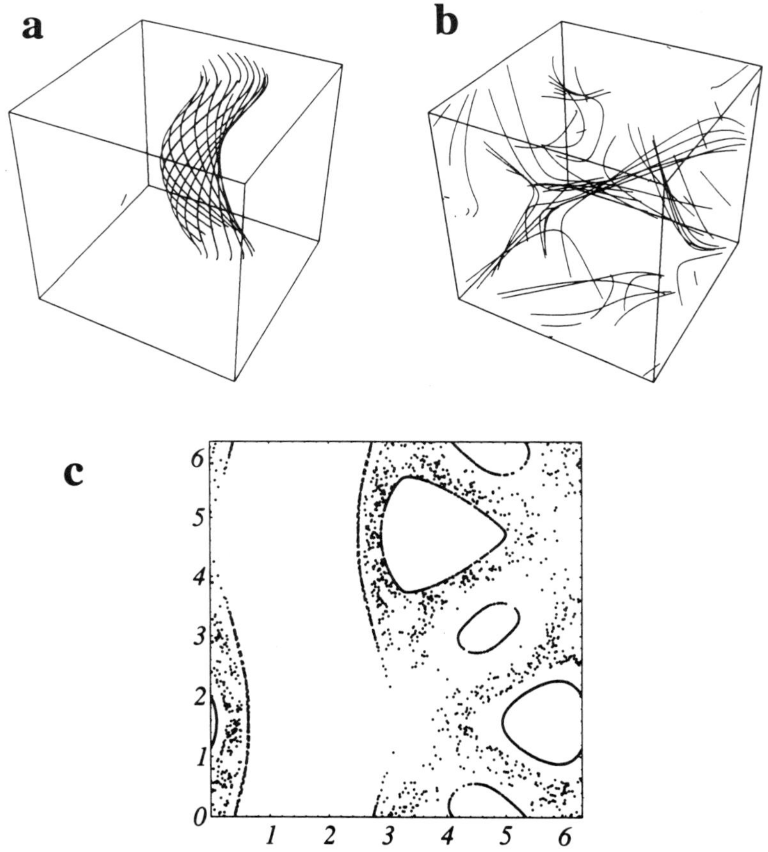

Fig. 1. Particle trajectories in the ABC flow (5) with $A = 1$, $B = C = \sqrt{2}$. A regular trajectory is shown in (a) and a chaotic trajectory is shown in (b). (c) A Poincaré section constructed by recording the intersections of several trajectories with the upper surface of the cube shown in (a) and (b).

where A, B, and C are parameters. There is no mixing of fluid between the regular streamtubes and the chaotic regions surrounding them. The ABC flow possesses the Beltrami property ($\boldsymbol{\omega} \times \mathbf{u} = 0$). In fact, the ABC flow has the feature that $\boldsymbol{\omega} = \mathbf{u}$ so streamlines are vortex lines. For $ABC = 0$ the flow is integrable, but for $ABC \neq 0$ the flow domain consists of regular streamlines and chaotic streamlines. Within the regular regions the variance of the distribution of non-diffusive tracer grows like t^2 in the direction of the streamline. The dispersion of a perfect tracer in the chaotic sea grows more slowly, linearly in time if the size of the regular regions is negligible. Consequently, in this mixed regime, the $t \to \infty$ behavior is dominated by the regular islands. In contrast, diffusive particles sample both types of streamlines and the dispersion of the tracer is intermediate between the ballistic spreading in islands and the diffusive spreading in the chaotic sea. This difference is illustrated in Fig. 2. The flow parameters are the same as those used to generate Fig. 1. Figure 2(a) shows the dispersion in the x-direction of 10 000 perfect particles placed uniformly within a cube of side 2π. As expected, the temporal evolution of the variance is quadratic in time. The rate of dispersion of diffusive particles is shown in Fig. 2(b) and is measured to be proportional to $t^{1.7}$. Here, as in the linear shear flow, there is a quantitative change in the rate of dispersion when the tracer is diffusive. However, in the mixed regime of chaotic advection, where regular and chaotic trajectories coexist, diffusion reduces the spreading rate. The explanation for this reduction is that particles spend some time on regular trajectories and some time on chaotic trajectories. The scaling exponent reflects the fraction of the flow domain occupied by regular trajectories.

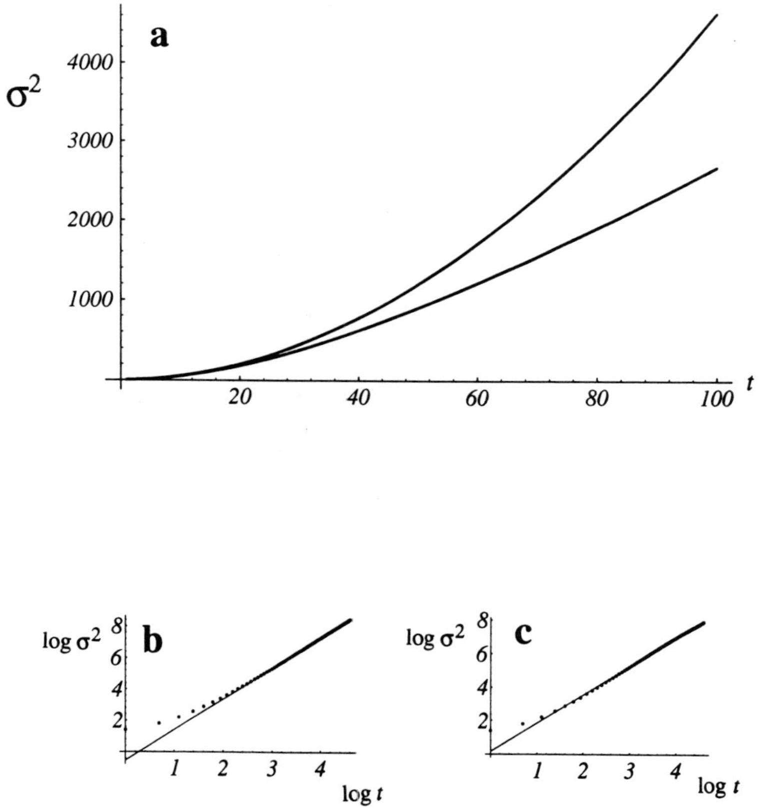

Fig. 2. Dispersion by the ABC flow (5) with $A = 1$, $B = C = \sqrt{2}$. In (a) the upper curve is for $D_{mol} = 0$ and the lower curve is for $D_{mol} = 0.01$. (b) and (c) are log–log plots of the particle variance in the x-direction versus time. In (b) $D_{mol} = 0$ and the slope of the curve is 2. In (c) $D_{mol} = 0.01$ and the slope of the curve is 1.7.

The *ABC* flow contains large islands at all parameter values. Thus, to explore the effect of diffusion on dispersion by flows that contain negligible islands it is necessary to investigate a flow that is a generalization of the *ABC* flow. A general formulation for a family of Beltrami flows (that included the *ABC* flow as a special case) has been given by Yakhot *et al*. [9] and has the form

$$u(\mathbf{x}) = \sum_{\mathbf{Q}\in\mathbf{S}} A(\mathbf{Q})(\mathbf{n} + i\mathbf{Q} \times \mathbf{n}) \exp(i\mathbf{Q}\cdot\mathbf{x}/l), \tag{6}$$

where $\mathbf{S}$ is a finite set of unit vectors $\mathbf{Q}$, $\mathbf{n}$ is a unit vector perpendicular to $\mathbf{Q}$, and the complex amplitudes are conjugate symmetric, $A(-\mathbf{Q}) = A^*(\mathbf{Q})$. The *ABC* flow consists of the six unit vectors $\mathbf{S} = \{(\pm1, 0, 0), (0, \pm1, 0), (0, 0, \pm1)\}$. By including the additional unit vectors $\pm\sqrt{3}(1/3, 1/3, 1/3)$ it is possible to create a flow with very small islands. The advection equations take the form

$$\begin{aligned} \dot{x}(t) &= A\sin z + C\cos y - 2D\sin(x + y + z)/\sqrt{3} \\ \dot{y}(t) &= B\sin x + A\cos z + D(\sin(x + y + z)/\sqrt{3} - \cos(x + y + z)) \\ \dot{z}(t) &= C\sin y + B\cos x + D(\sin(x + y + z)/\sqrt{3} + \cos(x + y + z)) \end{aligned} \tag{7}$$

where an additional parameter, D, appears. Dispersion by this flow with and without diffusion is contrasted in Fig. 3. In both cases the variance grows linearly in time.

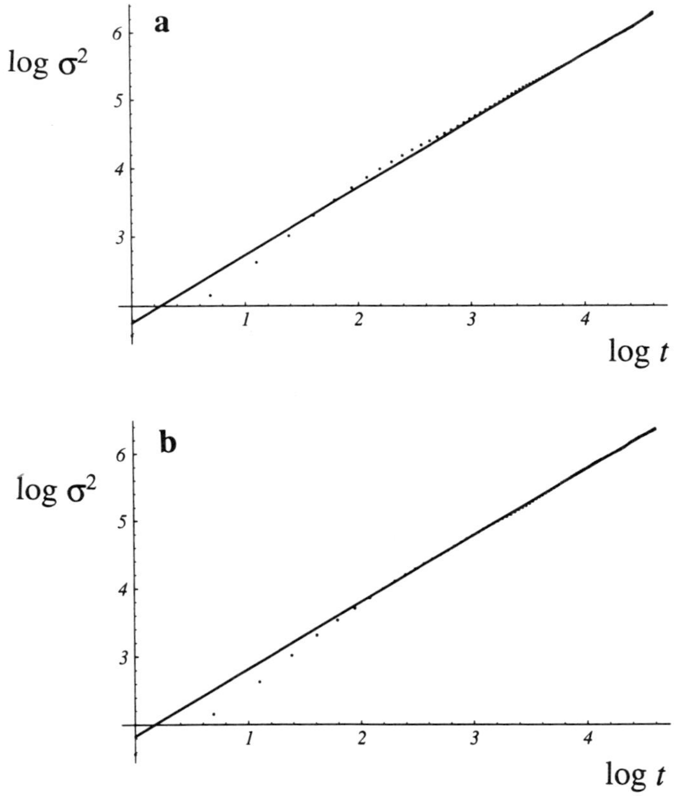

Fig. 3. Dispersion by the Beltrami flow (7) with $A = B = C = D = 1$. This is a log–log plot of the particle variance in the x-direction versus time. In (a) $D_{\mathrm{mol}} = 0$ and the slope of the curve is 1.0. In (b) $D_{\mathrm{mol}} = 0.01$ and the slope of the curve is 1.0.

3. CHAOTIC ADVECTION AND HOMOGENIZATION

In this section I will elaborate on a classical result that dates back to Prandtl [10]* and discussed in greater detail by Batchelor [11]. This result is frequently referred to as the 'Prandtl–Batchelor Theorem' and was discussed by these authors in terms of the vorticity and viscosity. Here we substitute the words 'tracer' for vorticity and (molecular) 'diffusivity' for viscosity. In words, the central feature of the Prandtl–Batchelor Theorem is that the steady-state distribution of a weakly diffusive tracer occupying the two-dimensional region within closed streamlines is uniform. More recently it has been shown by Rhines and Young [12] that the timescale on which this homogenization occurs is the diffusive timescale, $t \sim L^2/D_{\text{mol}}$, or in terms of the Péclet number, $t \sim (L/U)Pe$.

Below we briefly review the Prandtl–Batchelor result. The steady-state concentration of a passive tracer, $C(x, y)$, in a two-dimensional flow is described by

$$J(\psi, C) = D_{\text{mol}}(\partial_x^2 + \partial_y^2)C, \tag{8}$$

where the Jacobian, $J(a, b)$, is

$$J(a, b) = \partial_x a \partial_y b - \partial_x a \partial_y b$$

and the velocity components of the fluid are derived from the streamfunction $\psi(x, y)$ by

$$\dot{x}(t) = -\partial_y \psi, \qquad \dot{y}(t) = \partial_x \psi.$$

At this point we note that for perfect tracer $D_{\text{mol}} = 0$ and the solution of (8) has the form

$$C = C(\psi). \tag{9}$$

For small D_{mol} we can expand C in a power series in D_{mol}

$$C = C_0 + D_{\text{mol}} C_1 + \tfrac{1}{2} D_{\text{mol}}^2 C_2 + \dots$$

and use this expansion in (8) to find the first-order (in D_{mol}) expression

$$J(\psi, C_1) = (\partial_x^2 + \partial_y^2) C_0. \tag{10}$$

Integrating (10) over the area enclosed by a streamline gives

$$\iint J(\psi, C_1)\, \mathrm{d}x\, \mathrm{d}y = \iint (\partial_x^2 + \partial_y^2) C_0\, \mathrm{d}x\, \mathrm{d}y \tag{11a}$$

$$\iint \frac{\mathrm{d}C_1}{\mathrm{d}\psi}(\partial_x \psi \partial_y \psi - \partial_y \psi \partial_z \psi)\, \mathrm{d}x\, \mathrm{d}y = \oint \frac{\mathrm{d}C_0}{\mathrm{d}\psi}\, \mathrm{d}s = 0 \tag{11b}$$

$$\frac{\mathrm{d}C_0}{\mathrm{d}\psi} = 0. \tag{11c}$$

From equation (11c) it is clear that the concentration of diffusive tracer within the region of closed streamlines is independent of the streamfunction. This result is in contrast to the result for perfect tracer given by (9).

Rhines and Young [12] show that the homogenization process occurs in two steps. There is an initial step in which the tracer is homogenized along a streamline. This process is rapid and scales like $t \sim (L/U)Pe^{1/3}$. The second step is the diffusion of tracer across streamlines. This process is described by a diffusion equation with an effective diffusivity

---

*Apparently this result was first published in 1905 with the title "Über Flüssigkeitsbewegung bei sehr kleiner Reibung". This paper was reproduced in *Vier Abhandlungen zur Hydrodynamik und Aerodynamik* in 1927. In the following year an English translation appears as NACA Technical Memorandum No. 452. I am indebted to S. Leibovich for providing these references.

that is proportional to the molecular diffusivity. Consequently the timescale for homogenization by steady two-dimensional flows with closed streamlines is $t \sim (L/U)Pe$. For non-circular streamlines the effective diffusivity is greater than the molecular value and for circular streamlines the effective diffusivity is equal to the molecular value.

In two dimensions, chaotic advection can be achieved only by time-dependent flow fields and the concept of a streamline takes on a different meaning than the steady-state situation described above. However, the analog we are interested in is the homogenization of a tracer within a closed domain. Of course, for finite Péclet numbers the steady-state distribution is again a uniform distribution, but the time to attain this steady state is much shorter than for steady flow. In this case the mixing of tracers is accelerated by the chaotic trajectories. These trajectories enable tracer particles to sample a finite area of the domain. In fact, the scaling for the homogenization time is $t \sim (L/U)\ln Pe$, which diverges slowly as $Pe \to \infty$. This result can be derived in a heuristic fashion by considering the balance between the rate a material line is stretched, which for chaotic advection is exponential in time, and a characteristic diffusion length, which grows like $t^{1/2}$ [13]. For concreteness, consider the advection of a patch of fluid marked by red eye. Chaotic advection will stir this patch into long tendrils, and the flow domain will consist of striations of dyed and undyed fluid. In the absence of diffusion, the fractional area occupied by red fluid is a constant. However, the average distance between neighboring tendrils will diminish with time. On timescales short relative to a typical diffusion time, diffusive tracers will behave in essentially the same manner as non-diffusive tracers. Molecular diffusion will rapidly homogenize the fluid when it acts over short lengthscales. Letting A denote the area of the flow domain, the average spacing between striations is

$$l_{\text{sep}} = \frac{2A}{l}, \tag{12}$$

where l is the length of the idealized material contour bounding the red fluid in the absence of diffusion. The scaling for the growth of l by chaotic advection is $l = l_0 \exp(\lambda t)$, λ is the Lyapunov exponent of the advecting flow and l_0 is the initial length of the perimeter of red fluid. Equating (12) with the diffusion lengthscale $l_{\text{D}} \approx (D_{\text{mol}} t)^{1/2}$ gives the scaling for the homogenization time in terms of the Péclet number

$$t_{\text{hom}} \sim \ln Pe,$$

where the Péclet number is defined as $Pe = l_0^2 \lambda / D_{\text{mol}}$.

Finally, it is possible to contrast the steady-state concentration distribution for diffusive and perfect tracers advected by integrable and non-integrable flow fields. As noted above, the steady solution of the advection–diffusion equation for integrable flow fields is $C = C(\psi)$ for perfect tracers (see Fig. 4(a)) and $C = \text{const}$ for diffusive tracers. Of course, the chaotic advection of diffusive tracers yields the same solution as integrable advection, $C = \text{const}$. The solution for the chaotic advection of perfect tracers can have neither form. Because the velocity field is time-dependent, it is not possible to have a steady solution that is a function of ψ. In the absence of diffusion, the area of the domain occupied by red fluid is conserved by the incompressible flow. Thus it is always possible to find regions of clear fluid, and the concentration cannot be a constant independent of the spatial coordinates. However, given a finite lengthscale there will always be a finite time at which the fluids are mixed at lengthscales larger than this threshold. So perfect tracers, in the $t \to \infty$ limit, approache the diffusive solution (Fig. 4(c)).

This process of the effective homogenization of perfect tracers is illustrated in Fig. 5. This figure shows the logarithm of the concentration at fixed points in space plotted versus the instantaneous value of the streamfunction. The upper panel (Fig. 5(a)) is constructed

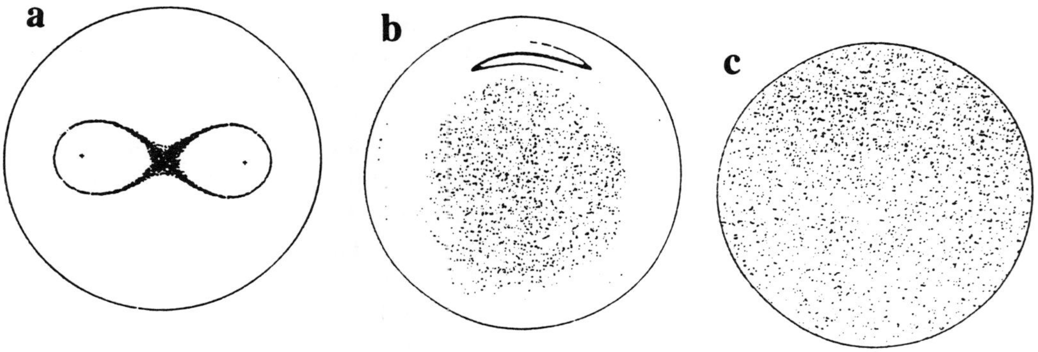

Fig. 4. The stirring of perfect particles by two point vortices located at fixed positions within a circular domain (see Aref [1] for details on this flow). In (a) the flow is steady and the particle distribution is a function of the streamfunction. In (b) the flow domain consists of a mixture of chaotic sea and islands and the particle distribution reflects this mixture. In (c) most particle trajectories are chaotic and the tracer appears homogenized.

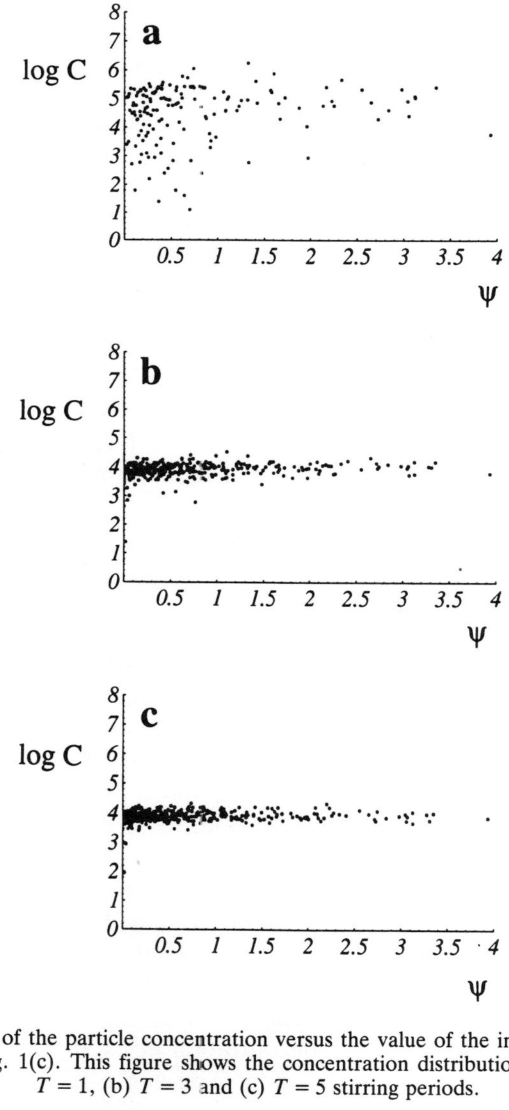

Fig. 5. Plots of the logarithm of the particle concentration versus the value of the instantaneous streamfunction of the flow used to generate Fig. 1(c). This figure shows the concentration distribution at three different times: (a) $T = 1$, (b) $T = 3$ and (c) $T = 5$ stirring periods.

after one stirring period and shows the concentration distribution is related to the streamfunction in some complicated way. In Fig. 5(b), after 3 stirring periods, it is apparent that the concentration is becoming more uniform, and by 5 stirring periods (Fig. 5(c)) the concentration appears to be independent of the streamfunction (at the level of coarse-graining in these concentration measurements).

In the above discussion I have assumed that almost all trajectories are chaotic. Of course, it is well known that non-integrable flows possess a mixture of chaotic and regular trajectories. For perfect particles in this mixed regime the nature of the local concentration field depends on whether the tracer is being advected in a regular manner, on KAM curves, or chaotically. By analogy with the distribution of perfect tracers in steady flows we expect the concentration distribution of perfect tracers in islands to be a function of the KAM tori (Fig. 1(b)). All of these cases are summarized in Table 1.

4. CHAOTIC ADVECTION AND SHEAR DISPERSION

The next classical result I wish to re-examine in the context of chaotic advection is shear dispersion. In a paper published in 1953, Taylor [14] observed that, after an initial transient, a passive scalar impurity flowing through a circular tube obeys a one-dimensional diffusion equation

$$\partial_t \bar{C}(x, t) + \bar{U}\partial_x \bar{C}(x, t) = D_{\text{eff}}\partial_x^2 \bar{C}(x, t) \tag{13}$$

with an effective diffusivity, D_{eff}, that is inversely related to the molecular value. In (13) x is the axial coordinate, $\bar{C}$ is the tracer concentration, $\bar{U}$ is the axial velocity component, and the overbar denotes an average over the cross-section of the tube. For the tube with circular cross-section studied by Taylor the averaging operator is

$$\overline{(\bullet)} = \frac{2}{a}\int_0^a (\bullet) r \, \mathrm{d}r \tag{14}$$

and the effective diffusivity is given by

$$D_{\text{eff}} = D_{\text{mol}} + \frac{U^2 a^2}{48 D_{\text{mol}}}, \tag{15}$$

where a is the radius of the tube. Equation (13) is an asymptotic result that is only valid at long times.

It is interesting to note Taylor's qualitative description of this process, "... since water moves at twice the mean speed near the centre of the pipe and the patch of colour at the mean speed, the clear water in the middle must approach the colour patch, absorb colour as it passes into it and then lose colour as it passes out, finally leaving the patch as perfectly clear water" (p. 187). This remark illustrates the difference between the kinematic template and the true behavior of the dye. In the absence of diffusion the dye would be advected as a fluid element. Owing to the velocity shear, the cross-stream concentration gradients and the separation of particles on different streamlines would grow

Table 1. The $t \to \infty$ scalings for the concentration field of diffusive and perfect tracers

Flow regime	Diffusive tracer	Perfect tracer
Regular	C = const	$C = C(\psi)$
Mixed	C = const	$C = C$(KAM curves) in islands, $C \to$ const elsewhere
Irregular	C = const	$C \to$ const

linearly in time. However, the motion of a diffusive tracer particle differs from that of a fluid element and the unconstrained growth of concentration gradients is arrested by diffusion. In fact, Taylor's result is due to the balance between the axial shear flow and radial diffusion. It is the radial diffusion that enables the patch of dye to sample the entire axial velocity profile. Consequently the centre of mass of the dye moves downstream with the sectionally-averaged velocity and the spread about the center of mass grows only like $t^{1/2}$. In other words, the center of mass of the tracer evolves like $\bar{x} \approx Ut$ and the variance about the mean position grows like

$$\sigma^2 \approx 2D_{\text{eff}}t. \tag{16}$$

I will return to (16) below.

The chaotic advection of non-diffusive tracers in a tube geometry would appear to contain the necessary ingredients for observing shear dispersion. In the absence of islands the chaotic streamlines permit each tracer particle to sample the entire axial velocity profile. Thus at long times we would expect to observe a cloud of tracer moving down the tube with a Gaussian distribution. The results of numerical experiments presented by Jones and Young [15] in a 'twisted pipe' geometry show that this does not occur. Instead, the distribution of tracers is skewed with a long tail of slow moving particles. Although the center of mass translates with the mean velocity, the variance grows like $t \ln t$ which is slightly faster than the linear rate obtained for the integrable advection of diffusive tracers. However, if we incorporate molecular diffusion with chaotic advection, we once again find $\sigma^2 \sim t$.

In this example, we see that there is a significant quantitative deviation between the kinematic template and the asymptotic evolution of a real tracer. Already at the level of the variance we see this deviation in the form of the logarithmic correlation. Furthermore, the skewness grows like t^2 rather than decaying like $t^{-1/2}$ as for diffusive tracers.

The explanation for both the $t \ln t$ growth of the variance in the absence of diffusion and the reappearance of the linear growth rate in the presence of molecular diffusion lies in the form of the velocity autocorrelation function. Owing to the no-slip boundaries along the tube the tail of the correlation function has the form

$$\mathscr{C}(t) = p/t, \tag{17}$$

where p is a coefficient with the dimension of time. The relationship between the correlation function and the variance is another important result due to Taylor [16]

$$\frac{d\sigma^2}{dt} = 2\overline{u'^2}\int_0^t \mathscr{C}(t')\,dt'. \tag{18}$$

Integration of (18) using the correlation function (17) gives an expression for the variance

$$\sigma^2 \approx 2U^2 pt \ln t \qquad \text{as} \quad t \to \infty \tag{19}$$

which agrees with the numerical experiments. The introduction of a small diffusivity changes this result in the following way. Molecular diffusion ultimately causes the correlation function to decay exponentially. As a consequence (18) no longer diverges with time. For large Péclet numbers the integral in (18) will still be dominated by the t^{-1} tail, but the integration is only carried out to some fixed value, t_{max}. The scaling for t_{max} has been worked out by Acrivos and Taylor [17] and has the form $t_{\text{max}} \sim Pe^{1/3}$. Thus the leading term of the long-time behavior of the variance for chaotic advection + diffusion is

$$\sigma^2 \sim (\tfrac{1}{3}D_{\text{mol}} Pe \ln Pe)t. \tag{20}$$

The effect of the Péclet number is quantitatively different for regular advection. Equation (16) can be rewritten in terms of *Pe* as

$$\sigma^2 \sim D_{\text{mol}} Pe^2 t \tag{21}$$

so the influence of the Péclet number as $Pe \to \infty$ is buffered by chaotic advection.

Finally, I conclude this section with a brief discussion of the scaling of the transient period. Shear dispersion occurs owing to the interplay between the velocity shear and the mixing of tracers across streamlines. The asymptotic diffusive behavior occurs after the tracer is well-mixed across the cross-section of the pipe. Thus the timescale before diffusive spreading of tracers is observed is the same as that required to achieve homogenization. Therefore, it is possible to estimate the duration of the pre-asymptotic transient by recalling the results for homogenization discussed in Section 2. In the straight tube considered by Taylor the tracer can only sample the axial velocity by diffusion, so in order for (13) to be valid

$$t \gg \frac{a}{U} Pe.$$

Even in curved tubes, where a secondary flow is established, the same scaling occurs because the streamlines of the flow are restricted to two-dimensional surfaces (streamtubes) and tracers can only cross these streamtubes by diffusion. However, chaotic advection enables tracers to wander seemingly randomly across the tube. As a consequence the relevant timescale is

$$t \gg \frac{a}{U} \ln Pe,$$

as was described in Section 2 above.

5. DISCUSSION

The central point of this article is that the effect of molecular diffusion on the dispersion process depends on the nature of the flow domain. In unbounded flow domains molecular diffusion permits tracer particles to sample regions of both regular motion and regions of chaotic motion. Consequently the dispersion obeys a power law with an exponent that is intermediate between that of a diffusive process ($\sigma^2 \sim t$) and ballistic spreading ($\sigma^2 \sim t^2$).

In bounded domains, on timescales that are long relative to a diffusive timescale, the distribution of tracer particles is independent of the style of advection. However, there is a quantitative difference in the timescale on which that asymptotic behavior is reached between chaotic advection and regular advection. The exponential divergence of nearby particles that are advected chaotically enhances the rate at which particles explore the flow domain. Consequently, the mixing of tracers is achieved more rapidly by chaotic advection. The same process is also responsible for the reduction of the pre-asymptotic transient in shear dispersion.

It is sometimes believed that the complex particle motion due to chaotic advection is equivalent to the regular advection of diffusive particles. However, in a non-integrable flow that possesses both trajectories that are confined to KAM tori and chaotic trajectories there is no mixing of perfect tracers between these regions. Thus, unlike the diffusive case, the tracer can never sample the entire flow domain. Furthermore, unlike diffusive processes, these regular structures allow long-time correlations in the particle motion. As a result the long-time behavior is dominated by the islands rather than the chaotic sea. Molecular diffusion provides a mechanism that eliminates the invariant structures, such as KAM tori, and truncates the correlation function. For diffusive tracers, the underlying template of chaotic motion is preserved, but the invariance of the KAM tori is removed. The result of chaotic trajectories and molecular diffusion, as we have seen, is to promote the mixing of tracers beyond that achieved by the interaction of integrable advection and diffusion.

REFERENCES

1. H. Aref, Stirring by chaotic advection, *J. Fluid Mech.* **143**, 1–21 (1984).
2. J. B. McLaughlin, Particle size effects on Lagrangian turbulence, *Phys. Fluids* **31**, 2544–2553 (1988).
3. L.-P. Wang, T. D. Burton and D. E. Stock, Quantification of chaotic dynamics for heavy particle dispersion in ABC flow, *Phys. Fluids* **A3**, 1073–1080 (1991).
4. H. Aref, Chaotic advection of fluid particles, *Phil. Trans. R. Soc. Lond.* **A333**, 273–288 (1990).
5. S. Chandrasekhar, Stochastic problems in physics and astronomy, *Rev. Mod. Phys.* **15**, 1–89 (1943).
6. M. C. Wang and G. E. Uhlenbeck, On the theory of Brownian motion—II, *Rev. Mod. Phys.* **17**, 323–342 (1945).
7. P. G. Saffman, The effect of wind shear on horizontal spread from an instantaneous ground source, *Q.J.R. Met. Soc.* **88**, 382–393 (1962).
8. T. Dombre, U. Frisch, J. M. Greene, M. Hénon, A. Mehr and A. M. Soward, Chaotic Streamlines and Lagrangian turbulence: The *ABC* flows, *J. Fluid Mech.* **167**, 353–391 (1986).
9. V. Yakhot, B. J. Bayly and S. A. Orszag, Analogy between hyperscale transport and cellular automation fluid dynamics, *Phys. Fluids* **29**, 2025–2027 (1986).
10. L. Prandtl, Über Flüssigkeitsbewegung bei sehr kleiner Reibung, in *Verhandlungen des dritten Internationalen Mathematikerkongress*. Teubner, Leipzig (1905).
11. G. K. Batchelor, On steady laminar flow with closed streamlines at large Reynolds number, *J. Fluid Mech.* **1**, 177–190 (1956).
12. P. B. Rhines and W. R. Young, How rapidly is a passive scalar mixed within closed streamlines? *J. Fluid Mech.* **133**, 133–145 (1983).
13. S. W. Jones, The enhancement of mixing by chaotic advection, *Phys. Fluids* **A3**, 1081–1086 (1991).
14. G. I. Taylor, Dispersion of soluble matter in solvent flowing slowly through a tube, *Proc. R. Soc. Lond.* **A219**, 186–203 (1953).
15. S. W. Jones and W. R. Young, Shear dispersion and anomalous diffusion by chaotic advection, *J. Fluid Mech.* Submitted (1994).
16. G. I. Taylor, Diffusion by continuous movements, *Proc. Lond. Math. Soc.* **20**, 196–212. (1921).
17. A. Acrivos and T. D. Taylor, Heat and mass transfer from single spheres in Stokes flow, *Phys. Fluids* **5**, 387–394 (1962).
18. G. E. Uhlenbeck and L. S. Ornstein, On the theory of Brownian motion, *Phys. Rev.* **36**, 823–841 (1930).

Resonant and Chaotic Advection in a Curved Pipe

P. E. HYDON*

Department of Applied Mathematical Studies, University of Leeds, Leeds LS2 9JT, UK

Abstract—The Lagrangian description is used to study the motion of fluid particles in a curved pipe, when laminar fully developed flow is generated by an oscillatory longitudinal pressure gradient. The (viscous) fluid is assumed to be incompressible, so that the transverse motion is governed by a Hamiltonian dynamical system, which is near-integrable when the flow is quasi-steady. Islands and chaotic regions are seen in a Poincaré section, in accordance wtih Kolmogorov–Arnol'd–Moser (KAM) theory. Resonance between the longitudinal and transverse motion causes ballistic longitudinal transport of particles whose trajectories lie on KAM tori within certain islands. Intermittent longitudinal transport is seen in the chaotic regions bordering such islands, as trajectories are trapped by cantori in the vicinity of both resonant and non-resonant KAM tori. The interaction of advection and molecular diffusion is discussed. When the Péclet number is very large, particles are transported longitudinally in an intermittent manner which resembles transport in chaotic regions. Here the trapping of trajectories is caused by islands, not cantori.

1. INTRODUCTION

The advection of a passive scalar through a curved pipe is a problem with many applications, such as the transport of gases through the airways of the lung and solute transport in chemical reactors. The interaction between advection and molecular diffusion has been studied over a wide parameter range (Eckmann and Grotberg [1], Sharp *et al.* [2]). In the current paper, attention is focused on the Lagrangian dynamics of fluid particles in oscillating flows, with the aim of understanding solute transport in such flows at very large Péclet number.

The pipe geometry is shown in Fig. 1; the pipe radius, a, and the radius of curvature, R, are taken to be uniform, and the pipe's centreline is an arc of a circle. (Chaotic advection by a steady flow in a twisted pipe has been studied by Jones *et al.* [3].) It is also assumed that the flow is laminar and fully-developed (and therefore independent of $\hat{s}$), and that the fluid is viscous and incompressible. The motion of a fluid particle is determined by the equations

$$\frac{d\hat{x}}{d\hat{t}} = \hat{u}\,(\hat{x}(\hat{t}),\, \hat{y}(\hat{t}),\, \hat{t}), \tag{1}$$

$$\frac{d\hat{y}}{d\hat{t}} = \hat{v}\,(\hat{x}(\hat{t}),\, \hat{y}(\hat{t}),\, \hat{t}), \tag{2}$$

$$\left(1 + \frac{\hat{x}}{R}\right)\frac{d\hat{s}}{d\hat{t}} = \hat{w}\,(\hat{x}(\hat{t}),\, \hat{y}(\hat{t}),\, \hat{t}), \tag{3}$$

where $(\hat{u}, \hat{v}, \hat{w})$ are the $(\hat{x}, \hat{y}, \hat{s})$ components of the Eulerian velocity field; a caret over a variable denotes that it is dimensional. If (1) and (2) can be solved to find the transverse

*Present address: Department of Mathematics and Statistics, University of Northumbria at Newcastle, Newcastle-Upon-Tyne NE1 8ST, UK.

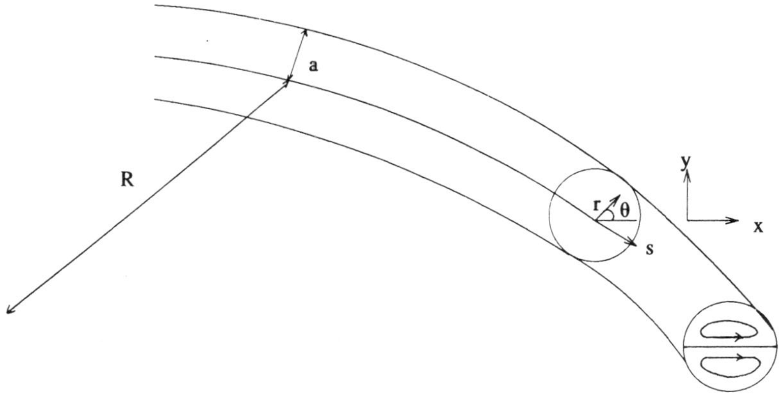

Fig. 1. Geometry of the curved pipe. Polar and Cartesian coordinates are used to describe motion in the transverse plane.

components of the particle's trajectory, then $\hat{s}(\hat{t})$ may be found from (3) by quadrature. The incompressibility condition is

$$\left(1+\frac{\hat{x}}{R}\right)^{-1}\left\{\frac{\partial}{\partial\hat{x}}\left[\left(1+\frac{\hat{x}}{R}\right)\hat{u}\right]+\frac{\partial}{\partial\hat{y}}\left[\left(1+\frac{\hat{x}}{R}\right)\hat{v}\right]\right\}=0. \tag{4}$$

Therefore, there exists a secondary streamfunction, $\hat{\psi}\,(\hat{x},\,\hat{y},\,\hat{t})$ such that

$$\hat{u}=\left(1+\frac{\hat{x}}{R}\right)^{-1}\frac{\partial\psi}{\partial\hat{y}};\qquad \hat{v}=-\left(1+\frac{\hat{x}}{R}\right)^{-1}\frac{\partial\psi}{\partial\hat{x}}. \tag{5}$$

Combining (1), (2) and (5), we see that the transverse components of the trajectory of a fluid particle are governed by the Hamiltonian dynamical system

$$\frac{d\hat{X}}{d\hat{t}}=\frac{\partial\hat{\psi}}{\partial\hat{y}},\qquad \frac{d\hat{y}}{d\hat{t}}=-\frac{\partial\hat{\psi}}{\partial\hat{X}}, \tag{6}$$

where

$$\hat{X}=\hat{x}\left(1+\frac{\hat{x}}{2R}\right). \tag{7}$$

Until now, the only requirement is that the flow is laminar and fully developed; no other constraints have yet been placed on either the curvature or the flow parameters. If the flow is steady, then (6) is integrable, and the problem is reduced to quadrature. Then (in general) each trajectory twists around a smooth secondary streamsurface (on which ψ is constant) whilst being advected longitudinally. We shall refer to the transverse components of a particle's trajectory as the *secondary orbit* of the particle (whether the flow is steady or not). When the secondary flow is steady, almost all secondary orbits are topologically equivalent to circles.

Equation (6) is usually non-integrable when the flow is unsteady, but some analytical progress may be made when the flow is near-integrable. We study quasi-steady flow which is driven by a periodic longitudinal pressure gradient. Here (6) is near-integrable, and KAM theory (see Arnol'd [4]) is applicable. Particles which lie in certain 'islands' are transported longitudinally by resonance between the secondary rotation and the longitudinal oscillation. (Resonant transport was first reported by Pedley and Kamm [5], who studied the interaction between longitudinal oscillations and a steady (integrable) secondary motion.) It is shown below that the set of all trajectories involved in resonant transport has non-zero measure; the proportion of particles which are transported by resonance is approximately the proportion of the pipe cross-section which is occupied by resonant islands (as visualized in a Poincaré section).

The paper concludes with a discussion of transport in those regions of the pipe where there is Lagrangian chaos, and of the main effects of molecular diffusion at large Péclet number.

2. EQUATIONS OF MOTION

From now on, it is assumed that $\delta \equiv a/R \ll 1$, and all but the leading order terms in an expansion of the governing equations in powers of δ are neglected. This assumption is made so that the velocity field is easy to calculate. However, it has been shown that the transverse motion is Hamiltonian for arbitrary δ, and so the behaviour which is described below will also be seen when δ is not very small.

Consider the flow corresponding to the oscillatory pressure gradient

$$\frac{\partial \hat{p}}{\partial \hat{s}} = -\rho K \cos \omega \hat{t}, \tag{8}$$

where ρ is the (uniform) fluid density, and K is an amplitude parameter. Using polar coordinates, with $\hat{x} = \hat{r}\cos\theta$, $\hat{y} = \hat{r}\sin\theta$, the Navier–Stokes equations reduce to

$$\frac{\alpha^2}{2\pi}\frac{\partial}{\partial t}\nabla^2\psi - \frac{1}{r}\frac{\partial \psi}{\partial r}\frac{\partial}{\partial \theta}\nabla^2\psi + \frac{1}{r}\frac{\partial \psi}{\partial \theta}\frac{\partial}{\partial r}\nabla^2\psi - G\left(\frac{\partial w}{\partial r} w \sin\theta + \frac{1}{r}\frac{\partial w}{\partial \theta} w\cos\theta\right) = \nabla^4\psi, \tag{9}$$

$$\frac{\alpha^2}{2\pi}\frac{\partial w}{\partial t} - \frac{1}{r}\frac{\partial \psi}{\partial r}\frac{\partial w}{\partial \theta} + \frac{1}{r}\frac{\partial \psi}{\partial \theta}\frac{\partial w}{\partial r} - \alpha^2\cos 2\pi t = \nabla^2 w, \tag{10}$$

where

$$\psi = \nu^{-1}\hat{\psi}, \qquad w = \frac{\omega}{K}\hat{w}, \qquad t = \frac{\omega}{2\pi}\hat{t}, \qquad r = a^{-1}\hat{r}, \qquad \nabla^2 = \frac{\partial^2}{\partial r^2} + \frac{1}{r}\frac{\partial}{\partial r} + \frac{1}{r^2}\frac{\partial^2}{\partial \theta^2} \tag{11}$$

(ν is the kinematic viscosity of the fluid). Note that the pressure gradient oscillation has unit period when time is nondimensionalized as above.

The dimensionless parameters occurring in (9) and (10) are

$$\alpha^2 = \frac{a^2\omega}{\nu}, \qquad G = \frac{2\delta a^2 K^2}{\omega^2\nu^2}. \tag{12}$$

α^2 is the 'unsteadiness parameter' (for quasi-steady flows, $\alpha^2 \ll 1$), and $\alpha^2\sqrt{G}$ is an 'unsteady Dean number.' The velocity vanishes at the solid boundary $r = 1$, and is bounded at $r = 0$, and therefore

$$\frac{\partial \psi}{\partial \theta} = \frac{\partial \psi}{\partial r} = w = 0 \text{ at } r = 1; \quad \frac{1}{r}\frac{\partial \psi}{\partial \theta}, \quad \frac{\partial \psi}{\partial r}, \quad w \text{ bounded at } r = 0. \tag{13}$$

A regular series expansion (in powers of α^2) is used to solve (9), (10) and (13). It is also assumed that $\alpha^2\sqrt{G} < 96$, which is equivalent to the limiting value of Dean's expansion parameter (Mullin and Greated [6]). Details of the solution are given in Hydon [7]. It turns out that the streamfunction may be written in the form

$$\psi = \frac{\alpha^2}{\pi} H(x, y, t)\cos^2 2\pi t, \qquad \text{where} \quad H(x, y, t) = H_l(x, y) + H_N(x, y, t); \tag{14}$$

here H_N is a small periodic perturbation (with period $\frac{1}{2}$) to the time-independent term H_I. In terms of the stretched time variable,

$$T = \int_0^t 2\cos^2 2\pi\tau \, \mathrm{d}\tau = t + \frac{1}{4\pi}\sin 4\pi t, \tag{15}$$

the equations of transverse particle motion are (to leading order in δ)

$$\frac{\mathrm{d}x}{\mathrm{d}T} = \frac{\partial H}{\partial y}, \qquad \frac{\mathrm{d}y}{\mathrm{d}T} = -\frac{\partial H}{\partial x}. \tag{16}$$

Note that (15) may be used to find $t(T)$, and that any function which is periodic in t with period $n/2$ (where n is a positive integer), is also periodic (with period $n/2$) in terms of T. Therefore the longitudinal velocity has period 1 in terms of T, and the perturbation H_N has period $\frac{1}{2}$. If H is replaced by H_I in (16), the resulting problem is autonomous (and therefore integrable); some trajectories are shown in Fig. 2. Each particle lies on a unique smooth closed curve, and the set of all such curves fills the transverse plane. Suppose that a particle is at (x_0, y_0) when $T = T_0$ and first returns to (x_0, y_0) when $T = T_0 + \bar{T}(x_0, y_0)$. The period of the unperturbed secondary orbit, $\bar{T}(x_0, y_0)$, is a well-defined smooth function of x_0 and y_0, and tends to infinity as $x_0^2 + y_0^2 \to 1$, or as $y_0 \to 0$.

Figure 3 is a Poincaré section which shows the effect of the perturbation H_N upon the secondary motion. Those secondary orbits whose period is commensurate with the period of the perturbation (i.e. rational) break into chains of islands, each surrounded by a region in which the secondary motion is chaotic. Secondary orbits whose period is 'sufficiently irrational' lie densely on KAM tori, and the projection of a KAM torus onto the Poincaré

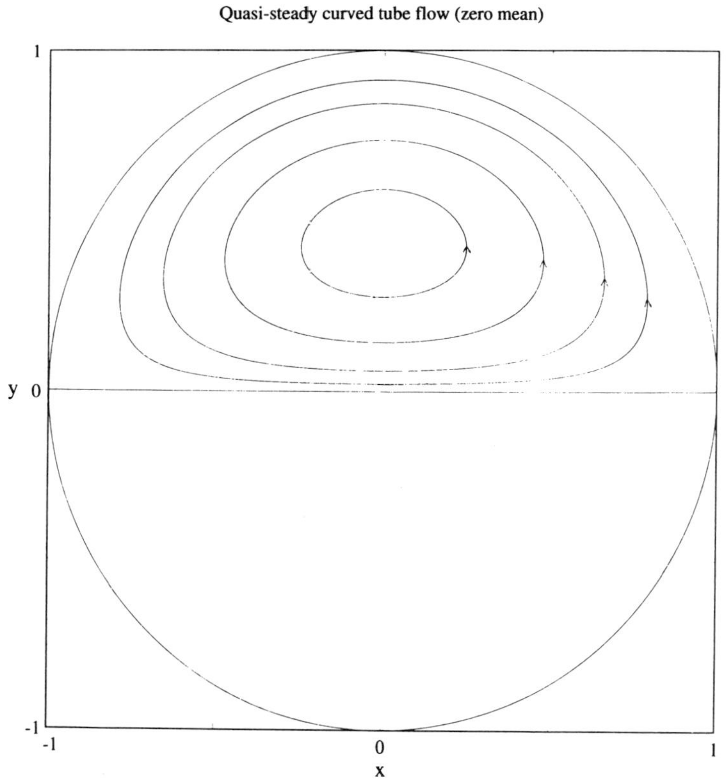

Fig. 2. Secondary orbits in the integrable approximation.

section is a smooth closed curve. Islands also consist of KAM tori and chaotic regions surrounding smaller islands, but Fig. 3 shows little of the detail at such a fine level (see Lichtenberg and Lieberman [8]).

The chaotic regions are not homogeneous, and a typical trajectory in a chaotic region spends long periods trapped close to those KAM tori which bound the region. This trapping is due to 'cantori' (Percival [9]), which have a Cantor set structure and act as partial barriers, preventing the trajectory from wandering freely through the chaotic region. As the amplitude of the perturbation term increases, KAM tori adjacent to the chaotic region are destroyed. Cantori are the remnants of these KAM tori, and they tend to trap trajectories at the edges of the chaotic region. A trajectory eventually passses through gaps in the cantori, and wanders through the chaotic region for a time, before being trapped by another group of cantori. This phenomenon has been described as 'stickiness' at the borders of the chaotic region (Shlesinger *et al.* [10]).

3. LONGITUDINAL TRANSPORT BY RESONANCE

Particles on certain trajectories are transported along the pipe by a resonant interaction between the secondary orbital motion and the longitudinal oscillation. Firstly, consider the unperturbed (integrable) problem (Fig. 4). The period of the secondary orbit shown is unity, as is the period of the longitudinal oscillation. When $T = 0$, the particle is at position A and is being advected in the direction of increasing s; the magnitude of the longitudinal velocity is relatively large, because the particle is in the centre of the pipe. When $T = \frac{1}{2}$, the particle has reached position B, and is being advected in the direction of decreasing s; its longitudinal velocity is smaller in magnitude than previously, because it is adjacent to the pipe wall. When $T = 1$, the particle reaches A once again, and the cycle recommences. However, the particle has been shifted some distance Δs along the pipe. Note that a particle located at position B at $T = 0$ would be shifted by a distance Δs in the opposite direction after one period, and so the effect of resonance is to stretch out the set of particles which lie on each resonant secondary orbit. A particle whose secondary orbit is resonant is shifted longitudinally by a fixed amount each period; this type of transport is commonly termed *ballistic*.

A necessary condition for resonance is that the period of the secondary orbit is of the form $n/(2m - 1)$, where n and m are positive integers [7]. Therefore, in the set of all trajectories, those which are resonant form a subset of measure zero. From this, one might infer that transport by resonance is not important in the current problem. However, the unperturbed (integrable) system is not an exact solution of the Navier–Stokes equations, and the effects of the perturbation must be considered.

The perturbation H_N causes the destruction of secondary orbits whose period is rational; in particular, each unperturbed secondary orbit with period $n/(2m - 1)$ is replaced by a chain of islands with a surrounding chaotic region. Henceforth, these islands will be termed 'resonant', because almost all particles whose trajectories lie on KAM tori within such islands are transported longitudinally by resonance (Fig. 5). Consider, for example, a particle trajectory which lies on a KAM torus inside one of the two large resonant islands formed by the break-up of the unperturbed secondary orbit of period 1. One such trajectory is shown (coloured pink) in Fig. 5. This trajectory is in the upper (right-hand) island at $T = 0, 1, 2, \ldots$ and is in the lower (left-hand) island at $T = \frac{1}{2}, \frac{3}{2}, \frac{5}{2} \ldots$ Longitudinal transport occurs by much the same mechanism as in the unperturbed case. The particle is advected slowly (quickly) in the direction of increasing (decreasing) s when it is in the upper (lower) island; the result is a net motion in the negative-s direction. Note that the particle's trajectory winds densely around the KAM torus; therefore the particle is

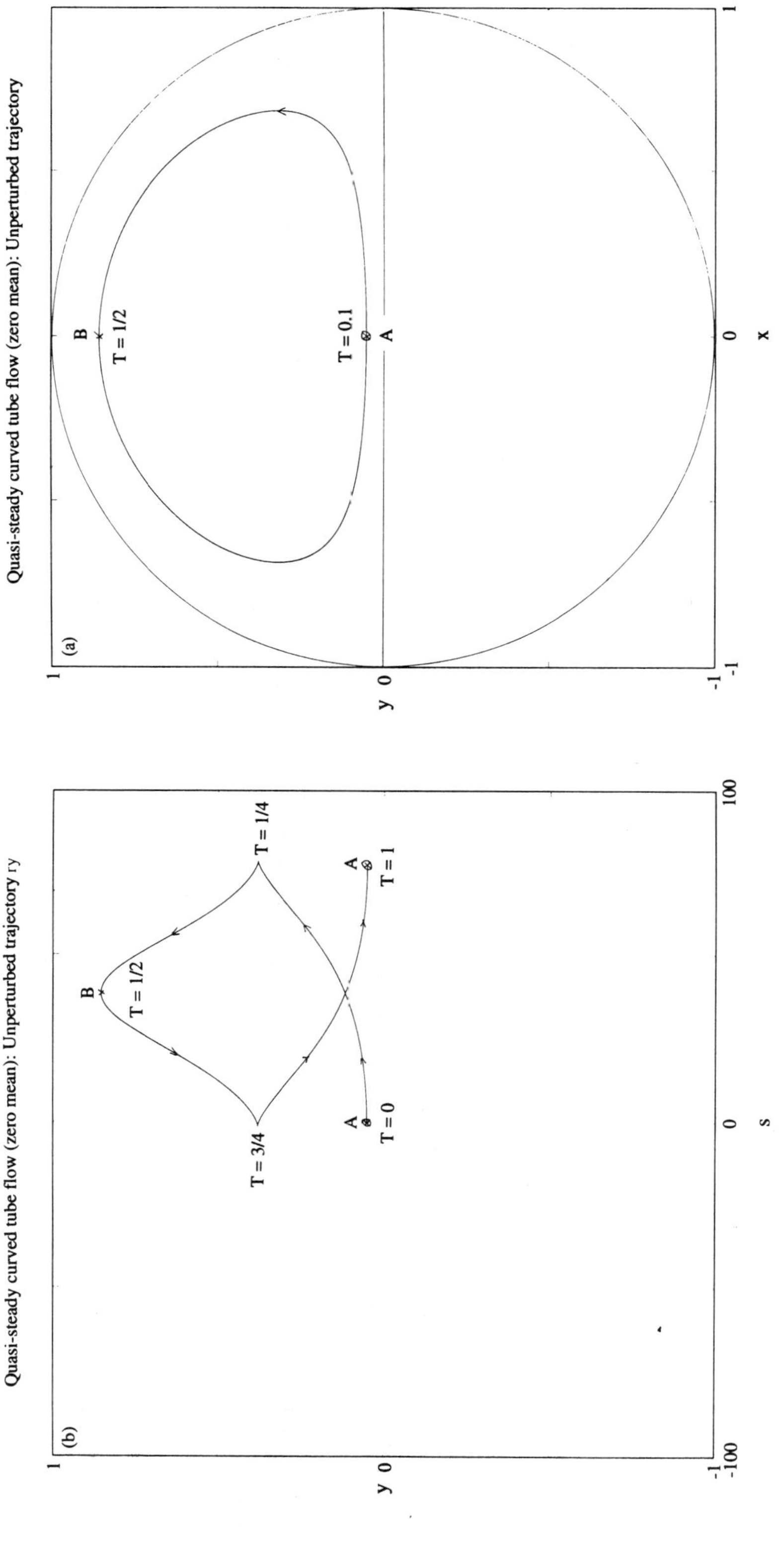

Fig. 4. Resonant longitudinal transport of a particle whose secondary orbit has period 1.

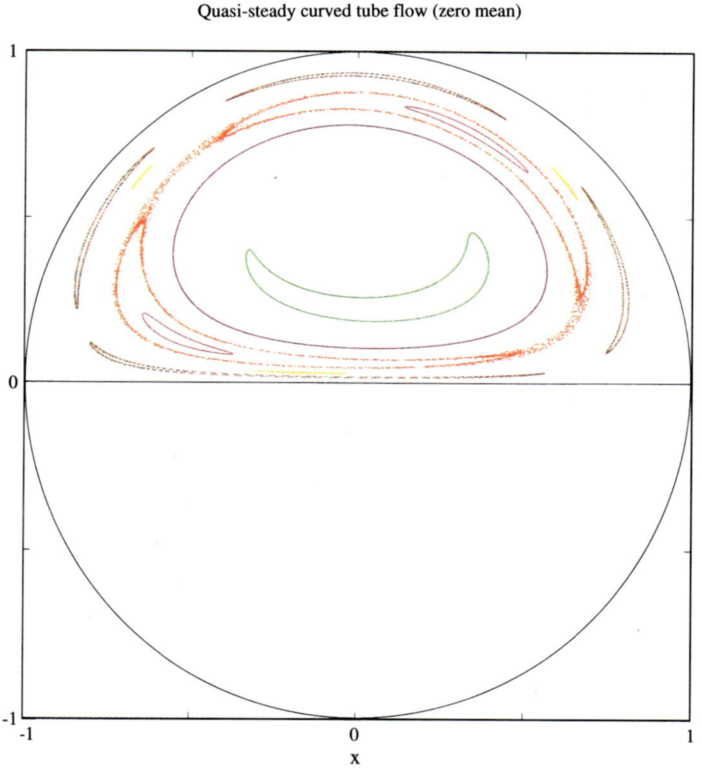

Fig. 3. Poincaré section at $T = k/2$, $k = 0, 1, 2 \ldots 4000$ showing the structure of the perturbed secondary orbits.

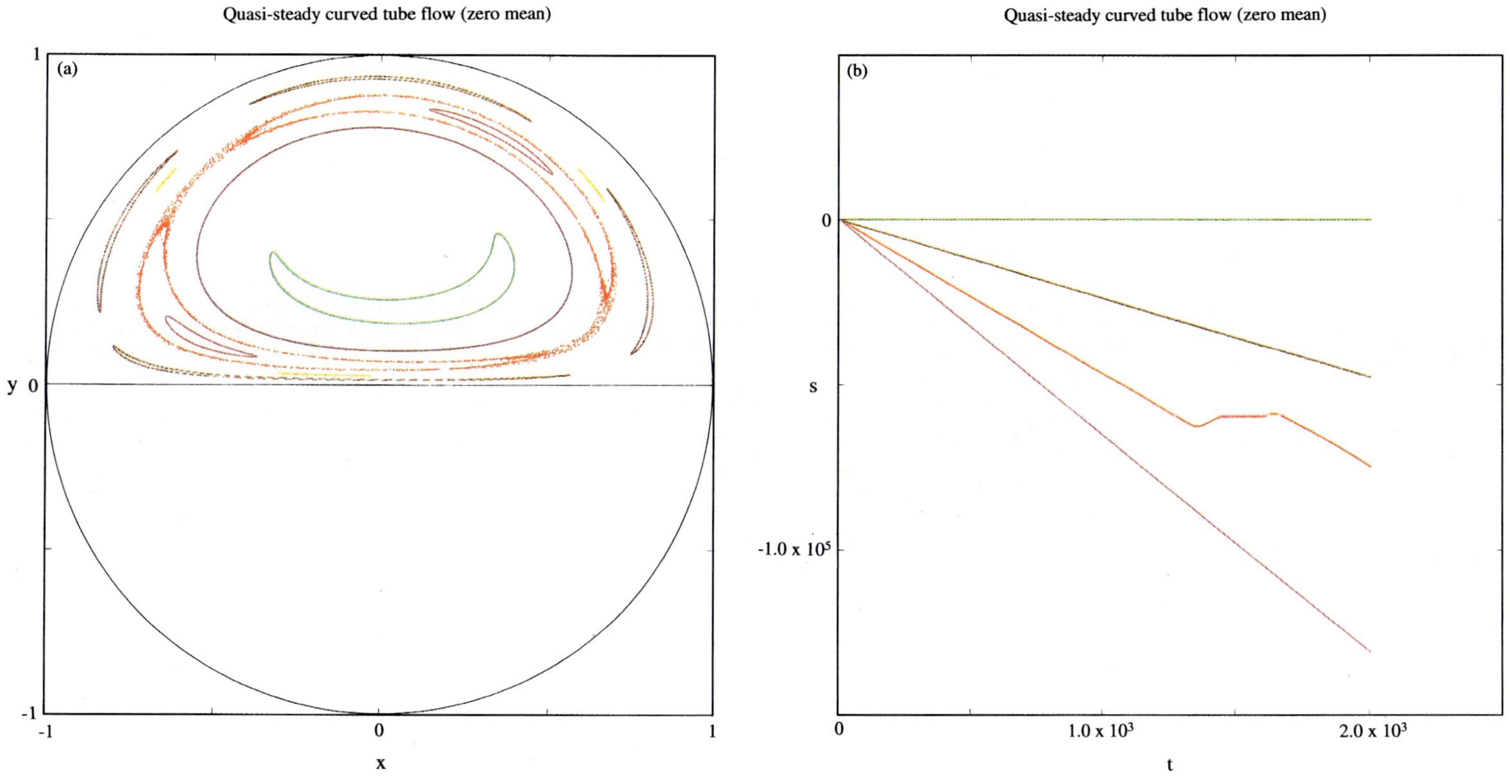

Fig. 5. Time series showing the dependence of longitudinal transport upon transverse position: resonant trajectories (pink & brown) lie inside resonant islands and the green, purple and yellow trajectories are non-resonant. The red trajectory in the vicinity of a chain of resonant islands appears to switch between resonance and non-resonance (see text).

not (in general) transported longitudinally by a fixed amount each period. Instead, the rate of longitudinal transport oscillates quasiperiodically about some mean value [7]. To see this, note that two incommensurate frequencies are involved: (1) the frequency of the longitudinal oscillation, and (2) the frequency of the transverse motion of the trajectory around the KAM torus.

The set of all KAM tori within a resonant island has non-zero measure, provided that the amplitude of the perturbation is not too great. Therefore, the set of all resonant trajectories has non-zero measure. Furthermore, it is shown below that a typical particle whose trajectory lies inside a resonant island, but not on a KAM torus, exhibits (modulated) ballistic longitudinal transport at approximately the same rate as a particle on the nearest KAM torus. The significance of this result is that a blob of marked fluid, placed within a resonant island at some initial time, may be advected arbitrarily far along the pipe (note that the blob will become more spread out with time). This is not possible for a blob placed elsewhere in the pipe cross-section.

4. CHAOS AND DIFFUSION

Consider the longitudinal transport of a particle whose trajectory lies in the chaotic region adjacent to a chain of resonant islands (Fig. 6). The trajectory is trapped behind cantori near to the boundaries of this region for much of the time, and the longitudinal transport of a trapped particle is similar to that experienced by a particle whose trajectory lies on the nearest KAM torus. To leading order, the chaotic region shown in Fig. 6(a) is bounded by four KAM tori; two of these are at the edge of islands, with resonant transport occurring in either the positive- or negative-s direction respectively. Resonant transport does not occur on the other two KAM tori. Therefore the trajectory is transported by resonance intermittently, with intervals when no resonant transport occurs (Fig. 6(b)). Trapping behind cantori has been modelled as a Levy walk (see [10]), and an ensemble of (the longitudinal components of) trajectories in a chaotic region has a variance which is of the order of T^{μ}, where $\mu \in [0, 2]$ ($\mu = 2$ corresponds to ballistic transport, which is the limiting rate at which particles are advected). This type of behaviour has become known as 'anomalous diffusion' (see Young and Jones [11] for examples).

Chaotic regions also occur within resonant islands; these are sandwiched between KAM tori on which resonant transport occurs. Therefore a particle whose trajectory wanders through one such chaotic region is intermittently trapped close to one or other of the bounding KAM tori. However, the average rate of resonant transport is approximately the same for all such KAM tori, because the longitudinal velocity profile is continuous and the chaotic region is thin. Therefore, a particle in the chaotic region is transported longitudinally at approximately the same average rate as a particle on the nearest KAM torus.

We have seen that longitudinal transport by advection has the rich structure predicted by KAM theory. Now we consider the effects of molecular diffusion. Diffusion allows fluid particles to cross both KAM tori and cantori, and therefore trapping of trajectories by the mechanism described above cannot occur. However, the KAM tori within islands occupy a set of non-zero measure, and a trajectory which has wandered into a resonant island (by diffusion) will take some time to emerge from the island. During much of this time, the secondary motion will be near to resonance with the longitudinal oscillation, subject to statistical fluctuations due to diffusion. Therefore the trajectory has episodes of near-resonance, when it is in the vicinity of a resonant island, and episodes of no resonance, when it has escaped from the vicinity of the resonant island. The effect is most obvious at very large Péclet number (Fig. 7). The Péclet number, Pe, is here defined as the timescale

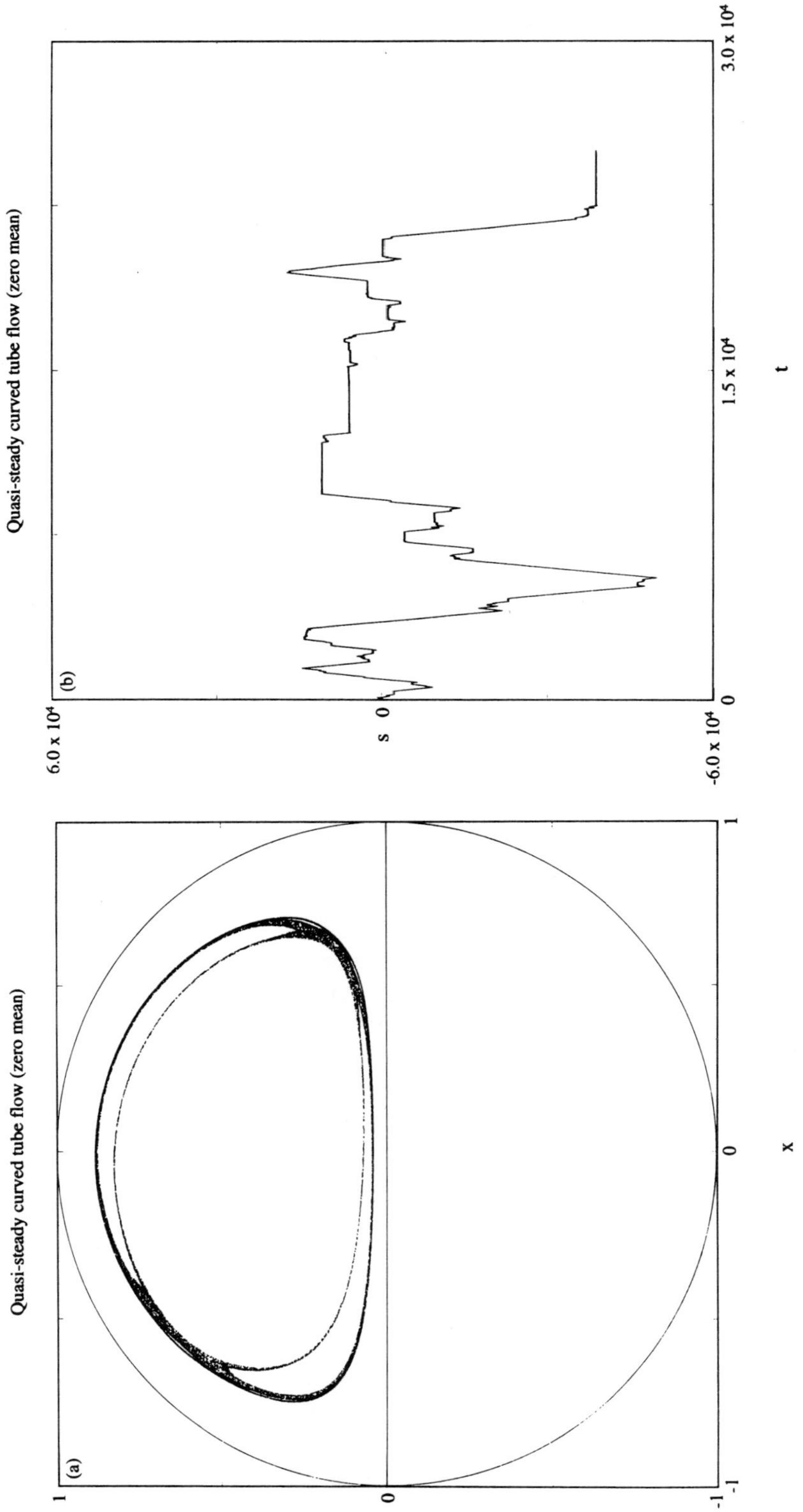

Fig. 6. Longitudinal transport in the chaotic region surrounding the period 1 resonance, shown as a time series at $T = k$, $k = 0, 1, 2 \ldots 25000$.

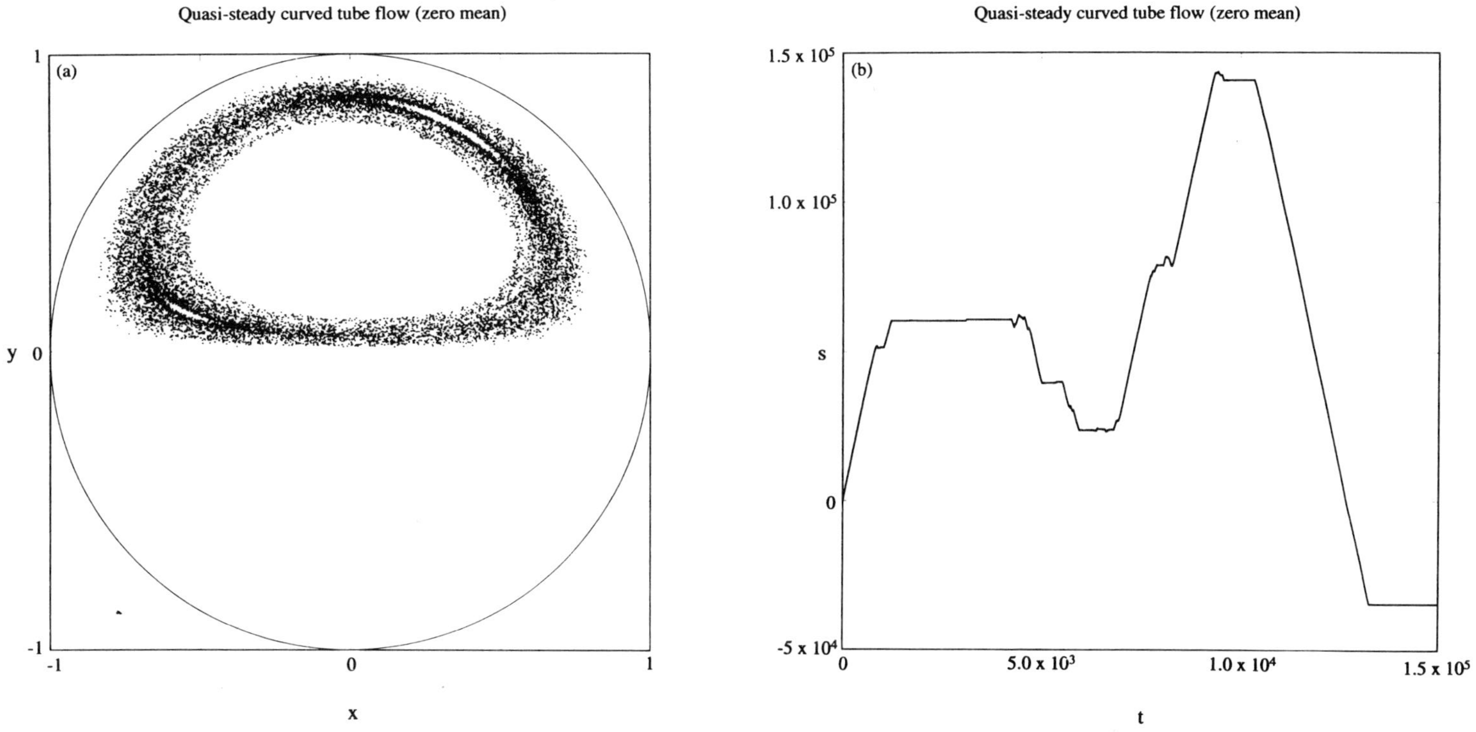

Fig. 7. Longitudinal transport at high Péclet number ($Pe = 1.25 - 10^5$), shown as a time series at $T = k$, $k = 0, 1, 2 \ldots 15000$.

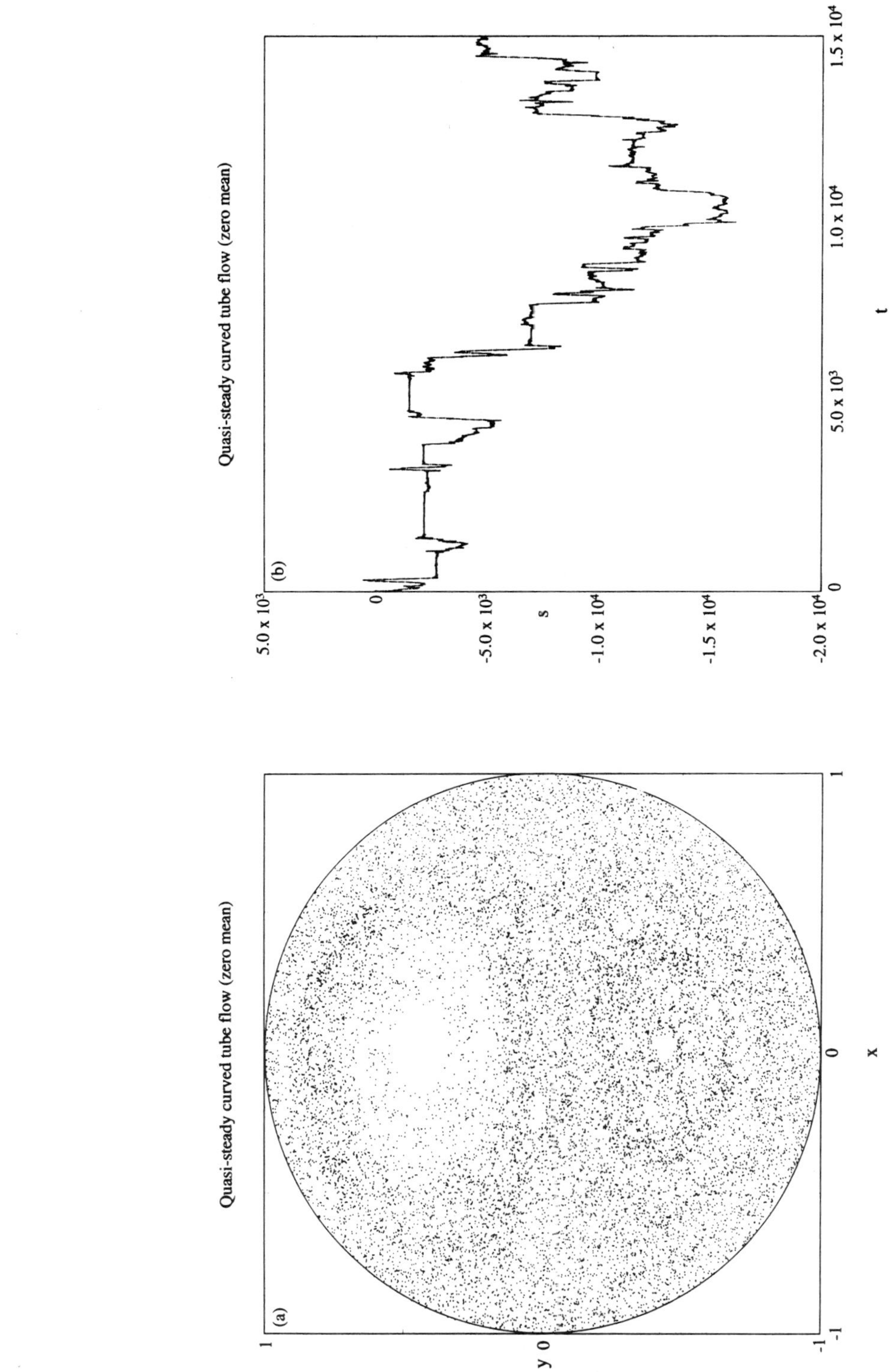

Fig. 8. As Fig. 7, but with $Pe = 1.25 \times 10^3$.

for transverse diffusion across the pipe, divided by the timescale for a fluid particle to be longitudinally advected the same distance, i.e.

$$Pe = \frac{aK}{8\omega D} = \frac{\nu}{8D}\sqrt{\frac{G}{2\delta}}, \tag{17}$$

where D is the molecular diffusivity of the particles.

This is strikingly reminiscent of the behaviour of trajectories in the chaotic region when there is no diffusion. Here, however, it is the islands which are the trapping regions. Consider a trajectory which is just outside a resonant island at some initial time. Diffusion may carry the trajectory into the island; the probability that the trajectory goes deep into the island is small, compared with the probability that the trajectory only briefly enters the island's border before being expelled by diffusion. When a trajectory does go far into the island, it experiences near-resonance for a considerable time. This is the explanation for the remarkable similarity between deterministic transport in the chaotic region in the absence of diffusion, and non-deterministic transport in the whole pipe when there is diffusion (with large Péclet number).

As the Péclet number decreases (Fig. 8), the underlying dynamical system becomes obscured by random fluctuations. When the diffusive timescale is comparable with the period of the pressure gradient oscillation, near-resonance does not (in general) occur between the secondary motion and the longitudinal oscillation.

The above discussion is qualitative, and describes effects seen in individual realizations of trajectories, as shown in Figs 7 and 8. These pictures have been produced by numerically integrating the Langevin equations

$$\frac{d\hat{x}}{d\hat{t}} = \frac{\partial\hat{\psi}}{\partial\hat{y}} + \eta_1(\hat{t}), \tag{18}$$

$$\frac{d\hat{y}}{d\hat{t}} = \frac{-\partial\hat{\psi}}{\partial\hat{x}} + \eta_2(\hat{t}); \tag{19}$$

here $\eta_1(\hat{t})$ and $\eta_2(\hat{t})$ are Gaussian random deviates with autocorrelation

$$\overline{\eta_i(\hat{t}_1)\eta_j(\hat{t}_2)} = 2D\delta_{ij}\delta(\hat{t}_1 - \hat{t}_2). \tag{20}$$

For simplicity, the Taylor approximation has been used (longitudinal diffusion is neglected), and therefore the longitudinal position of a particle is determined uniquely by the history of its transverse motion. Work is in progress to quantify the phenomena described above, and the results will be published shortly.

Acknowledgements—I thank Professor T. J. Pedley for discussing the ideas contained in this paper with me, and SERC and the Wellcome Trust for financial support.

REFERENCES

1. D. M. Eckmann and J. B. Grotberg, Oscillatory flow and mass transport in a curved tube, *J. Fluid Mech.* **188**, 509–527 (1988).
2. M. K. Sharp, R. D. Kamm, A. H. Shapiro, E. Kimmel and G. E. Karniadakis, Dispersion in a curved tube during oscillatory flow, *J. Fluid Mech.* **223**, 537–563 (1991).
3. S. W. Jones, O. M. Thomas and H. Aref, Chaotic advection by laminar flow in a twisted pipe, *J. Fluid Mech.* **209**, 335–357 (1989).
4. V. I. Arnold, *Mathematical Methods of Classical Mechanics*, 2nd ed. Springer, New York (1989).
5. T. J. Pedley and R. D. Kamm, The effect of secondary motion on axial transport in oscillatory tube flow, *J. Fluid Mech.* **193**, 347–367 (1988).
6. T. Mullin and C. A. Greated, Oscillatory flow in curved pipes—Part 2. The fully developed case, *J. Fluid Mech.* **98**, 397–416 (1980).

7. P. E. Hydon, Resonant advection by oscillatory flow in a curved pipe, *Physica D*, to appear (1994).
8. A. J. Lichtenberg and M. A. Lieberman, *Regular and Stochastic Motion*. Springer, New York (1983).
9. I. C. Percival, Variational principles for invariant tori and cantori, in *Nonlinear Dynamics and Beam-Beam Interaction*, edited by M. Month and J. C. Herera, pp. 302–310. American Institute of Physics, New York (1979).
10. M. F. Shlesinger, G. M. Zaslavsky and J. Klafter, Strange kinetics, *Nature Lond.* **363**, 31–37 (1993).
11. W. R. Young and S. W. Jones, Shear dispersion, *Phys. Fluids* **A3**, 1087–1101 (1991).

Application of Chaotic Advection to Heat Transfer

HSUEH-CHIA CHANG

Department of Chemical Engineering, University of Notre Dame, Notre Dame, IN 46556, USA

and

MIHIR SEN

Department of Aerospace and Mechanical Engineering, University of Notre Dame, Notre Dame, IN 46556, USA

Abstract—We review some of the work that has been done on the enhancement of convective heat transfer by chaotic mixing in fluids. Several devices that have been proposed for practical application are discussed. The physical processes of convective heat transfer are classified in terms of the thermal resistances involved. Techniques of reduction of the different resistances are presented with examples of their application. Analytical, numerical and experimental results are also presented.

1. INTRODUCTION

Enhancement of convective heat transfer has been a practical goal of thermal engineering for a long time. There are many different techniques that have been used effectively. Some, like external or internal fins, are based on increased heat transfer area. Others depend on increased mixing through the creation of turbulence or unsteady flow using inserts of various kinds, or by provoking mixing in recirculation zones and resonant enhancement of flow instabilities. There is some question, however, as to whether any of these methods can be used in high-viscosity liquids in which it is not possible to achieve turbulent or even unstable flow without excessive input of energy [1]. Laminar flow generally gives much poorer rates of heat transfer compared to turbulent flow. The problem is acute in fluids with high Prandtl number, *Pr*, i.e. fluids with high viscosity and low thermal diffusivity.

Consider first the simplest case of steady, laminar flow along a streamline. Heat transfer in a direction transverse to the streamlines is only due to conduction. This is a relatively slow molecular process for most fluids. Greater heat transport in a direction transverse to the main flow can be achieved by secondary flows which enable the fluid particles to be physically transported in that direction. Transverse mixing is thus one way in which heat flux can be increased. For flow in a pipe or in other internal flows, this is achieved by an exchange of fluid between the interior and the wall region. An internally inserted twisted tape, for example, provides some fluid movement normal to the flow direction. However, there is a cost in terms of increased pressure drop associated with the drag of the additional no-slip surfaces introduced by the insert. The pressure drop is increased as well as the power needed to pump the fluid. Chaotic mixing is presented here is an alternative to this.

There are two kinds of laminar mixing that can be identified, regular mixing and chaotic mixing. The former occurs due to fluid motion in which the pathlines are integrable. This is the case if, for example, the fluid particles in a finite domain return to the same point in a periodic manner. Another example is that of flow in a helical coil in which a pair of secondary flow vortices with closed streamlines are formed. Chaotic mixing occurs in flows in which the pathlines are not integrable. Early suggestions of the possibility of chaotic particle paths in laminar flows were put forward by Arnol'd [2] and Hénon [3]. The idea has been expanded by Aref [4, 5], Chien *et al.* [6], Chaiken *et al.* [7] and Khakhar *et al.* [8] and many others who have theoretically and experimentally studied velocity fields that lead to chaotic mixing. Two-dimensional flows must be time-dependent for this to happen. A common technique is for a steady flow to be subjected to a periodic perturbation. The concept can be extended to steady, three-dimensional flows along ducts in which the axial coordinate is then time-like as the fluid is advected downstream. Chaotic particle paths may be achieved if the duct geometry is periodically perturbed in the downstream direction. In this way chaotic mixing becomes useful from a practical point of view since many industrial devices involve flow in ducts. Transport of heat can then be enhanced in a relatively simple manner without the use of inserts or external energy.

Convective heat transfer from a fluid to the walls is due to the combined effects of thermal conduction and advection. Chaotic mixing is most effective in cases where conduction is weak compared to advection; that is in flows in which the Peclet number, *Pe*, is high. For a fluid of zero thermal diffusivity for which *Pe* is infinite, the temperature of a fluid particle remains constant as it moves with the flow. Heat transport within the fluid is entirely due to advection. The effect of finite diffusivity is to smooth out the temperature distribution. At a boundary, however, transfer or heat from fluid to boundary is by conduction alone.

Various ways of introducing chaotic mixing in practical devices have been proposed, most of which can be applied to heat transfer enhancement. This paper reviews of some of these techniques. We will begin by a description of some of these devices. Then we will classify the mechanisms of heat transfer and seek to understand the main methods of enhancement that can be employed. Physical and qualitative arguments are used throughout this review for the purposes of understanding the phenomena more than for quantitative predictions.

2. DEVICES FOR CHAOTIC MIXING

For chaotic mixing to occur, a flow field has to be chosen such that the particle paths are chaotic. In a two-dimensional region the flow must be time-dependent, either due to boundary conditions or to instability. This technique is thus suitable for heat transfer in a batch process. Chaotic mixing in three-dimensional flows can be steady and can be applied to continuous flow in a duct. Kusch and Ottino [9] have described some of the processes. The ones described below can be adapted to heat transfer applications, though the adaptation has not been carried through in all cases.

Khakhar *et al.* [10] have proposed an active, partitioned-pipe mixer to achieve chaotic mixing. A schematic of this device is shown in Fig. 1. This is a continuous-flow device consisting of a pipe partitioned into semicircular ducts by means of rectangular plate inserts at 90° to each other. The fluid to be mixed is forced through the pipe, while the entire pipe is rotated about its axis. The net result is that of chaotic mixing of the fluid particles. A simplified analysis of this mixer has recently been carried out by Ling [11].

A device is passive and does not need external sources of energy if mixing can be produced through internal forces such as inertia or buoyancy. One example is the Kenics mixer that is quite common in heat transfer applications. It is similar to the partitioned-

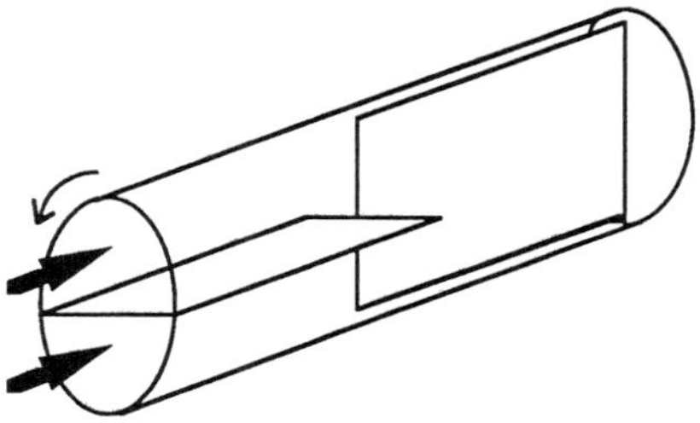

Fig. 1. Partitioned pipe mixer [10].

pipe mixer, except that the arrangement is stationary. Mixing is produced by twisting the plates to induce transverse motion.

Another useful device is the eccentric annulus mixer. In this the fluid occupies the space between two eccentric cylinders. If both cylinders rotate at a steady rate, but in opposite directions, the streamlines are all closed and mixing is regular. There is one streamline which connects a stagnation point to itself. If a time-periodic modulation of the rotational speed of the outer cylinder is introduced, mixing becomes chaotic. There is fluid advection across what was the stagnation streamline, thus increasing the heat flux across it.

A helically coiled pipe can also be adapted to produce chaotic mixing. A conventional, constant-axis coil has closed secondary flow streamlines which provide regular mixing. Chaotic mixing is produced in a relatively simple manner by changing the axis of the coil with spatial periodicity; the closed streamlines are then broken up. A wire-frame model of such an alternating axis coil is shown in Fig. 2(a). Each successive loop of the coil lies on a different plane, P or P′, which are mutually perpendicular, so that after completing each loop the coiling axis changes by 90°. The secondary flow produced also changes with coiling axis, as shown schematically in Fig. 2(b). The hydrodynamics of chaotic particle paths produced in such coils has been discussed by Lee *et al.* [12], Jones *et al.* [13] and Le Guer and Peerhossaini [14]. The notion of a heat exchanger based on this concept was mentioned by Peerhossaini and Le Guer [15]. We, along with our students, have studied heat transfer enhancement in this geometry extensively, and we rely heavily on that work in this review [16–19].

3. CLASSIFICATION OF HEAT TRANSFER PROBLEMS

Since there is a wide variety of designs in heat transfer equipment, heat transfer enhancement in each one must be dictated by the kind of flow pattern that exists. It is useful to classify them in some manner to be able to make general statements about the enhancement techniques. We will consider the two kinds of problems that are described below.

In the first kind we will include steady-state heat transfer between walls held at isothermal conditions. The fluid region can be three-dimensional in general, but a two-dimensional approximation may be used in specific cases. Since the temperature gradient is always imposed, one is typically interested in the heat transfer rate at large time. Transient performance in this kind of device is unimportant as long as the transients die rapidly enough. We also include in this group devices which work with time-periodic or aperiodic forcing, in which only a time-average behavior for large time is desired. Several examples of this kind of convective heat transfer problem are shown in Fig. 3. Figure 3(a)

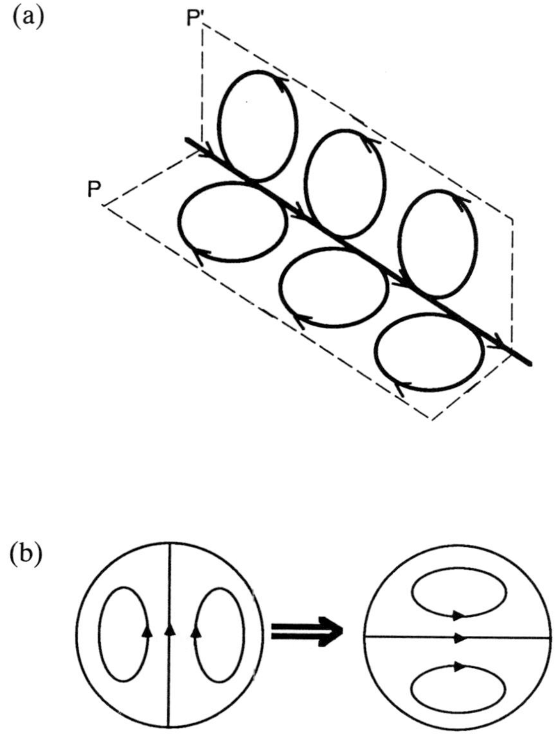

Fig. 2. Alternating axis coil mixer and secondary flow change [16, 18]. (a) Schematic of alternating axis (AA) coil. (b) Change in secondary flow due to change in coiling axis.

shows heat transfer between a cylinder and a cross-flowing fluid, the cylinder and the fluid being kept at different temperatures T_w and T_f, respectively. There is a large recirculation region behind the cylinder where the heat transfer rate is high. In Fig. 3(b) the recirculation is next to a wall and is due to a depression in the wall. The third example shown in Fig. 3(c) represents heat transfer between a wall and a surrounding fluid through a falling film. Recirculation of the closed streamlines within the large waves which cover the interfacial film increases the heat transfer [20]. These waves are fairly stationary; they move at constant speed and, if one is in the Lagrangian frame moving with the wave, the temperature within a wave crest quickly equilibrates and the subsequent heat transfer between the two constant temperature reservoirs, the wall and the surrounding fluid, is the heat transfer of interest. Other heat transfer processes belonging to this kind are heat transfer between a large-heat-capacitance solid object or liquid bubble with the surrounding fluid, between a solid plane and a constant-temperature fluid flowing over it, diffusion cells, heat transfer across Rayleigh–Bénard cells, and the eccentric annulus previously discussed.

The other kind of heat transfer devices are inherently unsteady. The fluid in the region of interest is initially at a temperature different from that of the walls, and due to heat transfer the fluid temperature eventually approaches that of the walls. Again, flow in such a device may be three- or two-dimensional. One example is the cooling of a hot fluid in a pan at room temperature, as schematically shown in Fig. 4(a). The external temperature T_{ext} is constant, while that of the fluid in the pan T_f varies with time. Three-dimensional flow will also be in this category if the third spatial dimension is time-like, as in flow in a

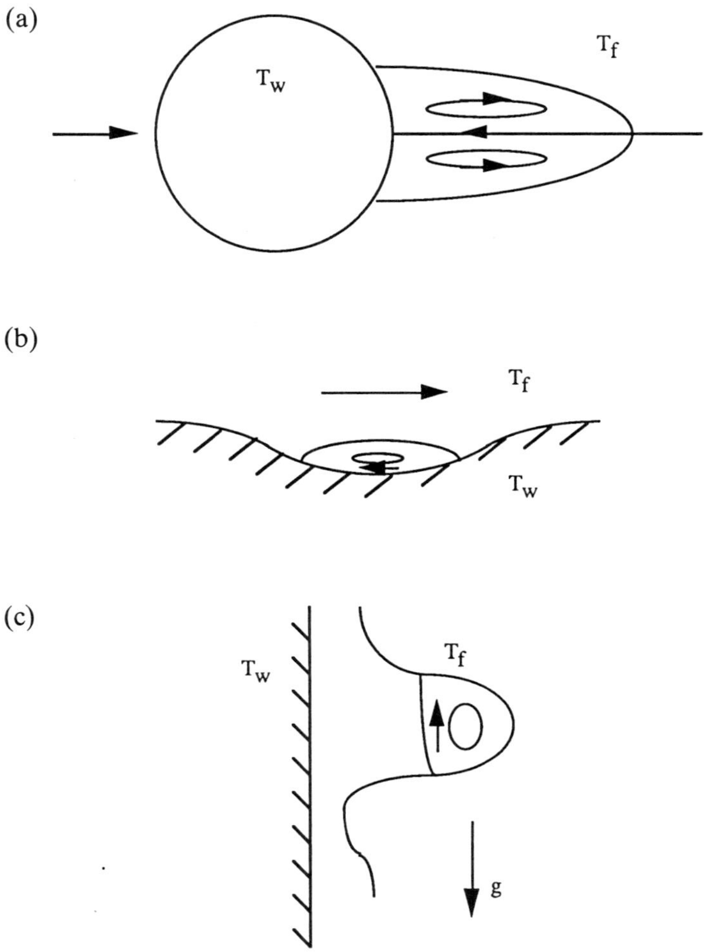

Fig. 3. Examples of heat transfer problems of the first kind. (a) Heat transfer between cylinder at T_w to exterior flow at T_f. (b) Heat transfer between wall with depression at T_w and exterior flow at T_f. (c) Heat transfer from wall at T_w to falling film at T_f.

duct or pipeline. One can then consider the axial coordinate to be similar to time, and think of the process as transient. In the Lagrangian frame of the fluid particle, the pertinent heat transfer is unsteady even though the flow is steady in an Eulerian sense. For example, if the heat transfer in the falling-film device shown in Fig. 4(b) is between the liquid and the external fluid, the film temperature T_f is a function of its vertical position. It will eventually reach the surrounding temperature T_{ext} after which there is no further heat transfer. Consequently, all pertinent heat transfer occurs during the transient stage in the Lagrangian frame. Another important example in this category is the heat-exchanger pipe, Fig. 4(c), where all heat transfer occurs before the fluids inside and outside the tube reach the same temperature.

Of course, the distinction between the two kinds of problems is neither sharp nor exclusive. Thus there are some devices that have both characteristics. The issue that we wish to address is that thermal resistances in the two kinds of problems are distributed differently. Thus the heat transfer enhancement techniques must also be different.

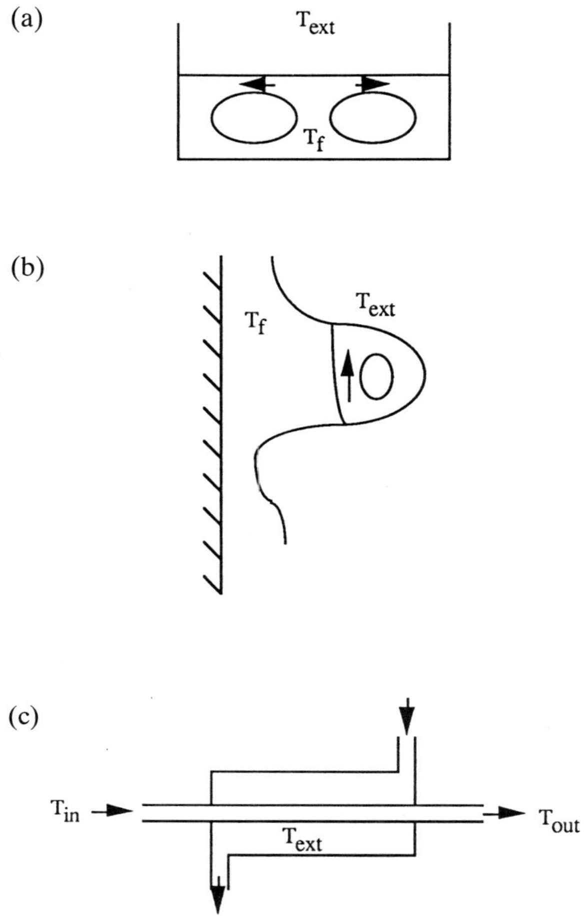

Fig. 4. Examples of heat transfer problems of the second kind. (a) Cooling of hot liquid in pan at T_f to exterior at T_{ext}. (b) Heat transfer between falling film at T_f and exterior fluid at T_{ext}. (c) Heat transfer between fluid in straight pipe and exterior at T_{ext}.

4. THERMAL RESISTANCES

To understand the variety of thermal resistances that can occur in convective heat transfer, we look at typical two-dimensional flow region shown in Fig. 5. Thermal boundary layers, a pair of closed streamlines vortices, and a stagnation streamline are shown. First, assume that we would like heat to be transferred from wall A to wall C, and that the other two walls, B and D, are adiabatic. This is a problem of the first kind. The heat flux Q can be written in terms of a total thermal resistance R_{th}, i.e. $Q = \Delta T/R_{th}$, where ΔT is the temperature difference between A and C. The objective of heat transfer enhancement techniques is to get a higher Q for the same ΔT, or a smaller ΔT for the same Q. In either case, we would like R_{th} to be as small as possible. The total thermal resistance can be qualitatively considered to be the sum of three separate resistances, $R_{th} = R_{th}^{bl} + R_{th}^{cr} + R_{th}^{sl}$, where R_{th}^{bl}, R_{th}^{cr}, and R_{th}^{sl} are the thermal resistances corresponding to the two thermal boundary layer regions marked as I, the two circulation regions II, and the region around the stagnation streamline marked as III, respectively. It is important to

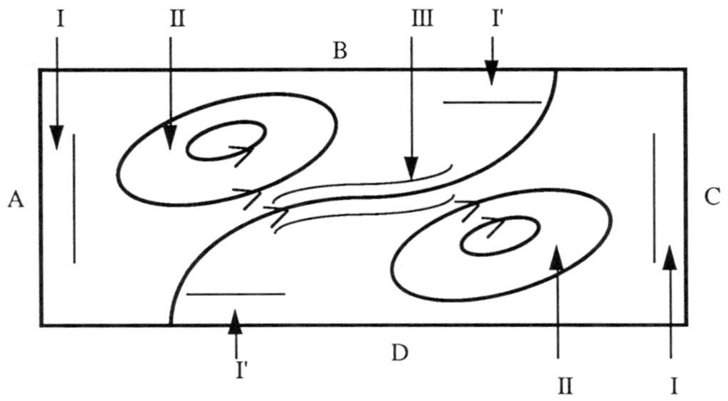

Fig. 5. Schematic of internal flow.

notice that, as drawn, these resistances are in series, so that the largest of the three resistances is the most critical.

One can also imagine other steady-state scenarios. Consider now heat transfer between walls B and D, with the other two adiabatic. Then, for the most part, circulation regions II and stagnation streamline region III are in series, while at the left and right extremes the latter does not exist. The combined resistance is also in series with the resistance of thermal boundary layer region I′.

Another situation arises when we consider the transient case in which the fluid is initially uniformly hot. Some of the walls are isothermal at a lower temperature, and others are adiabatic. Heat transfer in this instance is not between walls, but between the fluid and the isothermal walls. We can see that $R_{\mathrm{th}}^{\mathrm{sl}}$ does not play a critical role; the only resistances that affect the heat transfer are $R_{\mathrm{th}}^{\mathrm{cr}}$ and $R_{\mathrm{th}}^{\mathrm{bl}}$, and they are in series. This is a problem of the second kind.

It is important to identify and understand the critical thermal resistances in a given problem because the processes that affect each one of them are different. The design of an enhancement technique depends on which kind of resistance we would like to reduce. We will discuss them in more detail and introduce qualitatively, where applicable, the effect of chaotic advection in the high *Pe* limit.

4.1. *Thermal boundary layer*

Heat transfer through a laminar thermal boundary layer is mainly due to conduction, so that its resistance is given by $R_{\mathrm{th}}^{\mathrm{bl}} \sim \delta_{\mathrm{t}}/k$, where δ_{t} is the thermal boundary layer thickness, and k is the fluid thermal conductivity. The resistance can be decreased by modulation of the hydrodynamics in a downstream direction, so that the thermal boundary layer is constantly readjusting itself. There are heat transfer enhancement techniques which are based on this.

4.2. *Circulation region*

The thermal resistance of a circulation region often plays an important role in heat transfer enhancing techniques. The resistance is low when the advective time is much shorter than the thermal diffusion time, that is when *Pe* is high. Within the closed streamlines, rapid advection is used to mix the fluid and reduce temperature differences. A pertinent question concerns the time required to uniformize the temperature. There has been some controversy concerning this that has now been resolved [21]. Essentially, an initial temperature field will quickly be advected along the closed streamlines. No mixing occurs if there is no diffusion along the streamline. For weak thermal diffusion we have that mixing along each closed streamline is complete over a time $\tau_c \sim Pe^{1/3}(l/U)$ with $Pe = Ul/\kappa$, where l is the characteristic dimension of the vortex, U is the circulation velocity, and κ is the thermal diffusivity. However, a temperature gradient across streamlines still exists at this point, and gradient expulsion is complete when the cross-streamline gradient is removed by slow thermal diffusion. The time for this process is, of course, the thermal diffusive time $\tau_D = l^2/\kappa = Pe^{2/3}\tau_c$. Steady operation occurs for times much larger than τ_D if $Pe \gg 1$. After this characteristic time, the temperature differences within the circulation region as well as its thermal resistance become small. Heat flux enhancement for the eccentric cylinder can be estimated in this manner [22].

Reduction of the thermal resistance of the circulation region increases the heat transfer between two surfaces at different temperatures within which it is enclosed. If the heat transfer outside the circulation region is dominated by conduction, it is then a simple matter to estimate the flux enhancement due to recirculation to be of order

$$\eta \sim \frac{L}{L - l}, \tag{1}$$

where L is the distance separating the surfaces. The enhancement factor can be increased by enlarging the circulation region. In doing so the fraction of the total flow region occupied by low thermal resistance circulation is increased. In practice this can often be accomplished by breaking up of the circulation bubble by periodic forcing such that fluid elements can be transported across the previous boundary. From the perspective of heat transfer enhancement, however, this mechanism is akin to the advection-driven expulsion of temperature gradient in the unbroken bubble. Here, chaotic advection is used to expel temperature gradient to further increase l.

4.3. *Stagnation streamline region*

The stagnation streamline presents a thermal resistance that can be also be reduced. The basic ideas are conveniently approached for two-dimensional flows. Moreover, there are many physical situations in which this is a good approximation. Examples are long Rayleigh–Bénard rolls [23], long eccentric cylinders [22] and axisymmetric bubbles [24]. The theoretical advantage of the two-dimensional geometry is that the velocity field can be written in terms of the stream function and the trajectory of a fluid particle is described by a Hamiltonian system. There are many results in this area that can then be brought to bear on the problem.

When a Hamiltonian system with a continuous family of closed streamlines and a bounding homoclinic orbit or heteroclinic cycle is perturbed by small-amplitude periodic forcing, the usual approach is to use a Melnikov analysis to study the time-averaged dynamics. Steady heat transfer across the bounding streamline is enhanced by the presence of a homoclinic tangle. This occurs when the homoclinic orbit or heteroclinic cycle break up into stable and unstable manifolds which intersect an infinite number of times to give rise to infinitely many unstable periodic orbits as well as uncountably many non-periodic orbits. These orbits occupy a stochastic layer near the original bounding streamline and it is the chaotic advection within this stochastic layer which expels temperature gradient beyond the original boundary. Not all forcing will break up the boundary. Physically, if a heteroclinic orbit exists for the perturbation flow field that coincides with the original one, the latter remains unbroken. Mathematically, this means that the Melnikov function must have simple zeros—a condition that is satisfied generically by most perturbations.

Though most analyses to date have been two-dimensional, the above reasoning can be extended to three-dimensional circulations regions. Closed streamlines in a three-dimensional bubble belong to a surface, and the circulation region is bounded by a surface that is either closed or ends at a wall. The study of the geometry of this surface under perturbation is quite complicated for a completely three-dimensional flow. This is presumably just a technical difficulty and its response may be similar to that of two-dimensional flows. We may add that a Melnikov theory for n-dimensional bounding surfaces is now available [24] and the break up of three-dimensional bubbles can hence be analyzed, though with considerably more effort. It is probably simpler to design steady heat transfer devices with empirical testing guided by the two-dimensional estimates offered here.

5. PROBLEMS OF THE FIRST KIND

The eccentric annulus is a prototype of the first kind of problem and we will review its performance in some detail. For this geometry, using assumptions of low Reynolds number and small eccentricity, we can explicitly write down the velocity field on the basis of lubrication theory. The position of a fluid particle as a function of time can then be obtained by numerical integration. A Poincaré plot obtained by marking the position of a particle at periodic intervals is shown in Fig. 6 [22]. The difference between the three diagrams is the frequency of the perturbation. An optimum frequency for thickest stochastic layer can be clearly seen. Transport across the bounding streamline is achieved in a manner similar to the random walk process of diffusion.

Under periodic perturbation stochastic layers develop within the separation bubble. At small amplitude of perturbation the resonant orbits, which are closed streamlines whose turnover time is a rational multiple of the forcing period, yield even numbers of elliptic and hyperbolic fixed points for the Poincaré map. The stable and unstable manifolds of the latter undergo homoclinic entanglement by the classical KAM mechanism to produce the stochastic layers. At larger amplitudes, even the non-resonant orbits break up. However, such interior mixing due to chaotic mixing only serves to shorten the transient time τ_D of the unforced case. It does not affect the steady heat transfer after the transient since this is essentially independent of the length of the transient time—it does not matter how fast the temperature gradient is expelled.

The Melnikov analysis can also be used to estimate the size of the well-mixed region when periodic forcing is introduced. The details of the analysis is contained in [22], but we offer a more physical description here. Consider a particle following a closed streamline with turnover time τ_c near the bounding homoclinic orbit or heteroclinic cycle. It experiences a time-averaged perturbation due to a velocity normal to the unperturbed streamline $\sim O(\alpha U)$, where α is the normalized amplitude of the forcing or perturbation flow field. This perturbation pushes a fluid particle away from the original path by a distance $\alpha U \tau_c$ per circuit. Since α is small, this displacement is negligible except near the bounding homoclinic orbit or heteroclinic cycle where $\tau_c \to \infty$. The location of the edge of the stochastic layer, l_s, is approximately where the characteristic turnover time is $\tau_c = l/U$. This gives $l_s = \alpha l$.

We now have a circulation region that is of size $l + l_s$ instead of l. Modifying equation (1) we find the enhancement of heat transfer to be

$$\eta \sim \frac{L}{L - l - l_s} \sim \frac{L}{L - l - \alpha l}. \tag{2}$$

(a)

(b)

(c)

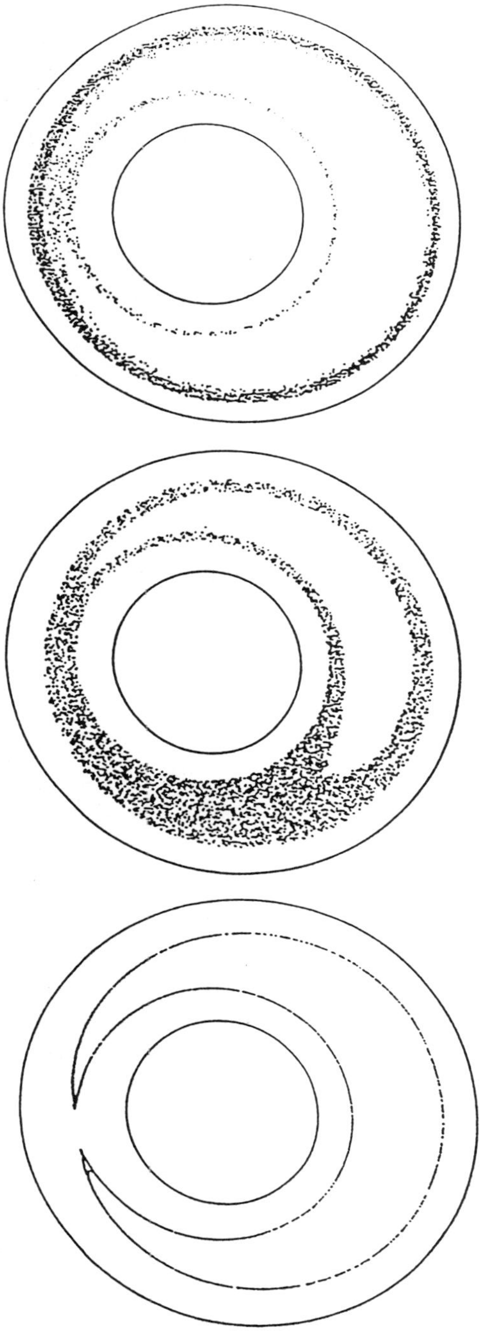

Fig. 6. Eccentric annulus mixer [22]. Frequency of perturbation increases from (a) to (c).

It is clear that the largest enhancement would occur if $L - l \sim O(\alpha l) \ll 1$. The original unperturbed circulation would then occupy almost the entire domain, and is already quite efficient in heat transfer. In the eccentric cylinder case, we were able to achieve no more than 50% enhancement over the unperturbed system.

The above estimate is only valid if the chaotic advection timescale is much shorter than the diffusion timescale. Otherwise, since diffusion is unable to expel the gradient outside the unperturbed boundary, nor should the perturbed flow field. The effective diffusion due to chaotic advection can again be quantitatively estimated using Melnikov theory [22]. We offer an order of magnitude estimate here. A displacement of $\alpha U \tau_c$ over time τ_c gives a stochastic diffusion coefficient $\kappa_s \sim (\alpha U \tau_c)^2/\tau_c$. For $\tau_c = l/U$, we get $\kappa_s \sim \alpha^2 Pe\kappa$.

For chaotic advection to be effective in a steady heat transfer device, we must have $\kappa_s \gg \kappa$, so that the amplitude of forcing must satisfy $\alpha \gg Pe^{-1/2}$. The dependence of the stochastic diffusion and heat transfer enhancement on forcing frequency is more subtle. It involves an interaction between the forcing period and the turnover time near the bubble boundary. If the forcing period τ_f is much longer than the turnover time $\tau_c = l/U$ at the outer edge of the stochastic layer, the displacement from the unperturbed trajectory essentially vanishes. On the other hand, if $\tau_f \ll \tau_c$, the effect also averages away in time since the nominal trajectory behaves as a flat base line for the fast zero-mean fluctuations. There is, hence, a resonance of sorts at the intermediate forcing frequency $\tau_f \sim l/U$. This leads to an optimum forcing frequency which was clearly demonstrated for the eccentric cylinder [22].

Similar arguments can be made for other problems of the first kind. If a circulation occupies the entire domain as in Rayleigh–Bénard instability, there will be thermal boundary layers of thickness $Pe^{-1/2}$ and $Pe^{-1/3}$ at the interface between cells and wall, respectively. In such cases, $L \sim l + Pe^{-1/3}$ and

$$\eta \sim Pe^{1/3}, \tag{3}$$

which is commonly used in Nusselt number correlations.

6. PROBLEMS OF THE SECOND KIND

In steady three-dimensional duct flows the axial coordinate is time-like. A problem typical of this kind is that of a helically coiled pipe. This particular example will be examined in some detail for several reasons: a conventional coil is often used in heat exchangers; it can be easily adapted to take advantage of chaotic mixing; it is also the only configuration that has been experimentally tested for heat transfer enhancement. In what follows we will often compare the performance of the straight pipe, the constant axis and alternating axis coils and refer to them by the symbols ST, CA, and AA respectively.

There is no recirculation region in an ST pipe with steady flow. The thermal boundary layer extends over the whole domain. Thus δ_t is large, and so is the thermal resistance. The velocity field in a pipe of circular cross-section can be written as $\mathbf{u} = w(r)\mathbf{e}_z$ where r is the radial coordinate and $\mathbf{e}_z$ is the unit vector in the axial direction. We shall use the center-line velocity for non-dimensionalization so that $w \sim O(1)$. An isothermal inlet condition $T = T_i$ is imposed and the core of the pipe remains at this temperature in the entrance region near the inlet. A developing thermal boundary layer near the wall of width

$$\delta_t \sim Pe^{-1/3} z^{1/3} \tag{4}$$

grows from the entrance; the pipe radius has been used as the characteristic length. The local azimuthally averaged heat flux at every z-station is then given by the Nusselt number

$$\langle Nu \rangle_\phi \sim Pe^{1/3} z^{-1/3}. \tag{5}$$

Heat transfer is most effective near the inlet. The thermal field becomes well developed when δ_t is of the order of the radius of the pipe. This defines the entrance length as

$$z_e \sim Pe^{1/3}. \tag{6}$$

Isotherms at any cross-section of a ST pipe are concentric circles. Moving downstream the core shrinks slowly as $Pe^{-1/3}z^{1/3}$. For large Pe, this implies that the core is locally a cylinder and the thermal field is nearly planar with weak z-dependence. A steady heat transfer process is approached if Pe is sufficiently high. To increase heat transfer we seek a mechanism to enlarge this region and to suppress the growth of boundary layers. Unlike problems of the first kind, however, the boundary of the well-mixed region does not coincide with the boundary of any closed circulation. Since the thickness of the boundary layer is small as before, we need simply to introduce a normal advective displacement at the core boundary. Advection within the core is again inconsequential. More importantly, since the core is axisymmetric, the axial velocity of all particles along the core boundary is identical, and one simply needs to be concerned with the transverse velocities provided the normal advections are weak i.e. if the pipe is nearly straight.

One way to improve heat transfer in a pipe is by coiling. The resulting CA coil has secondary motion with a closed circulation region and regular mixing as shown in one of the diagrams of Fig. 2(b). The thermal boundary layer thickness is much smaller and a circulation thermal resistance is created, while the total resistance is lower.

A simplified mathematical model of the problem shows the essential features of the process. In such an analysis we use the perturbation flow field provided by Dean [25]:

$$\mathbf{u} = w(r)\mathbf{e}_z + \delta^{1/4}Re^{-1/2}u(r, \theta)\mathbf{e}_r + \delta^{1/2}v(r, \theta)\mathbf{e}_\theta, \tag{7}$$

where $\delta \ll 1$ is the ratio between pipe radius and the coil radius, and Re is the flow Reynolds number. An orthogonal toroidal coordinate system is used which locally resembles cylindrical coordinates due to the weak coiling. This flow field is valid only when

$$Re\delta^{-1/2} \gg 1. \tag{8}$$

We also note that the pipe length of each coil in this configuration is of order δ^{-1}.

Nusselt number results of detailed numerical analysis near the entrance of a coiled pipe are shown in Fig. 7. Three distinct regions are observed along a coiled pipe [17]. In region 1, the thermal boundary layer grows as in a straight pipe. It is too thin to be affected by the transverse flow field and the local Nusselt number decays as before. In region 2, however, $\langle Nu \rangle_\phi$ begins to rise. The reason for this is apparent from the isotherms in Fig. 8 where the core is seen to go off-center with an eccentricity ϵ. It can be easily shown by lubrication theory that the Nusselt number of an eccentric core is higher than that of a concentric one of the same area according to the scaling

$$Nu(\epsilon) - Nu(0) \sim \epsilon^2. \tag{9}$$

An estimate of $\epsilon(z)$ can be obtained from the scaling for the radial velocity u in equation (7),

$$\epsilon \sim \delta^{1/4}Re^{-1/2}z. \tag{10}$$

The local enhancement of the CA configuration over the ST configuration should then be

$$\eta_{\mathrm{CA}}(z) - 1 \sim z^2\delta^{1/2}Re^{-1}. \tag{11}$$

Beyond region 2, when z approaches the entrance length z_e given by equation (6), the eccentric thermal boundary layer has grown to the center and the $\langle Nu \rangle_\phi$ again drops to the ST asymptote. The overall efficiency of a CA pipe over a ST pipe of the same length Z is then obtained by integrating equation (11) to give

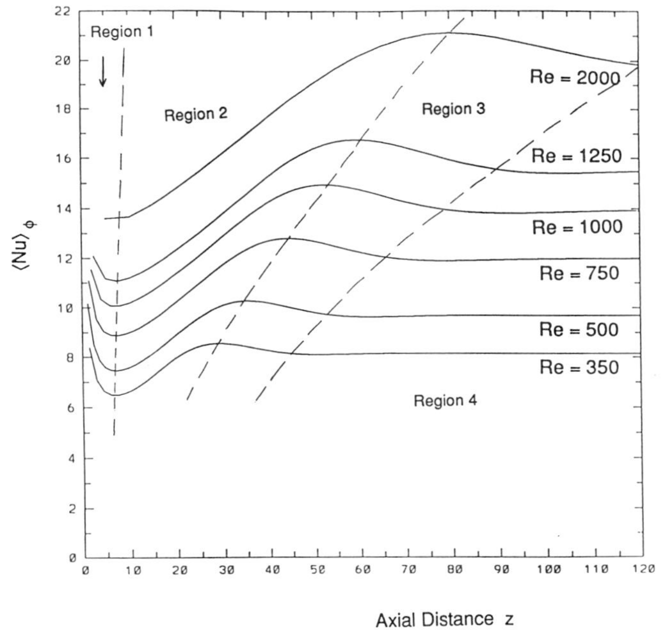

Fig. 7. Nusselt number in the thermal entrance region of a CA coil, $Pr = 1$, $\delta = 0.1$ [17].

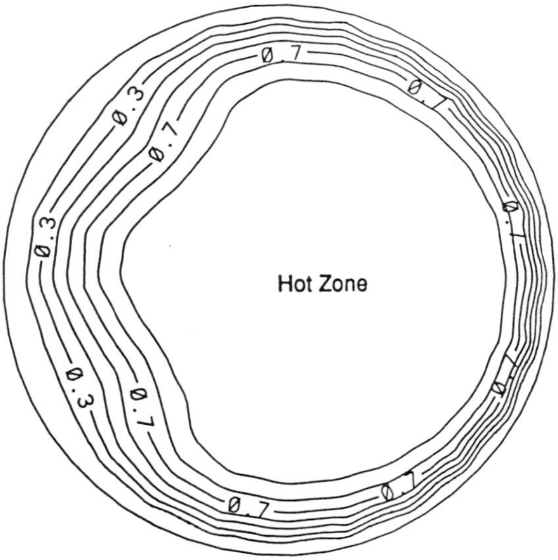

Fig. 8. Isotherms for CA coil at $z = 5$ for $Re = 750$, $\delta = 0.1$ [17].

$$\begin{aligned}\eta &\sim z_e^2 \delta^{1/2} Re^{-1}(z_e/Z) + 1 \\ &\sim Pe\,\delta^{1/2} Re^{-1}/Z + 1 \\ &\sim \delta^{1/2} Pr/Z + 1,\end{aligned} \tag{12}$$

which shows the insensitivity to Pe. It is clear that the CA pipe does not enhance heat transfer by enlarging the well-mixed core region which would suppress the thermal boundary layer. Instead, it does so by introducing core eccentricity.

The CA coil has two circulation regions as shown in one of the diagrams in Fig. 2(b), to which the chaotic advection concept can be applied in order to reduce their resistance. This is achieved by changing the coiling axis periodically every distance z_s in a downstream direction to produce the AA configuration. The eccentricity effect should average to nearly zero. It should also be noted that, since we are interested in heat transfer from the fluid to the wall, the thermal resistance of the stagnation streamline is not important.

For the resistance of the circulations to be effectively reduced, the switching length z_s should be far shorter than the entrance length z_e of equation (6) and, by definition, should be larger than the length of the coil $\sim O(\delta^{-1})$. Hence, there are a priori bounds

$$\delta^{-1} < z_s \ll Pe^{1/3} \tag{13}$$

to what the switching length should be for optimal enhancement. From equation (8), however, there is a condition required for weak coiling which combines with equation (13) to yield the condition for the validity of the analysis

$$Pe^{1/3} Re^2 \gg 1, \tag{14}$$

which is true for all high Pe flows. As long as equations (8) or (14) is satisfied, which requires the switching to be also weak as in the coiling such that the pipe is still almost straight, the core remains almost axisymmetric and it is only the transverse displacement at the core boundary that is pertinent. One can again neglect the axial velocity. The flow field (7) is now replaced by

$$\mathbf{u} = w(r)\mathbf{e}_z + [\delta^{1/4} Re^{-1/2} u(r, \theta)\mathbf{e}_r + \delta^{1/2} v(r, \theta)\mathbf{e}_\theta] + \alpha\mathbf{g}(r, \theta, z) \tag{15}$$

to leading order, where α is small and $\mathbf{g}(r, \theta, z) = \mathbf{g}(r, \theta, z + z_s)$, the perturbation flow field, is periodic in z. Since we are locally concerned only with the transverse displacement at the thermal core boundary, we need examine only the flow field near the boundary $r = r^*(z)$. The transverse displacement there over one switching length is given by the Poincaré map

$$\begin{aligned} r_{n+1} &= r_n + f_1(r, \theta) \\ \theta_{n+1} &= \theta_n + f_2(r, \theta) \end{aligned} \tag{16}$$

where

$$f_1 \sim -\frac{\partial H}{\partial \theta} + \alpha G_1(r, \theta, z)$$

$$f_2 \sim \frac{\partial H}{\partial r} + \alpha G_2(r, \theta, z)$$

and

$$\begin{aligned} -\frac{\partial H}{\partial \theta} &= \delta^{1/4} Re^{-1/2} u(r, \theta) z_s / w(r^*) \\ \frac{\partial H}{\partial r} &= \delta^{1/2} v(r, \theta) z_s / w(r^*) \\ G_i &= \int_0^{z_s} g_i(r, \theta, z')\,\mathrm{d}z'. \end{aligned} \tag{17}$$

The unperturbed system, $\alpha = 0$, can be written as a Hamiltonian because the nearly straight assumption stipulates that the transverse flow field is nearly divergence free with respect to the transverse divergence operator, and because w is evaluated at the core boundary r^*. Both conditions require the pipe to be nearly straight. It would be inappropriate to use this Hamiltonian Poincaré map approach if the coil is curved too rapidly i.e. if conditions (8) or (14) are violated. This is not to say strongly coiled pipes cannot enhance heat transfer. It simply means that the simplifying approach of the map (16) cannot be applied, and the connection with chaotic advection is not as clear. It is only in the case of almost straight pipes that one can treat the axial length z as a time-like variable.

Since $r^*(z)$ varies downstream and spans the entire pipe radius, one can actually study the map (16) by replacing r^* by r. In the CA case, the map generates two families of closed trajectories, each around an elliptic fixed point, separated by a heteroclinic orbit connecting two stagnation points at the wall. Under periodic forcing for which $\alpha \neq 0$, stochastic layers will again appear at the closed streamlines by the KAM mechanism and at the heteroclinic orbit. In contrast to problems of the first kind, however, the former now becomes very important, since the core boundary $r^*(z)$ traverses the entire cross-section, and the stochastic layer at every position serves to extend its boundary and suppress the thermal boundary layer.

Numerical analysis of the simplified Dean's hydrodynamics helps in understanding heat transfer in the AA coil. We also assume, for the moment, that a flow field is established immediately after every change in coil axis. Consider a material line which initially runs straight across the diameter of the pipe. As the fluid moves downstream the interface stretches due to the flow field. The velocity field can be integrated numerically to produce fluid particle trajectories. Figure 9(a) shows a comparison between the material lines at a given section in the CA and AA cases. The line in the latter case is much longer, so that diffusion of heat across it can be expected to be higher. The pathlines for any given initial set of fluid particles can also be computed. A set of about 400 fluid particles that are initially located within a small area disperse over the cross-section, as shown in Fig. 9(b). The Lyapunov exponent is found to have a value of 1.52, indicating that the trajectories are indeed chaotic.

The energy equation corresponding the Dean's flow field can also be solved numerically to determine the temperature field and the inner heat transfer coefficient. The two cases of CA and AA coils can be compared. Figure 10 shows the ratio of heat transfer coefficients as a function of Reynolds number. The AA coil has the greater heat transfer due to chaotic mixing, the effect which increases with Reynolds number. Since the axial velocity profile is the same in both coils, the pressure drop is also the same in both cases; the ratio of pressure drops is thus unity.

The simplified analysis outlined above makes several assumptions that can be relaxed in a detailed numerical computation [18]. For this purpose the Navier–Stokes and energy equations are solved. For the CA coil the fully developed axial gradient of the velocity is zero. This enables a parabolic march in the downstream direction from an initial state. In the AA coil the velocity repeats itself periodically in the downstream direction. The governing equations are solved using a control volume technique for periodic boundary conditions.

In the CA coil, the transversely averaged heat transfer coefficient goes to a constant value. For the AA geometry there is periodic variation along the length of the pipe. The ratios of the convective heat transfer coefficients and the pressure drops in the two coils are shown in Fig. 11. Once again chaotic mixing is seen to lead to better heat transfer without significant change in pressure drop.

One should be careful about using the Poincaré map approach for strongly coiled pipes. In this case, u, v and w are all functions of z to leading order and $\nabla_{\mathrm{T}} \cdot (u\mathbf{e}_r + v\mathbf{e}_z)$ does not vanish, where ∇_{T} is the transverse divergence operator. As a result, the map is not

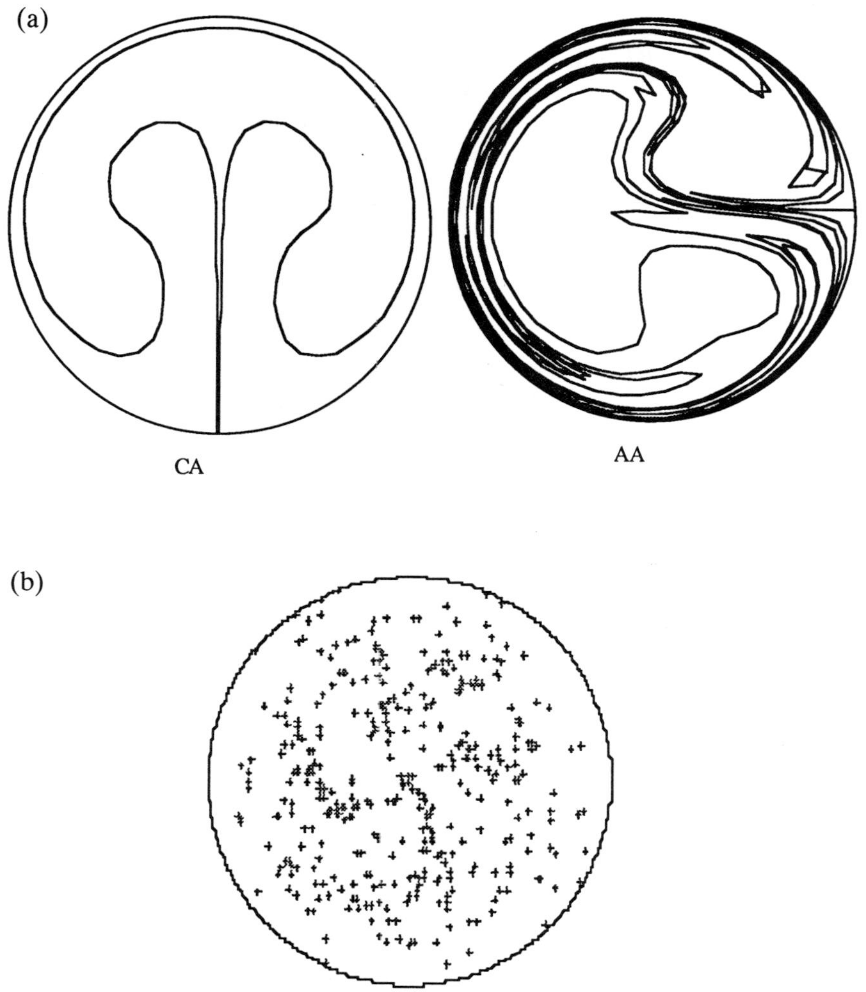

Fig. 9. Results of simplified numerical analysis. (a) Interfaces in cross-sectional plane of CA and AA coils [19]. (b) Dispersion of particles [16, 18].

conservative. However, since $\nabla \cdot \mathbf{u} = (\partial w/\partial z) + \nabla_T \cdot (u\mathbf{e}_r + v\mathbf{e}_z)$ vanishes for incompressible flow, we have

$$\begin{aligned}\int_\Omega \nabla_T \cdot (u\mathbf{e}_r + v\mathbf{e}_z)\,\mathrm{d}\Omega &= -\int_\Omega \frac{\partial w}{\partial z}\,\mathrm{d}\Omega \\ &= -\frac{\partial}{\partial z}\int_\Omega w\,\mathrm{d}\Omega \\ &= 0, \end{aligned} \tag{18}$$

indicating that the map is globally conservative. In equation (18), Ω is any cross-section area and $\int_\Omega w\,\mathrm{d}\Omega$ is just the constant volumetric flow rate. As a result, sources and sinks can coexist with chaotic orbits for the corresponding Poincaré map, as in the well-known damped pendulum example with a constant torque [26]. Domains of attraction of the sinks with fractal boundaries are expected and long chaotic transients should be the norm for

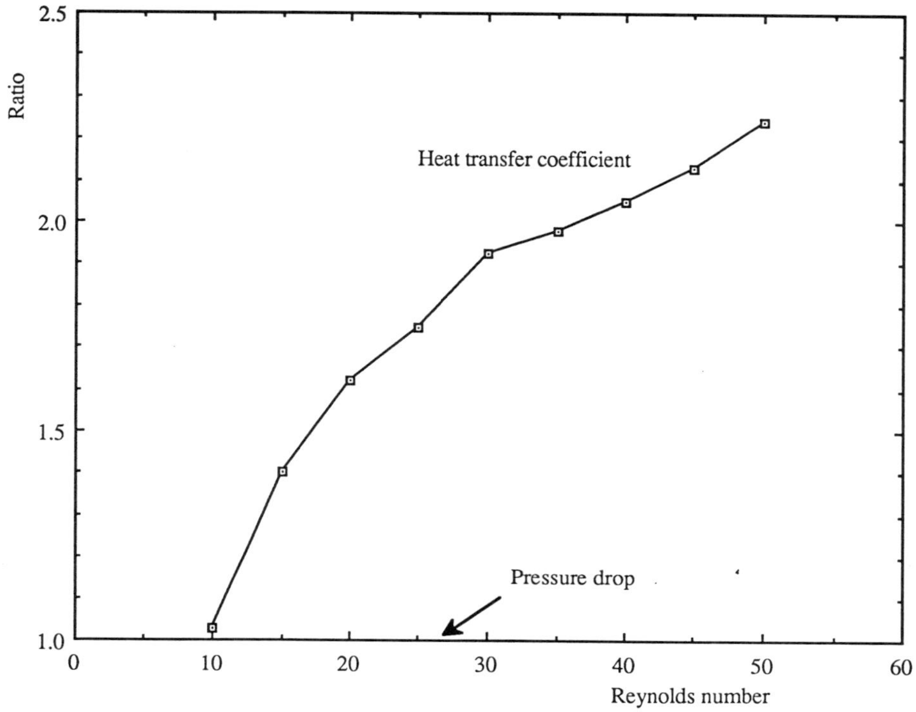

Fig. 10. Ratios obtained from simplified numerical model between AA and CA coils of heat transfer coefficients and pressure drops [16, 18].

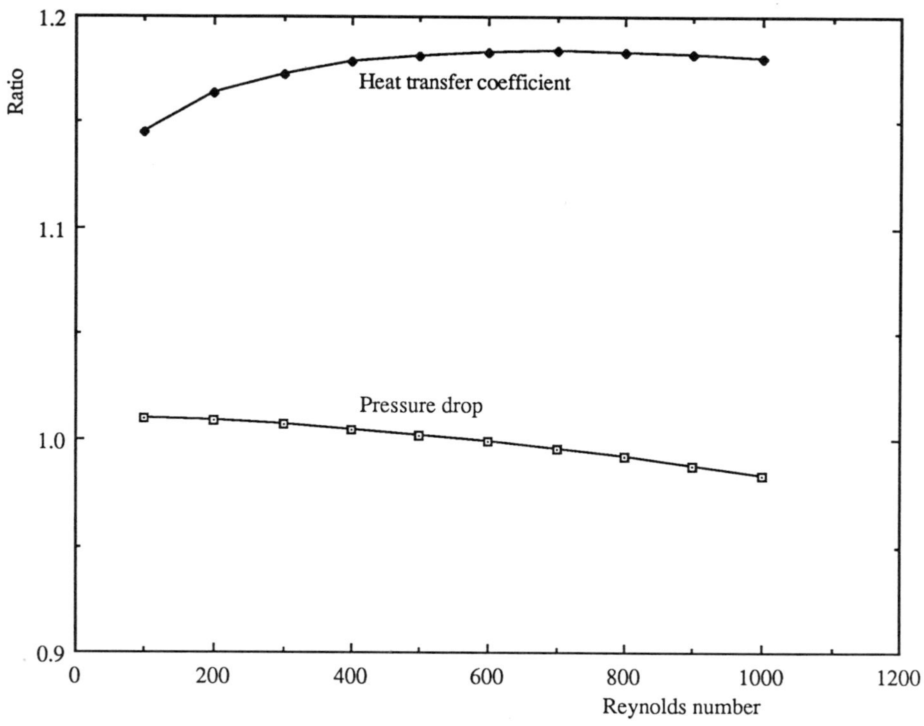

Fig. 11. Ratios obtained from exact numerical model between AA and CA coils of heat transfer coefficients and pressure drops [18].

these strongly coiled maps. One example is shown in Fig. 12 where the flow field is computed exactly for $\delta = 0.2$ and a switching angle of 30°. Four attracting points are observed. The kind of attractors obtained depends strongly on the switching angle, with fixed points, periodic attractors and chaos being obtained. Despite the rich dynamics one can extract from the Poincaré map of strongly coiled pipes, we have found weak correlation between the Lyapunov exponent and the heat transfer efficiency.

7. EXPERIMENTS WITH HEAT TRANSFER ENHANCEMENT

That chaotic mixing actually enhances heat transfer has been verified by comparative experiments on two different coils [16, 18]. Measurements were made in coils which were identical in every respect, except that one was of CA geometry and the other AA. This was achieved by making the coils out of the same number of copper U-tubes, but soldering them together differently. Great care was taken to minimize experimental errors by, for example, keeping the flow and measuring system the same, but changing only the coil being tested. An open-loop gravity-driven flow system was used to reduce flow instabilities and fluctuations. The experiments were conducted with cold water in the in-tube side and warm water in the over-tube side. Measurements were made of the mass flow rate and fluid temperatures entering and leaving the test coils. The overall heat transfer coefficient and flow Reynolds number were determined from this. The total thermal resistance to heat flow is that due to the inner and outer heat transfer coefficients, as well as the conduction in the pipe wall. The outer heat transfer coefficient was determined by measuring the transient response of the coils to a step change in the outside surface temperature by suddenly dipping the coil in a hot water bath and measuring the variation in wall temperature with time. A lumped mass analysis was used to find the outer heat transfer

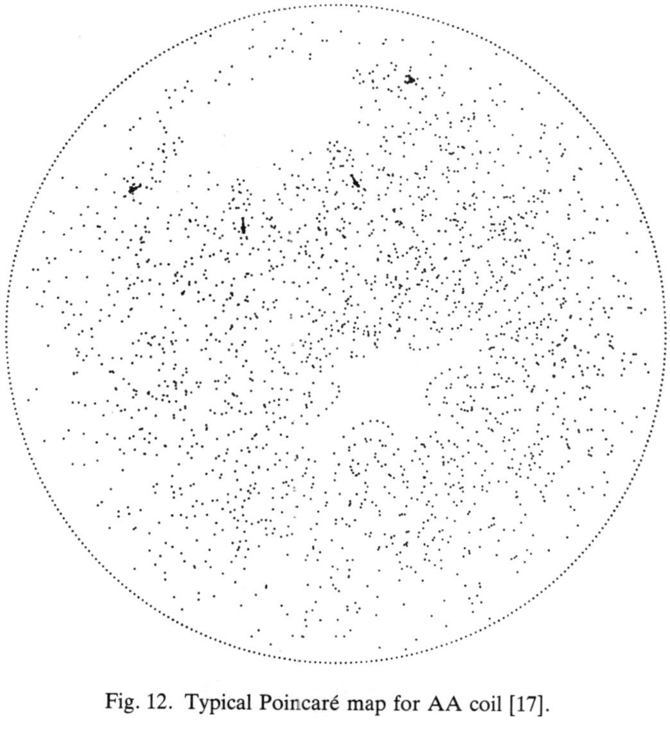

Fig. 12. Typical Poincaré map for AA coil [17].

coefficient. From the overall and outer heat transfer coefficients, the inner heat transfer coefficient was determined. The pressure drop in each coil was also measured. The Reynolds number range covered in the experiments included the transition Reynolds number of 7270. A careful error analysis that was carried out indicated that the difference between the two coils was larger than the uncertainty in the measurements.

The ratio of inner heat transfer coefficients in the CA and AA coils is shown in Fig. 13. The shape of this curve is qualitatively similar to that found from theoretical analysis. The ratio of pressure drops is also shown. Once again, there is increase in the heat transfer coefficient without a large penalty in terms of pressure drop. This effect is seen not only in the laminar regime as expected, but also in the turbulent regime.

Similar experiments have also been carried out by Peerhossaini *et al.* [27]. There are minor differences in the geometry of the coil used and in the flow field entering the coil. Their measurements confirm that the coil with chaotic mixing has a larger temperature drop compared to the conventional one. They have also used water as the working fluid. Experiments with fluids of higher *Pr* which should give larger factors of enhancement have yet to be performed.

8. CONCLUSIONS

Enhancement of heat transfer is especially important in the laminar flow of high-viscosity fluids since the generation of turbulence is prohibitive in terms of energy consumption, and in fluids of low thermal diffusivity where heat conduction is small. Heat transfer can be enhanced by reducing thermal resistances, which, in general terms, can be classified as due

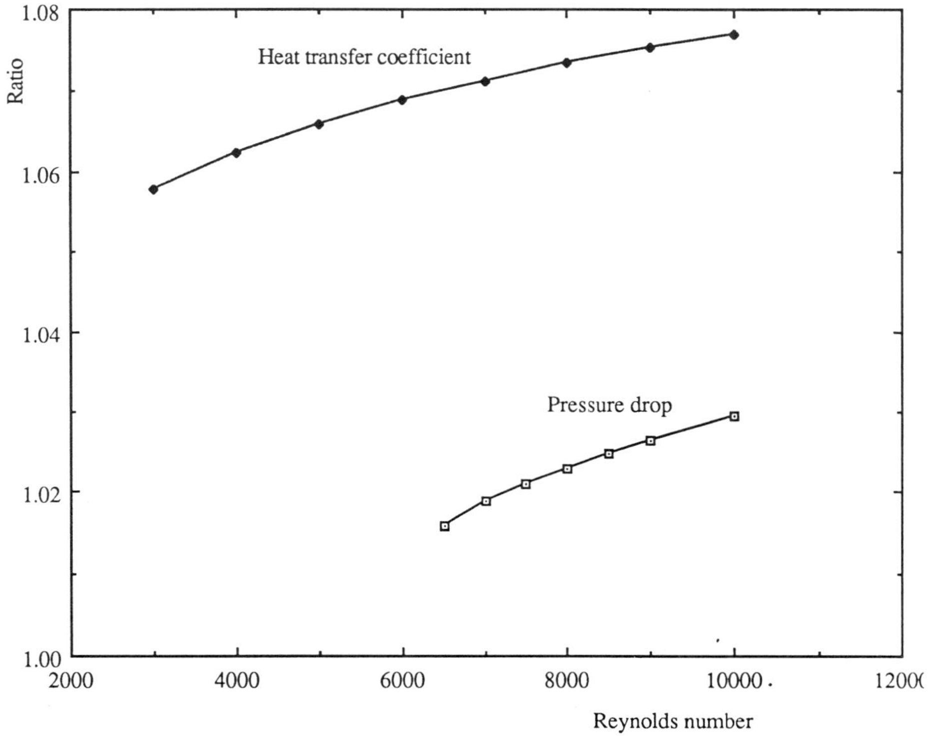

Fig. 13. Experimentally obtained ratios between AA and CA coils of heat transfer coefficients and pressure drops [16, 18].

to thermal boundary layers, recirculation regions, and stagnation streamlines. Different techniques apply for the reduction of each one of these resistances.

Mixing is the crucial factor for the reduction of thermal resistance in a recirculation region. Chaotic mixing leads to higher heat transfer compared to regular mixing, which in turn is better than no mixing. Time-dependent perturbations are needed for chaotic mixing in a two-dimensional domain, while it can occur for steady flows in ducts with spatial modulation. Perturbations can also be used to break up a stagnation streamline and reduce the resistance to heat flow across it.

The operation of a number of chaotic mixing devices have been reviewed here. Active or passive equipment can be designed either for batch or continuous-flow processes for the transfer of heat, mass or other scalars. Much of the previous discussion of rate enhancement is equally applicable to all scalars as long as the relevant diffusivity is low. Furthermore, since the rate of chemical reaction depends on the rate at which the reacting species come together, chaotic mixing has also been found to increase the effectiveness of a tubular chemical reactor [19]. Practical realization of chaotic mixing using these or similar devices is close to reality.

Acknowledgments—We are grateful to our students Dr. N. Acharya, Mr. S. Ghosh and Mr. D. R. Sawyers who carried out the work reported here. We are also very thankful to Dr. Ferol Fish and the Gas Research Institute for their support of our research on heat transfer enhancement by chaotic mixing under Grant number 5090-260-1971.

REFERENCES

1. M. F. Edwards, High-viscosity liquids—are we stuck in a rut?, *Heat Transfer Engng*. **2**(2), 5 (1980).
2. V. I. Arnol'd, Sur la topologie des écoulements stationnaires des fluides parfaits, *C. R. Acad. Sci. Paris* **261**, 17–20 (1965).
3. M. Hénon, Sur la topologie de lignes courant dans un cas particulier, *C. R. Acad. Sci. Paris* **A262**, 312–314 (1966).
4. H. Aref, Stirring by chaotic advection, *J. Fluid Mech.* **143**, 1–21 (1984).
5. H. Aref, Chaotic advection in a Stokes flow, *Phys. Fluids* **29**, 3515–3521 (1986).
6. W. L. Chien, H. Rising and J. M. Ottino, Laminar mixing and chaotic mixing in several cavity flows, *J. Fluid Mech*. **170**, 355–377 (1986).
7. J. Chaiken, R. Chevray, M. Tabor, Q. M. and Tan, Experimental study of Lagrangian turbulence in a Stokes flow, *Proc. R. Soc. Lond.* **A408**, 165–174 (1986).
8. D. V. Khakhar, H. Rising and J. M. Ottino, Analysis of chaotic mixing in two model systems, *J. Fluid Mech.* **172**, 419–451 (1986).
9. H. A. Kusch and J. M. Ottino, Experiments on mixing in continuous flows, *J. Fluid Mech.* **236**, 319–348 (1992).
10. D. V. Khakhar, J. G. Franjione and J. M. Ottino, A case study of chaotic mixing in deterministic flows: the partitioned pipe mixer, *Chem. Engng. Sci.* **42**, 2909–2926 (1987).
11. F. H. Ling, Chaotic mixing in a spatially periodic continuous mixer, *Phys. Fluids* **A5**, 2147–2160 (1993).
12. C. S. Lee, J. J. Ou and S. H. Chen, Quantification of mixing from the Eulerian perspective: flow through a curved tube, *Chem. Engng. Sci*. **42**, 2484–2486 (1987).
13. S. W. Jones, O. M. Thomas and H. Aref, Chaotic advection by laminar flow in a twisted pipe, *J. Fluid Mech*. **209**, 335–357 (1989).
14. Y. Le Guer and H. Peerhossaini, Order breaking in Dean flow, *Phys. Fluids* **A3**, 1029–1032 (1991).
15. H. Peerhossaini and Y. Le Guer, Chaotic motion in the Dean instability flow – A heat exchanger design, *Bull. Am. Phys. Soc.* **35**, 2229 (1990).
16. N. Acharya, M. Sen and H.-C. Chang, Heat transfer enhancement in coiled tubes by chaotic mixing, *Int. J. Heat Mass Transf*. **35**, 2475–2489 (1992).
17. N. Acharya, M. Sen and H.-C. Chang, Thermal entrance length and Nusselt numbers in coiled tubes, *Int. J. Heat Mass Transf*. **37**, 336–340 (1994).
18. N. Acharya, Experimental and numerical investigation of heat transfer enhancement in coiled tubes by chaotic mixing, Ph.D. dissertation, University of Notre Dame, Notre Dame, Indiana.
19. D. R. Sawyers, M. Sen and H.-S. Chang, Effect of regular and chaotic mixing on bimolecular chemical reaction in helical-coil reactors, to be published.
20. H.-C. Chang, E. Demekhin and D. Kopelevich, Nonlinear evolution of waves on a vertically falling film, *J. Fluid Mech*. **250**, 433–480 (1993).

21. P. B. Rhines and W. R. Young, How rapidly is a passive scalar mixed within closed streamlines, *J. Fluid Mech*. **133**, 133–145 (1983).
22. S. Ghosh, H.-C. Chang and M. Sen, Heat-transfer enhancement due to slender recirculation and chaotic transport between counter-rotating eccentric cylinders, *J. Fluid Mech.* **238**, 119–154 (1992).
23. J. B. Weiss and E. Knobloch, Mass transport and mixing by modulated travelling waves, *Phys. Rev.* **A40**, 2579–2589 (1989).
24. P. Holmes, Some remarks on chaotic particle paths in time-periodic, three-dimensional swirling flow, *Contemp. Math.* **28**, 393–406 (1984).
25. W. R. Dean, Note on the motion of fluid in a curved pipe, *Philos. Mag. J. Sci.* **IV** (7th series), 208–223 (1927).
26. F. M. A. Salam and S. S. Sastry, Dynamics of forced Josephson junction circuit: The region of chaos, *IEEE Trans. Circuits and Systems* **32**, 784–797 (1995).
27. H. Peerhossaini, C. Castelain and Y. Le Guer, Heat exchanger design based on chaotic advection, *Exp. Thermal & Fluid Sci.* **7**, 333–344 (1993).

Interacting Two-dimensional Vortex Structures: Point Vortices, Contour Kinematics and Stirring Properties

V. V. MELESHKO

Institute of Hydromechanics, 252057 Kiev, Ukraine

and

G. J. F. VAN HEIJST

J. M. Burgers Centre for Fluid Mechanics, Fluid Dynamics Laboratory, Department of Technical Physics, Eindhoven University of Technology, P.O. Box 513, 5600 MB Eindhoven, The Netherlands

Abstract—This paper addresses the stirring properties of interacting two-dimensional vortex structures. It is well known from previous studies that essential features of dipole–dipole and dipole–monopole interactions, like partner exchange, are captured by simple point-vortex models of these structures. In order to investigate the mixing induced by such interactions, the method of contour kinematics is adopted here, which allows us to trace fluid particles that are initially confined within some closed contour. For example, by tracing the deformations of the separatrix enclosing the 'atmosphere' of a point-vortex dipole during the interaction with another single point vortex or vortex pair, the contour kinematics method provides information about the processes of entrainment and detrainment that take place during the interaction. Comparison of the modelling results with both full numerical simulations and laboratory experiments on dye-visualised vortex interactions in a stratified fluid reveals a surprisingly good qualitative agreement.

1. INTRODUCTION

> Now I think hydrodynamics is to be the root of all physical science, and is at present second to none in the beauty of its mathematics.
>
> W. Thomson to G. G. Stokes
> 20 December 1857

The above words are written just before Helmholtz's [1] great discovery of the laws of vortex motions in an ideal fluid. After this seminal paper these words have appeared to become the motto of a constant interest in experimental and theoretical investigations of vortex dynamics. Not much later, a fundamental paper of W. Thomson (Lord Kelvin) [2] with an original vortex theory of matter appeared, which stimulated many prominent scientists, including Maxwell [3] and J. J. Thomson [4] to study various problems of vortex interactions. Many of these problems still remain difficult and challenging to modern vortex dynamics [5] and even today have not been solved completely. At the same time, important progress was made along the experimental line by applying the concept of colouring the fluid to render its motion apparent. That idea was suggested by the effect of beautiful phenomena of smoke-ring motion, first shown by Rogers [6], Ball [7], and especially Tait [8] in air. In water such experiments on vortex rings were performed by Reusch [9] and in much more detail by Reynolds [10, 11]. Reynolds noticed, that "he found that the vortex played a part in fluid motion which he never dreamt of; that, in fact, it was the key to almost all the problems of internal fluid motion" [10]. In subsequent years a great number of experiments with vortex rings confirmed Reynolds's observation that the ring catches initially some coloured volume of fluid, upon which it grows in size as it proceeds. While moving it both entrains continually some of the surrounding fluid and leaves some coloured fluid behind in its wake. For a single vortex ring moving steadily without expansion through an inviscid fluid the term 'atmosphere' for the volume of trapped fluid was introduced by Kelvin [2] (see Fig. 1). This feature was considered in some detail by Hicks [12] and was later also mentioned in the fundamental textbooks by Lamb [13] and Batchelor [14].

Since the pioneering work in the last century, the dynamics of vortex rings has continued to be an interesting research topic up to the present day. Along with theoretical and numerical studies, simulations of various aspects of vortex rings now play an important role in the process of gaining a better insight in their dynamics. A comprehensive review of this subject was recently published by Shariff and Leonard [15], who also provided a detailed list of references.

In recent times the analysis of two-dimensional (2D) coherent vortex structures is increasingly attracting the attention of many fluid dynamicists. The emergence of these structures is a characteristic feature of quasi-geostrophic or 2D flows, and coherent vortices have indeed been observed abundantly in the oceans [16] and the planetary atmospheres [17]. Well-known oceanic examples are the Gulf Stream and Agulhas rings, vortices shed from coastal currents, and the subsurface Meddies. These coherent monopolar and dipolar vortices have relatively long lifetimes, and are believed to play an important role in the

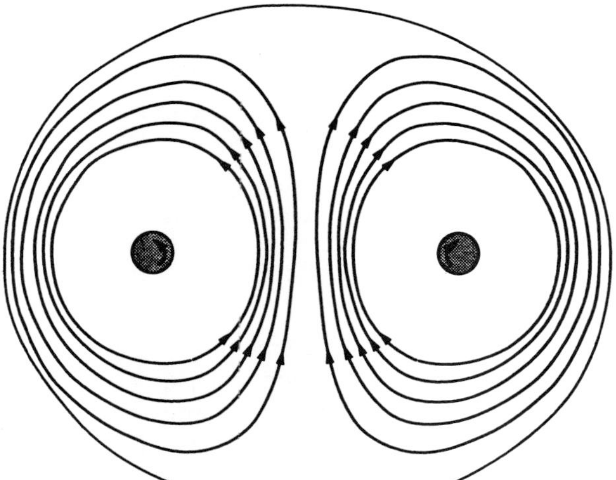

Fig. 1. Original drawing of the vortex pair atmosphere by W. Thomson (1867). "In a perfect fluid the bounding surface of the portion carried forward would necessarily be quite symmetrical on the anterior and posterior sides of the middle plane of the ring. The motion of the surrounding fluid must be precisely the same as it would be if the space within this surface were occupied by a smooth solid" [2].

transport of heat, salt and other properties. In this geophysical context the 2D motion is mainly due to the planetary rotation, the density stratification and also due to the dominantly 2D nature of the flow domain itself.

For 2D flow of an ideal unbounded fluid some important solutions to the Euler equations were known for many years, both for the case of distributed vorticity (e.g. the monopolar Rankine and Kirchhoff vortices, and the Chaplygin–Lamb dipole) and for concentrated (point) vortices. Relatively recently Stern [18] presented an exact solution for the modon, i.e. a dipolar coherent vortex solution to the barotropic vorticity equation with a linear gradient in the background vorticity, in geophysical fluid dynamics usually referred to as the β-plane approximation. Considerable theoretical progress has been made since Stern's pioneering paper, and a reasonably up-to-date review is given by Flierl [19].

Most of these stable coherent vortex structures were also observed experimentally in linearly stratified [20, 21] and rotating [22–24] fluids, in soap films flows [25] and in magnetohydrodynamic flows [26]. In the latter case the flow was kept two-dimensional by the presence of magnetic forces; in this respect it should be mentioned that apart from their geophysical relevance, coherent vortex structures are also studied within the context of plasma physics and astrophysics, in which 2D flow situations occur as well.

Numerical investigations of Chaplygin–Lamb dipoles or modons collisions under coaxial [27, 28] and collinear [29] initial movements also demonstrated the soliton-like character of these vortex structures. In view of their properties, these coherent vortices are sometimes referred to as the elementary particles of 2D turbulence.

An important question in the study of vortex structures concerns their stirring and mixing properties, in particular when smaller or larger perturbations are introduced by, for example, instability or interaction with other vortices. Recent experiments on interacting vortex dipoles [20, 21] have revealed the complicated nature of the advection of passive tracers during the interaction process, in which fluid initially trapped in the atmospheres of the original dipoles is left behind via detrainment, while some surrounding fluid is caught in the newly formed structures (entrainment).

Rather than studying these coherent vortices by full numerical simulation, in the present paper we will investigate the possibilities of modelling some essential features of the dynamics and the associated advection properties by a simplified model [30] in which the coherent vortex structures are represented by point vortices and their atmospheres. The main aim of the present paper is to further explore the validity of this point-vortex model to a quantitative analysis of the entrainment/detrainment processes in some typical cases of nonperiodic interactions of monopolar and dipolar vortex structures. As will be shown, the method of contour kinematics provides an extremely useful tool in studying the interaction-induced advection of passive tracers. We have confined ourselves to some typical cases of integrable motion of a vortex pair in combination with another single vortex or with another vortex pair. It is believed that studying the entrainment/detrainment that the vortex pair undergoes while interacting with the other vortex structures is a first—but essential—step in developing a better understanding of stirring processes in more general cases.

The paper is organized as follows: in Section 2 a general formulation is presented of advection properties of systems of point vortices, and the contour kinematics method is discussed. Some general characteristics of dipolar vortices (vortex pairs, Chaplygin–Lamb dipole, laboratory experiments) as well as definitions of entrainment/detrainment coefficients are given in Section 3. The interaction of a vortex pair with a monopole and with another vortex pair is discussed in Section 4. Results of the contour kinematics method obtained for both cases are compared with laboratory observations, and a good qualitative agreement is found. Finally, some general conclusions are given in Section 5.

2. FORMULATION OF THE ADVECTION PROBLEM

> The general idea of mixture is so familiar to us that the vast generalisation to which these ideas afford the key, remains unnoticed.
> O. Reynolds (1893)

2.1. *General statement of advection problem*

In many investigations of fluid motion the Eulerian representation is used and a velocity vector $\mathbf{u}$ is considered at any point of the fluid as a function of the coordinates $\mathbf{x}$ and time t. In other words, it is supposed that we are able to take a snapshot of the system at any time, and to observe the velocities, without having to track each individual particle during the flow evolution. In such an approach we measure the velocity by the volume which passes through a unit area rather than by the displacement travelled by a particle in a time unit. The interest in the processes of fluid stirring drives us to the Lagrangian approach based on the study of motion of an individual fluid particle $\mathbf{X}$ in a known Eulerian velocity field $\mathbf{u}(\mathbf{x}, t)$. That motion is described by the solution of the dynamical system

$$\dot{\mathbf{x}} = \mathbf{u}(\mathbf{x}, t) \tag{1}$$

with initial condition $\mathbf{x} = \mathbf{X}$ at time $t = 0$. Here, and in what follows, the dot denotes the time derivative. A particle moving according to these equations, is said to be subject to a passive advection.

Although the idea of such an approach was laid down in the last century by Rankine [31] and Maxwell [32], and was complemented by an elegant qualitative discussion by Reynolds [11] (he introduced the term 'folding' in order to note how the attenuated coloured layers are brought together*) and by simplified qualitative estimates by Welander [34] (who showed the importance of the mechanism of stretching and folding for the mixing process), up to recent times comparatively little has been done towards obtaining a clear view of the advection process in any but the simplest cases. To such complicated cases belong the processes of chaotic advection introduced by Aref [35]—chaotic trajectories of Lagrangian particles in a regular Eulerian velocity field. A general review of this topic is presented by Aref [36] and Ottino [37].

For an analysis of the stirring process we have to consider the initial 'coloured' region (dyed fluid in various experiments) and to solve equation (1) with different initial positions $\mathbf{X}$. At first sight the problem looks purely kinematical, but in order to know the velocity field $\mathbf{u}(\mathbf{x}, t)$ we have to solve some dynamical equations of fluid motion.

2.2. *Advection by point vortices*

It is interesting to analyse a special case of (1), when the velocity field contains no explicit time dependence, namely

$$\dot{\mathbf{x}} = \mathbf{u}(\mathbf{x}, \mathbf{x}_\alpha). \tag{2}$$

Here the functions $\mathbf{x}_\alpha(\alpha = 1, \ldots, N)$, obey the evolutionary equations

$$\dot{\mathbf{x}}_\alpha = \mathbf{u}_\alpha(\mathbf{x}_\alpha, \mu) \tag{3}$$

where $\mathbf{x}_\alpha = \mathbf{X}_\alpha$ at $t = 0$ with some parameter μ. Depending on the values of μ and N the solutions $\mathbf{x}_\alpha(t)$ of (3) may be either regular or chaotic.

---

*As Reynolds noted [33]: "it is a peculiarity of the motion of continuous fluid, otherwise we might infer that two people who had once shaken hands could never after be so much as a mile apart".

An important example of that approach is connected with two-dimensional motion of 'point' vortices in an unbounded plane. Here the Lagrangian viewpoint is particularly useful because, according to the Helmholtz theorem [1], the vortex lines are material, implying that point vortices move as fluid particles. This subject is of traditional interest in fluid mechanics [5, 13, 14].

The two-dimensional problem of the interaction of N point vortices of strength κ_α and position (x_α, y_α) in the unbounded plane (x, y) of an inviscid fluid consists in solving the following nonlinear system of first-order differential equations

$$\dot{z}_\alpha^* = \frac{1}{2\pi i}\sum_{\beta=1}^{N}{}' \frac{\kappa_\beta}{z_\alpha - z_\beta}, \qquad \alpha = 1, \ldots, N \tag{4}$$

with initial conditions $z_\alpha = z_\alpha^{(0)}$ at $t = 0$. Here $z_\alpha = x_\alpha + iy_\alpha$, the asterisk denotes complex conjugation, and the prime indicates omission of the singular term $\beta = \alpha$. The equations (4) may be also written in a real form as a system of $2N$ equations

$$\dot{x}_\alpha = -\frac{1}{2\pi}\sum_{\beta=1}^{N}{}' \frac{\kappa_\beta(y_\alpha - y_\beta)}{l_{\alpha\beta}^2}, \qquad \dot{y}_\alpha = \frac{1}{2\pi}\sum_{\beta=1}^{N}{}' \frac{\kappa_\beta(x_\alpha - x_\beta)}{l_{\alpha\beta}^2}, \tag{5}$$

where $l_{\alpha\beta}^2 = (x_\alpha - x_\beta)^2 + (y_\alpha - y_\beta)^2$ with the initial configuration of vortices given by $x_\alpha = x_\alpha^{(0)}$, $y = y_\alpha^{(0)}$ at $t = 0$.

It was already demonstrated by Kirchhoff [38] that the equations (5) define a Hamiltonian dynamical system

$$\kappa_\alpha \dot{x}_\alpha = \frac{\partial H}{\partial y_\alpha}, \qquad \kappa_\alpha \dot{y}_\alpha = -\frac{\partial H}{\partial x_\alpha}, \tag{6}$$

where the so-called Hamiltonian

$$H = -\frac{1}{8\pi}\sum_{\alpha,\beta=1}^{N}{}' \kappa_\alpha \kappa_\beta \ln l_{\alpha\beta}^2 \tag{7}$$

is a conserved quantity. In addition to H, the system (6) has three well-known first integrals, these being:

$$Q = \sum_{\alpha=1}^{N} \kappa_\alpha x_\alpha, \qquad P = \sum_{\alpha=1}^{N} \kappa_\alpha y_\alpha, \qquad I = \sum_{\alpha=1}^{N} \kappa_\alpha (x_\alpha^2 + y_\alpha^2). \tag{8}$$

A detailed analysis of the general properties of the system (5) is presented in Aref's survey [39].

The equation of the passive advection of an individual Lagrangian fluid particle induced by the point vortices may be obtained by considering a marker $Z = X + iY$ as a point vortex of zero strength. Its motion is then described by the equation

$$\dot{Z}^* = \frac{1}{2\pi i}\sum_{\beta=1}^{N} \frac{\kappa_\beta}{Z - z_\beta}, \tag{9}$$

with $Z = Z^{(0)}$ at $t = 0$, or equivalently by

$$\dot{X} = -\frac{1}{2\pi}\sum_{\beta=1}^{N} \frac{\kappa_\beta(Y - y_\beta)}{l_\beta^2}, \qquad \dot{Y} = \frac{1}{2\pi}\sum_{\beta=1}^{N} \frac{\kappa_\beta(X - x_\beta)}{l_\beta^2}, \tag{10}$$

where $l_\beta^2 = (X - x_\beta)^2 + (Y - y_\beta)^2$, with $X = X^{(0)}$, $Y = Y^{(0)}$ at $t = 0$. The equations (10) can be transformed into Hamiltonian form

$$\dot{X} = \frac{\partial \Psi}{\partial Y}, \qquad \dot{Y} = -\frac{\partial \Psi}{\partial X} \tag{11}$$

with

$$\Psi(X, Y, t) = -\frac{1}{4\pi}\sum_{\beta=1}^{N} \kappa_\beta \ln l_\beta^2. \qquad (12)$$

The implicit time dependence of Ψ is due to the time evolution of the functions $x_\beta(t)$ and $y_\beta(t)$, being the solutions to system (5).

The kinematical problem of passive advection is thus formally equivalent to the dynamical problem of a time-dependent, one degree of freedom, Hamiltonian system. It is important to stress that system (11) is, in general, nonautonomic and may consequently reveal chaotic behaviour [40].

It is well known, due to Gröbli [41], that for the rather complicated case of three vortices in an unbounded plane this problem is integrable, regardless of the values of strengths: the solution $z_\alpha(t)$ to equation (4) is obtained in a closed analytical form containing, in general, some quadratures. A formal proof of that fact, based on the Liouville theorem for integrals in involution, was mentioned briefly in Poincaré's lectures [42] on vortex dynamics. Hence, the advection by two interacting point vortices forms an integrable problem because a passive marker may be considered as a point vortex of zero strength (the so-called 'restricted' problem [40]). A similar simple statement does not hold for the advection by three vortices: the passive advection of a particle according to equation (2) now forms a non-integrable problem and a marker can move along a chaotic trajectory. Then the following question arises: if this is the case, what can we say about the behaviour of a cloud of fluid particles in the velocity field induced by a number of point vortices?

Thomson [2] was the first to show that a vortex pair (two point vortices of equal but oppositely-signed intensities) in its steady motion in an unbounded plane is also accompanied by an 'atmosphere' (sometimes referred to as the 'Kelvin oval' [43]), i.e. a fixed closed volume of fluid particles. Hicks [12] generalized this result and proved that when some vortex configuration moves steadily through an irrotational fluid it is possible to distinguish three definite regions of fluid motion: (i) a region of rotational motion, which conserves its identity, (ii) a region of irrotational cyclic motion surrounding the first, which also keeps its identity and volume and travels uniformly through the fluid with an undeformed boundary, (iii) a region of irrotational acyclic motion, outside the second region. The fluid in region (iii) remains at rest at infinity and is never displaced over more than only a small distance. Rom-Kedar *et al.* [44] showed that in the presence of small perturbations in the form of an oscillating external strain-rate velocity field at infinity the whole picture is radically changed: fluid is entrained and detrained across the boundary of the averaged atmosphere and chaotic particle motions occur in certain regions. Thus, it seems interesting to investigate how the initial atmosphere of a vortex pair develops during a mutual interaction with other point vortices.

2.3. *Numerical algorithm*

The present study discusses a few different vortex configurations for which the system (5) with associated initial conditions is integrated numerically in double precision by two different methods: (i) by the method of extrapolation with a variable step and order [45] and (ii) by the Adams–Bashforth predictor–corrector routine with a fourth-order Runge–Kutta starting scheme. For the problems addressed, both methods give the same results for the trajectories of the point vortices. The accuracy of the computations was investigated in several ways. Firstly, it was found that the invariants (7) and (8) were conserved to within 10^{-11} during the course of the simulation when altering the local accuracy from 10^{-11} to 10^{-13} in the first method and using a dimensionless time step $\Delta\tau = 0.001$ in the second one. Secondly, for the cases $N = 3$ (always) and $N = 4$ (for some cases of initial symmetry) the analytical solutions (in quadratures) to equations (5) may be obtained by using the invariants H, Q, P and I. Such analytically calculated trajectories of vortices were found to coincide completely with the ones obtained by direct numerical solution, at least for the timespans covered by the simulations. Using the analytical or numerical solutions $x_\alpha(t)$, $y_\alpha(t)$, the system (10) was solved numerically by means of the methods mentioned above.

Computationally, a material closed line surrounding the 'coloured' area is composed of a large number of markers (up to $M = 5000$) connected by small linear segments. If enough markers are used, the line appears smooth. Maintaining smoothness during the flow evolution can be difficult, since the line is likely to be stretched and folded. As a result any initial line may deform into a very complicated structure, and its length will increase considerably. We use the following algorithm for tracing the boundary: each time interval is divided into small time steps, usually $\Delta\tau = 0.1$ and an integration using the above-mentioned numerical scheme is performed. If at some moment $\tau_n = n\Delta\tau$ the distance between neighbouring points becomes more than a prescribed threshold value, or if the angle between neighbouring linear segments is less than 120° (e.g. folding begins), it is necessary to return to $\tau = 0$, add points in proper places on the boundaries, repeat the calculations only with these new points up to τ_n, and then proceed. This approach is here referred to as the method of contour kinematics.

The surface S enclosed by any deformed curve L was calculated using Stokes' theorem, namely

$$S = \iint_S \mathrm{d}x\,\mathrm{d}y = \frac{1}{2}\oint_L (-y\,\mathrm{d}x + x\,\mathrm{d}y) \approx \frac{1}{2}\sum_{i=1}^{M}(x_i y_{i+1} - y_i x_{i+1}). \tag{13}$$

This algorithm for calculating the area S is known to be very accurate, and thus provided a useful tool to check the overall accuracy of the advection calculations.

3. DIPOLAR VORTICES

> All scientific advance in discovering the laws of nature has been made by first simplifying problem and reducing it to certain ideal conditions, and this is what mathematicians have done in studying the motion of a liquid.
> H. S. Hele-Shaw (1899)

Within the context of 2D flows the dipolar vortex can be concisely defined as a compact symmetric arrangement of two regions of oppositely signed vorticity. Dipolar vortices are characterized by a translational motion, and they contain a net linear momentum. When embedded in a quiescent ambient fluid, the vortex dipole shows a steady translation in a direction defined by its symmetry axis. The self-propelling mechanism associated with the dipolar vorticity distribution provides an important agent for the transport of material properties. In view of their nonzero linear momentum, the generation of dipoles requires a local, linear forcing.

3.1. *Atmosphere of the vortex pair*

The simplest model of a vortex dipole is the vortex pair: two point vortices of strength $\pm\kappa$ separated at some distance, say $2b$. If their initial positions are on the y-axis with

$z_1^{(0)} = ib$ and $z_2^{(0)} = -ib$, this vortex pair moves along the x-axis with a constant velocity $\kappa/4\pi b$. In other words, the expressions

$$z_1(t) = \frac{\kappa}{4\pi b}t + ib, \qquad z_2(t) = \frac{\kappa}{4\pi b}t - ib \tag{14}$$

are the solutions to equations (4). In fact, this solution was already mentioned in Helmholtz's pioneering [1] paper.

More interesting is the motion of fluid particles surrounding a uniformly propagating vortex pair. Using equations (14) and the dimensionless co-moving coordinate system (ξ, η), with $\xi = x/b - \kappa t/4\pi b^2$, $\eta = y/b$, the system (10) describing particle advection may be written as

$$\frac{d\xi}{d\tau} = \frac{2(\xi^2 - \eta^2 + 1)}{[(\eta + 1)^2 + \xi^2][(\eta - 1)^2 + \xi^2]} - \frac{1}{2}, \quad \frac{d\eta}{d\tau} = \frac{4\xi\eta}{[(\eta + 1)^2 + \xi^2][(\eta - 1)^2 + \xi^2]}, \tag{15}$$

or, in autonomic Hamiltonian form

$$\frac{d\xi}{d\tau} = \frac{\partial \Psi_{VP}}{\partial \eta}, \qquad \frac{d\eta}{d\tau} = -\frac{\partial \Psi_{VP}}{\partial \xi} \tag{16}$$

where

$$\Psi_{VP}(\xi, \eta) = \frac{1}{2}\left(\ln \frac{(\eta + 1)^2 + \xi^2}{(\eta - 1)^2 + \xi^2} - \eta\right). \tag{17}$$

Here, and in what follows, the dimensionless time $\tau = \kappa t/2\pi b^2$ is introduced. Thus in the (ξ, η)-system the particle motion is steady and the streamlines $\Psi_{VP}(\xi, \eta) = C$ coincide with pathlines of fluid particles. Two types of streamlines are formed, viz. a system of closed lines and a system of non-closed lines, as shown in Fig. 2(a). The closed streamline $C = 0$ defines a fixed, closed volume of fluid, approximately an ellipse with axes $1.73b$ and $2.09b$ and area $S = 3.61\pi b^2$; this is the 'atmosphere', which moves permanently with the vortex pair. The equation of the atmosphere boundary (the lines Ψ_u, Ψ_l) was already obtained by W. Thomson [2] as

$$\xi^2 = 2\eta \coth \frac{\eta}{2} - 1 - \eta^2. \tag{18}$$

The inner streamline Ψ_0: $|\xi| \leq \sqrt{3}$, $\eta = 0$, also corresponds to $C = 0$, and acts as the symmetry axis. Thus, the atmosphere of the vortex pair consists of two symmetrical parts bounded by lines $\Psi_u \cup \Psi_0$ and $\Psi_l \cup \Psi_0$, respectively. These lines intersect in two hyperbolic fixed points P_- and P_+. The lines Ψ_u and Ψ_l are referred to as the stable manifold of P_- and the unstable manifold of P_+, respectively; the line Ψ_0 is the stable manifold of P_+, and the unstable manifold of P_-. It means that in the course of time all points lying on Ψ_u and Ψ_l will move towards P_- (and can reach it only after infinite time). In the same way, all points on Ψ_0 will move towards P_+. All fluid particles inside the atmosphere move in a periodic way along closed lines. It is useful to mark these lines by Ψ_ε, where the value ε $(0 \leq |\varepsilon| < 1)$ corresponds to the value η at $\xi = 0$.

Analysis of equations (15) shows that for $|\xi_0| < \sqrt{3}$ the motion of an initial particle $(\xi_0, 0)$ on the streamline Ψ_0 can be expressed implicitly as

$$\tau = \frac{4}{\sqrt{3}} \ln \frac{(\sqrt{3} + \xi)(\sqrt{3} - \xi_0)}{(\sqrt{3} - \xi)(\sqrt{3} + \xi_0)} + \xi_0 - \xi. \tag{19}$$

For large values of τ, equation (19) has the the following asymptotic behaviour

$$\xi(\tau) \approx \sqrt{3} + (\xi_0 - \sqrt{3}) \exp\left(-\tau \frac{\sqrt{3}}{4}\right). \tag{20}$$

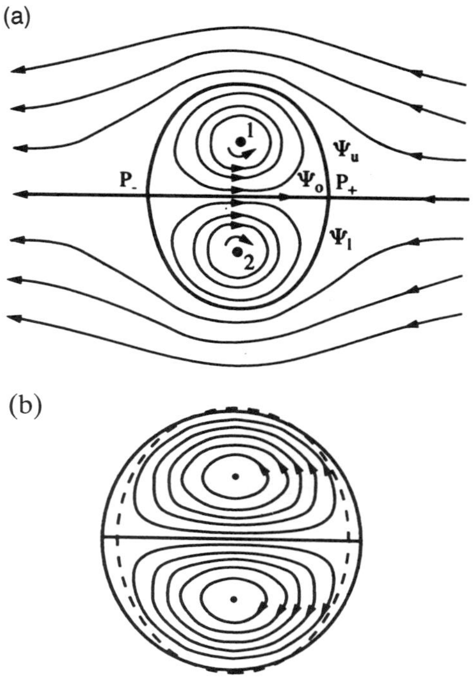

Fig. 2. (a) Flow pattern associated with a freely moving point–vortex pair: the streamlines are drawn in the comoving frame of the unperturbed flow, showing the atmosphere of the vortex pair. The fixed points are marked by P_+, P_-, the lower and upper atmosphere boundaries are marked by Ψ_l and Ψ_u, respectively; the common line is indicated by Ψ_0. (b) Chaplygin–Lamb dipole: circular vortex patch with oppositely signed vorticity in upper and lower semicircles. The broken line represents the boundary of the equivalent vortex pair atmosphere.

This result implies that theoretically any point ξ_0 on Ψ_0 (excluding P_-) can reach P_+ only after *infinite* time. In numerical simulations, however, any point on Ψ_0 has practically reached P_+ for $\tau \approx 10$. The length $l(\tau)$ of any straight-line segment on the separatrix Ψ_0 for large τ decreases as $l(\tau) \approx l(0)\exp(-\tau\sqrt{(3)}/4)$.

It is much more complicated to find analytical expressions for the motion on the separatrices Ψ_u and Ψ_l, but it is to be expected that in general they will have a time dependence similar to (20). It is worth noting that in spite of the rather simple expression (18) for the separatrices Ψ_u and Ψ_l, numerically tracing a fluid particle motion initially located on them is rather complicated: even the smallest computational error in determining its coordinates in the vicinity of P_- causes the particle to 'leap' onto the straight line Ψ_0, along which it then rapidly moves to point P_+, where it remains.

The solution to the equations (10) enables us to study the formation process of a vortex pair atmosphere from an arbitrary distribution of tracers. For the case of an initial marker distribution in a rectangular region covering the atmosphere, this process is shown in Fig. 3. Regardless of the form and size of the initial region, the form of the atmosphere asymptotically approaches the curve described by (18). Those fluid particles which are initially not situated inside the atmosphere will gradually form a long tail in the wake of the moving pair. In other words, during its motion the vortex pair catches only those markers lying at $\tau = 0$ inside the atmosphere, while the other fluid particles simply move around it as if the atmosphere were a rigid body. It can be shown that the length of a small vertical line $\xi_0 = 0$, $2.09 - \varepsilon/2 \leq \eta_0 \leq 2.09 + \varepsilon/2$ crossing the separatrix Ψ_u increases in time only linearly, $l(\tau) \sim \varepsilon\tau$. The absence of exponential stretching and the consequent poor mixing is an obvious result of complete integrability of the problem under consideration. Any fluid

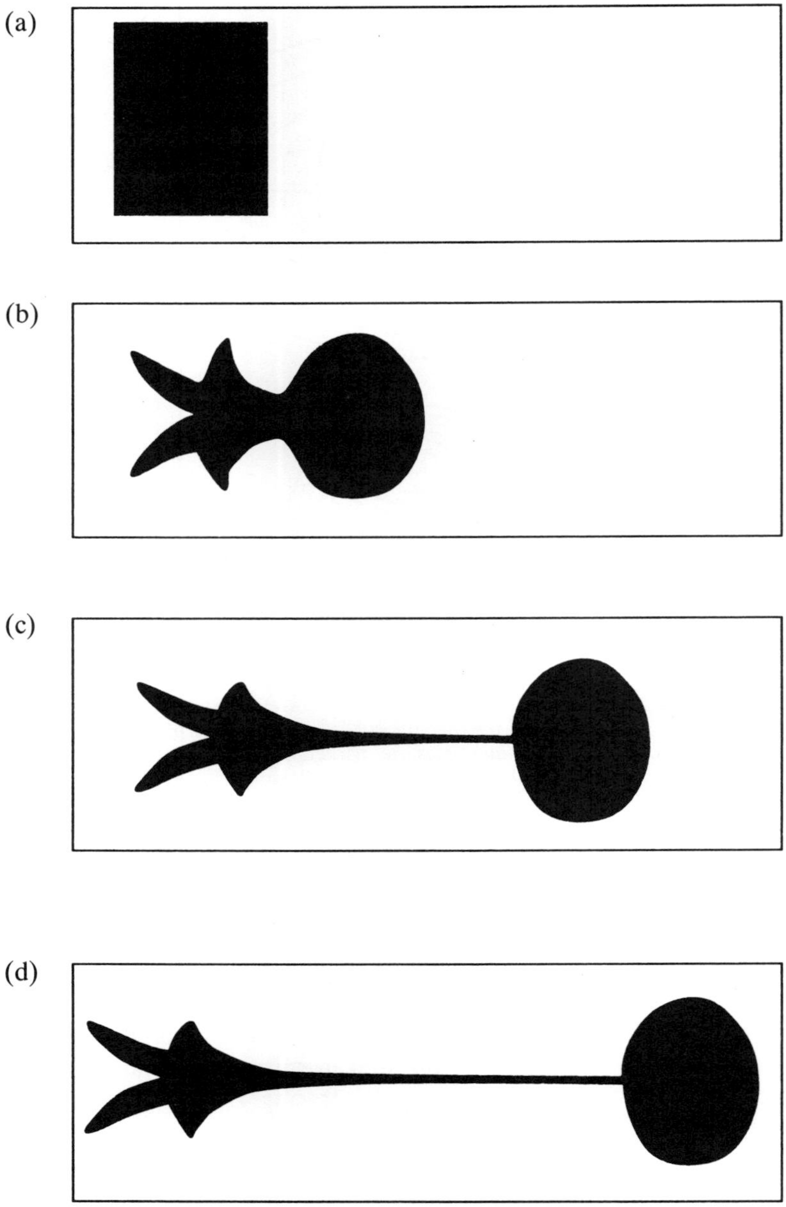

Fig. 3. The formation process of the vortex pair atmosphere is nicely illustrated by following passive tracers that are initially confined in a rectangular area covering the atmosphere. The panels show the calculated evolution of the tracer distribution for (a) $\tau = 0$; (b) 9; (c) 22; (d) 30 (from ref. [30]).

particle outside the atmosphere region (Fig. 4(a)) is not caught by the approaching vortex pair, but is pushed aside by its atmosphere. This 'pushing' effect of the translating dipole is comparable to that of a translating vortex ring. For the latter case the effect was nicely illustrated in a classical experiment by Northrup [46], who studied the motion of a watch chain suspended in a fluid due to an approaching vortex ring. A couple of his photographs are reproduced in Fig. 4(b), on which it can be clearly seen that the chain is slightly curved forward before the ring is actually touching it.

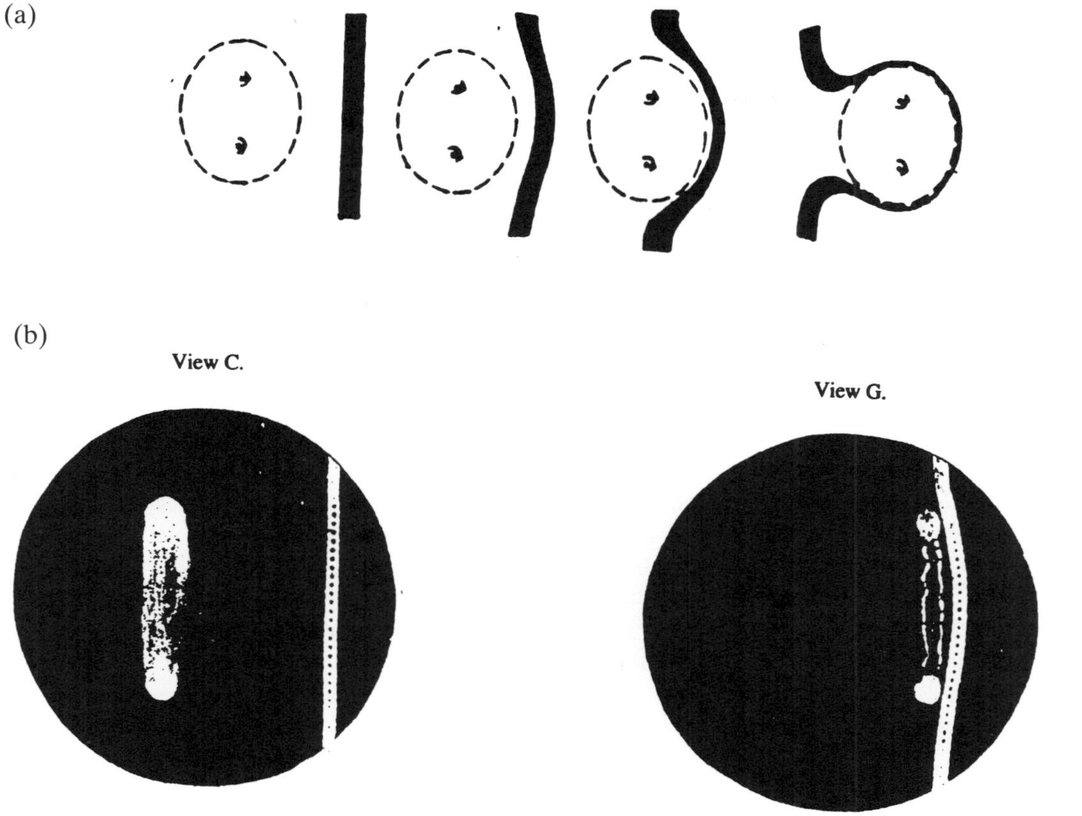

Fig. 4. (a) Distortion of a strip of marked fluid due to a translating vortex pair, calculated for $\tau = 0$, 2.5, 5 and 10. (b) Classical photographs by Northrup [46], showing that a watch chain suspended in a fluid is deformed by the approaching vortex ring.

3.2. *Detrainment and entrainment of vortex pair atmosphere*

In the previous examples the vortex pair atmosphere conserves its entity, due to the absence of any additional external excitation of the two vortices. The study of the temporal evolution of manifolds Ψ_0, Ψ_u, Ψ_l via the Poincaré map, a standard tool for studying non-integrable systems with a bounded phase space under periodical excitation, provides important information about chaotical behaviour of surrounding fluid [44]. The global perspective provided by the Poincaré section is extremely useful for assessing the quality of a stirring process. However, in the majority of the cases considered here (except for special cases of mutual trapping) the phase space of the freely moving vortex pair is unbounded, and the Poincaré section is not immediately available. Also, most fluid particles spend only a finite time in the 'chaotic' zone, rendering the classical theory of Lyapunov exponents inappropriate for such types of motion. If, after some interaction with another vortex pair or with a monopole vortex, the distance between the vortices in a freely propagating vortex pair is changed, a new atmosphere will be gradually created. The vortex pair may detrain some of the initially trapped fluid particles, may entrain some new surrounding fluid into the new atmosphere, or will do both. In order to characterize these processes quantitatively it is useful to introduce so-called detrainment and entrainment coefficients. Let us consider the deformation of the separatrices $\Psi_u \cup \Psi_0$ or $\Psi_l \cup \Psi_0$ enclosing the halves of the vortex pair atmosphere; the area enclosed by the separatrix associated with point vortex i is denoted by SI_i. At the end of the interaction process each separatrix will, in general, be deformed (the flow is two-dimensional, so that lines are mapped onto lines without intersections) and in some cases it appears that only a part SF_i of the initial area SI_i is eventually inside a freely moving vortex pair ij with new separatrix area SN_{ij}, see Fig. 5. It is then possible to introduce the *entrainment coefficient* E_{ij} and the *detrainment coefficient* D_{ij} of vortex pair ij as

$$E_{ij} = 1 - \frac{SF_i + SF_j}{SN_{ij}}, \qquad D_{ij} = 1 - \frac{SF_i + SF_j}{SI_i + SI_j}. \tag{21}$$

For the case when vortex j is an initial monopole, but after interaction with a vortex pair forms a new pair ij with vortex i of the original pair, the detrainment coefficient can be introduced in the following manner

$$D_i = 1 - \frac{SF_i}{SI_i}. \tag{22}$$

It should be noted that in these definitions the entrainment is related to the final atmosphere area of vortex pair ij after the interaction process is completed, whereas the detrainment is related to the initial areas of the vortices i and j. In general, the detrainment coefficient shows how much 'dyed' fluid is left in the wake of a vortex pair after interaction. The entrainment coefficient E_{ij} shows how much ambient fluid has eventually been trapped by the ultimate vortex pair ij.

3.3. *Chaplygin–Lamb dipole*

In some recent experiments [20, 21, 25, 26] of 2D coherent vortex structures the Chaplygin–Lamb dipole with a linear relationship between vorticity and stream function

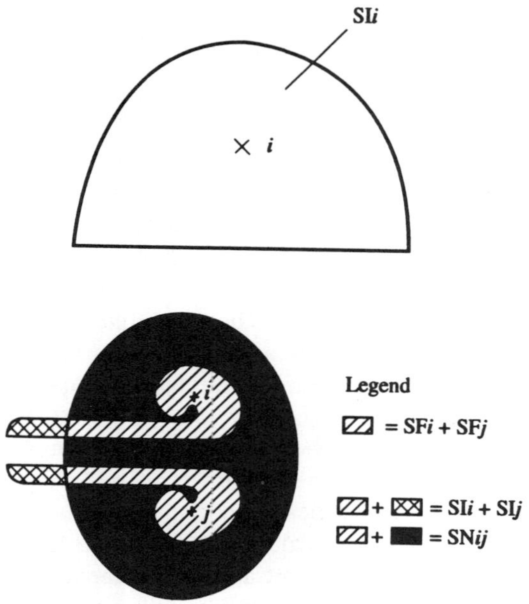

Fig. 5. Definition sketch of the entrainment and detrainment coefficients E_{ij} and D_{ij}, see equation (21).

inside a circular region was observed. The theoretical solution for such a distributed vortex structure was derived independently by Chaplygin [47] and Lamb [13] (a detailed survey is presented in [48]). This inviscid model corresponds to a steady solution of the 2D Euler equations, with the vorticity ζ being zero outside a circle of radius a, and being proportional to the stream function ψ inside that circular boundary. In terms of polar coordinates (r, θ) the solution is

$$\psi(r, \theta) = \begin{cases} -[2V_d/(kJ_0(ka))]J_1(kr)\sin\theta & \text{if } r \leqslant a \\ V_d(r - a^2/r)\sin\theta & \text{if } r \geqslant a \end{cases}$$
$$\zeta(r, \theta) = \begin{cases} -[2V_d k/(J_0(ka))]J_1(kr)\sin\theta & \text{if } r \leqslant a \\ 0 & \text{if } r \geqslant a \end{cases} \tag{23}$$

where ka is a root of the equation $J_1(ka) = 0$. By superposing a velocity V_d parallel to the x-axis, one obtains a circular vortex patch travelling with velocity V_d through a liquid which is at rest at infinity. The impulse of the vortex is given by $2\pi\rho a^2 V_d$, where ρ is the fluid density.

For the smallest root $ka = 3.83$, the solution (23) represents a circular vorticity region (see Fig. 2(b)) with a positive (upper half) and negative (lower half) distribution of vorticity, and without a vorticity jump across the boundary $r = a$. The intensity of the vorticity in each semi-circular area is equal to $\kappa = 6.83V_d a$ and the vortex centres (here defined as the points inside the circle that move with horizontal velocities equal to V_d) have coordinates $r = 0.48a$, $\theta = \pm\pi/2$.

The interaction behaviour of two Chaplygin–Lamb dipoles initially moving coaxially and collinearly (i.e. in parallel directions with some offset) has been studied numerically [27–29], and these studies revealed the existence of various types of interaction: either a *direct* or an *exchange scattering* of positive and negative vorticity regions, or a *mutual trapping* into a tripolar vortex structure.

A general characteristic of the collinear dipole–dipole interactions is that substantial entrainment and detrainment takes place, the latter resulting in patches or filaments of vorticity left behind in the wake of the newly formed dipoles. Of course, a full analytical description of such an interaction presents immense complications.

In order to proceed analytically, one may therefore explore the possibilities offered by point vortices in the modelling of the Chaplygin–Lamb-type vortex dipoles. As we have seen thus far, the Chaplygin–Lamb dipole and the point-vortex dipole have a few important characteristics in common: both types of dipoles translate steadily, both carry a certain impulse and both have an atmosphere. A suitable way of connecting the point–vortex pair to the Chaplygin–Lamb dipole is by two important physical quantities, viz. the total impulse (being $2\pi a^2 \rho V_d$ and $2b\kappa\rho$, respectively) and the translation velocity (being V_d and $\kappa/4\pi b$, respectively). Thus, we obtain $a = 2b$, $\kappa_d = 1.09\kappa$, $S_d = 1.11S$. Although the point–vortex dipole has a non-circular atmosphere, described by (18), for the above parameter values it deviates only slightly from the Chaplygin–Lamb atmosphere. This is illustrated in Fig. 2(b), in which the broken line represents the separatrix of the equivalent vortex pair.

In the study of advection properties of interacting dipolar and monopolar vortices the comparison between completely numerical simulations and point–vortex models is, of course, very useful. Additionally, one may compare the model results with laboratory observations of interacting vortices, such as reported by [20, 21, 25, 26]. Such a comparison is often bound to be qualitative, but may nevertheless provide important information.

3.4. *Experimental set-up for generating quasi-2D vortex structures*

In the laboratory, quasi-2D vortex structures were made in a stratified fluid. Dipolar vortices were generated by a pulsed horizontal injection of a small volume of fluid of matching density through a split-shaped nozzle. Such an injection results in a 3D-turbulent patch of motion, in which the vertical motion is quickly suppressed by the restoring action of the ambient density stratification—a process called gravitational collapse. The injected fluid soon occupies a flat, pancake-shaped region in which the motion is planar, i.e. quasi-2D. After some time one observes that the flow becomes organized in a regular dipolar vortex, which translates quasi-steadily through the fluid without any significant changes in shape. The flow evolution is conveniently examined by dye visualization, but quantitative information about the flow structure can be obtained by streak photography of tracer particles [20, 49]. These flow measurements have revealed that the vortex dipoles thus created can be described in good approximation by the inviscid 2D Chaplygin–Lamb model: the observed relation between vorticity and stream function is close to linear, the vorticity is confined to a region that is close to circular, and the translation speed agrees in good approximation with that predicted by the Chaplygin–Lamb model [49]. Monopolar vortices can be generated in different ways, e.g. by applying a small rotating solid sphere [50] or by applying a mixer, i.e. a rotating stirring device, by which rotational motion is induced in a relatively thin layer of the stratified fluid. This latter technique has been used here. Again, the evolution of the vortex structure could be investigated by addition of some dye or by tracking little tracer particles. In the next section we will discuss dipole–monopole and dipole–dipole interactions in which only dye visualization was used.

4. VORTEX PAIR INTERACTION

> The human mind is seldom satisfied, and is certainly never exercising its highest functions, when it is doing the work of a calculating machine. What the man of science, whether he is a mathematician or a physical inquirer, aims at is to acquire and develop clear ideas of the things he deals with. For this purpose he is willing to enter on long calculations, and to be for a season a calculating machine, if he can only at last make his ideas clearer.
>
> J. C. Maxwell (1870)

4.1. *Interaction of vortex pair with monopole*

The general problem of motion of three-point vortices was exhaustively analysed by Gröbli [41]. Although his analysis was repeated partially in [51, 52] and was well-known to scientists of the 19th century (although references to his pioneering work in textbooks [13] and [38] are rather scarce), up to recent times this work was forgotten completely. Some of its results have been re-derived independently—see the review papers by Aref [39] and Aref *et al.* [53] for details. In the general case the analytical formulae are rather complicated. However, the very interesting particular case of a vortex pair interacting with a single vortex when $\kappa_1 = -\kappa_2 = \kappa$, $\kappa_3 = \kappa$ provides more simple analytical expressions.

For this monopole–dipole configuration it is possible—for any initial positions of the three vortices—to take the origin of the Cartesian x, y-coordinate system in the centre of vorticity ($P = Q = 0$) and to choose the axes so that at $\tau = 0$ the vortices are situated in a manner shown in Fig. 6(a).

Analytical expressions for the polar coordinates $(r_\alpha, \theta_\alpha)$ of the vortex positions were obtained [41] in terms of elliptic integrals. The analysis of Gröbli's solution provided in

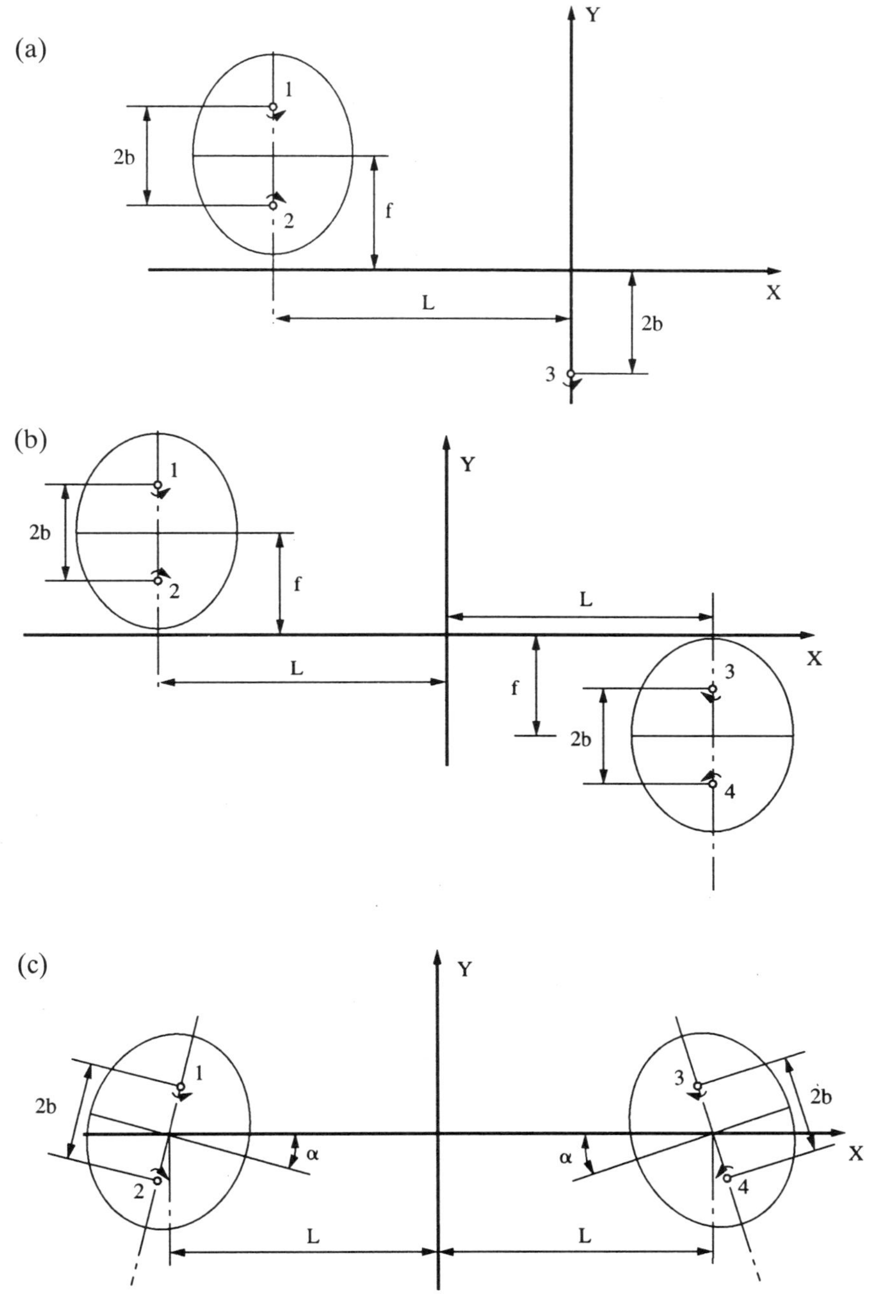

Fig. 6. Initial vortex positions: (a) vortex pair 12 with single vortex 3; (b) two identical collinear vortex pairs 12 and 34; (c) two identical vortex pairs 12 and 34 initially inclined under an angle α.

[54–56] shows that all the possible cases of interaction between a vortex pair and a single vortex may be classified by means of the two dimensionless parameters

$$\lambda = \frac{f + b}{4b} \frac{L^2 + (f + 3b)^2}{L^2 + (f + b)^2}, \qquad \zeta_0 = \frac{L^2 + (f + 3b)^2}{4b^2}. \tag{24}$$

There are three different types of interactions, depending on the values of λ, ζ_0 and the roots ζ_1, ζ_2 of the polynomial $\zeta^2 - 4(\lambda^2 - \lambda)\zeta + 4\lambda$:

(a) $\lambda < -1/4$ or $\lambda > 2$ corresponds with *direct scattering*: after interaction pair 12 moves to infinity and the interaction may be schematically denoted as $12 + 3 \Rightarrow 12 + 3$. The initial and final positions of vortex 3 do not coincide;

(b) $0 \leq \lambda < 2$ or $-1/4 < \lambda < 0$ (when $\zeta_0 > 1$) is an *exchange scattering*: after interaction, the new pair 23 moves to infinity, and the interaction scheme is $12 + 3 \Rightarrow 1 + 23$;

(c) $-1/4 < \lambda < 0$ ($\zeta_2 \leq \zeta_0 \leq \zeta_1$) corresponds with *mutual trapping*: the three vortices rotate around the origin for all time. In this case pair 12 is situated close to vortex 3 from the very beginning.

The various types of interactions that arise for different values of f and L are schematically indicated in the diagram of Fig. 7.

The limiting cases when $\lambda = -1/4$ and $\lambda = 2$ are also of interest. Here the freely propagating vortex pair 12 can initially be situated far away from vortex 3. During its motion this pair catches vortex 3 in such a way that a new steady configuration is formed. For these cases all the integrals are calculated in elementary functions and simple expressions for the vortex coordinates are obtained (see [30] for additional details). It turns out that this three-vortex configuration is *unstable*, however.

Let us now consider the advection of fluid initially confined in the vortex pair atmosphere during the interaction process. Figure 8(a) shows that during a *direct scattering* the inner regions $|\varepsilon| > 0.1$ of the pair 12 atmosphere move practically without any disturbance. Pair 12 leaves behind a long 'tail' and stirring takes place only in the region $0 \leq |\varepsilon| < 0.1$. In order to gain insight in the stirring at the edge of the atmosphere, let us consider Fig. 8(b, c) which show the deformation of two closed fluid lines $\Psi_u \cup \Psi_0$ and $\Psi_l \cup \Psi_0$ for some intermediate stages of the vortex interactions. In Fig. 8(b) one clearly observes a tendency for the formation of a tail, consisting of material of the atmosphere halves associated with vortices 1 and 2, and also a thin curved 'tongue', consisting of material of the lower atmosphere half, that is being wrapped around the other. It is also apparent from this figure that the material contained in the tail and the curved tongue

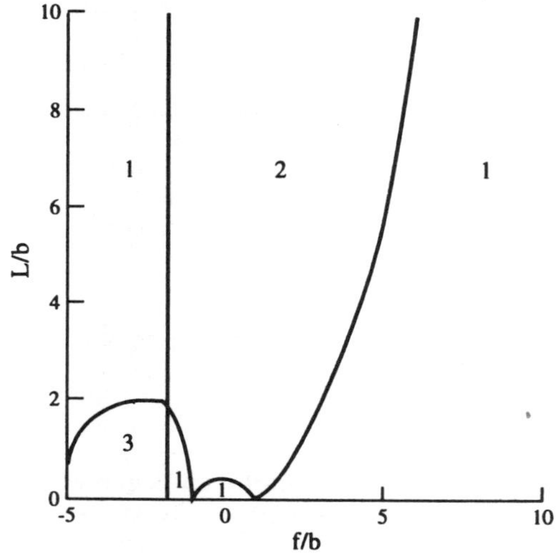

Fig. 7. Classification diagram for the interaction between a vortex pair and a monopole: the numbers refer to (1) direct scattering, (2) exchange scattering, (3) mutual trapping.

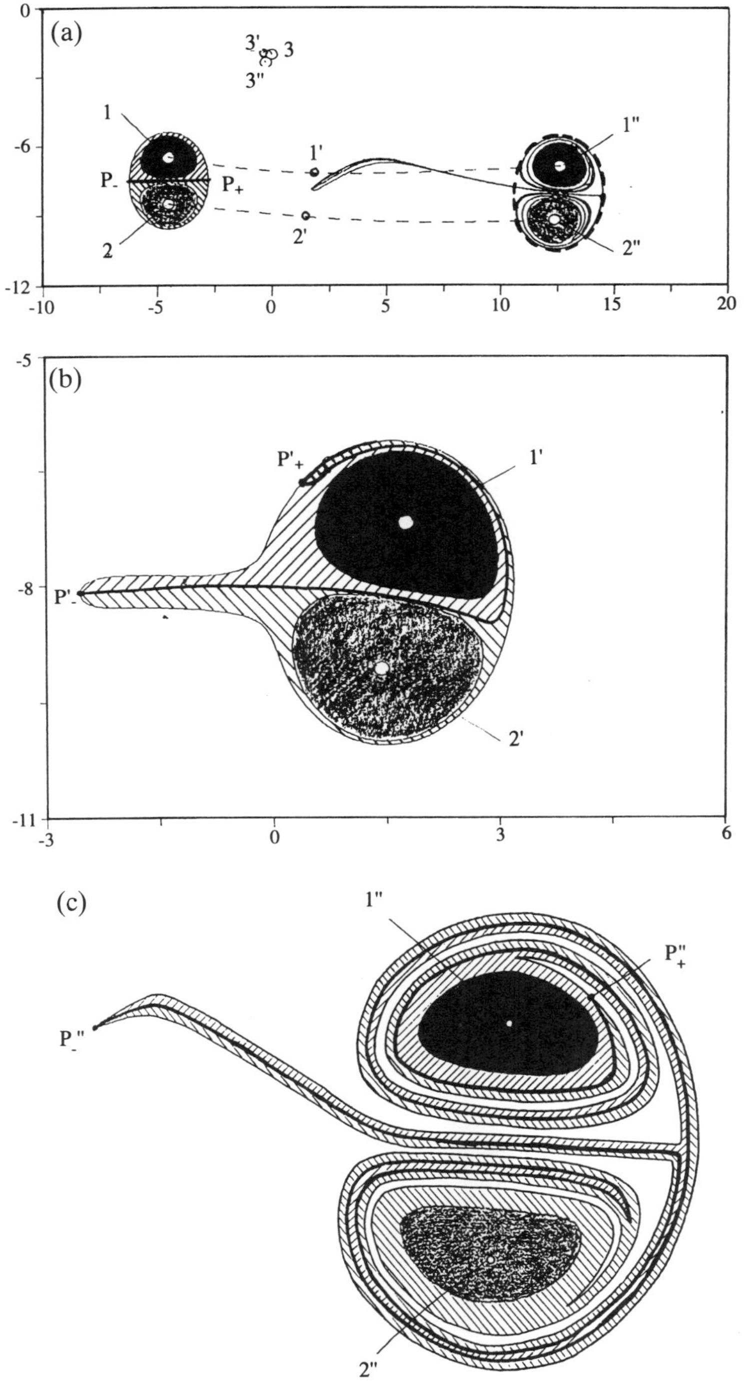

Fig. 8. Schematic representation of the interaction of vortex pair 12 with a single vortex 3 for the case $b = 1$, $f = -7.5$, $L = 4.5$, which corresponds with a *direct scattering*. The grey and black regions initially enclosed by fluid lines $\varepsilon = -0.1$ and $\varepsilon = 0.1$ indicate unstirred regions of the atmosphere; the common line Ψ_0 is represented by a bold line. The initial atmosphere regions $0 \leqslant \varepsilon \leqslant 0.1$ and $-0.1 \leqslant \varepsilon \leqslant 0$ are represented by ▨ and ▧, respectively. Graph (a) shows the initial positions (1, 2, 3 at $\tau = 0$), intermediate positions (1′, 2′, 3′ at $\tau = 9$) and 'final' positions (1″, 2″, 3″ at $\tau = 30$) of vortices and their trajectories (broken lines). The heavy broken line represents the atmosphere boundary of the vortex pair at the quasi-ultimate stage; (b) atmosphere deformation at the intermediate stage, $\tau = 9$; (c) schematic picture of the atmosphere at the quasi-final stage, $\tau = 30$, (from ref. [30]).

originates from the region $0 \leqslant |\varepsilon| < 0.1$; the cores of both atmosphere halves remain virtually unaffected. At later stages the inner fluid line $\Psi_{0.1}$ begins to deform slightly while the line $\Psi_{-0.1}$ conserves its initial form. A more complicated picture is shown in Fig. 8(c). This picture is typical for further considerations of curve topology. While being stretched and folded, the curves remain closed and preserve their topological structure, no matter how complicated their deformations are. For a correct interpretation of Fig. 8(c) it should be remarked that the lines that are whorled around the vortices 1 and 2 originate from an initial T-shaped element of tracers, consisting of a straight line segment $-\sqrt{3} \leqslant \xi_0 \leqslant -1.60$ of the initial boundary Ψ_0 and small segments ($|\eta_0| \leqslant 0.1$) of the boundaries Ψ_u and Ψ_l issuing from the hyperbolic point P_-. A feature that can not be seen from Fig. 8(c) is the uneven stretching of the lines that are wrapped around the core regions: it was found that infinitesimal fluid-line elements emanating from the point P_- show an exponential stretching for the finite time of the interaction, while line elements originating from the area around P_+ are only stretched to a small amount. Also, it is clear that some portion of the material originally in the region $0 \leqslant |\varepsilon| < 0.1$ is left behind in the thin, long tail, nicely showing the detraining effect due to the perturbing vortex 3. At the same time, this perturbation results in substantial entrainment into the dipolar structure, as can be seen from the unmarked spiral-shaped bands in the new atmosphere.

A typical case of an *exchange scattering* is shown in Fig. 9. Again, neither entrainment nor detrainment of surrounding fluid occurs in the inner regions $|\varepsilon| \geqslant 0.1$. One only

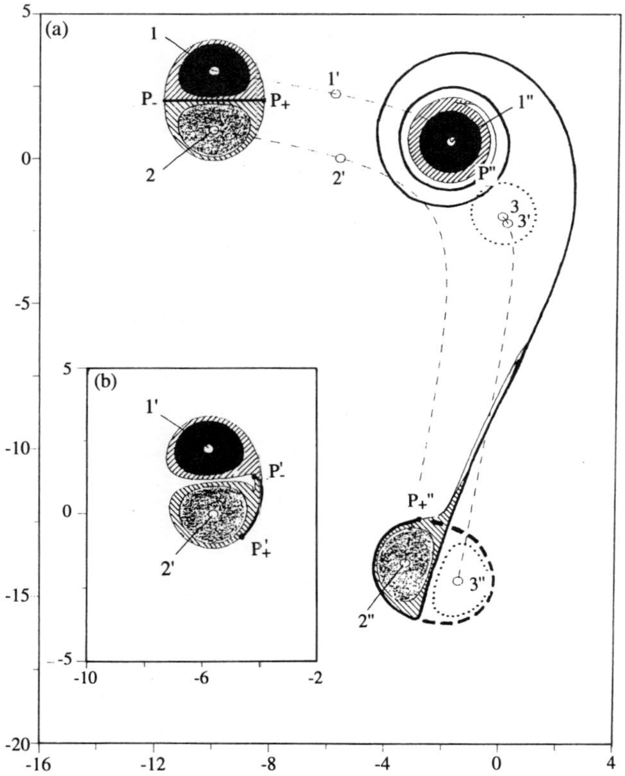

Fig. 9. Schematic representation of the interaction of vortex pair 12 with a single vortex 3 for the case $b = 1$, $f = 2$, $L = 10$, which corresponds with an *exchange scattering*. Symbols and graphical indications are the same as in Fig. 8. Graph (a) shows the initial positions (1, 2, 3 at $\tau = 0$), intermediate positions (1′, 2′, 3′ at $\tau = 10$) and 'final' positions (1″, 2″, 3″ at $\tau = 50$) of the vortices as well as their trajectories. The dotted lines indicate fluid initially surrounding vortex 3, which is finally trapped into the new atmosphere. Graph (b) presents a detailed picture of the atmosphere deformation at the intermediate stage $\tau = 10$, (from ref. [30]).

observes a change in shape of the inner region ($\varepsilon \geqslant 0.1$) associated with vortex 1: at the stage in which it is practically at rest, it is transformed into a circular region. At the same time the region initially enclosed by line $\Psi_{-0.1}$ (denoted by grey tones) is observed to be present inside the new atmosphere of vortex pair 23 without any appreciable changes in its shape. A noteworthy feature is the considerable elongation of line Ψ_0 during this interaction. The whole deformation process of the fluid lines Ψ_u, Ψ_l, Ψ_0 is qualitatively the same as shown in Fig. 8(c). Some additional details of the curve topology at an intermediate stage are presented in Fig. 9(b).

The new atmosphere of the vortex pair 23 also contains some fluid initially surrounding vortex 3, as indicated by the dotted line in Fig. 9(a). In order to examine how the fluid in the immediate vicinity of vortex 3 was advected throughout the interaction process, we traced particles initially located on a curve of radius $r_0 = 1.3b$ centred on the initial vortex 3 position (this radius was chosen such that the circle has an area slightly less than the semi-atmosphere of the initial vortex pair 12). It is seen that at the end of the interaction process all the fluid initially contained in this circle now forms the new half of the atmosphere of pair 23.

This scenario of partner exchange was also observed in laboratory experiments in a stratified fluid (see Section 3.4), in which a dipolar vortex was directed at a monopolar vortex. The sequence of events during the interaction is nicely shown by the photographs of Fig. 10. In this experiment the initial monopole vortex is visualized by orange dye, while the dipolar vortex (generated by a pulsed injection of a small fluid volume through the orifice visible on the right) is coloured with green dye. The monopolar vortex was produced by locally stirring the fluid in anticlockwise direction. It is observed that the dipole, when approaching the monopole, gradually widens and splits up into two parts: one half (with clockwise circulation) pairs with the monopole and forms a new dipole, while the other half is left behind. Although the vortices in the laboratory experiment have continuous spatial vorticity distribution (the dipole has the characteristics of the Chaplygin–Lamb dipole, while the monopole has a Gaussian structure), their interaction behaviour is very similar to that predicted by the simple point–vortex model. This similarity does not only concern the trajectories of the individual vortex centres and the partner exchange, but

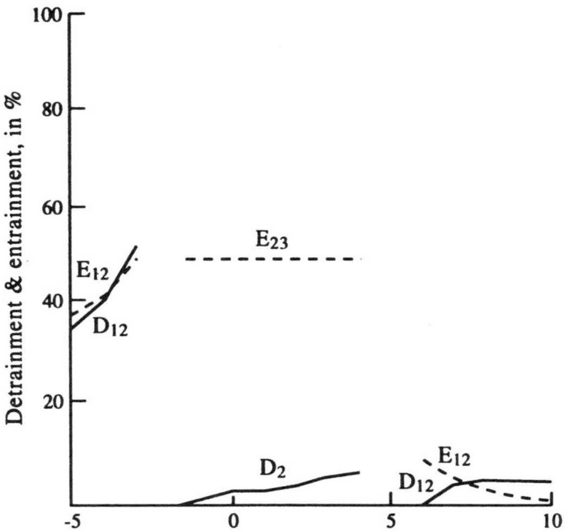

Fig. 11. Graphical presentation of the calculated entrainment and detrainment coefficients (E, D) as a function of the parameter f/b for the monopole–dipole interaction, with the dipole initially located at a large distance ($L = 10b$) from the vortex monopole. Note that for $-2 < f/b < 6$ the interaction is of the type *exchange scattering*, while outside that interval *direct scattering* occurs.

the similarity is also visible in the distribution of the passively advected dye (compare Figs 9 and 10). Apparently, the advection properties of the continuous-vorticity flow structures are adequately described by the contour kinematics method.

Calculated values of the entrainment and detrainment coefficients for the case when initially the vortex pair is situated far away ($L = 10b$) from the monopole vortex 3 are shown graphically as a function of f/b in Fig. 11. It is interesting to note that for the *direct scattering* when $f \geqslant 6b$ the detrainment coefficient is almost zero; in that case there is only some entrainment due to an increase of the final distance between vortices in the vortex pair after interaction. Relatively large values of the entrainment and detrainment coefficients are obtained for the case of direct scattering when $f \leqslant -2b$.

In the case of *mutual trapping* (when $-1/4 < \lambda < 0$) such an advection analysis is not so useful because in this case an initial atmosphere can not be defined properly—pair 12 is always situated close to vortex 3. Moreover, when $L = 0$ and $-3b < f < -2b$ the distance between vortices 1 and 3 is less then $2b$. For this case it is more convenient to use the Poincaré section technique in order to analyse the stirring behaviour of passive particles. This problem is beyond the scope of the present paper and will be addressed elsewhere.

4.2. *Interaction of two vortex pairs*

The interaction of two vortex pairs 12 ($\kappa_1 = -\kappa_2 = \kappa$) and 34 ($\kappa_3 = -\kappa_4 = a\kappa$) having initial separation distances between vortices $2b$ and $2d$, respectively, and which are freely propagating parallel to the x-axis, has been considered in detail by Eckhardt and Aref [57] and recently in [58] for $a = 1$. A system of four vortices generally possesses chaotic trajectories. It was shown [59, 60] that integrable types of motion are obtained either when $Q = P = 0$ (that is, if $b = ad$) or when the vortex system during its motion has an additional axial or central symmetry, which is the case for so-called *coaxial* and *collinear* pairs. Equations (5) for two coaxial or collinear pairs permit the integration in quadratures. This was done long ago by Gröbli [41] and Goryachev [61] for the case of equal intensities of the pairs, i.e. $a = 1$. Depending on the initial vortex positions these types of symmetry lead to various forms of interaction: direct scattering, exchange scattering or mutual trapping.

4.2.1. *Collinear pair with centre of symmetry.* The initial positions of the vortex pairs are shown in Fig. 6(b). The central symmetry of the problem is conserved during the flow evolution and the solution to (5) can be found in a closed form using some quadratures [56, 61, 62]. All possible cases of vortex pairs interaction are characterized by the two dimensionless parameters

$$h = \frac{2b}{|f|}\,\frac{L^2 + f^2}{\sqrt{[(L^2 + (f + b)^2)(L^2 + (f - b)^2)]}}$$

$$\zeta_0 = \frac{\sqrt{[(L^2 + (f + b)^2)(L^2 + (f - b)^2)]}}{2|f|b} \tag{25}$$

and the roots ζ_1, ζ_2 of the polynomial $\zeta^2 - h\zeta + 1$. One can distinguish the following cases:

(a) $0 < h < 2$ is a *direct scattering*: $12 + 34 \Rightarrow 12 + 34$;
(b) $h > 2$ and $\zeta_0 > \zeta_1 > \zeta_2$ is an *exchange scattering*: $12 + 34 \Rightarrow 14 + 23$;
(c) $h > 2$ and $1/h < \zeta_0 < \zeta_2$ is a *mutual trapping*;
(d) $h = 2$, $\zeta_0 > 1$ is an *asymptotical trapping* into a steady rotating, rhombic configuration. Note that such a trapping is unstable: any small change in the value of h during the numerical integration of (5) leads this configuration to dissolve into two

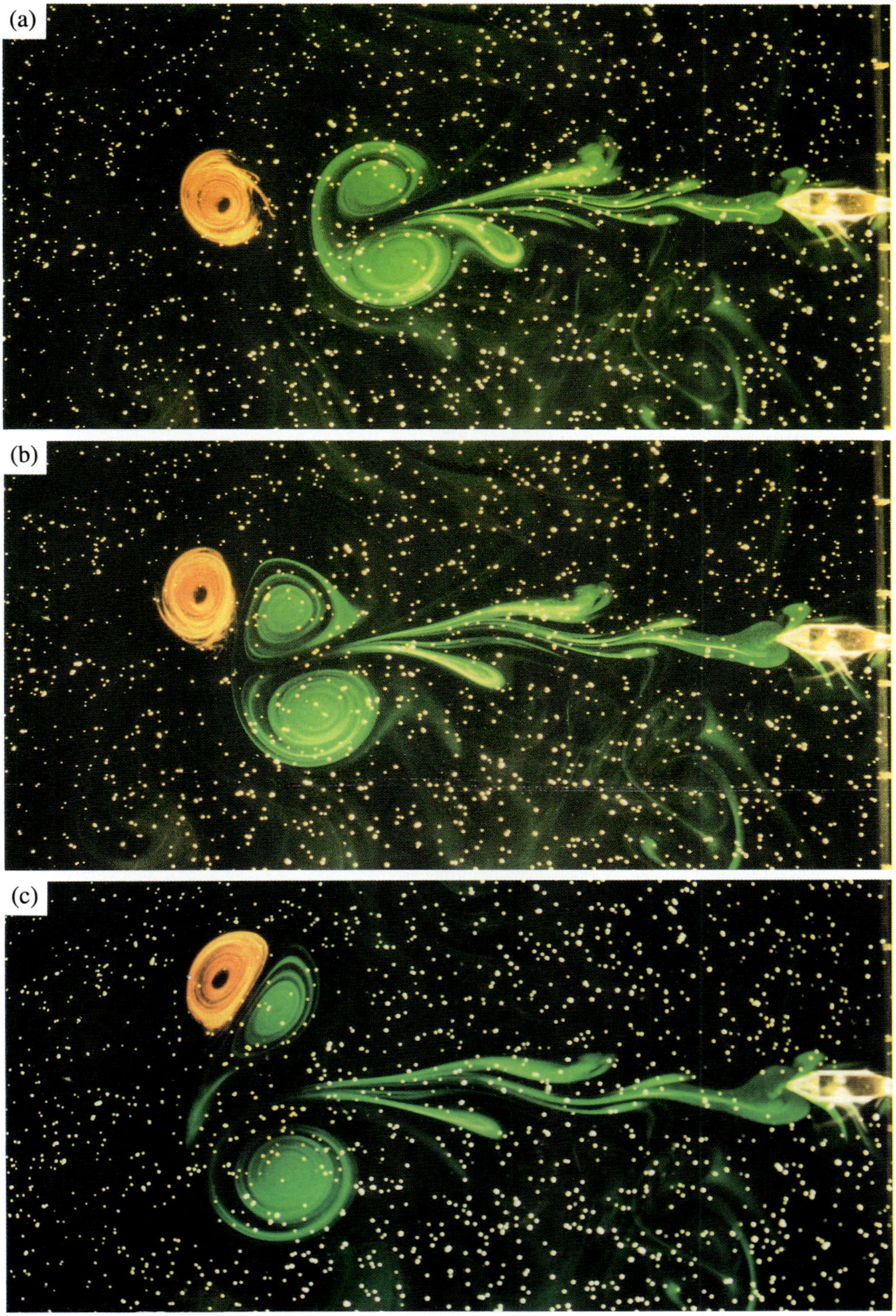

Fig. 10. Typical sequence of photographs showing an exchange scattering between a dipolar vortex and a monopolar vortex as observed in the laboratory. The monopolar vortex (orange dye) was created by applying a mixer during some short time interval, while the dipolar vortex was generated by a horizontal squirt of dyed fluid (green) of matching density through the orifice visible on the right-hand side of the pictures.

vortex pairs freely propagating to infinity. Thus, the analysis of the advection processes in that case should be provided by means of the exact analytical representations of the vortex trajectories (see [30] for additional details).

The interaction classification of collinear pairs appears to be quite useful also for the considerably more complicated case of two collinear Chaplygin–Lamb dipoles. In the analogy between Chaplygin–Lamb dipole and the vortex pair with its atmosphere as described in Section 3.3, numerical results [29] of exchange scattering, direct scattering and mutual trapping (tripole formation) correspond to values $h = 3.9$, 1.3 and 1.95, respectively. These values agree reasonably well with those of the point–vortex model. Moreover, in the case of exchange scattering the new vortex pair 14 moves away under practically the same angle to the x-axis as the new dipole. These points of correspondence indicate the rather good possibilities of this simple model to predict and to track experimental results [20].

The behaviour of the fluid initially confined in the vortex pair atmospheres is quite fascinating. The case of a *direct scattering* is illustrated in Fig. 12. The most intensive stirring takes place when the pairs fulfil a rotation around the origin. The offset of the dipole axes is chosen such that vortices 1 and 3 are situated nearer to each other than vortices 2 and 4. This implies that the upper part of the atmosphere of vortex pair 12 undergoes a more intensive stirring than the lower half. Only a rather small inner region $\varepsilon \geqslant 0.8$ conserves its entity during the interaction. Some fluid inside the region $0 \leqslant \varepsilon \leqslant 0.5$ gets trapped by vortex 3 into the atmosphere of vortex pair 34. A small part of the boundary line Ψ_u forms a long tail connecting the fluid that is trapped in the atmosphere of vortex pair 34 with that remaining inside the original atmosphere of vortex pair 12, see Fig. 12. The stirring of the lower part of the initial atmosphere of pair 12 is less intensive. The boundary line Ψ_l is seen to curl around vortex pair 12 alone and is not trapped into the atmosphere of pair 34. The detrainment coefficient for vortex pair 12 has been calculated to be $D_{12} = 0.24$. The calculations reveal that after the *direct scattering* interaction, only the closest dipole halves (which have equally-signed circulations) have exchanged material, while the outer dipole halves remain relatively unaffected, except for some detrainment leading to the formation of tails. The entrainment of surrounding fluid into the new atmospheres is $E_{12} = 0.22$. It is tempting to compare the results obtained by the contour kinematics method with those obtained by the full numerical simulation of the collinear interaction of Chaplygin–Lamb dipoles [29]. Three stages of the collinear interaction of such dipolar vortices for an initial offset $1.5a$ (with a the dipole radius) are shown in Fig. 13, in which the contours are now iso-vorticity lines. The correspondence between the flow evolutions in Figs 12 and 13 is striking.

For the case of an *exchange scattering* intensive stirring takes place only in the regions $0 < |\varepsilon| < 0.1$ of both pairs. In the final stage of interaction, shown in Fig. 14, the boundary $\Psi_u \cup \Psi_0$ of the original pair 34 not only surrounds its own inner region $\varepsilon > 0.1$ in the new pair 23, but it is also partially wrapped around the atmosphere boundary of the new vortex pair 14. The internal atmosphere regions $|\varepsilon| \geqslant 0.1$ practically do not show any stirring at all and their shapes at the final and initial stages are almost identical. For this particular case, the entrainment and detrainment coefficients of the newly formed vortex pair are $E_{14} = 0.13$ and $D_{14} = 0.15$, respectively. A comparison between the contour kinematics results and dye-visualizations of a dipole collision created in a laboratory experiment [20] is shown in Fig. 15 for four different stages of the interaction process. The initial conditions used for the numerical simulations were taken from the experiments. This ‘tuning’ gave rise to slightly different radii and intensities, as well as a somewhat different inclination to the horizontal axis for the left and right dipoles. In spite of the absence of the lobes visible in the dye distributions of the laboratory experiment (the result from the formation of the

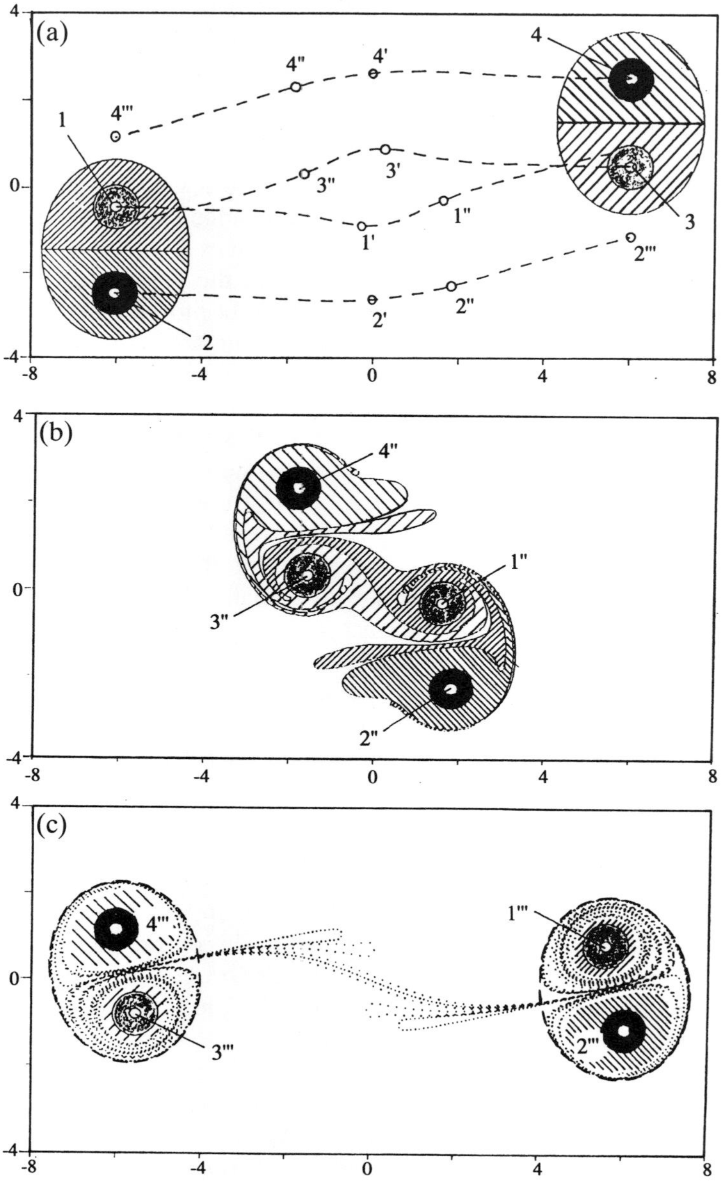

Fig. 12. Schematic representation of the interaction of collinear vortex pairs 12 and 34 in the *direct scattering* régime, with $b = 1$, $f = 1.5$, $L = 6$. The grey and black regions initially enclosed by fluid lines $\varepsilon = 0.5$ and $\varepsilon = -0.5$ indicate unstirred regions of both dipole atmospheres. The hatched areas are defined by $0 \leqslant \varepsilon \leqslant 0.5$ and show the intricate details of the entrainment and detrainment during the interaction process. Graph (a) shows the initial vortex positions (1–4 at $\tau = 0$), the positions at two intermediate stages ($1'$–$4'$ at $\tau = 12$ and $1''$–$4''$ at $\tau = 15$) and at the 'final' positions ($1'''$–$4'''$ at $\tau = 24$); broken lines indicate the trajectories of the individual vortices. Graph (b) shows the situation at $\tau = 15$, while graph (c) shows the distribution of markers (initially on the fluid lines $\varepsilon = 0$ and $\varepsilon = \pm 0.5$) at the 'final' stage $\tau = 24$. In the latter graph the broken lines delineate the boundaries of the new dipole atmospheres, (from ref. [30]).

dipoles; these lobes were not included in the initial stage of the contour kinematics simulation, see Fig. 15(a)), the numerical results based on the point–vortex model extended with the contour kinematics calculation show an excellent agreement with the experiments. This close correspondence is somewhat surprising, since in the experiments

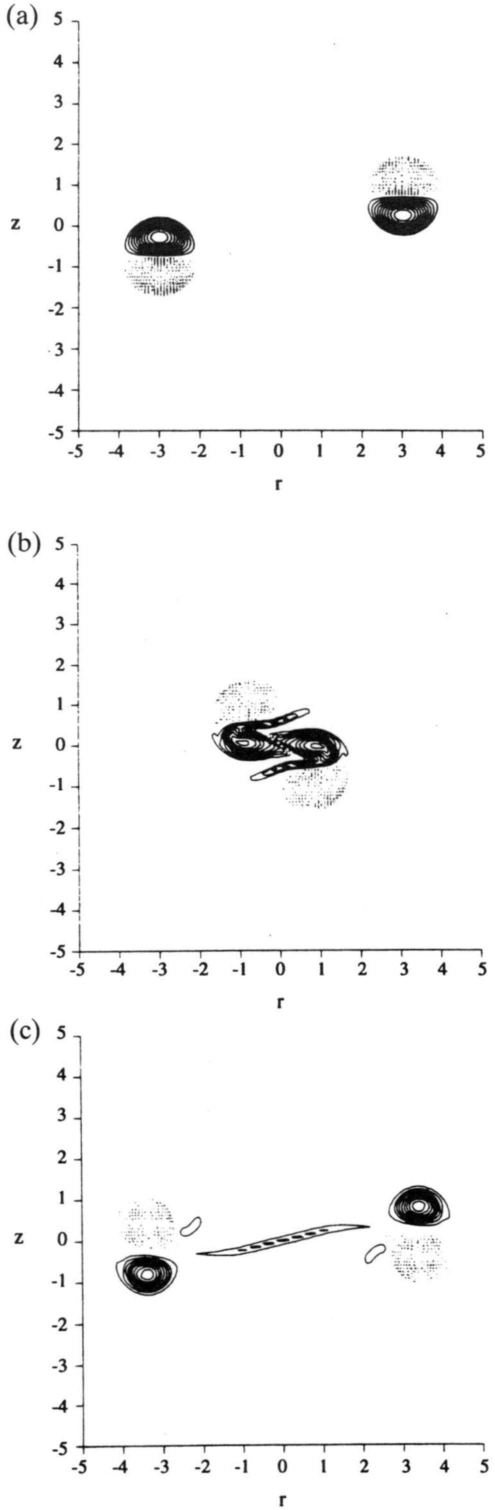

Fig. 13. Numerical simulations of two collinear Chaplygin–Lamb dipoles with the axes offset at a distance $1.5a$, where a is the dipole radius, for (a) $\tau = 0$, (b) $\tau = 4$ and (c) $\tau = 8$. The contours are iso-vorticity lines, (from ref. [29]).

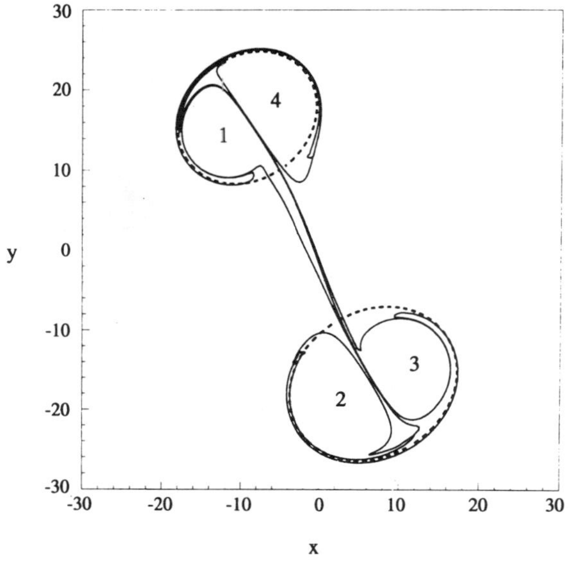

Fig. 14. Graph showing the contours $\varepsilon = 0$ and $\varepsilon = \pm 0.1$ of the initial atmospheres of vortex dipoles 12 and 34 after *exchange scattering* at $\tau = 11.5$. The initial configuration is given by $b = 5$, $f = 1.1$, $L = 10$.

the flow structures have an essentially distributed vorticity, whereas such a continuous distribution is entirely absent in the point–vortex model. Nevertheless, the perfect agreement between the distributions of the passive scalars in the laboratory experiment and in the simulations provides important evidence for the power and effectiveness of the contour kinematics method.

4.2.2. *Oblique collisions of equal pairs.* For the case of symmetrical location of two equal pairs approaching each other under the same angle (Fig. 6(c)) the problem is formally equivalent to the motion of two coaxial vortex pairs 14 and 23 which initially move towards each other. Such a kind of problem was also considered by Gröbli [41], who gave the full analytical solution depending on one dimensionless parameter, characterizing the initial geometry of the problem. On the basis of that solution it is possible to show (for details see [56]) that an *exchange scattering* $12 + 34 \Rightarrow 14 + 23$ will take place for any value of the angle α. One of the interesting results of the analytical solution is that for $L \gg b$ the distances d_{14} and d_{23} between the new vortex pairs 14 and 23 are

$$d_{14} = 2b[\sqrt{(1 + \sin^2 \alpha)} - \sin \alpha], \qquad d_{23} = 2b[\sqrt{(1 + \sin^2 \alpha)} + \sin \alpha], \tag{26}$$

apparently independent of L.

Results of the contour kinematics method for some typical cases of oblique interactions are presented in Figs 16, 17 and 18 for $\alpha = 0°$, $\alpha = 45°$ and $\alpha = 90°$, respectively.

The case $\alpha = 0°$ (see Fig. 16) corresponds to the well-known situation of a 'head-on' collision of two vortex pairs or, equivalently, to the perpendicular collision of one vortex pair with a solid plane wall with free-slip boundary conditions. As in all previous cases, intensive stirring takes place only near the boundaries of the atmospheres. After the actual interaction process is completed, the initial atmospheres are distorted in such a way that they form a long tail along the symmetry axis, thus connecting the original atmosphere halves that are now confined in the newly formed dipoles moving in opposite directions. The values of the entrainment and detrainment coefficients of the newly formed vortex

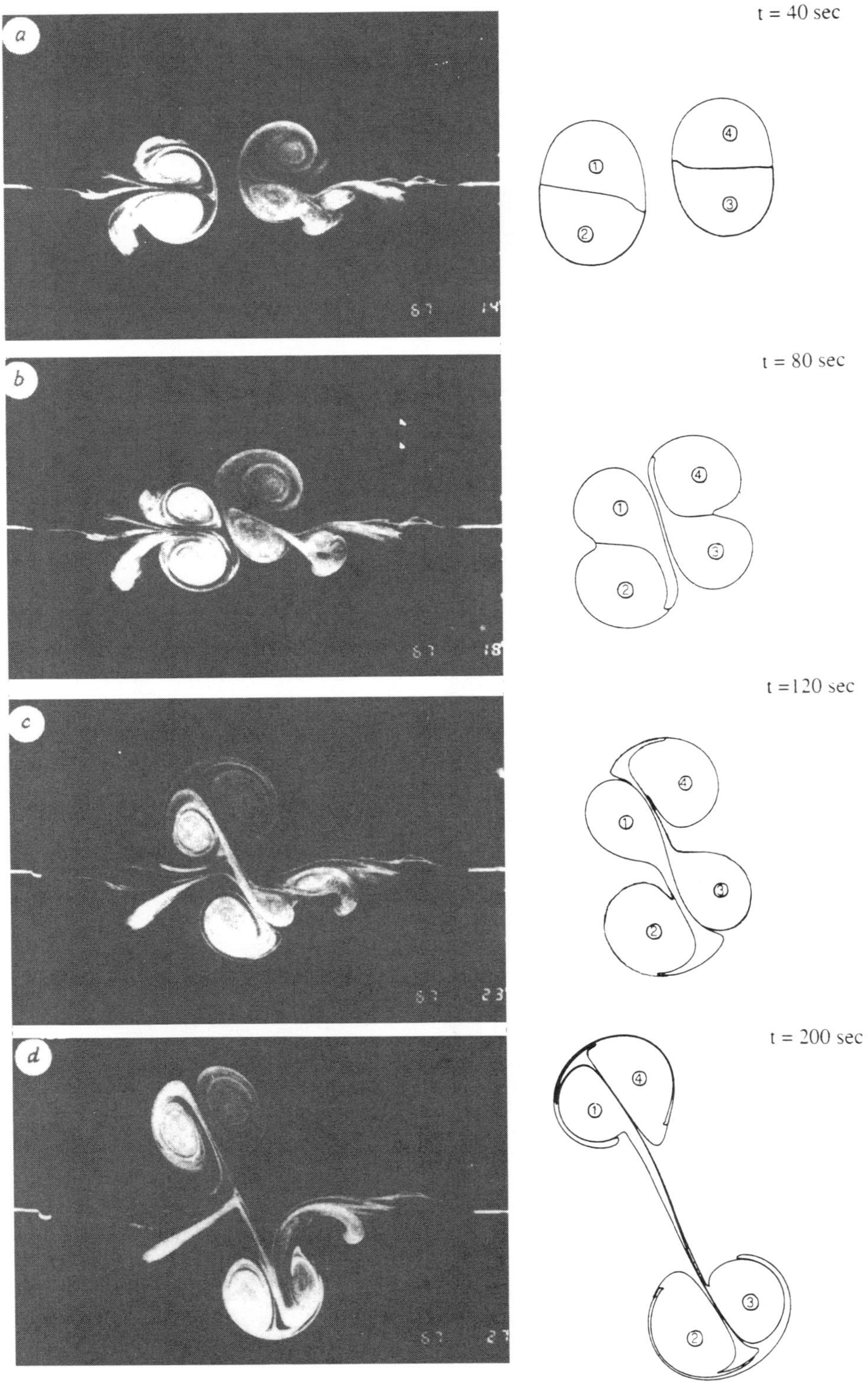

Fig. 15. Comparison between dye-interaction experiments on colliding dipoles [20] and the contour kinematics simulation for approximately similar initial conditions.

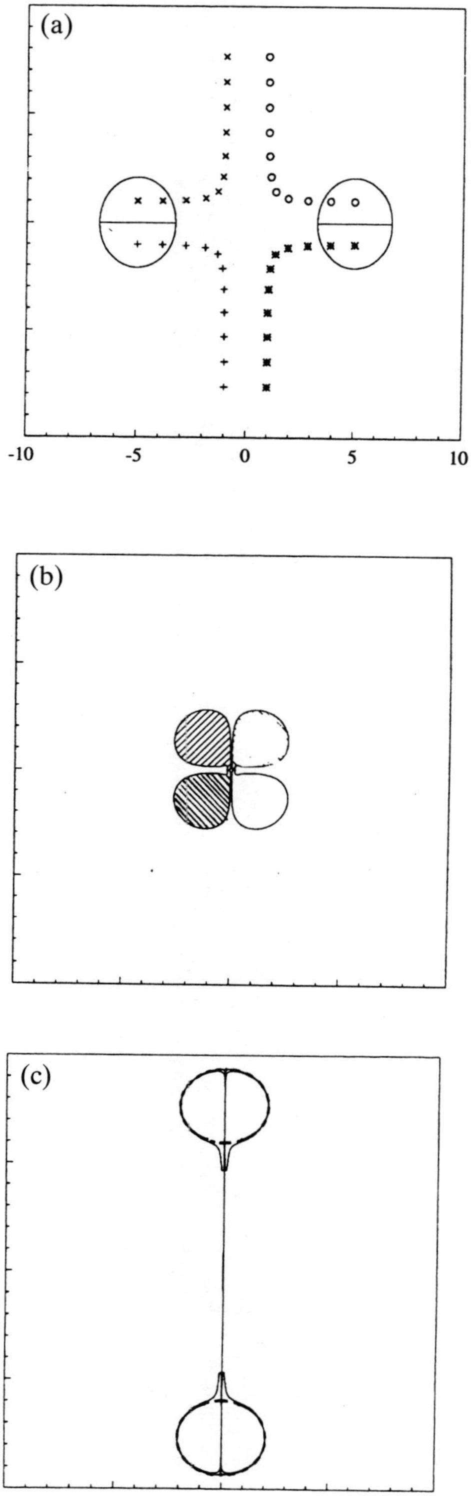

Fig. 16. Graph showing the evolution of the initial atmospheres of two frontally colliding ($\alpha = 0°$) vortex pairs at (a) $\tau = 0$, (b) $\tau = 60$ and (c) $\tau = 150$. The crosses and dots in (a) represent the positions of the individual vortices at time intervals $\delta\tau = 15$ during the interaction. Initial parameters: $b = 1$, $L = 5$.

pairs are $E_{23} = 0.04$ and $D_{12} = 0.06$, respectively. Due to the symmetry of the problem the coefficients E_{14} and D_{34} have the same values.

For the oblique collision with $\alpha = 45°$ (see Fig. 17) the newly formed pair 14 does not entrain any fluid, $E_{14} = 0$, but due the separation distance d_{14} being less than the initial distance d_{12}, the pair 14 detrains some fluid. On the other hand, pair 23 detrains practically nothing of its initial area but instead entrains a considerable amount of surrounding fluid ($E_{23} = 0.74$).

For the interaction with $\alpha = 90°$ (see Fig. 18) it is necessary to continue the calculations during a longer timespan in order to capture the entire interaction process. It appears that the motion of the boundaries surrounding vortices 1 and 4 become extremely complicated due to the intensive curling of the lines, and therefore only markers are used in the diagram of Fig. 18(c). In this case the larger pair 23 has entrained a significant amount of surrounding fluid into its new area ($E_{23} = 0.83$), while the smaller pair 14 shows considerable detrainment.

The oblique interactions have the following characteristics: vortices 2 and 3 do not show any detrainment, $D_2 \approx 0$ and $D_3 \approx 0$, and the detrainment of pair 12 (or 34) is entirely connected with that of vortex 1 (or 4) and changes with the initial angle α i.e. the coefficient $D_{12}(= D_{34})$ depends on α. On the other hand, the entrainment of surrounding fluid into the 'small' new pair 14 is very small, $E_{14} \approx 0$. The entrainment coefficient E_{23} also depends on the value of α. The dependence of E_{12} and E_{23} on α are shown graphically in Fig. 19. It is worth noting that both curves can be closely approximated by the following empirical formulas

$$D_{12} = 0.46 - 0.40\exp(-\alpha/25), \qquad E_{23} = 0.88 - 0.84\exp(-\alpha/25)$$

where the angle α is measured in degrees.

Finally, it should be noted that the results described here show good qualitative agreement with laboratory experiments on oblique dipole collisions in a stratified fluid. It is beyond the scope of the present paper, however, to discuss this comparison in detail; these results will be published elsewhere.

5. CONCLUSIONS

> If assumptions must be made, it is better to assume as little as possible.
>
> M. Faraday (1843)

In spite of its 'classical' character, two-dimensional fluid dynamics is still a most fascinating field for experimental, theoretical and numerical researchers. In his fundamental paper, Helmholtz [1] not only stated the main laws of vortex dynamics for the general case of motion in an inviscid incompressible fluid, but also formulated the simplified 2D model of point vortices in an unbounded plane. Since that time the interest in this model periodically increased and decreased. The point–vortex model was widely used by many scientists at the end of the last century and beginning of the present century in the study of different aspects of 2D motion [13, 63, 64]. Besides offering some mathematical challenges, which resulted in the development of solution methods for such types of problems (like, for example, the conformal mapping method), the theory of 2D flows permitted the explanation of lift forces on an airofoil in inviscid fluid. The model of a singular point vortex in a uniform flow carrying its own atmosphere is also used by J. J. Thomson [65] as a quantitative analogy of the fundamental conception of Faraday's force lines in electromagnetic theory. Starting with the papers by Onsager [66] and Welander [34] the point–vortex model also found an entrance into statistical mechanics and kinetic

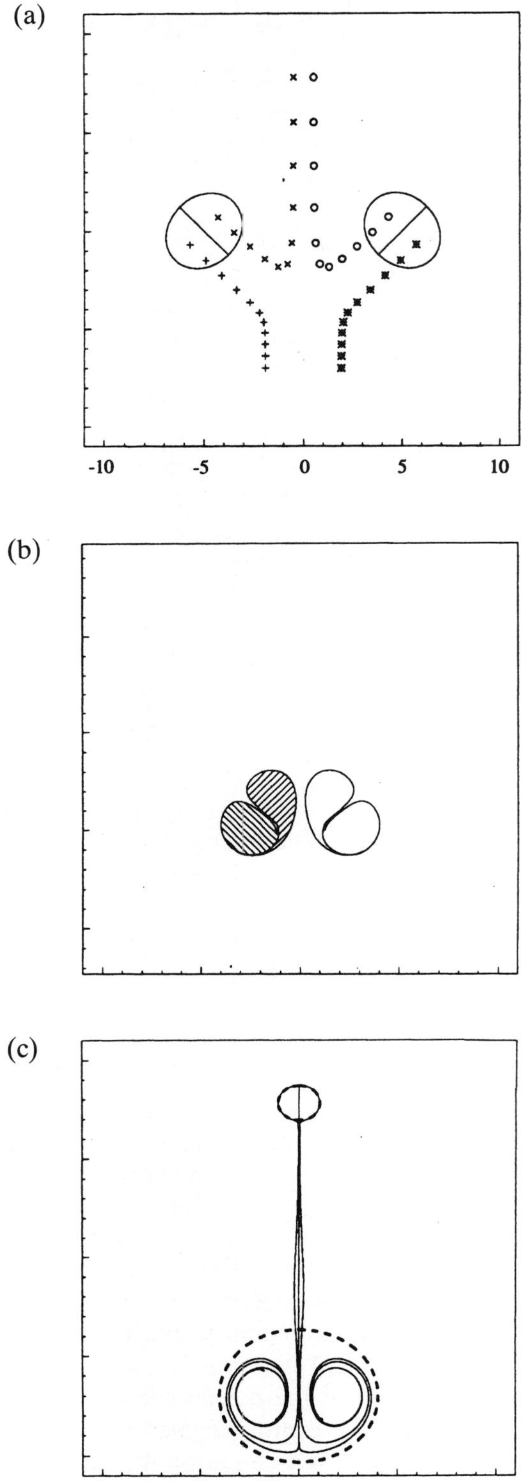

Fig. 17. Graph showing the evolution of the initial atmospheres of two obliquely colliding ($\alpha = 45°$) vortex pairs at (a) $\tau = 0$, (b) $\tau = 60$ and (c) $\tau = 150$. The crosses and dots in (a) represent the positions of the individual vortices at time intervals $\delta\tau = 15$ during the interaction. The boundaries of the new atmospheres are denoted by broken lines in (c). Initial parameters: $b = 1$, $L = 5$.

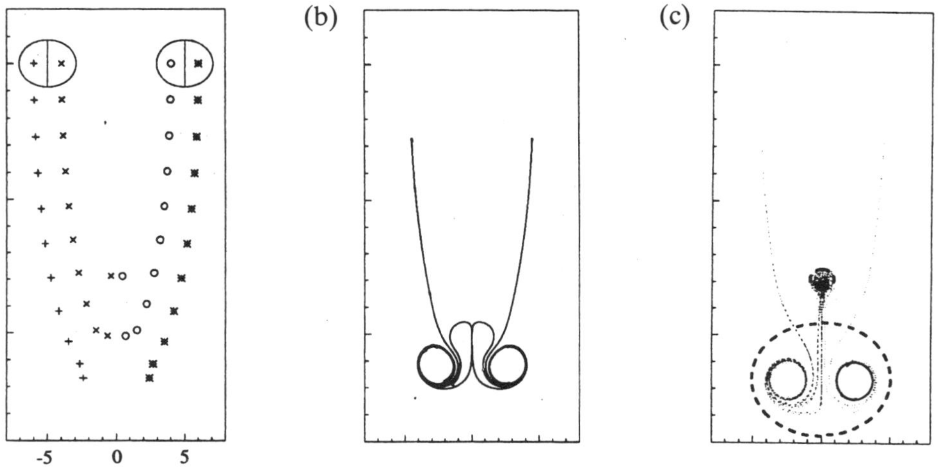

Fig. 18. Graph showing the evolution of the initial atmospheres of two 'parallel' colliding ($\alpha = 90°$) vortex pairs at (a) $\tau = 0$, (b) $\tau = 315$ and (c) $\tau = 350$. The crosses and dots in (a) represent the positions of the individual vortices at time intervals $\delta\tau = 35$ during the interaction. The boundaries of the new atmospheres are denoted by broken lines in (c). Initial parameters: $b = 1$, $L = 5$.

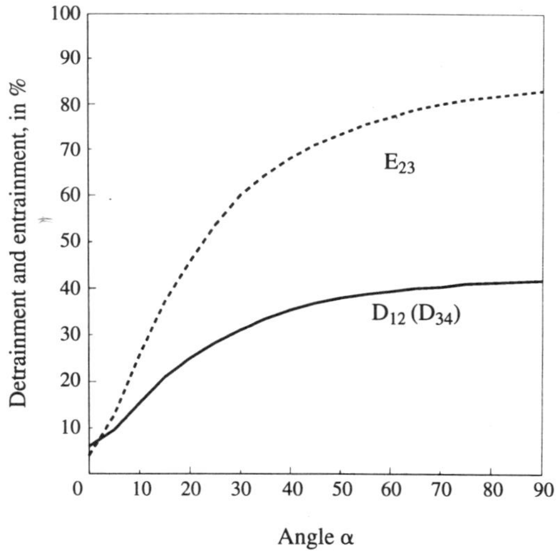

Fig. 19. Graphical representation of the entrainment and detrainment coefficients E_{23} and $D_{12}(D_{34})$ as a function of the angle α.

theory—see review papers [67, 68]. In meteorology, the use of a point-vortex model has also been found in early studies to be successful in describing some large-scale atmospheric processes [34].

With the advent of powerful computers it recently became possible to perform complete numerical simulations of evolving 2D flow, either viscous or inviscid (see e.g. [25, 28, 29, 69, 70]). Although the results obtained in this way are impressive, in view of the considerable hardware requirements it is desirable to know to what extent some relatively simple cases of localized vorticity, can be modelled in a more simple approach. This question has intrigued quite some investigators during the last one or two decades. Of course, 2D flows with continuous vorticity distributions can perhaps be modelled in a better way by taking a cloud of point–vortices (rather than one or two). Although this approach has yielded some successes (see e.g. [71], in which the problem of vortex merger is addressed), one of the major computational problems is the decreasing timestep when individual vortices become too close. This problem is eliminated in the so-called cloud-in-cell method that was first described by Christiansen [72], who applied it to the problems of vortex merger and the von Kármán vortex street formation. By introducing a strength modulation (in order to satisfy the constraints of conservation of volume and potential vorticity acting on a vortex tube with an arbitrarily assigned cross-sectional area) this cloud-in-cell method has also been applied successfully to describe the motion of vortices on a β-plane [73] or in a rotating fluid with bottom topography [74]. Another approach consists of modelling continuous distributions of vorticity by patches of uniform vorticity, and by calculating only the deformation of their boundaries. This technique is generally known as the contour dynamics method, and various aspects of this approach are reviewed in ref. [75].

In the present paper we have studied both analytically and numerically 2D inviscid flows governed by a vortex pair in the presence of another point vortex or another vortex pair in an unbounded plane. Attention was focussed mainly on the induced advection of passive scalars, and in particular on the stirring process inside the atmosphere, for some integrable cases of point–vortex motion.

In spite of this regular motion, in the case of a single vortex pair interacting with either a single or another vortex pair, intensive stirring is observed in the atmosphere, in particular in the regions near its boundaries. In most cases, a freely propagating vortex pair leaves the interaction scene, sometimes after a partner exchange, sometimes consisting of the original vortices. The size of the atmosphere of such an ultimately surviving vortex pair is essentially dependent on the initial conditions of the vortex interaction. In general, entrainment and/or detrainment takes place, their magnitudes being defined by the coefficients introduced in Section 3.2. This definition was based on the adopted approach of the *contour kinematics* method, which consists of tracing some arbitrarily defined material contours that are passively advected by the flow induced by the point vortices.

The results obtained with this contour kinematics technique have been compared for a few cases with both full numerical solutions of the 2D Euler equations and dye-visualized laboratory observations of interacting vortex structures: the contour kinematics calculations showed perfect agreement. In view of this successful result, we may have confidence in the effectiveness of the approach adopted, and also in predicting the advection properties of more complicated 2D flows.

Acknowledgments—This paper was written while one of the authors (VVM) was visiting the Fluid Dynamics Laboratory in Eindhoven, within the framework of a collaboration project that is financially supported by the Netherlands Organization for Scientific Research (NWO). The authors gratefully acknowledge this support. Also we are grateful to Shura Gurzhi for some useful calculations and to the students René Kieft and Patrick Janssen for performing the vortex interaction experiments.

REFERENCES

1. H. Helmholtz, Über Integrale der hydrodynamischen Gleichungen, welche den Wirbelbewegungen entsprechen, *J. reine angew. Math.* **55**, 25–55 (1858).
2. W. Thomson, On vortex atoms, *Phil. Mag.* (ser. 4) **34**, 15–24 (1867).
3. J. C. Maxwell, Atom, in *Encyclopedia Britannica*, 9th ed. London (1878).
4. J. J. Thomson, *A Treatise on the Motion of Vortex Rings*. Macmillan, London (1883).
5. P. G. Saffman, *Vortex Dynamics*. Cambridge University Press, Cambridge (1992).
6. W. B. Rogers, On the formation of rotating rings by air and liquids under certain conditions of discharge, *Amer. J. Sci.* (ser. 2) **26**, 246–258 (1858).
7. R. Ball, Account of experiments upon the retardation experienced by vortex rings of air when moving through air, *Trans. R. Irish Acad.* **25**, 135–155 (1871).

8. P. G. Tait, *Recent Advances in Physical Sciences*. Macmillan, London (1876).
9. E. Reusch, Über Ringbilding in Flüssigkeiten, *Ann. Phys. Chem.* **110**, 309–316 (1860).
10. O. Reynolds, On vortex motion, *Proc. R. Instn. Great Britain* **8**, 272–279 (1877).
11. O. Reynolds, Study of fluid motion by means of colored bands, *Proc. R. Instn. Great Britain* **14**, 129–138 (1893), or *Nature*, **50**, 161–164 (1894).
12. W. M. Hicks, The mass carried forward by a vortex, *Phil. Mag.* (ser. 6) **38**, 597–612 (1919).
13. H. Lamb, *Hydrodynamics*, 3rd ed. Cambridge University Press, Cambridge (1906).
14. G. K. Batchelor, *An Introduction to Fluid Dynamics*. Cambridge University Press, Cambridge (1967).
15. K. Shariff and A. Leonard, Vortex rings, *Ann. Rev. Fluid Mech.* **24**, 235–279 (1992).
16. A. R. Robinson, editor *Eddies in Marine Science*. Springer, Berlin (1983).
17. R. Scorer, *Environmental Aerodynamcis*. Ellis Horwood, Chichester (1978).
18. M. E. Stern, Minimal properties of planetary eddies, *J. Marine Res.* **33**, 1–13 (1975).
19. G. R. Flierl, Isolated eddy models in geophysics, *Ann. Rev. Fluid Mech.* **19**, 493–530 (1987).
20. G. J. F. van Heijst and J. B. Flór, Dipole formation and collisions in a stratified fluid, *Nature*, **340**, 212–215 (1989).
21. S. I. Voropayev and Y. D. Afanasyev, Two-dimensional vortex–dipole interactions in a stratified fluid, *J. Fluid Mech.* **236**, 665–689 (1992).
22. R. C. Kloosterziel and G. J. F. van Heijst, An experimental study of unstable barotropic vortices in a rotating fluid, *J. Fluid Mech.* **223**, 1–24 (1991).
23. G. R. Flierl, M. E. Stern and J. A. Whitehead, The physical significance of modons: laboratory experiments and general integral constraints, *Dyn. Atmos. Oceans*, **7**, 233–264 (1983).
24. E. J. Hopfinger and G. J. F. van Heijst, Vortices in rotating fluids. *Ann. Rev. Fluid Mech.* **25**, 241–289 (1993).
25. Y. Couder and C. Basdevant, Experimental and numerical study of vortex couples in two-dimensional flows, *J. Fluid Mech.* **173**, 225–251 (1986).
26. J.-M. Nguyen Duc and J. Sommeria, Experimental characterization of steady two-dimensional vortex couples, *J. Fluid Mech.* **192**, 175–192 (1988).
27. H. Aref, Coherent features by the method of point vortices, Woods Hole Ocean. Inst. Tech. Rep. 80–53, pp. 233–249 (1980).
28. J. C. McWilliams and N. J. Zabusky, Interactions of isolated vortices—I: Modons colliding with modons, *Geophys. Astrophys. Fluid Dyn.* **19**, 207–227 (1982).
29. P. Orlandi and G. J. F. van Heijst, Numerical simulation of tripolar vortices in 2D flow, *Fluid Dyn. Res.* **9**, 179–206 (1992).
30. V. V. Meleshko, M. Yu. Konstantinov, A. A. Gurzhi and T. P. Konovaljuk, Advection of a vortex pair atmosphere in a velocity field of point vortices, *Phys. Fluids*, **A4**, 2779–2797 (1992).
31. W. J. M. Rankine, On plane water lines in two dimensions, *Phil. Trans. R. Soc. Lond.* **163**, 345–377 (1864).
32. J. C. Maxwell, On the displacement in a case of fluid motion, *Proc. Lond. Math. Soc.* **3**, 82–87 (1870).
33. O. Reynolds, The motion of fluids, *Nature,* **21**, 342–344 (1880).
34. P. Welander, Studies on the general development of motion in a two-dimensional, ideal fluid, *Tellus*, **7**, 141–156 (1955).
35. H. Aref, Stirring by chaotic advection, *J. Fluid Mech.* **143**, 1–21 (1984).
36. H. Aref, Chaotic advection of fluid particles, *Phil. Trans. R. Soc. Lond.* **A333**, 273–288 (1990).
37. J. P. Ottino, Mixing, chaotic advection, and turbulence, *Ann. Rev. Fluid Mech.* **22**, 207–253 (1990).
38. G. Kirchhoff, *Vorlesungen über Mathematische Physik. Mechanik*. Teubner, Leipzig (1876).
39. H. Aref, Integrable, chaotic and turbulent vortex motion in two-dimensional flows, *Ann. Rev. Fluid Mech.* **15**, 345–389 (1983).
40. H. Aref, Chaos in the dynamics of a few vortices—fundamentals and application, in *Theoretical and Applied Mechanics*, edited by N. Olhøff and F. I. Niordson, pp. 43–68. North-Holland, Amsterdam (1985).
41. W. Gröbli, Specielle Probleme über die Bewegung geradliniger paralleler Wirbelfäden, *Vierteljahr. Naturfor. Geselsch. Zürich* **22**, 37–81, 129–165 (1877).
42. H. Poincaré, *Théorie des Tourbillons*. G. Carré, Paris (1893).
43. J. M. Robertson, *Hydrodynamics in Theory and Application*. Prentice-Hall (1965).
44. V. Rom-Kedar, A. Leonard and S. Wiggins, An analytical study of transport mixing and chaos in an unsteady vortical flow, *J. Fluid Mech.* **214**, 347–394 (1990).
45. E. Hairer, S. P. Nørset and G. Wanner, *Solving Ordinary Differential Equations—I: Nonstiff Problems*. Springer, Berlin (1987).
46. E. P. Northrup, An experimental study of vortex motions in liquids, *J. Franklin Inst.* **172**, 211–226, 345–368 (1911) or *Nature*, **88**, 463–468 (1911).
47. S. A. Chaplygin, One case of vortex motion in fluid. *Trans. Phys. Sect. Imp. Moscow Soc. Friends of Natural Sciences*, **11**, N2, 11–14 (1903).
48. V. V. Meleshko and G. J. F. van Heijst, On Chaplygin's investigations of two-dimensional vortex structures in an inviscid fluid, *J. Fluid Mech.* **272**, 157–182 (1994).
49. J. B. Flór and G. J. F. van Heijst, An experimental study of dipolar vortex structures in a stratified fluid, *J. Fluid Mech.* (1993).
50. J. B. Flór, W. S. S. Govers, G. J. F. van Heijst and R. van Sluis, Formation of a tripolar vortex in a stratified fluid, *Appl. Sci. Res.* **51**, 405–409 (1993). (Submitted.)

51. J. C. van den Berg, De Wervelbeweging, Ph.D. thesis, Leiden (1888).
52. N. Quint, De Wervelbeweging, Ph.D. thesis, Amsterdam (1888).
53. H. Aref, N. Rott and H. Thomann, Gröbli's solution of the three-vortex problem: a case study of graduate work in the 1870's, *Ann. Rev. Fluid Mech.* **24**, 1–20 (1992).
54. H. Aref, Motion of three vortices, *Phys. Fluids* **22**, 393–400 (1979).
55. T. P. Konovalyuk, Classification of interactions of a vortex pair with a point vortex in an ideal fluid, *Fluid Mechanics—Sov. Res.* **21**, 113–122 (1992).
56. V. V. Meleshko and M. Yu. Konstantinov, *Dynamics of Vortex Structures*. Naukova Dumka, Kiev (1993).
57. B. Eckhardt and H. Aref, Integrable and chaotic motions of four vortices. II. Collision dynamics of vortex pairs, *Phil. Trans. R. Soc. Lond.* **A326**, 655–696 (1988).
58. T. Price, Chaotic scattering of two identical point vortex pairs, *Phys. Fluids* **A5**, 2479–2483 (1993).
59. B. Eckhardt, Integrable four-vortex motion, *Phys. Fluids* **31**, 2796–2801 (1988).
60. N. Rott, Constrained three- and four-vortex problems, *Phys. Fluids* **A2**, 1477–1480 (1990).
61. D. N. Goryachev, *On Some Cases of Motion of Rectilinear Vortices*. Moscow University Press, Moscow (1898).
62. H. Aref, Point vortex motion with a center of symmetry, *Phys. Fluids* **25**, 2183–2187 (1982).
63. F. Auerbach, Wirbelbewegung, in *Handbuch der Physik* (2nd edn), edited by A. Winkelmann, Vol. 1, pp. 1047–1074 (1908).
64. A. E. H. Love, Hydrodynamik: Theoretische Ausführungen, in *Encyklopädie der Mathematischen Wissenschaften*, edited by F. Klein and C. Müller, Vol. 4, pp. 84–147 (1901).
65. J. J. Thomson, *Electricity and Matter*. Constable, Westminster (1904).
66. L. Onsager, Statistical hydrodynamics, *Nuovo Cimento* **6** (Suppl.), 279–287 (1949).
67. R. H. Kraichnan and D. Montgomery, Two-dimensional turbulence, *Rep. Prog. Phys.* **43**, 549–619 (1980).
68. H. Aref, Stochastic particle motion in laminar flows, *Phys. Fluids* **A3**, 1009–1016 (1991).
69. J. C. McWilliams, The emergence of isolated coherent vortices in turbulent flow, *J. Fluid Mech.* **146**, 21–43 (1984).
70. B. Legras, P. Santangelo and R. Benzi, High-resolution numerical experiments for forced two-dimensional turbulence, *Europhys. Lett.* **3**, 811–818 (1988).
71. K. Tsuboi and Y. Oshima, Merging of two-dimensional vortices by the discrete vortex method, *J. Phys. Soc. Japan* **54**, 2137–2145 (1985).
72. J. P. Christiansen, Numerical simulation of hydrodynamics by the method of point vortices, *J. Comput. Phys.* **13**, 363–379 (1973).
73. N. J. Zabusky and J. C. McWilliams, A modulated point-vortex model for geostrophic, β-plane dynamics, *Phys. Fluids* **25**, 2175–2180 (1982).
74. O. U. Velasco Fuentes and G. J. F. van Heijst, Experimental study of dipolar vortices on a topographic β-plane, *J. Fluid Mech.* **259**, 79–106 (1994).
75. D. I. Pullin, Contour dynamics methods, *Ann. Rev. Fluid Mech.* **24**, 89–115 (1992).

An Analytical Study of Chaotic Stirring in Tidal Areas

S. P. BEERENS, H. RIDDERINKHOF and J. T. F. ZIMMERMAN*

Netherlands Institute for Sea Research, P.O. Box 59, 1790AB Den Burg (Texel), The Netherlands

Abstract—Chaotic advection is studied in a flow representative of tidal areas. The flow consists of a residual flow, represented by a lattice of eddies, perturbed by a tidal flow. The physical background of the flow is given by means of a dynamical model for tide–topography interaction. Lagrangian advection in this flow can be described in terms of perturbed Hamiltonian systems. For small perturbations analytical techniques, like Melnikov's method, provide mixing coefficients. But also in the limit of large perturbations analytical results can be achieved. In this paper the method of *orbit expansion* is presented. The coordinates are transformed into a system, moving with the perturbation. By integration over the period of the perturbation, one obtains an (first-order) approximation of the Poincaré map. The next order can be obtained by a new coordinate transformation, this time moving with both the perturbation and the lower-order displacement. Again the moving system is integrated over a period of the perturbation. In this way an analytical approximation of the Poincaré map can be constructed. Using this approximate map one can find analytical expressions for the mixing coefficients. This method is applied successfully to a model of a tidal area. It can explain the non-monotonic dependence of the mixing on the topographic wavenumber.

1. INTRODUCTION

Mixing in tidal areas seems often to be more intense than theories [1, 2, 3, 4] predict. An explanation can be that large-scale mixing is merely a consequence of chaotic Lagrangian trajectories of waterparcels in a deterministic tidal velocity field, rather than of some sort of turbulent diffusion [5]. In this paper we present an analytical study of chaotic stirring in a tidal flow.

Tidal areas are characterized by their shallowness and their strong tidal currents. The nonlinear interaction of tidal currents and bottom topography produces currents [6], which give nonzero contributions to the tidally averaged currents. These so-called residual currents are important for the transport and mixing properties of the tidal flow [7, 4]. In first approximation, the flow in tidal areas can be considered to be the superposition of a time-dependent part (the tidal flow) and a time-independent part (the residual flow). It was first recognized in [8], that the advection problem in this type of flows, studied in the Lagrangian way, can be written in terms of perturbed 2D Hamiltonian systems. From dynamical system theory [9, 10] it is known, that these systems can be chaotic. The work of Aref [11] and Ottino [12] has shown that this *chaotic advection* is relevant for mixing in fluids.

Ridderinkhof and Zimmerman [5] applied these ideas in a Lagrangian study of particle motion in a realistic model and in a simplified model of a tidal area. Their simple kinematical model can be interpreted as a strongly perturbed Hamiltonian system, which hinders the direct application of analytical methods. Numerical experiments showed that

*Also at Institute for Marine and Atmospheric Research, Utrecht University, The Netherlands.

the amount of chaotic mixing depends on two parameters; λ, the ratio of the tidal excursion and the residual eddy diameter and ν, the ratio of residual velocity and the tidal velocity amplitude. They found an as yet unexplained oscillatory dependence of the mixing on the parameter λ.

In Section 2 of this paper we will introduce the current field which is used in our kinematical model. A dynamical model for tide–topography interaction is used to derive an expression for the (dynamical) relation between the parameters λ and ν. Section 3 gives a brief summary of some central concepts in chaotic mixing and applies a method to derive analytical expressions for chaotic mixing coefficients to our kinematical model. This method can only be applied if unrealistic parameter values are taken (weak perturbations). In Section 4 a similar approach is followed to derive analytical expressions for a strongly perturbed Hamiltonian system, which is representative for a tidal area. The discussion and conclusions are given in Section 5.

2. A MODEL FOR THE EULERIAN CURRENT FIELD IN TIDAL AREAS

In previous studies of chaotic mixing in tidal areas [5, 8] a kinematical model has been applied in which the underlying Eulerian current field is assumed to be a superposition of a tidal and a residual current field. The tidal field is assumed to be a spatially uniform oscillating current in one direction and the residual current field is assumed to be an infinite sequence of clockwise and anticlockwise rotating eddies (see Fig. 1). The (dimensionless) velocity field in the kinematical model, which is also used in the next sections reads:

$$\begin{aligned} u &= \pi\lambda\nu\Psi_y(x, y) + \pi\lambda\cos 2\pi t, \\ v &= -\pi\lambda\nu\Psi_x(x, y) \end{aligned} \tag{1}$$

where the streamfunction $\Psi(x, y)$ is given by:

$$\Psi(x, y) = \frac{1}{\pi}\sin\pi x \sin\pi y. \tag{2}$$

The dimensionless parameters of the model are:

$$\lambda = \frac{L_T}{L_E} = \frac{\text{tidal excursion}}{\text{eddy diameter}}, \qquad \nu = \frac{u}{U} = \frac{\text{velocity in residual eddies}}{\text{tidal velocity amplitude}}. \tag{3}$$

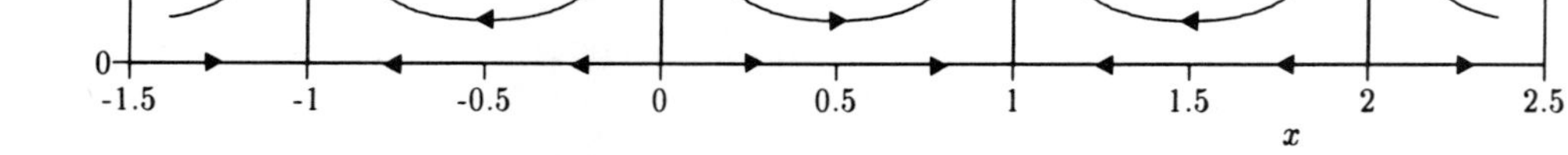

Fig. 1. Streamlines of the unperturbed flow. The streamlines of the residual velocity field as given by (2). The field is an alternating sequence of cyclonic and anti-cyclonic eddies. The centers of the cells are elliptic points and the corners are hyperbolic points.

The difference with previously applied models [5, 8] is that we omitted the time dependence in the eddy field to ease analytical calculations. It is clear that the parameters λ and v determine the quality of the solutions, if Lagrangian trajectories of particles are studied.

Since residual eddies are generated by the nonlinear interaction of tide and bottom topography [6, 13, 14], the parameters v and λ are not independent in a dynamical sense. Generation of residual eddies is most conveniently described in terms of vorticity dynamics. In the next subsections we will use solutions of the vorticity equation for a shallow tidal area with an irregular bottom topography to derive a response function, which gives the residual vorticity as a function of topographic lengthscale. Results are used to derive a relation between the parameters v and λ of our basic kinematical model.

2.1. *Vorticity generated by bottom topography*

Starting from the 2D shallow water equations one can derive the vorticity equation for tidal currents interacting with small-amplitude bottom topography. This equation can be derived by scaling as in [14] or in a more formal way by a multiple scale expansion as in [15]. From scaling emerge three parameters:

$$\beta = \pi L_{\mathrm{T}}/L_{\mathrm{W}}, \quad \delta = h_0/H, \quad \phi = L_{\mathrm{B}}/L_{\mathrm{W}}, \tag{4}$$

where L_{T} is the tidal excursion, L_{W} the tidal wavelength, L_{B} the topographic lengthscale, H is the height of the water column and h_0 is the amplitude of the topography.

These parameters are assumed to be small and of the same order of magnitude. With this assumption the vorticity equation becomes:

$$\partial_t \eta + \mathbf{U}\cdot\nabla\eta + \frac{r}{H}\eta = \frac{f}{H}\mathbf{U}\cdot\nabla h - \frac{r}{H^2}\mathbf{U} \wedge \nabla h, \tag{5}$$

where $\eta = \nabla \wedge \mathbf{u}$ is the vorticity, $\mathbf{u}$ is the topographically induced velocity, which is assumed to be much smaller than $\mathbf{U}$, the unperturbed tidal velocity ($\mathbf{u} \ll \mathbf{U}$). h is the bottom topography profile, f is the Coriolis parameter and r is the friction parameter in a linearized bottom friction law. The vorticity equation can be written in dimensionless form using the following scaling:

$$t' = \sigma t, \qquad \mathbf{x}' = \frac{2\mathbf{x}}{L_{\mathrm{T}}}, \qquad \eta' = \frac{\eta}{\sigma}, \qquad \mathbf{U}' = \frac{2\mathbf{U}}{\sigma L_{\mathrm{T}}},$$

$$h' = \frac{h}{H}, \qquad f' = \frac{f}{\sigma}, \qquad r' = \frac{r}{H\sigma},$$

where σ is the tidal frequency and $L_{\mathrm{T}}/2$ is the tidal excursion amplitude. The dimensionless form of the vorticity equation reads (dropping the primes):

$$\partial_t \eta + \mathbf{U}\cdot\nabla\eta + r\eta = f\mathbf{U}\cdot\nabla h - r\mathbf{U} \wedge \nabla h. \tag{6}$$

The r.h.s. of (6) contains the production terms due to planetary vortex stretching, resulting in clockwise circulation around hills on the northern hemisphere, and bottom friction torques which produce both clockwise and anticlockwise circulations, depending on the polarization of the tidal current. On the l.h.s. are the local derivative of the vorticity and the advective term, which accounts for the transfer of tidal vorticity to the residual field, and the dissipation term due to bottom friction (turbulent dissipation is neglected).

2.2. *Solutions of the vorticity equation*

Solutions of (6) can be found by introducing Fourier transformations of the space-dependent variables η and h:

$$\eta(\mathbf{x}, t) = \iint \hat{\eta}(\mathbf{k}, t) e^{-i\mathbf{k}\cdot\mathbf{x}}\, d\mathbf{k}, \qquad h(\mathbf{x}) = \iint \hat{h}(\mathbf{k}) e^{-i\mathbf{k}\cdot\mathbf{x}}\, d\mathbf{k}. \tag{7}$$

The integrals are over all of the 2D wavenumber space and $\mathbf{k} = (k, l)$ is the nondimensional wavenumber of the bottom topography. The solutions of the transformed equation can be found in [14] and in [15] for a circularly polarized tide and in [16] for a linear polarization. The general solution can be written as

$$\hat{\eta}(\mathbf{k}, t) = \hat{h}(\mathbf{k}) \sum_{p=-\infty}^{\infty} c_p(\mathbf{k}) e^{2\pi p t}. \tag{8}$$

We see that the vorticity is proportional to the bottom component times a weight-function $c_p(\mathbf{k})$, which is a complicated function of wavenumber $(\mathbf{k})$.

Since we are only interested in the residual vorticity we will only consider the case $p = 0$. Furthermore we will take the topographic wavenumber in the x-direction to be the same as in the y-direction $(k = l)$ and assume a linearly polarized tidal current. In that case the residual vorticity $\hat{\hat{\eta}}(k)$ reads

$$\hat{\hat{\eta}}(k) = \hat{h}(k) \frac{f + r}{i2\sqrt{2}} \sum_{m=-\infty}^{\infty} J_m(k) \left(\frac{J_{m+1}(k)}{i(1 + m) + r} - \frac{J_{m-1}(k)}{i(1 - m) + r} \right), \tag{9}$$

where the $J_m(k)$ are Bessel functions of the first kind, order m. The residual vorticity has a maximum for $k = \mathcal{O}(1)$, which is a typical value for tidal areas. This is not a coincidence, but reflects that the dominant wavelength of sandbanks may be the wavelength with the fastest growth rate, i.e. the wavelength with maximal vorticity response [17].

2.3. *v as a function of λ*

Having derived an expression for the residual vorticity generated by a uniform tidal current, we are now able to express the parameter $v = u/U$ in the parameter λ. The residual velocity $\hat{\hat{u}}(k)$ is related to the residual vorticity $\hat{\hat{\eta}}(k)$ according to:

$$\hat{\hat{u}}(k) = \frac{i\hat{\hat{\eta}}(k)}{k}. \tag{10}$$

The relation between the kinematical parameter λ and the topographical wavenumber k follows from scaling of k:

$$k = \pi L_T/L_B = \pi L_T/2L_E = \pi\lambda/2. \tag{11}$$

Combining (9), (10) and (11) the expression for v as function of λ becomes

$$v(\lambda) = \hat{h}(\lambda) F(\lambda, f, r), \tag{12}$$

where the weight-function $F(\lambda, f, r)$ is given by

$$F(\lambda, f, r) = \frac{(f + r)}{\pi\lambda\sqrt{2}} \sum_{m=-\infty}^{\infty} J_m(\pi\lambda/2) \left(\frac{J_{m+1}(\pi\lambda/2)}{i(1 + m) + r} - \frac{J_{m-1}(\pi\lambda/2)}{i(1 - m) + r} \right).$$

The function $F(\lambda, f, r)$ is shown in Fig. 2 for parameter values $f = 0.8$ and $r = 1.0$, which

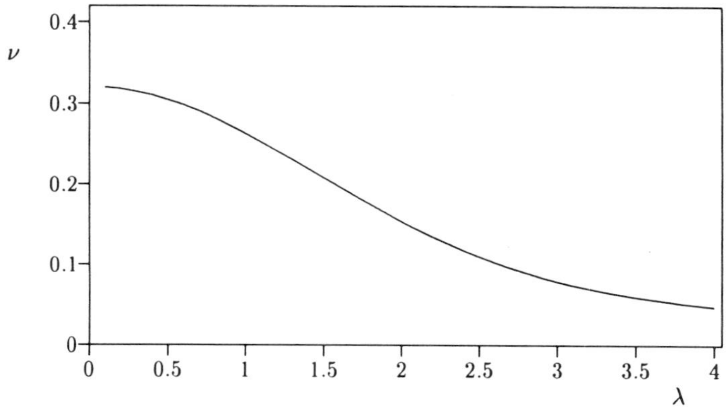

Fig. 2. The parameter ν as a function of λ. The scaled residual velocity (ν) as a function of (λ) as calculated in Section 2.3. The residual velocity decreases with increasing topographic wavenumber.

are representative for a tidal area like the Wadden Sea. From this figure it is clear that the value of ν is always substantially less than one.

If the set of equations (1) is used for a kinematical model, one should be aware that its parameters are not independent in a dynamical sense, as equation (12) shows. From a dynamical point of view it is more appropriate to work with the independent parameters λ and another parameter related to the amplitude of the topography. The parameter ν is a function of these two parameters and therefore not independent, but since it is the ratio of the strength of the two velocity fields that we use, we will (for computational convenience) treat ν and λ as independent parameters and later on add relation (12) for more realistic computations.

3. BASIC CONCEPTS OF CHAOTIC MIXING

Different techiques have been developed to analyse chaotic mixing in flow fields which are similar to the flow field introduced in the previous section [12, 18]. The basic idea underlying these techniques is that the advection problem in 2D incompressible fluids can be formulated in terms of Hamilton's formalism, where the streamfunction (Ψ) plays the role of the Hamiltonian. The trajectories of advected particles are given by:

$$\dot{x} = f_1(x, y) = \Psi_y(x, y) \quad \text{and} \quad \dot{y} = f_2(x, y) = -\Psi_x(x, y). \tag{13}$$

In our application we express the residual field $\mathbf{f}(\mathbf{x})$ in terms of a streamfunction and use the tidal field $\mathbf{g}(t)$ as a perturbation of the Hamiltonian system:

$$\dot{\mathbf{x}} = \mathbf{f}(\mathbf{x}) + \mathbf{g}(t). \tag{14}$$

In the following subsections we will briefly summarize the ideas behind some qualitative and quantitative methods that can be used to analyse these systems. We will apply these methods to study particle spreading in our kinematical model (1). Numerical solutions of particle trajectories in our model are obtained by applying a fourth order Runge–Kutta method.

3.1. *Poincaré sections*

System (14) will in general be chaotic. Qualitative insight in the behaviour of chaotic dynamical systems can be obtained by means of *Poincaré sections*. Poincaré sections can be constructed by plotting the successive intersections of the trajectory of a particle in phase-space and a lower-dimensional subspace. In case of time periodic perturbations, like our tidal flow, a Poincaré section can be constructed in a straightforward way. Following a particle, we plot its position after each period of the perturbation. We will call the Poincaré map of the tidal advection problem *the Tidal Map* (**T**),

$$\mathbf{x}_{n+1} = \mathbf{T}(\mathbf{x}_n). \tag{15}$$

Given the position of a particle ($\mathbf{x}_n$), **T** gives the position one tidal period later ($\mathbf{x}_{n+1}$). An orbit of the Tidal Map is defined as the set of iterates of **T**

$$\{\mathbf{x}_0, \mathbf{T}(\mathbf{x}_0), \mathbf{T}^2(\mathbf{x}_0), \cdots, \mathbf{T}^n(\mathbf{x}_0)\} \cap \{\mathbf{T}^{-1}(\mathbf{x}_0), \cdots, \mathbf{T}^{-n}(\mathbf{x}_0)\}, \tag{16}$$

where $\mathbf{x}_0$ is an arbitrary initial position.

Figure 3 shows plots of the Tidal Map for fixed $\nu = 0.3$ and different values of λ. In these plots one can distinguish between regions where orbits are smooth and regular, and regions where orbits are chaotic. Particles on regular orbits move around the elliptic fixed point in the middle of their cell. Particles in these regions are trapped. We will refer to these regions as *trapped regions*. The chaotic orbits are concentrated in the regions near the hyperbolic points in the corners of the cells. Particles in these regions move irregularly, changing cell every now and then. These regions are well mixed and we will refer to these regions as *mixing regions*.

As the parameter λ is raised, many interesting features occur. At certain parameter values the elliptic fixed points in the middle of the cells bifurcate. These bifurcations can be related to the action of **T**. We will address this subject later on, in Section 4.2.

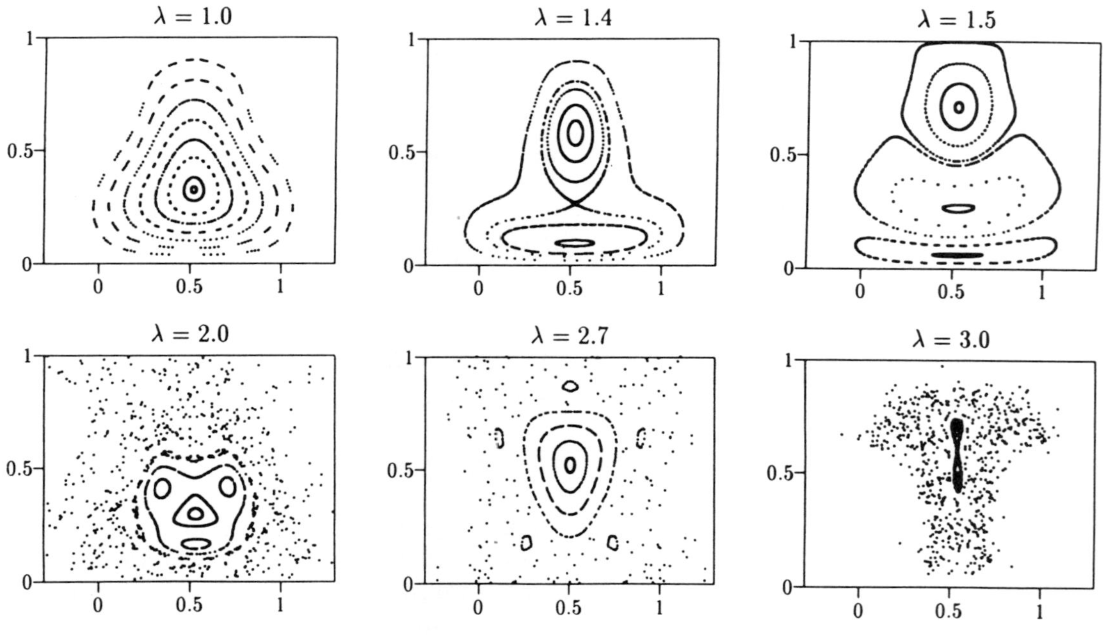

Fig. 3. Plots of the Tidal Map.

3.2. *Lobe dynamics*

Chaotic dynamics in fluids is often associated with *hyperbolic points* of the flow [9, 18]. Suppose that we have an unperturbed Hamiltonian system containing hyperbolic points. (In our model we have hyperbolic points in the corners of the cells, see Fig. 1.) The hyperbolic points are connected by *manifolds*. The unstable manifold is the set of orbits, which starts in a hyperbolic point and the stable manifold is the set of orbits, which ends in a hyperbolic point. In the unperturbed system these manifolds coincide, but due to the perturbation they get disconnected. In that case they have infinitely many intersection points (intersection points of the stable and unstable manifold are called *heteroclinic points*). The manifolds form a complicated structure of *lobes* (see Fig. 4). The dynamics of these lobes are important for the transport across the cell boundaries [19].

The complicated structure in Fig. 4 can be understood in the following way. If the stable manifold of a hyperbolic point and the unstable manifold of another hyperbolic point have one intersection point (heteroclinic point), then they will have infinitely many intersection points. The manifolds are invariants of **T**. So the image of a heteroclinic point under **T** will lie both on the stable and on the unstable manifold. This means that the image of a heteroclinic point will again be a heteroclinic point. The orbit constructed in this way is called the *heteroclinic orbit*. The lobes get longer and thinner as they reach the hyperbolic point. This is due to the fact that **T** is *area-preserving*. The lobes are mapped onto each other by **T**, so all lobes will have the same area. If we follow the heteroclinic orbit towards the hyperbolic point, the heteroclinic points come closer together and the lobes get longer and thinner. The lobes grow parallel to the unstable manifold of the hyperbolic point. As

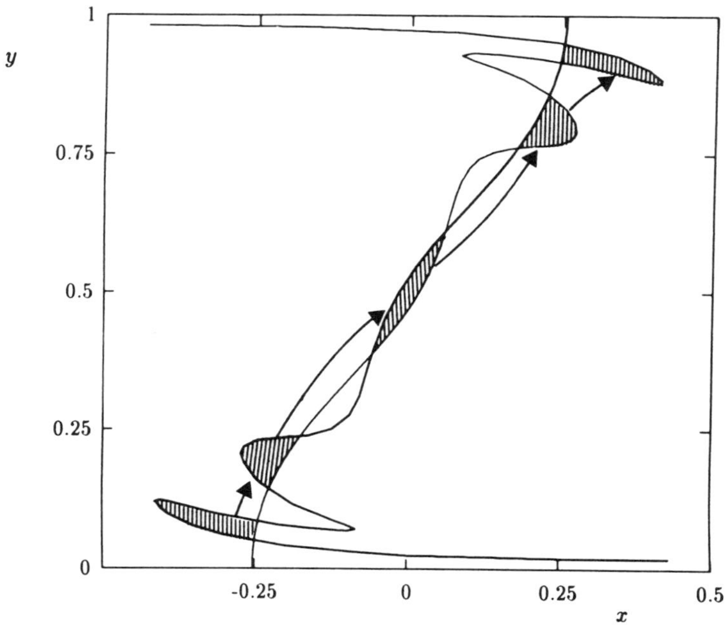

Fig. 4. The stable and unstable manifold. A typical picture of the lobe structure near the cell-boundary in the Tidal Map for $v = 0.2$ and $\lambda = 1.9$. Depicted are the unstable manifold of the hyperbolic fixed point $(-0.25, 0)$ and the stable manifold of the hyperbolic fixed point $(0.25, 1)$. The arrows denote the action of the Tidal Map. The lobes enclosed by the manifolds are mapped onto each other. Near the central heteroclinic point $(-0.06, 0.39)$, material is turned over from one to the other cell. The lobes, which establish the exchange, are called the turnstile lobes. This figure has been obtained by application of the software package DSTOOL [20].

they get longer, they will intersect more and more with other lobes. Secondary intersection points occur and the dynamics gets very complicated [21]. The system is chaotic in these regions.

The lobes are important for mixing and transport in the flow. Near the central heteroclinic point material is transported across the boundary of the cells. The amount of exchanged material during one iteration is exactly the area of a lobe. In Fig. 4 we see how the exchange process works. The lobes near the central heteroclinic act as a *turnstile*. They hand over material across the cell boundary. In the next subsection we will discuss quantitative aspects of this exchange.

3.3. *Quantification of chaotic mixing*

Welander already noticed in 1955 in a pioneering paper [22] on this subject, that the quantification of mixing in stirred fluids is far from trivial. There are at least two problems in defining mixing or dispersion coefficients in chaotically advecting flows. Firstly, in the generic situation, not all of the fluid moves along chaotic orbits. Thus regions that are non-mixing coexist with regions where mixing occurs. Secondly, even for that part of the fluid which is not trapped in non-mixing regions, the orbits often have a more complicated character than a simple random walk (due to memory effects), so that an unambiguous definition of a dispersion coefficient can be problematic.

In this paper we will use the quantification of chaotic mixing as suggested by Ryrie [23]. She studied a class of spatially periodic flows and found that chaotic mixing in these types of flows can be described using two coefficients. The first one (α) is the rate of exchange between neighbouring cells.

$$\alpha = a/A, \tag{17}$$

where a is the area of the turnstile lobe and A is the total area of the cell. But since chaotic mixing is localized in the areas near the hyperbolic points, one cannot suffice with just the rate of exchange. Large unmixed areas can occur, even if the exchange rate is high. One should also know how much fluid is participating in the mixing process. This quantity is contained in the parameter γ, which is the area of the mixing region A_{mix} relative to the total area,

$$\gamma = A_{\text{mix}}/A. \tag{18}$$

A dispersion coefficient can now be defined by assuming that the fraction of the cell fluid that participates in transfer across the cell boundaries mixes instantaneously over the mixing region of the cell during each iteration. Then particles in the mixing region are performing a random walk on the large scale of the cell lattice [23] and consequently a (dimensionless) dispersion coefficient D for these particles can be defined as:

$$D = \frac{1}{2}\frac{\mathrm{d}\sigma^2}{\mathrm{d}t} = \frac{1}{2N}\frac{d\sum_{n=1}^{N}(i_n - i_0)^2}{\mathrm{d}t}, \tag{19}$$

where i_n is the cell-number of the nth particle originally present in the mixing region of cell zero. The sum is over all N particles in the ensemble and t is the time divided by the period of the perturbation.

The dispersion coefficient D and the exchange rate α are related in the following way: The chance that a particle leaves a cell with unit area during one iteration is equal to 2α (α for each neighbour, see (17)). The squared cell-number increment will be 1 for each particle that leaves the cell, thus

$$D = \frac{1}{2N} \frac{\mathrm{d}\sum_{n=1}^{N}(i_n - i_0)^2}{\mathrm{d}t} = \frac{1}{2N} \frac{2N\alpha}{1} = \alpha. \tag{20}$$

This is under the assumption that mixing in the mixing region of a cell is instantaneous. But actually this is not a very good assumption. After entering a cell through the turnstile, it will take some iterations (tidal periods), before the particle is in a lobe which is long enough to reach the other end of the cell [24]. Until that time it has no chance of leaving the cell. Therefore the dispersion coefficient will be less than α. We can regard α as the upper limit of D. For the reality of tidal areas, however, the assumption can yet be a good one, as in these areas there is always a 'background' of small scale (turbulent) mixing, which tends to homogenize sharp gradients in watermasses and thus will quickly destroy the complex structures of lobe dynamics.

3.4. *Application to weakly perturbed Hamiltonian systems*

The above-defined mixing coefficients α and γ can be determined analytically for weakly perturbed Hamiltonian systems. In our system this means that the tidal perturbation should be much smaller than the unperturbed residual current. This limit ($\nu \to \infty$) is physically unrealistic for studies of tidal areas (see Section 2), but may find applications elsewhere. Nevertheless we will work out this limit, since a similar approach will be used in Section 4, where an analytical method for strongly perturbed Hamiltonian systems is presented.

3.4.1. *The Melnikov function.* In case of weakly perturbed Hamiltonian systems, one can derive an analytical expression for the mixing coefficient α [23]. This expression is found using Melnikov's function, which is a measure for the distance between the stable and unstable manifold in weakly perturbed Hamiltonian systems [19, 25]. The main problem in the computation of the Melnikov function is to find a parameterization of the unperturbed manifold ($\mathbf{q}_0 = \mathbf{q}_0(t)$). If this is possible the Melnikov function is defined in the following way:

$$M(t) = \int_{-\infty}^{\infty} \mathbf{f}(\mathbf{q}_0(\tau)) \wedge \mathbf{g}(t + \tau)\,\mathrm{d}\tau, \tag{21}$$

where $\mathbf{f}(\mathbf{x})$ is the unperturbed Hamiltonian system and $\mathbf{g}(t)$ is the (time-dependent) perturbation. For system (1) the unperturbed manifold (connecting the hyperbolic points $H = (0, 1)$ and $O = (0, 0)$) is parameterized by [26]:

$$\mathbf{q}_0(t) = \begin{cases} x_0 = 0 \\ y_0 = (2/\pi)\arctan\exp(-\pi^2\lambda\nu t) \end{cases} \tag{22}$$

where t goes from $-\infty$ (H) to ∞ (O).

For the calculation of the Melnikov integral (21) we need an expression for $\sin(\pi y_0)$, rather than for y_0. This can be found using trigonometric relations.

$$\sin(\pi y_0) = 2\sin(\tfrac{1}{2}\pi y_0)\cos(\tfrac{1}{2}\pi y_0) = \frac{2\exp(-\pi^2\lambda\nu t)}{1 + \exp(-2\pi^2\lambda\nu t)} = \operatorname{sech}(\pi^2\lambda\nu t), \tag{23}$$

where we used: $\tan(\tfrac{1}{2}\pi y_0) = \exp(-\pi^2\lambda\nu t)$.

Using (23) the Melnikov integral (21) of system (1) becomes

$$M(t) = \pi^2\lambda^2\nu\int_{-\infty}^{\infty} \operatorname{sech}\pi^2\lambda\nu\tau\cos 2\pi(t + \tau)\,\mathrm{d}\tau, \tag{24}$$

which can be solved ([27], p. 503) to give

$$M(t) = \pi\lambda \operatorname{sech}(1/\lambda\nu)\cos 2\pi t. \tag{25}$$

With the Melnikov function one can compute the distance (d) between the stable and the unstable manifold, as this distance (d) is given by:

$$d(t) = \frac{\|M(t)\|}{\|\mathbf{f}(x_0(t),\, y_0(t))\|}. \tag{26}$$

3.4.2. *Chaotic mixing coefficients.* Now that we know the distance between the manifolds, we can calculate the area of the lobes enclosed between two intersection points of the manifolds, which is equal to the exchange rate α:

$$\alpha = \int_{l_i}^{l_{i+1}} d(l)\, dl = \int_{t_i}^{t_{i+1}} \|M(t)\|\, dt = \lambda \operatorname{sech}(1/\lambda\nu), \tag{27}$$

where we used that $\|\mathbf{f}\| = dl/dt$ (l is the length of the arc and the subindices i and $i+1$ denote intersection points). We can also find an analytical expression for the other mixing coefficient, γ. The particles in the mixing region are contained in the lobes. These particles follow hyperbolic orbits, changing lobe every iteration. The total mixing region will be contained in the area enclosed by the hyperbolic curve connecting the maxima of the Melnikov distance (26) and the coordinate axes (see Fig. 5). This area can be computed if we assume the hyperbolic curve to be

$$y = X/2x, \tag{28}$$

where X is the Melnikov distance near the central heteroclinic point. Because of the

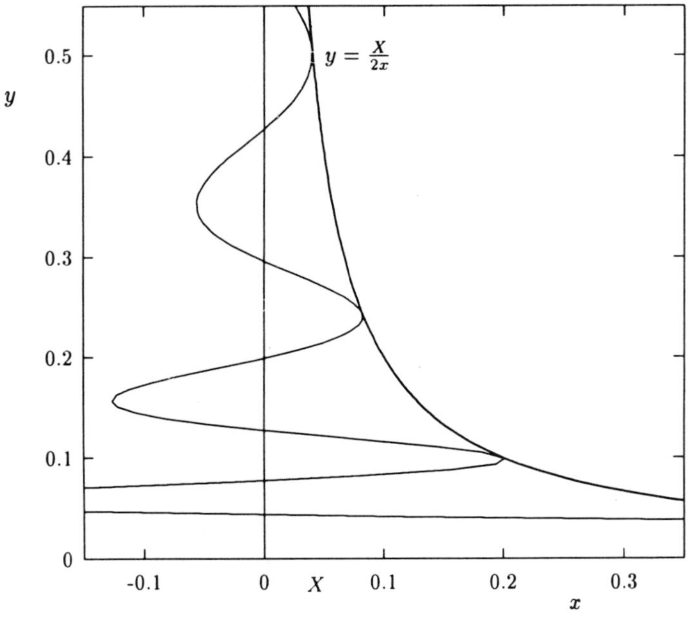

Fig. 5. Schematically the situation near a hyperbolic point. A schematical picture of the situation near the hyperbolic point $O = (0,\ 0)$, which is used to quantify the mixing coefficient γ. The unstable manifold of the hyperbolic fixed point $H = (0,\ 1)$ meanders downwards. The local maxima of the manifold are connected by a curve, which we assume to be a hyperbola (28).

cosine-like shape of the lobes, one may argue that the mixing region is a factor π smaller than this area. Thus for γ we have:

$$\gamma = \frac{2}{\pi}X\left(1 + \ln\frac{1}{2X}\right). \tag{29}$$

γ is four times the area enclosed by the hyperbola $yx = \frac{1}{2}X$, the x-axis from 0 to $x = \frac{1}{2}$ and the y-axis from 0 to $y = \frac{1}{2}$, divided by π. X is the Melnikov distance (26) at $(x, y) = (0, \frac{1}{2})$

$$X = \frac{1}{v}\operatorname{sech}\frac{1}{\lambda v}. \tag{30}$$

The above computations work only for small perturbations ($v \gg 1$). Only then can we use Melnikov's method, which results from a first-order perturbation in $1/v$. In that case the assumption we made to compute γ (the orbit connecting the Melnikov maxima being a hyperbola) is also correct.

A comparison of α and γ given by (27) and (29) with numerical results is given in Fig. 6 for fixed $v = 10$. The numerical results are obtained by releasing 10 000 particles distributed homogeneously over an unperturbed cell. γ is obtained directly by counting the number of particles that are trapped and α is calculated indirectly from a numerical determination of the dispersion coefficient and relation (20). (An exact calculation of α by determining the area of the lobes is very laborious.)

Figure 6 shows that the analytical expressions for both α and γ work quite well, but only for small values of the parameter λ. This is due to the fact that the pararmeter which

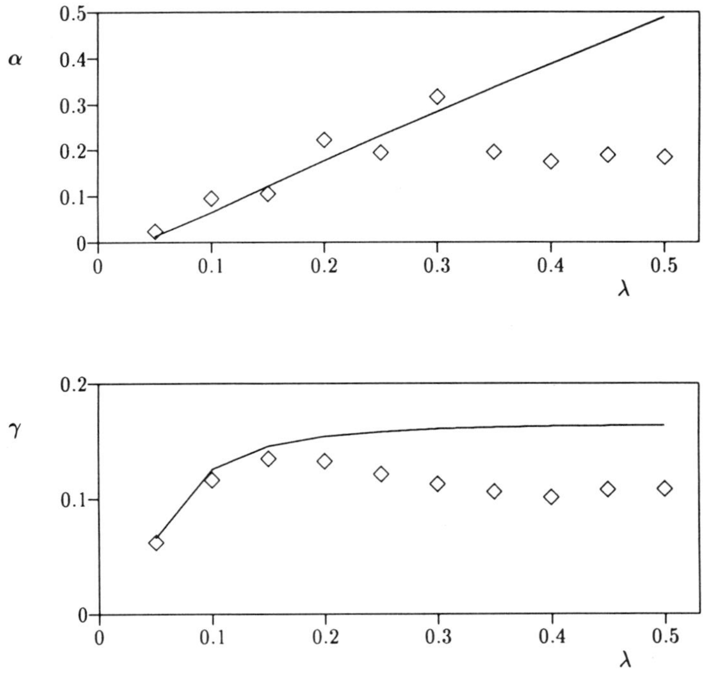

Fig. 6. The mixing coefficients α and γ vs λ for $v = 10$. The mixing coefficients α and γ versus the parameter λ in the case of a small perturbation ($\frac{1}{v} = 0.1$). The diamonds are the numerical results, the smooth curves are the analytical functions (27) and (29).

measures the disturbance of the Hamiltonian system is not just ν, as was assumed above, but actually is a combination of λ and ν, as can be seen in the expression for the Melnikov function (25).

4. MIXING IN A TIDAL FLOW: A STRONGLY PERTURBED HAMILTONIAN SYSTEM

Realistic values of the parameters λ and ν are such that our kinematical model for mixing in tidal areas can be considered as a strongly perturbed Hamiltonian system (see Section 2). Transport in such a system is governed by lobe dynamics. However, Melnikov's method cannot be applied for a quantification of the chaotic transport, unlike the case of weakly perturbed Hamiltonian systems. Therefore we have to develop other tools. We will use the method of *orbit expansion* [3] to derive an analytical approximation of the Tidal Map (15) in the limit $\nu \to 0$. Subsequently, this expression for particle displacement is used to quantify the lobe dynamics, giving analytical expressions for α and γ. Results are compared with numerical experiments.

4.1. *Orbit expansion*

The method of orbit expansion has been developed to estimate the tidally averaged Lagrangian displacement in tidal areas. In this method it is assumed that, over a single tidal period a particle trajectory is mainly the tidal excursion, with some correction to it, because of the residual velocity field. (This assumption is correct if the tidal current is much stronger than the residual current.) These two contributions to the trajectory are called the zeroth and first-order displacements. Higher orders can be generated by the following iterative process. For the trajectory of a particle $\mathbf{Y}(t)$ we write: $\mathbf{Y}(t) = \mathbf{Y}(0) + \sum_{n=0}^{\infty}\mathbf{Y}_n(t)$, where the iterates $\mathbf{Y}_n(t)$ are given by

$$
\begin{aligned}
\mathbf{Y}_0(t) &= \int_0^t (\mathbf{g})(\tau)\,\mathrm{d}\tau \\
\mathbf{Y}_1(t) &= \int_0^t (\mathbf{f})(\mathbf{Y}_0(\tau))\,\mathrm{d}\tau - \mathbf{Y}_0(t) \\
\mathbf{Y}_2(t) &= \int_0^t (\mathbf{f})(\mathbf{Y}_0(\tau) + \mathbf{Y}_1(\tau))\,\mathrm{d}\tau - (\mathbf{Y}_0(t) + \mathbf{Y}_1(t)) \\
&\vdots \\
\mathbf{Y}_n(t) &= \int_0^t (\mathbf{f})\left(\sum_{i=0}^{n-1}\mathbf{Y}_i(\tau)\right)\mathrm{d}\tau - \sum_{i=0}^{n-1}\mathbf{Y}_i(t).
\end{aligned}
$$

4.2. *Analytical approximation of the Tidal Map*

Using the method of orbit expansion the Tidal Map can be approximated by a series of maps, which is constructed analytically, where $\mathbf{T}$ itself is chaotic. Integrating the successive iterates over a tidal period we obtain

$$
\mathbf{T}(\mathbf{x}_n) = \mathbf{x}_n + \mathbf{M}_0(\mathbf{x}_n) + \mathbf{M}_1(\mathbf{x}_n) + \mathbf{M}_2(\mathbf{x}_n) + \cdots, \tag{31}
$$

where $\mathbf{M}_i = \mathbf{Y}_i(T)$.

Since the tide is a purely periodic signal in our model (1) the zeroth order displacement $\mathbf{M}_0$ vanishes,

$$\mathbf{M}_0 = \int_0^1 \pi\lambda \cos(2\pi\tau)\,d\tau = 0.$$

The first order is obtained by substituting $\mathbf{Y}_0(t)$ into the residual field, which is equivalent to transforming the coordinates into a frame moving with the periodic tide, and integrating over a tidal period. The coordinate transformation is given by:

$$x = x' - \lambda\pi \int^t \cos 2\pi\tau\,d\tau = x' - (\lambda/2)\sin 2\pi t, \qquad y = y', \tag{32}$$

where x and y are the new (moving) coordinates and x' and y' are the original coordinates. Substituting (32) into (1) we get

$$\begin{aligned}\dot{x} &= \pi\lambda v \sin \pi(x + (\lambda/2)\sin 2\pi t)\cos \pi y \\ \dot{y} &= -\pi\lambda v \cos \pi(x + (\lambda/2)\sin 2\pi t)\sin \pi y.\end{aligned} \tag{33}$$

This can be rewritten in terms of (time-dependent) streamfunctions in the following way

$$\begin{aligned}\dot{x} &= 2\kappa v \Psi_y(x, y)\cos(\kappa \sin 2\pi t) + 2\kappa v \Xi_y(x, y)\sin(\kappa \sin 2\pi t) \\ \dot{y} &= -2\kappa v \Psi_x(x, y)\cos(\kappa \sin 2\pi t) - 2\kappa v \Xi_x(x, y)\sin(\kappa \sin 2\pi t),\end{aligned} \tag{34}$$

where κ is shorthand for $\pi\lambda/2$ (11). The streamfunctions Ψ and Ξ are given by

$$\Psi(x, y) = (1/\pi)\sin \pi x \sin \pi y, \tag{35}$$

$$\Xi(x, y) = (1/\pi)\cos \pi x \sin \pi y. \tag{36}$$

Note that the transformation induces an extra velocity field associated with the streamfunction Ξ. Integrating (34) over a tidal period gives the first-order displacement $\mathbf{M}_1$,

$$\mathbf{M}_1(\mathbf{x}_n) = \begin{pmatrix} 2\kappa v J_0(\kappa)\Psi_y(\mathbf{x}_n) \\ -2\kappa v J_0(\kappa)\Psi_x(\mathbf{x}_n) \end{pmatrix}, \tag{37}$$

where $J_0(\kappa) = \int_0^1 \cos(\kappa \sin 2\pi t)\,dt$ (see [27], p. 405). We see that the extra velocity field due to the streamfunction Ξ does not contribute to $\mathbf{M}_1$, because $\int_0^1 \cos(\kappa \sin 2\pi t)\,dt = 0$. The first-order displacement is a 2D 'sine-map'. Its action is directed along the streamlines of the residual field and the amplitude depends on v and λ. The behaviour of the displacement in parameter space is almost periodic because of the Bessel function $J_0(\pi\lambda/2)$.

The same procedure is followed for the second-order displacement map, $\mathbf{M}_2$. Now we transform the coordinates into a frame moving both with the tide and the first-order Lagrangian velocity, which is assumed to be independent of time. This transformation is given by:

$$\begin{aligned}x &= x' - u_1 t - (\lambda/2)\sin 2\pi t, \\ y &= y' - v_1 t,\end{aligned} \tag{38}$$

in which (u_1, v_1) are the components of the first-order tidally averaged Lagrangian velocity: $(u_1, v_1) = (1/T)\mathbf{M}_1(\mathbf{x})$.

These coordinates are substituted into (1). The trigonometric functions, resulting from the substitution of (38) into (1), are written in forms with single arguments and integrated over a tidal period. After some manipulation the integrals can be written in the following standard type (see pp. 400–401 of ref. [27]):

$$\begin{aligned}
\int_0^{1/2} \cos(\kappa \sin(2\pi t)) \sin(\pi \upsilon_1 t)\,\mathrm{d}t &= -(\upsilon_1/4\pi)(1 - \cos(\pi \upsilon_1/2)) s_{-1,\upsilon_1/2}(\kappa) \\
\int_0^{1/2} \cos(\kappa \sin(2\pi t)) \cos(\pi \upsilon_1 t)\,\mathrm{d}t &= -(\upsilon_1/4\pi) \sin(\pi \upsilon_1/2)) s_{-1,\upsilon_1/2}(\kappa) \\
\int_0^{1/2} \sin(\kappa \sin(2\pi t)) \cos(\pi \upsilon_1 t)\,\mathrm{d}t &= (1/2\pi)(1 + \cos(\pi \upsilon_1/2)) s_{0,\upsilon_1/2}(\kappa) \\
\int_0^{1/2} \sin(\kappa \sin(2\pi t)) \sin(\pi \upsilon_1 t)\,\mathrm{d}t &= (1/2\pi) \sin(\pi \upsilon_1/2) s_{0,\upsilon_1/2}(\kappa)
\end{aligned} \tag{39}$$

where $s_{\mu,\nu}(\kappa)$ is Lommel's function ([27], p. 985).

In order to derive a more workable expression for the second-order displacement, we assume (half) the amplitude of the first-order displacement ϵ to be small compared to the cell diameter,

$$\epsilon = \kappa \nu J_0(\kappa) \ll 1. \tag{40}$$

This is correct not only for small ν, but also for values of λ which are close to $2n - \frac{1}{2}(n \in \mathbb{N})$, where the Bessel function $J_0(\pi\lambda/2)$ has its zeros. If we expand the trigonometric expressions in the integrals up to first order in ϵ, e.g. $\sin(\pi \upsilon_1/2) \approx -\epsilon\pi\Psi_x$ etc., we find the relatively simple result

$$\mathbf{M}_2(\mathbf{x}_n) = \epsilon \begin{pmatrix} \Psi_x^2(\mathbf{x}_n) s_{0,\epsilon\Psi_x}(\kappa) - \Psi_y^2(\mathbf{x}_n) s_{0,\epsilon\Psi_y}(\kappa) \\ \Psi_x(\mathbf{x}_n)\Psi_y(\mathbf{x}_n)(s_{0,\epsilon\Psi_y}(\kappa) - s_{0,\epsilon\Psi_x}(\kappa)) \end{pmatrix}. \tag{41}$$

The second-order map is also proportional to ϵ, but has some extra 'periodicity' due to the Lommel functions. Its spatial structure is more complicated than that of the first-order map. The motion is no longer along the streamlines of the residual field, but now cross-cell boundary transport, i.e. transport in the x-direction, is possible. The second-order displacement in the y-direction is small for all parameters and for all $\mathbf{x}_n$.

The third-order map can be constructed in similar way. It is very complicated and each component consist of over 70 terms which are products of derivatives of both streamfunctions (35) and (36). All terms are proportional to ϵ^2 times a product of Lommel functions with complicated second indices. However, a second-order approximation already gives adequate estimates for the mixing coefficients, as we will see in the next subsection.

4.3. *Analytical expressions for chaotic mixing coefficients*

In this subsection we will derive analytical expressions for the chaotic mixing coefficients α and γ, using the approximate map (31) constructed in the previous subsection. Using the first- and second-order map (37) and (41) we can make estimates of the displacements under $\mathbf{T}$. Especially, we are interested in the displacements near the cell boundary, where the exchange takes place.

The exchange coefficient α is equal to the area of the turnstile lobes. We will try to find an expression for the area of one of the turnstile lobes (the lobe containing the point $\mathbf{Q} = (0, \frac{1}{2})$). We take the motion of the point $\mathbf{Q} = (0, \frac{1}{2})$ to be representative for the motion of all points in the turnstile lobe (see Fig. 7). The tidally averaged displacement $\mathbf{X}$ of the point $\mathbf{Q}$ can be approximated by

$$\mathbf{X} = \mathbf{M}_1(\mathbf{Q}) + \mathbf{M}_2(\mathbf{Q}). \tag{42}$$

This is the simplest possible approximation, which produces any nonzero mixing coefficients. Due to the addition of $\mathbf{M}_2$, a horizontal cross-boundary displacement occurs, where

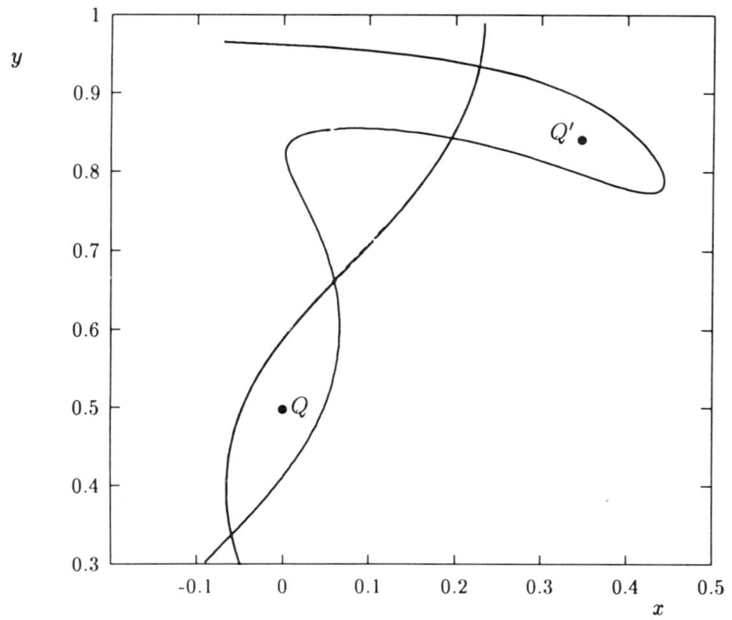

Fig. 7. The situation near the turnstile lobes. This DSTOOL-figure depicts the situation near the turnstile lobes for $v = 0.3$ and $\lambda = 2.0$. The point $Q = (0, \frac{1}{2})$ is mapped onto $Q' = (X, Y + \frac{1}{2})$. If we assume the lobes to be cosine-shaped, the displacement X is an estimate for the amplitude and Y for the wavelength of the cosine. With these estimates we can calculate the area of the lobes (45) and the area of the mixing region (46).

$\mathbf{M}_1$ produces only vertical displacement of $\mathbf{Q}$. The explicit form of $\mathbf{X}$ reads using (37) and (41)

$$\mathbf{X} = \begin{pmatrix} X \\ Y \end{pmatrix} = \begin{pmatrix} \pi\lambda v J_0(\pi\lambda/2) s_{0,\epsilon}(\pi\lambda/2)/2 \\ \pi\lambda v J_0(\pi\lambda/2) \end{pmatrix}. \tag{43}$$

Assuming that the displacement of $\mathbf{Q}$ is representative for all the points in the turnstile lobe and that the lobes are cosine shaped, the area of the turnstile lobe can be estimated by:

$$a = \frac{1}{\pi} XY, \tag{44}$$

where X is an estimate for the amplitude and Y for the wavelength of the cosine (see also Fig. 7). Using the same approach as in Section 3 for the determination of γ, we now have the following expressions for the mixing coefficients

$$\alpha = \frac{1}{\pi} XY, \tag{45}$$

$$\gamma = \frac{2}{\pi} X \left(1 + \ln \frac{1}{2X}\right). \tag{46}$$

Comparison with numerical results obtained by the same method as in Section 3.4.2, are given in Fig. 8, which shows the mixing coefficients α and γ versus λ for fixed $v = 0.1$. The minima in the numerical results are reproduced exactly by the analytical curves. The maxima show a small phase difference, but the amplitudes are rather good.

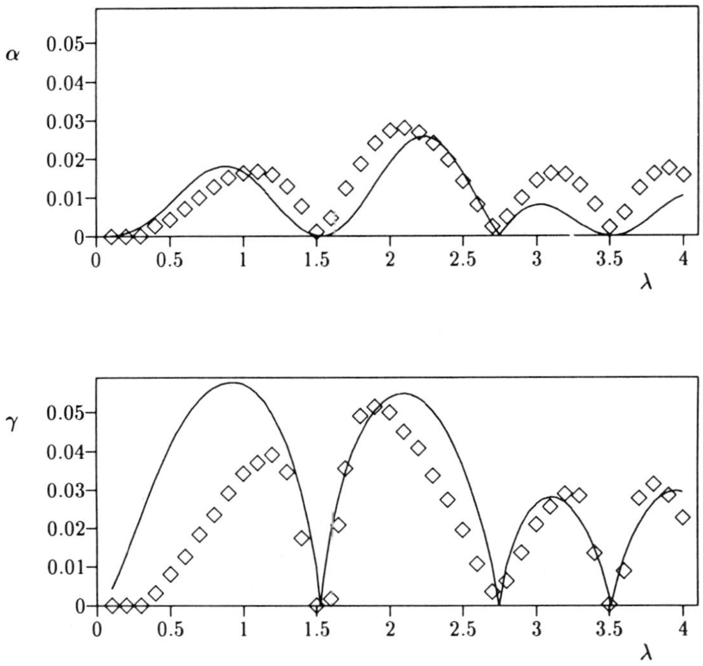

Fig. 8. The mixing coefficients α and γ vs λ for $v = 0.1$. In these figures we see the mixing coefficients α and γ versus the parameter λ in case of large perturbation ($\frac{1}{v} = 10$). The mixing depends on λ in a non-monotonic way. The analytical curves (smooth lines), described by (45) and (46), reproduce this behaviour quite well and not only for small λ.

If we also take into account the dynamical relation (12) between the parameters v and λ, which is certainly relevant in the range of λ-values representative for tidal areas ($\lambda \in [0, 4]$), we get slightly different results. This is shown in Fig. 9. Due to the decreasing λ-dependency of v, the maxima at shorter topographic wavelength become lower. The chaotic mixing is highest near $\lambda = 1$ and $\lambda = 2$.

The analytical approximation of **T** explains the minima occurring in Figs 8 and 9. The averaged tidal Lagrangian displacement (action of **T**) is proportional to the Bessel function $J_0(\pi\lambda/2)$, which is zero for $\lambda \sim 2n - \frac{1}{2}$, $n \in \mathbb{N}$. The mixing coefficients show minima not only for zeros of $J_0(\pi\lambda/2)$, but also for $\lambda \sim 2n + \frac{1}{2}$, $n \in \mathbb{N}$, where the Lommel function $s_{0,0}(\pi\lambda/2)$ has its zeros. At these parameter values the action of **T** is not zero, $J_0(\pi\lambda/2)$ is even close to its maximum, but near the turnstile the action of **T** is directed parallel to the cell boundary. In that case no exchange takes place, although the system is far from stable. These minima are not very stable and disappear at higher values of v.

Minima in the action of **T** correspond with bifurcations of the central elliptic point as shown in the Poincaré sections in Fig. 3. The second plot of Fig. 3 ($\lambda = 1.4$) gives the situation just before the first zero of $J_0(\pi\lambda/2)$, which is proportional to the action of **T** (see equation (37)). The central elliptic point bifurcates into two elliptic points and one hyperbolic point. A little further ($\lambda = 1.5$) the central elliptic point reappears, due to a bifurcation of the hyperbolic point. So near a minimum in the action of **T** bifurcations take place. At $\lambda = 2.0$ (fourth plot) there are a lot of chaotic orbits and not only inside the cell. Here a maximum in cross-boundary transport is present (γ has a maximum close to $\lambda = 2.0$), because the action of **T** in the x-direction at the cell boundary has its maximum. $\lambda = 2.7$ (fifth plot) is near a zero of the Lommel function; there we see that chaotic orbits

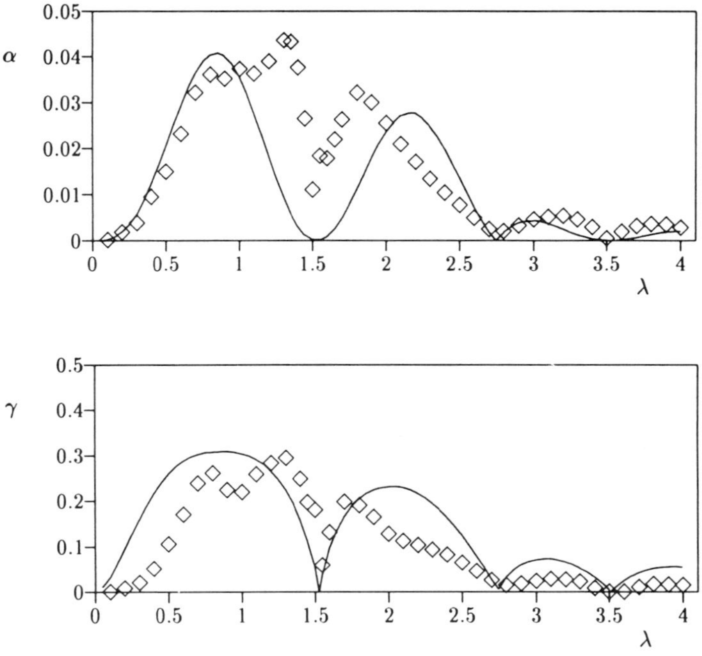

Fig. 9. The mixing coefficients α and γ vs λ for $\nu = \nu(\lambda)$. In these figures we see the mixing coefficients α and γ versus the parameter λ taking into account the dynamical relation (12) between ν and λ. Due to this relation the maxima become smaller at higher wavenumber.

inside the cell still exist, but the cross-boundary transport has almost vanished. At $\lambda = 3.0$ we see again an increase of γ (sixth plot).

For a more detailed study of chaotic mixing at realistic values of ν we need to analyse the higher-order terms of (31), which might be the subject of later studies.

5. DISCUSSION AND CONCLUSIONS

In this paper mixing in a spatially periodic flow perturbed by a time-dependent flow (1) is studied. Both in the limit of weak perturbation and in the limit of strong perturbation analytical expressions for the mixing coefficients are derived.

In the case of small perturbations Melnikov's method can be used to find characteristics of the lobe structure at the cell boundary. With this information chaotic mixing coefficients can be derived by the method of Ryrie [23]. The Melnikov method results from a first-order perturbation. Comparison with numerical results shows that the method works only for very small perturbations. A more general disadvantage of Melnikov's method is that deriving the necessary parameterization of the unperturbed manifolds is problematic in more complex flows.

In the case of strong perturbations, which is representative for tidal areas with irregular bottom topography, we can apply the method of orbit expansion, resulting in an analytical approximation of the Poincaré map. Using this approximation, displacements near the cell boundary can be studied and chaotic mixing coefficients can be estimated from analytical expressions. These analytical expressions gave similar results as numerical experiments and can explain the non-monotonic dependence of the mixing in the kinematical model on the ratio of the tidal excursion and the lengthscale of the residual eddies (λ), as found in [5].

A more realistic kinematical model for tidal areas is obtained if its parameters are determined from a dynamical model for tide–topography interactions. Such a model show that for a fixed amplitude of the tidal velocity, the magnitude of the residual velocity decreases for increasing topographic wavenumber. This results in a decreasing intensity of chaotic mixing for increasing parameter λ.

Experiments with a realistic numerical model of the Wadden Sea, a shallow tidal area in the North of the Netherlands, show that chaotic mixing occurs all over the area. Attributing a dispersion coefficient to the particle spreading gives $D \approx 100\ \mathrm{m^2\,s^{-1}}$ [5]. Characteristic values of the parameters v and λ for this area are $v \approx 0.3$ and $\lambda \approx 2$. For these parameter values our simple model gives (substituting $L_E = 10$ km and $T = 45\,000$ s) that about 30% of the area is chaotic and the dispersion coefficient is about $D = 200\ \mathrm{m^2\,s^{-1}}$. Thus, the kinematical model underestimates the area of the mixing region, whereas the spreading expressed in terms of dispersion coefficients is almost similar to the results of the realistic numerical model, taking into account that our method gives the upper limit for the dispersion coefficient.

The difference between the occurrence of global chaos in the realistic tidal model and the local chaos in our kinematical model is most presumably caused by the (over) simplifications of the current field. The kinematical model consists of only one topographic wavelength (spatial mode), whereas the realistic tidal model has a realistic topography with many modes. This will probably contribute to more chaos. Other contributions may be in temporal modes. In correspondence with the dynamical model for tide–topography interaction, the introduction of more temporal components (tidal constituents) in our kinematical model might lead to such a transition for local to global chaos. Numerical experiments with our kinematical model show that the area of the mixing region (γ) has a maximal value at $v \approx 1$ (see Fig. 10), for fixed $\lambda = 1$. (Note that Fig. 10 shows that for both limits $v \to 0$ and $v \to \infty$ the area of the region mixing goes to 0, which we also find in the analytical expressions which we derived for both limits.) A value for v larger than 0.3 is physically unrealistic in tidal areas. However, the introduction of more tidal constitutents (e.g. overtides) does increase the ratio of the generated spatially nonuniform currents and the unperturbed (spatially uniform) tidal current (8). These effects can possibly be parameterized by a higher effective value of v and therefore the introduction of more modes in the kinematical model is expected to increase the chaotic area. This will be the subject of future research.

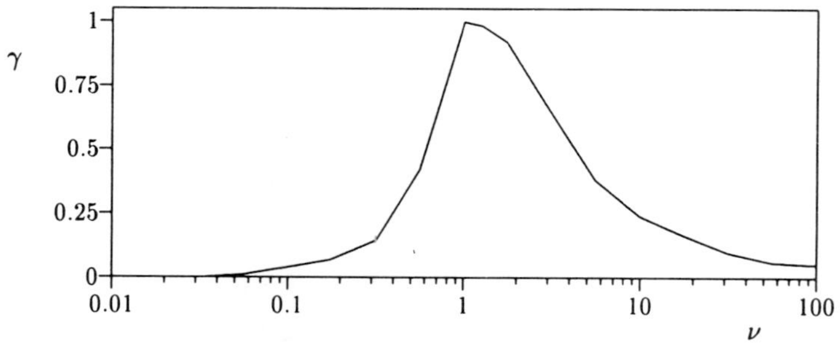

Fig. 10. The area of the mixing region γ vs v for fixed $\lambda = 1$. Results of numerical experiments on the area of the mixing region (γ) versus v on a logarithmic scale. γ has a maximum for $v \approx 1$, when both fields are of equal strength. γ disappears both in the limit $v \to 0$ (strong perturbation) and $v \to \infty$ (weak perturbation).

Acknowledgement—The authors would like to thank Dr. L. R. M. Maas for helpful comments and suggestions. One of the authors (SPB) was supported by the Netherlands Organization for the Advancement of Research (MFO/NWO) under contract 752-365-038.

REFERENCES

1. H. B. Fischer *et al.*, *Mixing in Inland and Coastal Waters*. Academic Press, New York (1979).
2. J. T. F. Zimmerman, Mixing and flushing in the Western Dutch Wadden Sea 1, *Neth. J. Sea Res.* **10**, 149–191 (1976).
3. J. T. F. Zimmerman, Mixing and flushing in the Western Dutch Wadden Sea 2, *Neth. J. Sea Res.* **10**, 397–439 (1976).
4. J. T. F. Zimmerman, The tidal whirlpool: a review of dispersion by tidal and residual currents, *Neth. J. Sea Res.* **20**, 133–154 (1986).
5. H. Ridderinkhof and J. T. F. Zimmerman, Chaotic stirring in a tidal system, *Science* **258**, 1107–1111 (1992).
6. J. T. F. Zimmerman, Dynamics, diffusion and geomorphological significance of tidal residual eddies, *Nature Lond.* **290**, 549–555 (1981).
7. H. Ridderinkhof, Tidal and residual flows in the Western Dutch Wadden Sea (I & II) *Neth. J. Sea Res.* **22**, 1–22 and 185–198 (1988).
8. R. Pasmanter, Dynamical systems, deterministic chaos and dispersion in shallow tidal flow, in *Physical Processes in Estuaries* pp. 42–52. Springer, Heidelberg (1988).
9. M. Tabor, *Chaos and Integrability in Nonlinear Dynamics: An Introduction*. Wiley, New York (1989).
10. S. Wiggins, *Introduction to Applied Nonlinear Dynamical Systems and Chaos*. Springer, New York (1980).
11. H. Aref, Stirring by chaotic advection, *J. Fluid Mech.* **143**, 1–21 (1984).
12. J. M. Ottino, *The Kinematics of Mixing: Stretching, Chaos and Transport*. Cambridge University Press, Cambridge (1989).
13. H. Ridderinkhof, Tidal and residual flows in the Western Dutch Wadden Sea (III) *Neth. J. Sea Res.* **24**, 9–26 (1989).
14. J. T. F. Zimmerman, Vorticity transfer by tidal currents over an irregular topography, *J. Mar. Res.* **38**, 601–630 (1980)
15. L. R. M. Maas, J. T. F. Zimmerman and N. M. Temme, On the exact shape of the horizontal profile of a topographically rectified tidal flow, *Geophys. Astrophys. Fluid Dynamics* **38**, 105–129 (1987).
16. J. T. F. Zimmerman, Topographic generation of residual circulation by oscillatory (tidal) currents, *Geophys. Astrophys. Fluid Dynamics* **11**, 35–47 (1978).
17. S. J. M. H. Hulscher, H. E. de Swart and H. J. de Vriend, The generation of offshore sandbanks and sand waves, *Cont. Shelf. Res.* **13**, 1183–1204 (1993).
18. S. Wiggins, *Chaotic Transport in Dynamical Systems*. Springer, New York (1992).
19. V. Rom-Kedar, Transport in two-dimensional maps, PhD thesis (1989).
20. J. Guckenheimer, M. R. Myers, F. J. Wicklin and P. A. Worfolk, *DSTOOL: A Dynamical System Toolkit with an Interactive Graphical Interface.* Center for Applied Mathematics, Cornell University (1990).
21. D. Beigie, L. Leonard and S. Wiggins, A global study of enhanced stretching and diffusion in chaotic tangles. *Phys. Fluids A* **3**(5), 1039–1050 (1990).
22. P. Welander, Studies on the general development of motion in two-dimensional ideal fluids, *Tellus* **7**, 141–156 (1955).
23. S. C. Ryrie, Mixing by chaotic advection in a class of spatially periodic flows, *J. Fluid. Mech.* **236**, 1–26 (1990).
24. V. Rom-Kedar, A. Leonard and S. Wiggins, An analytical study of transport, mixing and chaos in an unsteady vortex flow, *J. Fluid. Mech.* **214**, 347–394 (1988).
25. A. J. Lichtenberg and M. A. Lieberman, *Regular and Stochastic Motion*. Springer, New York (1983).
26. A. L. Bertozzi, Heteroclinic orbits and chaotic dynamics in planar fluid flows, *SIAM J. Math. Anal.* **19**, 1271–1294 (1988).
27. I. S. Gradshteyn and I. W. Ryzhik, *Table of Integrals, Series and Products*. Academic Press, New York (1965).

Stretching and Alignment in Chaotic and Turbulent Flows

M. TABOR and I. KLAPPER

Program in Applied Mathematics, University of Arizona, Tucson, AZ 85721, USA

Abstract—The kinematical and dynamical issues involved in stretching and alignment in chaotic and turbulent flows are examined. Dynamical systems tools, such as Lyapunov exponents, are discussed in the context of fluid mechanical problems and their connection with stretching rates in turbulent flows. Formalisms for the stretching and alignment of passive scalars, passive vectors and nonpassive vectors are developed and the interplay between kinematics and dynamics emphasized. This enables us to compare and contrast the behavior of line elements, gradients of scalar fields, vorticity and magnetic field lines.

1. INTRODUCTION

Over the past ten years the study of chaotic advection, namely the chaotic motion of particles in deterministic dynamical systems corresponding to (usually) simple fluid flows, has become a popular topic [1–4]. It provides an opportunity to bring the whole panoply of dynamical systems methods to the 'useful' physical context of mixing in fluids. In the case of two-dimensional flow fields it has the added appeal of being amenable to experimental study and, at the theoretical level, castable in the geometrically appealing Hamiltonian framework. Many of these ideas are discussed in detail in other articles in this issue.

The bulk of the mathematical tools brought to bear on this problem are, however, restricted to deterministic flow fields, and one must ask how useful chaotic advection studies are in the context of turbulent flows in which the velocity field is effectively stochastic. In other words, what ideas and insights that we have learnt from chaotic advection studies can be carried over to high Reynolds number flows? Which mathematical tools and geometrical characterizations of the particles' orbits are still useful in turbulent regimes? Our aim is to look for those descriptions of an orbit which are useful as the transition is made from deterministic dynamical systems to what are, effectively, stochastic flows governed by the Navier–Stokes equations. That we can still hope to do this follows from the fact that in the case of fluid flows the description of the fluid in terms of fluid particles, namely the Lagrangian description, is governed by ordinary differential equations. The complexity lies in the immensely complicated spatio-temporal behaviour that the velocity field develops as it becomes more turbulent. The processes that are a consequence of having a given velocity field, irrespective of its origin, is the subject matter of kinematics. The processes that govern the velocity field itself is the topic of dynamics—in the case of (turbulent) fluids this dynamics is governed by the Navier–Stokes equation. At this level it could be argued that chaotic advection (in deterministic flow fields) has little to do with fluid mechanics and is simply an example of what might be termed 'kinematic chaos' [5].

The Lagrangian description of fluids provides a very natural bridge between fluid dynamics and dynamical systems. We shall see that some of the ideas of dynamical systems theory can, with care, be carried over to turbulent flows. Furthermore, some of the concepts used in the description of stretching and alignment in turbulence may also be useful in the study of dynamical systems. These close links come about from the use of the

Lagrangian description in which the phase space that the fluid particles explore is just the three-dimensional configuration space (x, y, z) in which the fluid lives. Thus the fluid particle motion is described by a system of three first-order ordinary differential equations (odes). It is the right-hand sides of these odes, namely the dynamics, that determines if the flow is either a (simple) deterministic flow, a highly turbulent one (i.e. high Reynolds number) or some intermediate regime. Considered as a dynamical system many of the properties of the orbits, e.g. sensitivity to initial conditions, are determined by the tangent map, namely the local linearization of the flow about the orbit in question. In fluid mechanics the time derivative of the tangent map is usually referred to as the velocity gradient tensor: its symmetrized and anti-symmetrized forms being, of course, the rate of strain and vorticity tensors, respectively. It is the behavior and properties of the tangent map—at all Reynolds numbers—that will be a central theme of this article.

In comparing and contrasting the behaviour and characterization of trajectories in deterministic and stochastic flows we will address basic issues such as the rate of stretching and the degree of alignment of various types of quantities in both types of flow field. Many discussions of mixing concentrate on passive scalars, whereas here we will discuss the entire gamut of quantities ranging from passive scalars to passive vectors to nonpassive vectors. In the next section we begin by reviewing some standard ideas involved in the calculation of the largest Lyapunov exponent; namely the exponent of line stretching. This leads on to a discussion of the other exponents and, in particular, an efficient approach to calculating the rate of area stretching in three-dimensional volume-preserving flows. This latter result will be used later on in our discussion of scalar gradient stretching. The discussion of stretching in chaotic flows is followed in Section 2.3 by a comparison with the description of line stretching in turbulent flows. This introduces the key concept of the 'persistence' of a straining field, namely its rate of rotation relative to the timescales of the stretching itself. This idea is developed more fully in Section 3 and various quantities that might profitably characterize the nature of the straining field are described. The central result of this section is a presentation of an exact, kinematic, description of alignment dynamics of a passive line element in unsteady flow fields. This formalism can be extended to nonpassive quantities including vorticity. In examining the stretching and alignment of vorticity in Section 4 another key theme emerges: namely the interplay of kinematics and dynamics in turbulent flows. Although kinematics can describe the detailed dynamics in fine-scale structures, a complete description is not possible without a coupling to the dynamics (i.e. the Navier–Stokes equations) which controls the creation and destruction of these structures. Of particular significance is the nature of pressure fluctuations which, by virtue of their nonlocality, are inherently non-Lagrangian. The properties of turbulent pressure fields are very much an open topic in turbulence theory to date. Section 4 concludes with a brief description of the stretching and alignment of passive vectors, namely scalar gradients, and another nonpassive vector, namely magnetic field in the fully coupled equations of magneto-hydrodynamics.

Some of the discussion presented here is contained in an earlier review on the characterization of classical trajectories at all Reynolds numbers [6]. However, here we include some new ideas on characterizing straining fields and the stretching and alignment of vectorial quantities such as scalar gradients and magnetic fields.

2. STRETCHING IN DYNAMICAL SYSTEMS

2.1. *The equations of motion*

Since we are interested in dynamical systems corresponding to incompressible fluids we only consider volume preserving flows, i.e.

$$\nabla.\mathbf{u} = 0 \tag{1}$$

where the divergence operator is either two or three-dimensional. Here $\mathbf{u} = (u, v, w)$ denotes the three-component velocity field (vector field), and $\mathbf{x} = (x, y, z)$ the three-dimensional configuration space (phase space). In the case of tensorial quantities we use the obvious notation u_i and x_i, $i = 1, 2, 3$, for the three components of velocity and space, respectively.

A major difference between the two and three-dimensional cases is that the dynamics of fluid particles in two-dimensional flows is Hamiltonian whereas the three-dimensional case is not. In the former case, a stream function $\psi = \psi(x, y, t)$ can be introduced such that the two velocity components are given by the standard relations $u = \partial\psi/\partial y$ and $v = -\partial\psi/\partial x$. The stream function plays the role of a Hamiltonian [7] and the fluid particle motions are governed by the dynamical system

$$\dot{x} = \frac{\partial\psi}{\partial y}, \qquad \dot{y} = -\frac{\partial\psi}{\partial x} \tag{2}$$

in which x and y play the role of canonical variables. For steady flows the Hamiltonian is autonomous i.e. $\psi = \psi(x, y)$, leading to the well-known result that the flow is integrable and all the trajectories lie on smooth curves (KAM tori in the language of Hamiltonian dynamics and streamlines in the fluid dynamical language). For unsteady flows i.e. explicitly time-dependent ψ, the motion is typically nonintegrable, and the particle dynamics is generically chaotic. As described in other articles in this volume, chaotic advection phenomena can be analyzed and predicted in detail by drawing on the extensive understanding of Hamiltonian dynamics.

In three dimensions the fluid particle trajectories are governed by the non-Hamiltonian dynamics

$$\begin{aligned} \dot{x} &= u(x, y, z, t) \\ \dot{y} &= v(x, y, z, t) \\ \dot{z} &= w(x, y, z, t). \end{aligned} \tag{3}$$

Unlike the two-dimensional case chaotic motion is still possible for stationary flows i.e. when the r.h.s. of (3) is time-independent.

The velocity gradient tensor, $\mathbf{A}$, of the flow (3) is

$$A_{ij} = \frac{\partial u_i}{\partial x_j}. \tag{4}$$

In the dynamical systems context it is almost exclusively studied in the Lagrangian frame, i.e. along particle orbits, whereas in the fluid dynamics it is studied in both the Lagrangian and Eulerian frame. Furthermore, in fluid dynamics more significance is attached to the symmetrized and anti-symmetrized forms of $\mathbf{A}$. The symmetrized form is the rate-of-strain tensor $\mathbf{S}$

$$S_{ij} = \frac{1}{2}\left(\frac{\partial u_i}{\partial x_j} + \frac{\partial u_j}{\partial x_i}\right). \tag{5}$$

S has three real eigenvalues, s_i, $i = 1, 2, 3$, which for an incompressible flow, must sum up to zero i.e.

$$s_1 + s_2 + s_3 = 0.$$

This means that $s_1 \geqslant 0$ and $s_3 \leqslant 0$. The intermediate eigenvalue s_2 can have either sign—a fact that proves to be very significant in stretching and alignment issues in turbulence. The anti-symmetrized form of $\mathbf{A}$ is the vorticity tensor $\mathbf{\Omega}$,

$$\Omega_{ij} = \frac{1}{2}\left(\frac{\partial u_i}{\partial x_j} - \frac{\partial u_j}{\partial x_i}\right) \tag{6}$$

which has two, complex-conjugate, pure imaginary eigenvalues and one zero eigenvalue whose eigenvector is the vorticity vector $\omega_k = -\frac{1}{2}\varepsilon_{ijk}\Omega_{ij}$. (We note that this definition of vorticity is one half the standard definition $\omega = \nabla \wedge \mathbf{u}$ and, strictly speaking, should be identified as the angular velocity [8]. However, for internal consistency we will, unless otherwise stated, use the above form as our definition of vorticity.)

In dynamical systems studies these symmetric and antisymmetric forms of **A** are generally not considered separately and the notion of $\boldsymbol{\Omega}$—so central to fluid dynamics—being an important quantity, is rarely considered. For the most part all antisymmetric information, which corresponds to the imaginary parts of the eigenvalues of **A**, is ignored.

Linearization of the flow field (3) (or its n-dimensional generalization $\dot{x}_i = u_i(x_1, \ldots, x_n)$), gives the standard tangent mapping

$$\frac{\mathrm{d}\delta\mathbf{x}}{\mathrm{d}t} = \mathbf{A}.\delta\mathbf{x}, \tag{7}$$

where $\delta\mathbf{x} = (\delta x_1, \ldots, \delta x_n)$ and **A** is evaluated along the particular orbit about which the flow is linearized. Formally, the fundamental matrix solution of equation (7) from, say, time zero to time t is the tangent map associated with the orbit with initial condition $\mathbf{x}_0 = \mathbf{x}(t = 0)$, namely

$$\mathbf{M}(t, 0; \mathbf{x}_0) = T\colon \exp\left(\int_0^t \mathbf{A}(\mathbf{x}(t'))\,\mathrm{d}t'\right), \qquad \mathbf{M}(0, 0; \mathbf{x}_0) = \mathbf{I} \tag{8}$$

where T: denotes the time-ordered product and **I** is the identity matrix. In the context of fluid mechanics **M** is often called the Jacobian matrix.

A canonical model of a stationary three-dimensional flow field is the Arnold–Beltrami–Childress (ABC) system

$$\begin{aligned} u &= A \sin z + C \cos y \\ v &= B \sin x + A \cos z \\ w &= C \sin y + B \cos x, \end{aligned} \tag{9}$$

which has, among other contexts, proved useful in the study of the kinematic dynamo. The flow (9) is a solution to the (time-independent) Euler equations

$$(\mathbf{u}.\nabla).\mathbf{u} = -\frac{1}{\rho}\nabla p, \tag{10}$$

where p is the pressure and ρ the fluid density with $\nabla.\rho = 0$. (As a solution to the Euler equations (9) might be thought of as a fully 'turbulent', i.e. $Re = \infty$, flow but since it resolves just one spatial scale it should not be considered as such!) The ABC system has the special property of being a Beltrami flow, i.e. a flow for which the vorticity is parallel to the velocity, that is $\boldsymbol{\omega} = \lambda\mathbf{u}$, where $\boldsymbol{\omega} = \nabla \wedge \mathbf{u}$ is the vorticity vector. A detailed study of (9) by Dombre *et al*. [9] reveals an expected generic mixture of regular and chaotic orbits and a complicated stable/unstable fixed point structure. We also note that any positive measure region of a steady flow satisfying Euler's equation (10) and containing a dense streamline has the Beltrami property [10]. This can be seen by writing (10) in the form

$$\mathbf{u} \times \boldsymbol{\omega} = -\frac{1}{\rho}\nabla(p + \tfrac{1}{2}|\mathbf{u}|^2).$$

Dotting both sides with **u** yields $\mathbf{u}.\nabla(p + \frac{1}{2}|\mathbf{u}|^2) = 0$ which implies that the quantity $p + \frac{1}{2}|\mathbf{u}|^2$

is constant along streamlines. The presence of a dense streamline (which can be taken to correspond to a chaotic one) implies that $p + \frac{1}{2}|\mathbf{u}|^2$ is constant throughout the volume and hence $\mathbf{u} \times \boldsymbol{\omega} = 0$, i.e. the region has the Beltrami property.

2.2. *Lyapunov exponents for dynamical systems*

Although the calculation of Lyapunov exponents, especially the largest one, is standard knowledge, we recall it here to (i) highlight the many interesting open questions that still remain and (ii) to set the stage for a comparison with similar-looking quantities used to quantify line stretching in turbulent flows.

Working with the tangent mapping (7) one can introduce the norm

$$d(t) = \sqrt{\left\{\sum_i [\delta x_i(t)]^2\right\}} \tag{11}$$

and then define the mean rate of divergence, σ, of two initial points separated by a distance $d(0)$, namely

$$\sigma = \lim_{t\to\infty} \frac{1}{t} \ln \frac{d(t)}{d(0)}. \tag{12}$$

This is the largest Lyapunov exponent, $\sigma_1(\mathbf{x}_0)$, which, for typical points, is dependent on the initial condition $\mathbf{x}_0$. A positive Lyapunov exponent implies great sensitivity to initial conditions and is an accepted hallmark of a chaotic orbit. It should, of course, be emphasized that there is no guarantee that the limit $t \to \infty$ in (12) exists for all $\mathbf{x}_0$. When the limit does exist this suggests that there is also, remarkably, some asymptotic direction, termed the Lyapunov (eigen)vector, along which the stretching occurs. It is also possible to calculate relatively short-time averages in (12): such quantities can give some indication of local instabilities but do not guarantee actual chaos.

A direct computation of the time average (12) is not the most practical way of computing σ_1 since exponential growth of d(t) will lead to overflow. As is well known, the standard way to proceed is to compute the separation for some small time interval, τ, and then rescale the separation and start again—this is illustrated in Fig. 1(a). The separation after the jth time step is defined as

$$d_j = |\delta\mathbf{x}^{(j-1)}(\tau)|,$$

where

$$\delta\mathbf{x}^j(0) = \frac{\delta\mathbf{x}^{(j-1)}(\tau)}{d_j}.$$

One then defines the 'finite time' quantity $\sigma^{(N)}$,

$$\sigma^{(N)} = \frac{1}{N\tau}\sum_{j=1}^{N} \ln(d_j) \tag{13}$$

and claims that

$$\sigma_1 = \lim_{N\to\infty} \sigma^{(N)}. \tag{14}$$

The quantity (13) represents a time average over the d_js, but each d_j in the sum is precisely correlated with its preceding one. Below we will describe a quantity representing a mean stretching rate in turbulence which looks like (13) but, as will be explained, the similarity is superficial.

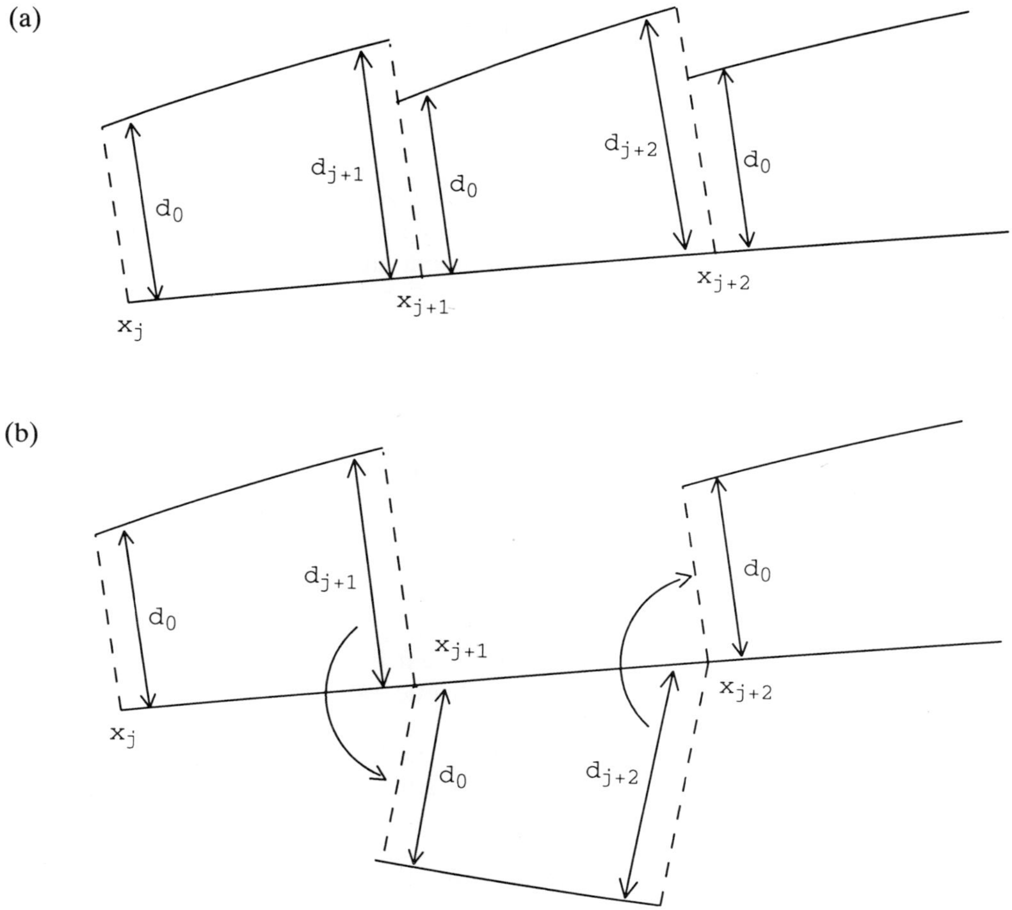

Fig. 1. (a) Computation of largest Lyapunov exponent: after each period τ the distance from the reference orbit is rescaled back to d_0. (b) Model of computation of largest Lyapunov exponent in a stochastic flow field: after each time step the separation vector is rescaled and randomly reoriented.

For a n-dimensional system there are, in fact, a spectrum of Lyapunov exponents: $\sigma_1, \ldots, \sigma_n$—each with an associated eigen-direction ξ_i. Briefly, for every initial condition $\mathbf{x}_0$ (belonging to a set of measure 1—this set is sometimes referred to as the set of regular points) there exists an orthonormal set of vectors $\xi_i(t)$ such that

$$\sigma_i = \lim_{t\to\infty} \frac{1}{t} \ln \|\mathbf{M}(t, 0; \mathbf{x}_0)\xi_i(t)\| \tag{15}$$

exists, where $\mathbf{M}$ is the tangent map (8). By noting that

$$\|\mathbf{M}(t, 0; \mathbf{x}_0)\xi_i(t)\|^2 = \xi_i(t)^\dagger \mathbf{M}^\dagger(t, 0; \mathbf{x}_0)\mathbf{M}(t, 0; \mathbf{x}_0)\xi_i(t) \tag{16}$$

we see that $\exp(\sigma_i^2 t)$ and ξ_i are the eigenvalues and eigenvectors, respectively, of $\lim_{t\to\infty} \mathbf{M}^\dagger\mathbf{M}$. Since this is a real symmetric matrix (hence real eigenvalues and orthogonal eigenvectors) one can write

$$\mathbf{M}^\dagger\mathbf{M} = \mathbf{Q}^2, \tag{17}$$

where **Q** is another real symmetric matrix. Therefore we can write

$$\mathbf{M} = \mathbf{RQ} \tag{18}$$

where **R** is a real unitary matrix [11]. Thus we note that the 'rotational' information contained in **R** is lost in the calculation of the eigenvalues and eigenvectors of $\mathbf{M}^{\dagger}\mathbf{M}$. Some preliminary studies of this rotational information is given in our earlier review.

Calculating the full spectrum of exponents, rather than just the largest one, is a computational challenge: the basic method being due to Benettin *et al*. [12]. For a volume-preserving flow one must, of course, have

$$\sum_{j=1}^{N} \sigma_j = 0. \tag{19}$$

Now, for three-dimensional incompressible flows it is possible to extract the second Lyapunov exponent efficiently from a calculation of the growth of an infinitesimal *area* element and subsequently, by using (19), the third exponent as well. Since this calculation contains instructive differences from the calculation of the first, line-stretching, exponent and will also be used in our later discussion of scalar gradient stretching (Section 4.2), we describe it here. Denoting an infinitesimal area element by $\delta\mathbf{y} \times \delta\mathbf{z}$, it follows, by volume conservation, that

$$0 = \frac{\mathrm{d}}{\mathrm{d}t}[\delta\mathbf{x}.(\delta\mathbf{y} \times \delta\mathbf{z})] = \left(\frac{\mathrm{d}}{\mathrm{d}t}\delta\mathbf{x}\right).(\delta\mathbf{y} \times \delta\mathbf{z}) + \delta\mathbf{x}.\frac{\mathrm{d}}{\mathrm{d}t}(\delta\mathbf{y} \times \delta\mathbf{z}). \tag{20}$$

Using (7) we can thus write

$$\delta\mathbf{x}.\frac{\mathrm{d}}{\mathrm{d}t}(\delta\mathbf{y} \times \delta\mathbf{z}) = -(\mathbf{A}.\delta\mathbf{x}).(\delta\mathbf{y} \times \delta\mathbf{z}) = -\delta\mathbf{x}.(\mathbf{A}^T.(\delta\mathbf{y} \times \delta\mathbf{z})), \tag{21}$$

and hence, by the arbitrariness of $\delta\mathbf{x}$,

$$\frac{\mathrm{d}}{\mathrm{d}t}(\delta\mathbf{y} \times \delta\mathbf{z}) = -\mathbf{A}^T.(\delta\mathbf{y} \times \delta\mathbf{z}). \tag{22}$$

Integration of this equation requires the inverse of the transpose of the propagator **M** defined in (8), which we shall denote as $\mathbf{M}^{-T}$. The asymptotic growth rate of a typical $(\delta\mathbf{y} \times \delta\mathbf{z})$ is the largest eigenvalue of $\lim_{t\to\infty} \mathbf{M}^{-T}$, which is the largest eigenvalue of $\lim_{t\to\infty} \mathbf{M}^{-1}$. Denoting this eigenvalue as λ', we note that it is the inverse of the smallest eigenvalue, λ_3, of $\lim_{t\to\infty} \mathbf{M}$. Thus, using (19),

$$\ln \lambda' = -\ln \lambda_3 = \ln \lambda_1 + \ln \lambda_2 = \sigma_1 + \sigma_2. \tag{23}$$

Before proceeding to our discussion of stretching rates in turbulent flows it is worth pausing to ask the following questions.

(i) In the definition of exponents does the limit $t \to \infty$ exist? This question has been answered by Oseledec [13] who proves that there are initial conditions—the so called regular points—for which the limit does indeed exist. For a volume-preserving flow these regular points have Lesbegue measure 1. In an ergodic region (with respect to Lebesgue measure) the Lyapunov exponents are constant except on set of measure zero.

(ii) How well does the limit $t \to \infty$ (or in practice $N \to \infty$) converge? Little is known about this—often $\sigma^{(N)}$ is observed to be a highly oscillatory function of N. Whether these oscillations contain any useful information is an interesting question.

(iii) What is the meaning of the eigenvectors ξ_i? Although they must say something about the local geometry of the set on which the orbit in question lies, what this something is is not understood. A better understanding of this issue would be useful for the dynamo problem, where the (asymptotic) direction in which magnetic field lines point is of fundamental importance.

2.3. *Stretching rates for turbulent flows*

The study of the stretching of line and surface elements in turbulent flows has a long history (see, for example, Monin and Yaglom [14]) that has tended to proceed independently of advances in the understanding of dynamical systems. The first major treatment was given by Batchelor [15] who made two fundamental assumptions. The first was that a finite material element can be considered as a collection of infinitesimal elements since in a turbulent flow the stretching of elements separated by more than a few Kolomogorov length scales (i.e. the smallest scale in the turbulent cascade) should be completely decorrelated. As long as one is averaging over many different material elements this appears to be a reasonable assumption.

However, because of the non-uniform stretching of different infinitesimal elements, a small measure of such elements can (and, in fact, generally do) dominate the growth of a finite element. In other words, the growth rate of a *finite* element is in general larger than that of an *infinitesimal* one. This idea is nicely illustrated by the simple but instructive dynamical systems example of Farmer *et al.* [16]. Consider the non-uniform Baker's map on the unit square:

$$\left.\begin{aligned} x_{n+1} &= \alpha x_n \\ y_{n+1} &= y_n/\alpha \end{aligned}\right\} \text{ if } y_n < \alpha$$

$$\left.\begin{aligned} x_{n+1} &= (1-\alpha)x_n + \alpha \\ y_{n+1} &= (y_n - \alpha)/(1-\alpha) \end{aligned}\right\} \text{ if } y_n > \alpha$$

where $0 < \alpha < 0$. The map breaks the (unit) square into two pieces of dimension $1 \times \alpha$ and $1 \times (1-\alpha)$, stretches them by factors of α^{-1} and $(1-\alpha)^{-1}$, respectively, and then pastes them together back into a square (see Fig. 2). Using this construction we can now compare the growth rates of typical infinitesimal and finite line elements. In the limit of the number of iterations, N, of the mapping becoming large a typical point visits the region $y < \alpha$ $N\alpha$ times whereas it visits the region $y > \alpha$ $N(1-\alpha)$ times. (Here typical means a measure 1 set of points.) Thus a typical infinitesimal vector at this typical point is stretched by a factor $(1/\alpha)^{\alpha N}$. $(1/1-\alpha)(1-\alpha)N$. Thus the Lyapunov exponent for the map is

$$\sigma = \alpha \ln(1/\alpha) + (1-\alpha)\ln(1/1-\alpha).$$

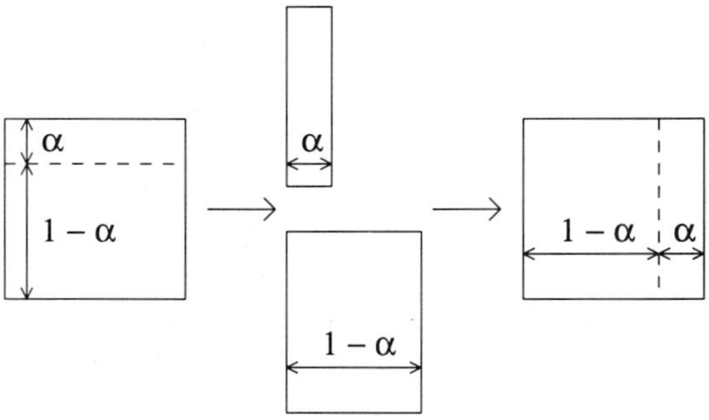

Fig. 2. Action of the non-uniform Baker's map.

Now, by contrast, consider the fate of a typical finite line segment. For simplicity assume that the line segment is a vertical line extending from the top to the bottom of the square. It easy to see that the length of this line element is doubled at each iteration and thus has the growth rate ln (2). Plotting σ as a function of α and comparing it with ln (2) (see Fig. 3) we see that the latter dominates the former except at $\alpha = 1/2$ where the growth rates of the infinitesimal and finite line elements both have the same value of ln (2). We note that when $\alpha = 1/2$, stretching is exactly uniform throughout the square. Hence, in general, a finite line element grows exponentially faster than an infinitesimal one.

The second assumption of Batchelor was that the infinitesimal elements $\mathbf{l}$ do, in fact, stretch exponentially. On the basis of this a *local* stretching rate was introduced

$$\bar{\xi} = \overline{\frac{1}{|\mathbf{l}|}\frac{\mathrm{d}|\mathbf{l}|}{\mathrm{d}t}} = \overline{\frac{\mathrm{d}}{\mathrm{d}t}\ln|\mathbf{l}(t)|}, \tag{24}$$

where the overbar represents an average over different realizations of the (turbulent) velocity field. In particular it was claimed that

$$\lim_{t\to\infty} \bar{\xi} > 0. \tag{25}$$

At first sight this looks like the description of the largest Lyapunov exponent given in the previous section with (24) and (25) being analogous to (13) and (14), respectively. However, in (24) the average is over different realizations of the flow field rather than along a single trajectory (this would be like removing the correlations between the successive d_js in (13)—a point on which we will elaborate below) and, furthermore, there is no analogue of Oseledec's theorem that guarantees that the limit $t \to \infty$ in (25) exists. Thus the idea of a stretching rate in turbulence being, somehow, the equivalent of a Lyapunov exponent has to be treated with caution.

At the time of Batchelor's investigations little was known about the validity of (25) and it was not until much later that Cocke [17, 18] demonstrated for Gaussian random fields which, to a first approximation, provide a model of turbulent flows, that:

(i) $\ln|\mathbf{l}(T)| > \ln|\mathbf{l}(0)|$ for $T > 0$ i.e. two points do separate on average for a finite time but that

(ii) $\lim_{t\to\infty} \bar{\xi} > 0$ i.e. there is, in fact, no long-time exponential separation.

A simplified version of Cocke's proof is given by Orzsag [19].

Although we must now bear in mind that $\bar{\xi}$ is only a finite time quantity it is still a useful

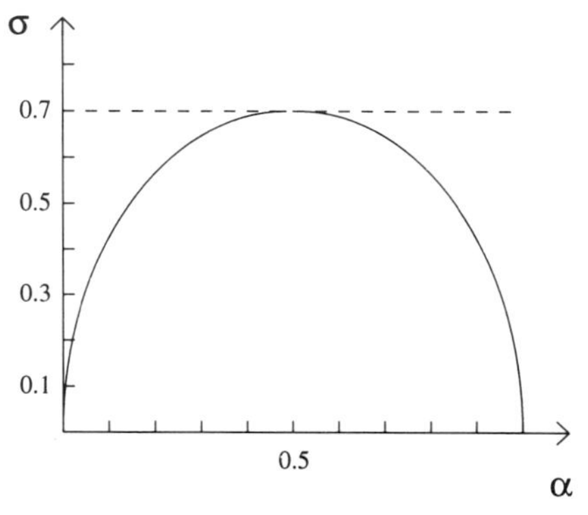

Fig. 3. Plot of $\sigma = \alpha \ln(1/\alpha) + (1-\alpha)\ln(1/1-\alpha)$ vs α. The horizontal line corresponds to $\sigma = \ln(2)$. The two curves meet at $\alpha = 1/2$.

one to study. The stretching of an infinitesimal line element, $\mathbf{l}$, moving along a particular Lagrangian path, is governed by the equation

$$\frac{d\mathbf{l}}{dt} = \mathbf{A}.\mathbf{l} \tag{26}$$

which, on decomposing $\mathbf{A}$ into its symmetric and anti symmetric parts, can also be written as

$$\frac{d\mathbf{l}}{dt} = \mathbf{S}\mathbf{l} + \boldsymbol{\omega} \wedge \mathbf{l}, \tag{27}$$

where $\mathbf{S}$ is the rate of strain and $\boldsymbol{\omega}$ the vorticity vector. Using (24) the local stretching rate ζ can be expressed as

$$\zeta = \sum_{ij} S_{ij} \hat{l}_i \hat{l}_j \quad i, j = 1, 2, 3 \tag{28}$$

where the $\hat{l}_i$ are the unit vector components associated with $\mathbf{l}$. The natural question to ask is for an estimate of $\bar{\zeta}$ in turbulent flows. Here Batchelor and Townsend [20] suggested that

$$\bar{\zeta} = \bar{s}_1, \tag{29}$$

i.e. the (ensemble) mean of the largest eigenvalue of the rate-of-strain tensor.

Before discussing the significance of this estimate we explore the relationship between $\bar{\zeta}$ and Lyapunov exponents and ask under what circumstances the calculation of the largest exponent, i.e. the long-time average of the stretching rate of an infinitesimal line element, could yield a number related to the average (eigenvalue) of the rate-of-strain tensor. This connection might seem difficult to make, since the former is the average of a nonlocal quantity (the tangent map) whereas the latter contains only local information. Recalling our earlier discussion of the largest exponent we write

$$\sigma = \lim_{N\to\infty} \frac{1}{N} \sum_{n=1}^{N} \ln(d_n), \tag{30}$$

where

$$d_n = |\mathbf{l}_n| / |\mathbf{l}_{n-1}| \tag{31}$$

is the ratio of the lengths of the (infinitesimal) line element $\mathbf{l}$ at discrete time steps n and $n-1$. The time average in (30) is the average of the d_ns along the particular orbit that $\mathbf{l}$ is following. We consider the tangent mapping (26) in discrete form and write

$$\frac{\mathbf{l}_n - \mathbf{l}_{n-1}}{\Delta t} = \mathbf{A}_{n-1}.\mathbf{l}_{n-1}, \tag{32}$$

and further decompose it into its symmetric and antisymmetric parts, namely

$$\mathbf{l}_n = (\mathbf{S}_{n-1}.\mathbf{l}_{n-1} + \boldsymbol{\omega}_{n-1} \wedge \mathbf{l}_{n-1})\Delta t + \mathbf{l}_{n-1}. \tag{33}$$

Forming the inner product $\mathbf{l}_n.\mathbf{l}_n$ gives

$$\mathbf{l}_n.\mathbf{l}_n = ((\mathbf{l}_n, \mathbf{S}_{n-1}.\mathbf{l}_{n-1}) - \boldsymbol{\omega}_{n-1}.(\mathbf{l}_n \wedge \mathbf{l}_{n-1}))\Delta t + \mathbf{l}_n.\mathbf{l}_{n-1}, \tag{34}$$

which can be rewritten as

$$|\mathbf{l}_n|^2 = (|\mathbf{l}_n||\mathbf{l}_{n-1}|(\hat{l}_n, \mathbf{S}_{n-1}\hat{l}_{n-1})) - |\omega_{n-1}||\mathbf{l}_n||\mathbf{l}_{n-1}| \cos\theta_n \sin\varphi_n)\Delta t + |\mathbf{l}_n||\mathbf{l}_{n-1}| \cos\varphi_n, \tag{35}$$

where $\hat{l}_n$ denotes unit vector, φ_n is the angle between $\mathbf{l}_n$ and $\mathbf{l}_{n-1}$ and θ_n is the angle between $\boldsymbol{\omega}_{n-1}$ and the vector $(\mathbf{l}_n \wedge \mathbf{l}_{n-1})$. Dividing both sides of (35) by $|\mathbf{l}_n||\mathbf{l}_{n-1}|$ gives

$$d_n = |\mathbf{l}_n/\mathbf{l}_{n-1}| = ((\hat{l}_n, \mathbf{S}_{n-1}.\hat{l}_{n-1}) - |\boldsymbol{\omega}_{n-1}| \cos\theta_n \sin\varphi_n)\Delta t + \cos\varphi_n. \tag{36}$$

Thus the stretching increment d_n can be seen to be made up of a part, $\zeta_n = (\hat{l}_n, \mathbf{S}_{n-1}.\hat{l}_{n-1})$, which is determined solely by the rate-of-strain tensor and a part, $|\boldsymbol{\omega}_{n-1}| \cos\theta_n \sin\varphi_n$, which is determined by the vorticity. The angle φ_n is the angle between the successive elements $\mathbf{l}_n$ and $\mathbf{l}_{n-1}$. If at each step we were to choose their relative orientations to be random (see Fig. 1(b)), the ensemble average of (36) would be

$$d_n = \langle \zeta_n \rangle_\varphi, \tag{37}$$

which, inserted into the limit (30) would yield a 'Lyapunov exponent', which would be some (averaged) function of the eigenvalues of **S**.

Returning to the estimate (29) we note that it implicity assumes that the straining field is 'persistent', namely that the combined effect of the vorticity and the rotation of the principle axes of the straining field is sufficiently small such that the material element has time to align with the strain axis with largest eigenvalue and hence experience maximum stretching. This idea was consistent with an earlier experiment of Townsend [21] on the decay of heat spots in grid turbulence. (He found that these spots experienced maximum cooling—which was consistent with his theoretical model in which the strain axes were stationary, i.e. the strain was persistent.) Batchelor and Townsend gave the estimate

$$\bar{s}_1 = 0.43/\tau_k,$$

where τ_k is the Kolmogorov turnover time given by $(\nu/\varepsilon)^{1/2}$, where ε is the rate of energy dissipation and ν the kinematic viscosity.

Pursuing, for a moment, the idea of estimates for stretching rates in turbulence it is clear that two, finitely close, particles will only separate significantly over regions of space for which the associated *velocity gradient* (i.e. straining field) is well correlated. The largest scale over which this can occur in (fully developed) turbulent flow is, almost certainly, the Kolmogorov microscale—this is the idea underlying the Batchelor–Townsend estimate. As a function of Reynolds number, *Re*, the microscale scales as $Re^{-3/4}$. Of course, as this scale becomes smaller, the associated gradient increases. This gradient (the microscale frequency) scales as $Re^{1/2}$. Considering that a mean stretching rate is a long-time average of particle separation the competition between these two effects as a function of *Re* is clearly very complicated. Just because a velocity field becomes more turbulent does not necessarily mean that the overall stretching (of finitely separated particles) will increase.

For many years the idea that the strain is persistent was firmly entrenched in the lore of fluid dynamics (see e.g. [14]). Some fairly recent pieces of evidence have now challenged this idea.

(i) A study of vorticity alignment by Ashurst *et al.* [22] in large-scale numerical simulations of the Navier–Stokes equations showed that the vorticity tended to align with the intermediate strain axis rather than the direction with the largest eigenvalue. In ideal fluids the stretching of vorticity (discussed in detail later) looks formally similar to that of line stretching, and one would therefore expect vorticity alignment with the axis with largest eigenvalue if the strain were persistent. Other numerical experiments all now seem to confirm the observations of Ashurst *et al.* We will eventually develop an alignment theory that explains these observations.

(ii) Girimaji and Pope [23] studied Lagrangian orbits and line stretching in spectral code simulations of homogeneous isotropic turbulence at Taylor scale Reynolds numbers, Re_λ, of up to 90, and found that the mean strain rate on a material line to be about one-third of the Batchelor and Townsend result. We also note their finding that the average direction cosine between the (infinitesimal) line element(s) and the strain axis with largest eigenvalue was around 0.9 (for persistent strain it should be 1.0). These results, coupled with other numerical tests of various timescales, strongly indicated that the strain was not persistent.

(iii) An exact analytical solution to the Euler equations due to Vieillefosse [24, 25] showed that the vorticity actually blows up in finite time in the direction of the intermediate strain axis, i.e. the axis associated with the eigenvalue s_2. Although his model makes some strong, and probably unrealistic, assumptions about the pressure field, it is very suggestive. A similar analysis has also been carried out more recently by Majda [26]. More general explanations for the stretching of vorticity along the intermediate direction have been given by Dresselhaus and Tabor [27] and She *et al*. [28].

3. ALIGNMENT AND THE CHARACTERIZATION OF STRAINING FIELDS

3.1. *Persistence of strain and alignment*

As mentioned in Section 2.2, the calculation of Lyapunov exponents discards fundamental information about rotation. However, certain, *instantaneous*, rotational information can be obtained from the dynamics in a straightforward way. The value of this information is that it can provide an estimate of the relative strengths of instantaneous 'stretching' and 'rotation'; domination of the former over the latter implying that stronger local stretching can take place. Consider, first, the stretching of a line element $\mathbf{l}$ in a 2D flow by a constant (in time) flow gradient $\nabla \mathbf{u} = \mathbf{A} = \mathbf{S} + \mathbf{\Omega}$ and work in the coordinate frame in which $\mathbf{S}$ is diagonal with eigenvalues s_x and s_y (where we assume $s_x \geq 0$ and $s_x + s_y = 0$) and corresponding eigenvectors ξ_x and ξ_y. The vorticity $\boldsymbol{\omega} = \nabla \times \mathbf{u}$ points out of the plane and is taken to rotate anti clockwise. Imagine, now, a typical situation of $\mathbf{l}$ (in the first quadrant) with components (l_x, l_y) making an angle θ with the ξ_x axis (see Fig. 4). The vorticity will try to rotate $\mathbf{l}$ anticlockwise whereas the straining action will try to rotate it clockwise by dragging it towards the stretching axis ξ_x. We now determine the angle θ at which these two competing effects balance thereby maintaining $\mathbf{l}$ in a steady direction in which it is stretched. In a time step Δt, $\mathbf{l}$ will have moved in the plane by the amounts

$$l_x(t + \Delta t) = l_x(t) + (s_x l_x(t) - \omega l_y(t))\Delta t$$
$$l_y(t + \Delta t) = l_y(t) + (s_y l_y(t) + \omega l_x(t))\Delta t.$$

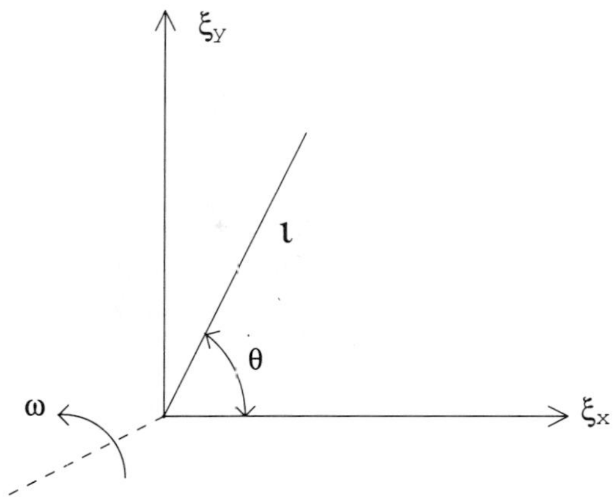

Fig. 4. Line element $\mathbf{l}$ lying in the plane of the eigenvectors of the two dimensional (diagonalized) rate-of-strain tensor $\mathbf{S}$.

If the angle θ is to remain constant after Δt we must have

$$\tan\theta = \frac{l_y}{l_x} = \frac{l_y(t+\Delta t)}{l_x(t+\Delta t)} = \frac{l_y}{l_x}\left(\frac{1+(s_y+\omega l_x/l_y)\Delta t}{1+(s_x-\omega l_y/l_x)\Delta t}\right), \tag{38}$$

where l_x denotes $l_x(t)$ and l_y denotes $l_y(t)$. This implies that

$$1 = \frac{1+(s_y+\omega l_x/l_y)\Delta t}{1+(s_x-\omega l_y/l_x)\Delta t}, \tag{39}$$

and hence

$$s_y - s_x = -\omega\left(\frac{l_y}{l_x}+\frac{l_x}{l_y}\right) = -\frac{2\omega}{\sin 2\theta}. \tag{40}$$

Treating this as an equation for θ it will have a solution $\theta = \sin^{-1}(2\omega/s_x - s_y)$ only if

$$\frac{4\omega^2}{(s_y - s_x)^2} \leqslant 1$$

or

$$(s_y - s_x)^2 - 4\omega^2 \geqslant 0. \tag{41}$$

Note that this condition involves the difference (squared) of the eigenvalues of **S** which, for the area-preserving case ($s_y = -s_x = -s$) reduces to

$$s^2 - \omega^2 \geqslant 0. \tag{42}$$

We note that the solution of (40) (if it exists) also gives the directions of the eigenvectors of **A**—which has eigenvalues $\pm(s^2-\omega^2)^{1/2}$. The angle between these eigenvectors may be shown to be $\pi/2 - 2\theta$.

Now, if the axes of **S** are rotating with the (steady) rate ω' the same argument given above carries through with ω simply replaced by $(\omega - \omega')$ (i.e. the net rotation in the rate-of-strain frame) and leads to the modified condition

$$s^2 - (\omega - \omega')^2 \geqslant 0. \tag{43}$$

For unsteady flows in which s_x, s_y, ω, ω' are all fluctuating in time, the analysis is considerably more complicated and will be considered below. However, at a more qualitative level we might well expect that if the (now fluctuating) quantity $s^2 - (\omega - \omega')^2$ is persistently greater than zero there will be stretching and if it is persistently less than zero there will be rotation.

Denoting the l.h.s. of (42) by μ^2, we can, in fact write

$$\mu^2 = s^2 - \omega^2 = \frac{1}{2}\,\text{Trace}\,\mathbf{A}^2, \tag{44}$$

which is easily computed and, furthermore, for the 2D steady (i.e. **S** fixed in time) case can easily be shown to be the Gaussian curvature of the associated streamlines [29]. Clearly if $\mu^2 < 0$ the motion is rotation dominated whereas for $\mu^2 > 0$ the motion is strain dominated with a tendency for local stretching. More generally though, the criterion for local strain or rotation domination must be determined from the quantity

$$\mu^2 = s^2 - (\omega - \omega')^2 \tag{45}$$

for which, unfortunately, the simple computation in terms of Trace $\mathbf{A}^2$ is no longer possible. We note, however, that the simpler quantity (44) has a nice fluid-mechanical interpretation since taking the divergence of the Navier–Stokes equation gives

$$\nabla.(\mathbf{u}.\nabla)\mathbf{u} = -\frac{1}{\rho}\nabla^2 p = \text{Trace } \mathbf{A}^2 = \mu^2, \tag{46}$$

where in 3D (recalling that $\boldsymbol{\omega}_k = -\frac{1}{2}\varepsilon_{ijk}\Omega_{ij}$)

$$\mu^2 = \sum_{i=1}^{3}(s_i^2 - \omega_i^2) + 3\omega^2, \tag{47}$$

where $\omega = |\boldsymbol{\omega}|$. However, it is important to note that in 3D, unlike the 2D case, the negativity of μ^2 does not mean that local stretching cannot take place since, whatever the local nature of **A**, it will always have (generically) at least one eigenvalue with a positive real part. In that sense, there will always be some degree of local stretching in 3D.

For a given initial condition μ^2 shows rapid fluctuations as a function of time and a more useful quantity may be the long-time average of the real and imaginary parts of μ itself (i.e. $\mu = (\text{Trace } \mathbf{A})^{1/2}$) along a trajectory [29]. The ratio

$$\chi = \frac{\mu_{\text{Re}}}{\mu_{\text{Im}}} \tag{48}$$

gives a simple measure of the mean strain-rotation ratio experienced by particle orbits. (The inverse of this quantity was introduced long ago in fluid mechanics by Truesdell [30].) However, as just mentioned, χ may not be useful as far as determination of actual stretching behavior is concerned.

Following the earlier terminology of Frank and Mackley [31] we had termed the quantity (47) the ‘persistence of strain’ since it gives a simple measure of strain domination along a particular orbit. However, as far as a particular, infinitesimal, line element moving along that orbit is concerned the quantity is only a true measure of the strain persistence if the straining field is constant in time. As already mentioned this is typically not the case and the strain axes themselves will also be rotating. Thus in order to obtain a true measure of the straining field felt by the line element it is necessary to work in the (rotating) frame of the rate-of-strain tensor. This is at the heart of a complete description of stretching and alignment and can provide what is perhaps a more meaningful definition of a ‘persistence of strain’ quantity.

Before we can introduce this quantity it is necessary to develop a precise stretching and alignment formalism [27]. In order to understand exactly what is going on in the stretching and alignment process it is necessary to perform an analysis in the orthonormal basis (ξ_1, ξ_2, ξ_3) in which the rate-of-strain tensor is diagonal. Denoting **X** as the matrix whose columns are the positively oriented orthonormal eigenvectors ξ_i (i.e. $\det \mathbf{X} = +1$ for all time) we can write

$$\mathbf{S}(t) = \mathbf{X}^\dagger \begin{pmatrix} s_1 & 0 & 0 \\ 0 & s_2 & 0 \\ 0 & 0 & s_3 \end{pmatrix} \mathbf{X},$$

where the eigenvalues of **S**, (s_1, s_2, s_3) are ordered so that $s_1 \geqslant s_2 \geqslant s_3$. This moving coordinate system will be referred to as the ‘strain coordinates.’ In addition we will assume that the flow is incompressible, i.e. $s_1 + s_2 + s_3 = 0$ so that $s_1 \geqslant 0$ and $s_3 \leqslant 0$ for all time. By transforming **l** into the strain coordinate system, the local separation rate (28) becomes

$$\zeta_{\text{line}} = \sum_{i=1}^{3} s_i \lambda_i^2, \tag{49}$$

where $\lambda = \mathbf{X}.\mathbf{l}$ is the material element in strain coordinates. The λ_i are the cosines of the angles between **l** and ξ_i. Thus equation (49) shows the intimate connection between the local stretching rate of a material element and its alignment with the strain axes. Furthermore we note the largest value attained by (49) corresponds to $\lambda = (\pm 1, 0, 0)$, i.e. $\zeta_{\text{line}} = s_1$ which is just the persistent strain estimate of Batchelor and Townsend.

The time evolution of the λ_i is determined from the relationship

$$\frac{\mathrm{d}}{\mathrm{d}t}\lambda_i = \frac{\mathrm{d}}{\mathrm{d}t}\langle \mathbf{l}, \xi_i \rangle = \left\langle \frac{\mathrm{d}}{\mathrm{d}t}\mathbf{l}, \xi_i \right\rangle + \left\langle \mathbf{l}, \frac{\mathrm{d}}{\mathrm{d}t}\xi_i \right\rangle. \tag{50}$$

The first of the two terms on the r.h.s. of (50) can be computed using equation (27). In order to evaluate the second term, expressions for the evolution of the ξ_i must be developed. Since the eigenvectors are orthonormal, $\langle \xi_i, \xi_j \rangle = \delta_{ij}$ it follows that $\langle \mathrm{d}\xi_i/\mathrm{d}t, \xi_j \rangle = 0$ for all time. Thus we can write

$$\frac{\mathrm{d}}{\mathrm{d}t}\xi_i = \boldsymbol{\omega}' \wedge \xi_i, \tag{51}$$

where the vector $\boldsymbol{\omega}'$ is the instantaneous axis of rotation of strain axes and we shall refer to it as the *strain rotation vector*. It will be seen to play a crucial role in the alignment equations. The components of $\boldsymbol{\omega}'$, in the basis of strain coordinates, can be determined by differentiating (with respect to time) the characteristic equation $\mathbf{S}.\xi_j = s_j \xi_j$, using equation (51) and taking the inner product of both sides with ξ_i. The result is

$$\omega'_k = \frac{1}{2}\varepsilon_{ijk} \frac{\left\langle \xi_i, \frac{\mathrm{d}}{\mathrm{d}t}\mathbf{S}\xi_j \right\rangle}{s_i - s_j}, \tag{52}$$

where ε_{ijk} is the usual third-order totally antisymmetric tensor. (For two dimensions one may easily verify that the $\boldsymbol{\omega}'_k$ remain finite should the eigenvalues become degenerate [27]. A more general proof of this, applicable to both two and three dimensions, is given in the Appendix.) At this stage the definition of $\boldsymbol{\omega}'$ is purely kinematical in origin—the dynamical content is in $\mathrm{d}\mathbf{S}/\mathrm{d}t$ which, in the fluid dynamical context, is governed by the Navier–Stokes equation.

Using the above results equation (50) becomes

$$\frac{\mathrm{d}}{\mathrm{d}t}\lambda = \begin{pmatrix} s_1 & 0 & 0 \\ 0 & s_2 & 0 \\ 0 & 0 & s_3 \end{pmatrix} \lambda - \zeta_{\text{line}}\lambda + (\boldsymbol{\omega} - \boldsymbol{\omega}') \wedge \lambda, \tag{53}$$

where ζ_{line} is given by equation (49). Thus the second term in (53) is nonlinear. The rotational terms, $\boldsymbol{\omega} - \boldsymbol{\omega}'$, which we will sometimes refer to as the effective rotation, $\mathbf{R}$, represent the net rotation seen by the element in the strain basis. The first two terms in (53) cause the material element to align with the principal strain axis with the largest strain rate, namely ξ_1. This claim can be justified by the following argument. Setting $\mathbf{R} = 0$ and setting $\gamma_i = s_i - \zeta_{\text{line}}$, equation (53) becomes

$$\frac{\mathrm{d}}{\mathrm{d}t}\lambda_i = \gamma_i \lambda_i. \tag{54}$$

Thus the components of the material element are instantaneously either growing or shrinking. Furthermore the incompressibility condition, the ordering $s_1 \leqslant s_2 \leqslant s_3$ and the condition $\zeta_{\text{line}} \leqslant s_1$ immediately give the inequalities

$$\gamma_3 \leqslant 0 \leqslant \gamma_1 \text{ and } \gamma_1 \geqslant \gamma_2 \geqslant \gamma_3. \tag{55}$$

Thus λ_1 always grows to 1 and λ_3 always shrinks to zero. The complexity lies in the behaviour of λ_2 and it can be shown that it will grow if s_2 is positive and shrink if s_2 is negative.

The above, exact, formalism again suggests that our original 'persistence of strain' quantity μ^2 (47) should be modified by replacing the local vorticity contribution $(\sum \omega_i^2)$ by the local rotation rate felt in the strain basis, namely

$$\mu^2 = \sum_{i=1}^{3}(s_i^2) - |\omega - \omega'|^2, \tag{56}$$

as suggested by our more heuristic arguments given earlier for a steadily rotating strain frame in 2D. However, we are again faced with the question of the extent to which (56) can give a useful criterion for local stretching in three, as opposed to two, dimensions. To proceed, we anticipate a result described in Section 4 which describes how vorticity itself has a tendency to stretch and align in the direction of the intermediate strain axis i.e. ξ_2. This being so, the flow will appear, locally, to be relatively 'planar' i.e. vorticity sticking out in the ξ_2 direction and the flow swirling around in the (ξ_1, ξ_3) plane. This suggests that a 2D persistence of strain criterion can be introduced in the form (cf. (45))

$$\mu_{2D}^2 = (s_1 - s_3)^2 - |\omega_2 - \omega_2'|^2, \tag{57}$$

where ω_2 and ω_2' are the rotational components in the ξ_2 direction. Thus most of the material element stretching will occur in the (ξ_1, ξ_3) plane if $\mu_{2D}^2 > 0$. Ratios of this quantity and the 3D version (56) might also be useful characterizations of the flow.

3.2. *Discriminant characterization of strain*

We have seen that the quantity $\mu^2 = \text{Trace } \mathbf{A}^2$ gives certain gross information about the eigenvalues of $\mathbf{A}$. In order to obtain more detailed information we need to consider the characteristic equation associated with $\mathbf{A}$, namely

$$\lambda^3 + a_2\lambda^2 + a_1\lambda + a_0 = 0. \tag{58}$$

Using standard identities [32] and the fact that the flow is incompressible we can show that

$$a_2 = -(\lambda_1 + \lambda_2 + \lambda_3) = \text{Trace } \mathbf{A} = 0 \tag{59}$$

$$a_1 = -(\lambda_1\lambda_2 + \lambda_2\lambda_3 + \lambda_1\lambda_3) = -\frac{1}{2}\text{Trace } \mathbf{A}^2 \tag{60}$$

$$a_0 = -\lambda_1\lambda_2\lambda_3 = -\frac{1}{3}\text{Tr}\,\mathbf{A}^3. \tag{61}$$

The discriminant for the cubic equation can now be written as

$$D = \frac{1}{36}\left[(\text{Tr}\,\mathbf{A}^3)^2 - \frac{1}{6}(\text{Tr}\,\mathbf{A}^2)^3\right]. \tag{62}$$

If $D < 0$ there is one real root and two complex conjugate roots; if $D = 0$ all roots are real and at least two are equal; if $D > 0$ all roots are real and distinct. There are various ways of looking at this quantity and we suggest that the ratio

$$R = \frac{(\text{Tr}\,\mathbf{A}^3)}{(\text{Tr}\,\mathbf{A}^2)^{3/2}} \tag{63}$$

can be studied with profit: if $-1/\sqrt{6} < R < +1/\sqrt{6}$ there is one real root and a pair of complex conjugate roots: if $R = \pm 1/\sqrt{6}$ the roots are degenerate; and if $R > \pm 1/\sqrt{6}$ there are three real and distinct roots. It is interesting to compare R (a quantity that derives from 'deterministic' considerations of a flow field) with the similar-looking skewness factor (which derives from 'statistical' considerations) which takes the form

$$S = \frac{\langle(\partial u/\partial x)^3\rangle}{\langle(\partial u/\partial x)^2\rangle^{3/2}}, \tag{64}$$

and gives a measure of the ratio of enstrophy production (i.e. vorticity stretching) to dissipation. Dimensionally R and S are the same which means (if the numerator and denominator of R are ensemble averaged) that at the level of Kolmogorov theory they should show the same Reynolds number scaling. Although one can easily show that the numerators of these two quantities are the same, their denominators are different: representing pressure fluctuations and dissipation respectively. This will result in different intermittency exponents [33].

4. STRETCHING AND ALIGNMENT OF VECTORIAL QUANTITIES

4.1. *Vorticity and the interplay between kinematics and dynamics*

Alignment equations analogous to (53) can also be derived for the vorticity. By taking the curl of the Euler equations

$$\frac{d\boldsymbol{\omega}}{dt} = \frac{\partial\boldsymbol{\omega}}{\partial t} + (\mathbf{u}.\nabla)\boldsymbol{\omega} = (\boldsymbol{\omega}.\nabla)\mathbf{u} = \mathbf{A}.\boldsymbol{\omega}, \tag{65}$$

which is identical in form to the evolution for material lines (26). For an infinitesimal vorticity element in the strain basis one can introduce the direction cosines $\mu = (\mu_1, \mu_2, \mu_3)$ representing the angles between the unit vector associated with $\boldsymbol{\omega}$ and the strain axes and define a local stretching rate

$$\zeta_{\text{vort}} = \sum_{i=1}^{3} s_i \mu_i^2. \tag{66}$$

Following the same steps used in the derivation of (53) one obtains

$$\frac{d}{dt}\mu = \begin{pmatrix} s_1 & 0 & 0 \\ 0 & s_2 & 0 \\ 0 & 0 & s_3 \end{pmatrix} \mu - \zeta_{\text{vort}}\mu + \boldsymbol{\omega}' \wedge \mu. \tag{67}$$

An important difference between (53) and (67) is, of course, the absence of the vorticity term $\boldsymbol{\omega}$ since, by definition, μ and $\boldsymbol{\omega}$ are parallel.

Comparison of the two alignment equations leads to some interesting conclusions. In the case of both line and vorticity stretching we suppose that the alignment is with some combination of ξ_1 and ξ_2, that is we can write

$$\mu \approx \mu_1\xi_1 + \mu_2\xi_2, \tag{68a}$$

and

$$\lambda = \lambda_1\xi_1 + \lambda_2\xi_2, \tag{68b}$$

where, on the basis of Ashurst *et al*. [22] we expect $\mu_2 > \mu_1$ in (68a) and, from Girimaji and Pope [23], we expect $\lambda_1 > \lambda_2$ in (68b). The difference between (53) and (67) lies in the term $\boldsymbol{\omega} \wedge \lambda$, which from (68) can be estimated as

$$\boldsymbol{\omega} \wedge \lambda \approx (\lambda_1\mu_2 - \lambda_2\mu_1)\xi_3.$$

From the above comments about the relative sizes of the $\mu_{1,2}$ and $\lambda_{1,2}$ we can expect this term to be non-negligible (whereas if both $\boldsymbol{\omega}$ and $\mathbf{l}$ aligned with ξ_1 we would expect it to be smaller). Since the contribution to $\boldsymbol{\omega} \wedge \lambda$ is through ξ_3 this will result in fluctuations in λ_3 and hence, since $s_3\lambda_3^2$ is negative definite, a reduction in ζ_{line}. Thus we speculate that

$$\zeta_{\text{line}} \leqslant \zeta_{\text{vort}}.$$

The alignment equation (53) is purely kinematical in origin and hence may be of value in analyzing the alignment and rotation questions we raised earlier for dynamical systems in general. The vorticity alignment equation (67), however, has an essential piece of dynamics built in (namely the feedback of $\boldsymbol{\omega}$ into s_i and $\boldsymbol{\omega}'$) since the starting point of its derivation, equation (65), makes use of the Euler equations. In order to make further comparisons of line and vorticity stretching in turbulence it is necessary to introduce the Navier–Stokes dynamics which enters into the alignment equations through expression (52) for the strain rotation $\boldsymbol{\omega}'$. By taking the antisymmetric and symmetric gradients, respectively, of the Navier–Stokes equations one can derive the general evolution for vorticity

$$\frac{d\boldsymbol{\omega}}{dt} = \mathbf{S}.\boldsymbol{\omega} + \nu\nabla^2\boldsymbol{\omega} \tag{69}$$

and strain

$$\frac{d\mathbf{S}}{dt} = -(\mathbf{S}^2 + \boldsymbol{\Omega}^2) - \frac{1}{\rho}\mathbf{P} + \nu\nabla^2\mathbf{S}, \tag{70}$$

where $\mathbf{P}$ is the pressure Hessian, i.e. $\mathbf{P}_{ij} = \partial^2 p/\partial x_i x_j$ (In equation (70) $\boldsymbol{\Omega}$ denotes, of course, the vorticity tensor as opposed to the vorticity vector $\boldsymbol{\omega}$). Using (70) in (52) leads to

$$\omega'_k = \varepsilon_{ijk}\frac{-\omega_i\omega_j - (1/\rho)P_{ij} + \nu D_{ij}}{s_i - s_j} \tag{71}$$

where the italic symbols denote projection onto the rate of stain axes i.e. $\omega_i = \omega.\xi_i$; $P_{ij} = (\xi_i, \mathbf{P}\xi_j)$ and $D_{ij} = (\xi_i, (\nabla^2\mathbf{S})\xi_j)$. Thus overall we can write

$$\boldsymbol{\omega}' = \boldsymbol{\omega}'_\omega + \boldsymbol{\omega}'_P + \boldsymbol{\omega}'_\nu, \tag{72}$$

where $\boldsymbol{\omega}'_\omega$, $\boldsymbol{\omega}'_P$, $\boldsymbol{\omega}'_\nu$ represent the contributions of, respectively, vorticity, pressure and dissipation to the strain rotation. It is important to note that in $\boldsymbol{\omega}'_\omega$ the contribution is quadratic in the vorticity.

Before proceeding a few words about the nature of the pressure field, $p = p(\mathbf{x}, t)$ are in order. From (46) we recall that the Laplacian of the pressure is the difference of the local strains and vorticity, i.e. $\nabla^2 p = \sum_i[(\omega_i^2 - \omega^2) - s_i^2]$. Treating the r.h.s. as a 'charge' the equation can be solved by standard techniques to give

$$p(\mathbf{x}, t) = \frac{1}{4\pi}\int\frac{\sum_i(\omega_i^2 - \omega^2) - s_i^2}{\|\mathbf{x} - \mathbf{x}'\|}\,d\mathbf{x}', \tag{73}$$

emphasizing the long-range nature of pressure (the integrand is slowly decaying) and its quadratic dependence on the strain and vorticity of nearby regions. The P_{ij} needed to evaluate the $\boldsymbol{\omega}'_P$ contribution to $\boldsymbol{\omega}'$ are found by suitable differentiation of (73). The interplay between the local quadratic contributions of vorticity to $\boldsymbol{\omega}'$, through $\boldsymbol{\omega}'_\omega$ and the corresponding nonlocal ones, through $\boldsymbol{\omega}'_P$, is a fascinating and poorly understood problem.

Equation (72) tells us that strain rotation is itself quadratically dependent on vorticity—both locally and nonlocally. Considerations such as these and identification of the term $(\boldsymbol{\omega} - \boldsymbol{\omega}') \wedge \lambda$ as the agent for misalignment in (53) suggests various ways of estimating the different contributions to the non-persistence of strain. We have proposed [27] that the following dimensionless quantities could provide useful measures of these contributions:

$$Q_\omega(t) = \|\boldsymbol{\omega} \wedge \boldsymbol{\lambda}\|/S$$
$$Q'_\omega(t) = \|\boldsymbol{\omega}'_\omega \wedge \boldsymbol{\lambda}\|/S$$
$$Q'_P(t) = \|\boldsymbol{\omega}'_P \wedge \boldsymbol{\lambda}\|/S$$
$$Q'_v(t) = \|\boldsymbol{\omega}'_v \wedge \boldsymbol{\lambda}\|/S$$

where $S = \|\mathbf{S}\boldsymbol{\lambda} \wedge \boldsymbol{\lambda}\|$. Thus $Q_\omega(t)$ measures the contribution of 'pure' vorticity, $Q'_\omega(t)$ measures the effect of vorticity through strain rotation, $Q'_P(t)$ the effects of pressure and $Q'_v(t)$ the effects of viscous terms.

The equations for vorticity and strain evolution (69) and (70) can be rewritten in the strain basis and, after setting $\nu = 0$ and simplifying, take the form

$$\frac{\mathrm{d}s_j}{\mathrm{d}t} = -s_j^2 + (\omega^2 - \omega_j) - P_{jj} \qquad j = 1, 2 \tag{74}$$

$$\frac{\mathrm{d}\omega_i}{\mathrm{d}t} = (s_i + \alpha_i)\omega_i + \beta_i \qquad i = 1, 2, 3 \tag{75}$$

where

$$\alpha_i = \sum_{j \neq i} \frac{\omega_j^2}{s_j - s_i} \tag{76}$$

and

$$\beta_i = -(\boldsymbol{\omega}'_P \wedge \boldsymbol{\omega})_i = \sum_{j \neq i} \frac{P_{ji}}{s_j - s_i}\omega_j. \tag{77}$$

In (74) we need only consider the evolution of s_1 and s_2 since $s_3 = -(s_1 + s_2)$. From (75) and (76) we see a significant concept, namely that in vorticity stretching the rotation of the strain axes due to vorticity itself actually plays the role of a *nonlinear rotation* which acts to always rotate the vorticity towards the strain axis with smaller strain. Since this contribution is quadratic in the vorticity it plays a significant role.

Study of equations (74) and (75) leads to the following qualitative description of vortex stretching when viewed from the Lagrangian frame. Here we reproduce the argument given in Dresselhaus and Tabor [27] (a similar scenario is briefly summarized in She *et al*. [28]). Suppose that at some time t the intermediate strain is approximately zero i.e. $s_2(t) \approx 0$, and that the (local) vorticity is weak, i.e. $\boldsymbol{\omega} \approx 0$. At this instant the nonlinear stretching rates α_i are small and the vorticity will begin to stretch in the ξ_1 direction. From equation (74) it follows that, since $s_2 \approx 0$, the stretched ξ_1 component of vorticity will cause s_2 to grow and become positive resulting in an increase in ω_2. As a consequence of this sequence of events α_1, which is always negative, will become large and will quickly act to nullify the previous growth in ω_1. Since s_2 is still small compared to s_1 the ω_2 stretching will proceed for a longer time than the initial stretching of ω_1 due to s_1. This relative long-time persistent alignment with ξ_2 survives until ω_2 itself becomes large, at which point rotation of vorticity into the ξ_3 direction becomes significant (due to the quadratic contribution of ω_2 to the always positive nonlinear stretching rate α_3). The net result is that now α_2 becomes negative and large. This suppresses the previous stretching of the intermediate vorticity ω_2. The alignment with the strong contracting direction causes a quick but not total destruction of the previous vorticity build up. It should be emphasized that the relatively small magnitude of s_2 implies that the vortex stretching in the intermediate direction is the most long-lived of the stretchings described in this scenario. This sequence of events is observed to be robust in a variety of simulations [34]—even in the presence of random forms of the pressure terms and is consistent with vortex stretching results in turbulence simulations [22, 28].

4.2. *Stretching of scalar gradients*

The advection equation for a passive scalar field $\theta = \theta(\mathbf{x}, t)$ in a volume preserving flow is

$$\frac{\mathrm{d}}{\mathrm{d}t}\theta = \kappa\nabla^2\theta \tag{78}$$

where κ is the scalar diffusivity. Taking the gradient of this equation gives us the scalar gradient evolution equation

$$\frac{\mathrm{d}}{\mathrm{d}t}\nabla\theta = -(\nabla\mathbf{u})^{-T}(\nabla\theta) + \kappa\nabla^2(\nabla\theta). \tag{79}$$

We note that in the absence of scalar diffusivity (79) is identical to the area element advection equation (22) and thus has the same solution, namely

$$\nabla\theta(\mathbf{x}, t) = \mathbf{M}^{-T}\nabla\theta(\mathbf{x}(0), 0). \tag{80}$$

This should not be unexpected since $\nabla\theta$ is a vector normal to surfaces of constant θ (a material surface when $\kappa = 0$) and of strength inversely proportional to the distance between adjacent surfaces of constant θ. Thus in volume-preserving flows $\nabla\theta$ is proportional (with a time-independent constant of proportionality) to any area element $\mathbf{b} \times \mathbf{c}$ tangent to the surface of constant θ.

We also note that by analogy with (53) we can write

$$\frac{\mathrm{d}}{\mathrm{d}t}\gamma = -\begin{pmatrix} s_1 & 0 & 0 \\ 0 & s_2 & 0 \\ 0 & 0 & s_3 \end{pmatrix}\gamma - \zeta_{\mathrm{area}}\gamma + (\omega - \omega') \wedge \gamma, \tag{81}$$

where γ is the projection of $\nabla\theta$ in the rate of strain basis, i.e. $\gamma_i = \xi_i.\nabla\theta$, and

$$\zeta_{\mathrm{area}} = -\sum_{i=1}^{3} s_i\gamma_i^2 \tag{82}$$

measures the rate of growth/shrinkage of an area element (corresponding to a particular θ surface). If $\nabla\theta$ is aligned with the principle eigenvector of $\mathbf{M}^{-T}$ then $\gamma \approx (0, 0, 1)$ and hence $\zeta_{\mathrm{area}} \approx -s_3$, which corresponds to the maximal area growth rate.

As in the case of infinitesimal area elements $\nabla\theta(\mathbf{x}, t)$ grows, in general, at the rate $\sigma_1 + \sigma_2$ and converges to the direction of the principal eigenvector of M^{-T}. From our area element interpretation it is clear that this direction must be perpendicular to the principal and intermediate eigen-directions of $\mathbf{M}$. (This fact can also be seen directly by comparison of the Jordan normal form conjugating matrices of $\mathbf{M}$ and $\mathbf{M}^{-T}$.) Now if we assume that vorticity aligns with the intermediate strain axis it follows that the direction of the vorticity is also an eigen-direction for $\mathbf{S}$ and $\mathbf{A}$ (either the principal or intermediate direction of $\mathbf{A}$ depending on the sign of $\mu_{2\mathrm{D}}^2$). If $\mathbf{A}$ does not rotate rapidly for a moderately long time then this direction should also be the principal or intermediate eigen-direction of $\mathbf{M}$ (see Fig. 5). In this case $\nabla\theta$ should be perpendicular to the principal and intermediate eigen-directions

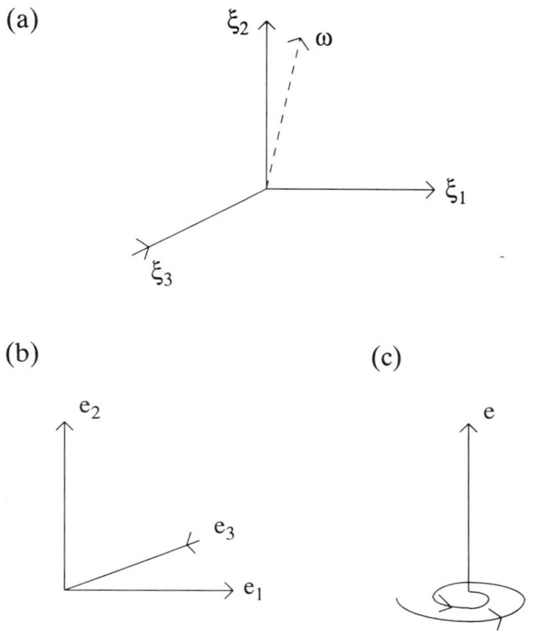

Fig. 5. (a) Sketch of orthogonal eigenvectors (ξ_1, ξ_2, ξ_3) of the three-dimensional rate-of-strain tensor **S** with the vorticity vector $\boldsymbol{\omega}$ approximately aligned with the intermediate axis ξ_2. Arrows indicate that eigenvalues associated with ξ_1 and ξ_2 are both positive and the eigenvalue associated with ξ_3 is negative. (b) Sketch of (non-orthogonal) eigenvectors ($\mathbf{e}_1$, $\mathbf{e}_2$, $\mathbf{e}_3$) of velocity gradient tensor **A** with positive (real) eigenvalues associated with $\mathbf{e}_1$ and $\mathbf{e}_2$. If $\mu_{2D}^2 > 0$ $\mathbf{e}_2$ is the intermediate or principle eigenvector of **A**. If **A** is slowly varying in time $\boldsymbol{\omega}$, ξ_2 and $\mathbf{e}_2$ will all be approximately aligned with each other. (c) Sketch of 'eigen-spaces' of **A** with pair of complex eigenvalues. If $\mu_{2D}^2 < 0$ **e** is the principle eigenvector of **A** and if **A** is slowly varying in time $\boldsymbol{\omega}$, ξ_2 and **e** will all be approximately aligned with each other.

of **M** and thus the intermediate strain direction—a prediction which appears to be supported numerically [22].

We remark that rapid growth of $\nabla\theta$ in a flow **u** with $\kappa = 0$ provides an indication of efficient mixing. More specifically, if we call exponential growth of the quantity $\int|\nabla\theta|\,\mathrm{d}x$ a necessary (although not sufficiant) condition for efficient mixing, then a recent result [35] shows that **u** must be chaotic (or turbulent).

4.3. *Stretching of magnetic field lines*

The equations of incompressible magneto-hydrodynamics are

$$\frac{\mathrm{d}}{\mathrm{d}t}\mathbf{u} = -\nabla p + \mathbf{j} \times \mathbf{B} \tag{83}$$

$$\frac{\mathrm{d}}{\mathrm{d}t}\mathbf{B} = \mathbf{A}.\mathbf{B} + \eta\nabla^2\mathbf{B}, \tag{84}$$

where **B** is the magnetic field, $\mathbf{j} = \nabla \times \mathbf{B}$ is the electric current and $\nabla.\mathbf{u} = \nabla.\mathbf{B} = 0$ (see e.g. [36]). We immediately see that the magnetic induction equation (84), in the limit $\eta = 0$, is identical to the (infinitesimal) line element advection equation (2.7). Thus, in this limit, magnetic field lines are material lines and the magnetic field increases or decreases in magnitude with the expansion or contraction of material lines. However, **B** is by no means passive, since the velocity field **u** and hence the velocity gradient **A**, are closely tied to **B** through the Lorentz force $\mathbf{j} \times \mathbf{B}$.

Let us first suppose that **B** is small and hence that the Lorentz force, which is quadratic in **B**, can be neglected. In this case, equation (83) for **u** decouples from **B** and hence **u** can be proscribed independently of **B**. This is called the kinematic dynamo approximation and is frequently used in dynamo studies (see e.g. [37]). Under the kinematic approximation when $\eta = 0$, equation (84) is truly an equation for the advection of a passive vector field. (Thus, given a specific deterministic velocity field, the study of kinematic dynamo effects becomes, effectively, an exercise in dynamical systems.) When η is small but nonzero, highly non-trivial field cancellations can occur especially when **u** is chaotic. An understanding of this process is the aim of kinematic fast-dynamo theory. As with the scalar gradient case described above, the exponential growth of $\int |\mathbf{B}|\, dx$ in the limit $\eta \to 0$ implies that **u** is chaotic or turbulent [35].

When $\mathbf{j} \times \mathbf{B}$ is not neglected the problem becomes considerably more complicated. In keeping with our philosophy of studying gradient quantities we introduce both the symmetric and antisymmetric parts of $\nabla\mathbf{B}$, namely $\mathbf{C} = \frac{1}{2}(\nabla\mathbf{B} + \nabla\mathbf{B}^T)$ and $\mathbf{J} = \frac{1}{2}(\nabla\mathbf{B} - \nabla\mathbf{B}^T)$. The latter is, of course, the tensor associated with the current i.e. $j_k = -\frac{1}{2}\varepsilon_{ijk}\mathbf{J}_{ij}$. The physical interpretation of **C** is not so obvious. By working with these quantities a system of equations, analogous to equations (69) and (70) can be derived, namely

$$\frac{d\boldsymbol{\omega}}{dt} = \mathbf{S}.\boldsymbol{\omega} - \mathbf{C}.\mathbf{j} + \mathbf{B}.\nabla\mathbf{j} + \nu\nabla^2\boldsymbol{\omega} \tag{85}$$

$$\frac{d\mathbf{S}}{dt} = -(\mathbf{S}^2 + \boldsymbol{\Omega}^2) + (\mathbf{C}^2 + \mathbf{J}^2) + \mathbf{B}.\nabla\mathbf{C} - (1/\rho)\mathbf{P} + \nu\nabla^2\mathbf{S} \tag{86}$$

$$\frac{d\mathbf{C}}{dt} = \boldsymbol{\Omega}.\mathbf{C} - \mathbf{C}.\boldsymbol{\Omega} - \mathbf{S}.\mathbf{J} - \mathbf{J}.\mathbf{S} + \mathbf{B}.\nabla\mathbf{S} + \eta\nabla^2\mathbf{C} \tag{87}$$

$$\frac{d\mathbf{J}}{dt} = \boldsymbol{\Omega}.\mathbf{J} - \mathbf{J}.\boldsymbol{\Omega} + \mathbf{S}.\mathbf{C} - \mathbf{C}.\mathbf{S} + \mathbf{B}.\nabla\boldsymbol{\Omega} + \eta\nabla^2\mathbf{J} \tag{88}$$

together with the conditions

$$\mathrm{Tr}\,\mathbf{S} = \mathrm{Tr}\,\mathbf{C} = 0 \tag{89}$$

and

$$(1/\rho)\mathrm{Tr}\,\mathbf{P} = \mathrm{Tr}\,[-(\mathbf{S}^2 + \boldsymbol{\Omega}^2) + (\mathbf{C}^2 + \mathbf{J}^2)]. \tag{90}$$

Using (86) the strain rotation (71) now takes the form

$$\omega_k' = \varepsilon_{ijk}\frac{-\omega_i\omega_j + j_ij_j + (\mathbf{C}^2)_{ij} - (1/\rho)P_{ij} + (\mathbf{B}.\nabla\mathbf{C})_{ij} + \nu D_{ij}}{s_i - s_j} \tag{91}$$

where the italic j_i corresponds to **j** projected in the rate of strain basis **J** i.e. $j_i = \mathbf{j}.\xi_i$ and the quantities $(*)_{ij}$ correspond to the inner products $(\xi_i, {}^*\,\xi_j)$.

Preliminary work (Klapper and Tabor [38]) involving the numerical integration of (85–88) indicates certain behaviors that are consistent with the results of large scale numerical simulations (Brandenburg *et al*. [39]), namely:

(i) there is a tendency for vorticity to align in the intermediate strain direction;

(ii) there is also a tendency for the electric current to align with the intermediate strain—apparently because current is generated by shearing of magnetic field and the strongest shearing generally occurs in the plane of the principle and least strains (see Fig. 6);

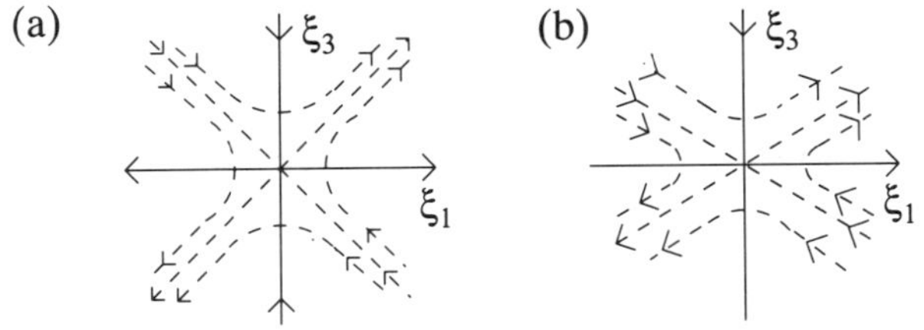

Fig. 6. (a) Sketch of the 'shearing plane': ξ_1 and ξ_3 are the eigenvectors of **S** corresponding to the maximal stretching and contracting directions. Dashed lines represent **B** lines corresponding, for illustrative purposes, to an x-type null point. (b) Same as (a) at a later time. The **B** field lines have been sheared by the straining action of **S**.

(iii) depending on the relative strengths of the vorticity and the strain in the plane perpendicular to the vorticity, magnetic field either aligns with current in the intermediate strain direction or anti-aligns with current by converging to the plane perpendicular to the intermediate strain direction.

This last observation is important since it determines whether the Lorentz force $\mathbf{j} \times \mathbf{B}$ is significant or not.

Acknoweldgements—This work was supported by the AFOSR under grant AFOSR-90-0284 and the DOE under grant DE-FG03-93-ER25174. I.K. thanks the NSF for a Postdoctoral Fellowship. The authors would like to thank J-D. Fournier for past discussions on Lyapunov exponents.

REFERENCES

1. H. Aref, Stirring by chaotic advection, *J. Fluid Mech.* **143**, 1–21 (1984).
2. J. Chaiken, R. Chevray, M. Tabor and Q. M. Tan, Experimental study of Lagrangian turbulence in a Stokes flow, *Proc. R. Soc. Lond.* **A408**, 165–174 (1986).
3. J. M. Ottino, *The Kinematics of Mixing: Stretching, Chaos and Transport*. Cambridge University Press, Cambridge (1989).
4. V. Rom-Kedar, A. Leonard and S. Wiggins, An analytical study of transport, mixing and chaos in an unsteady vortical flow, *J. Fluid Mech.* **214**, 347–394 (1990).
5. J. T. Stuart and M. Tabor, The Lagrangian picture of fluid dynamics, *Phil. Trans. R. Soc.* **A333**, complete volume (1990).
6. M. Tabor, Stretching and Alignment in General Flow Fields: Classical Trajectories from Reynolds Number Zero to Infinity, in *Topological Aspects of the Dynamics of Fluids and Plasmas*, edited by H. K. Moffatt, G. M. Zaslavsky, P. Comte and M. Tabor, pp. 83–110, NATO ASI Series E 218. Kluwer Academic Press, Dordrecht (1992).
7. M. Tabor, *Chaos and Integrability in Nonlinear Dynamics: An Introduction*. Wiley, New York (1989).
8. G. K. Batchelor, *An Introduction to Fluid Dynamics*. Cambridge University Press, Cambridge (1967).
9. T. Dombre, U. Frisch, J. M. Greene, M. Hénon, A. Mehr and A. M. Soward, Chaotic streamlines in the ABC flows. *J. Fluid Mech.* **167**, 353–391 (1986).
10. V. I. Arnold, Sur la géométrie différentielle des groupes de Lie de dimension infinite et ses applications a l'hydrodynamique des fluids parfait, *Ann. Inst. Fourier* **16**, 319–361 (1966).
11. I. Goldhirsch, P.-L. Sulem and S. A. Orszag, Stability and Lyapunov stability of dynamical systems: a differential approach and a numerical method, *Physica D* **27**, 311–337 (1987).
12. G. Benettin, L. Galgani, A. Giorgilli and J.-M. Strelcyn, Lyapunov characteristic exponents for smooth dynamical systems and for Hamiltonian systems; a method for computing all of them, *Meccanica* **15**, 9–21 (1980).
13. V. I. Oseledec, A multiplicative ergodic theorem. Lyapunov characteristic numbers for dynamical systems, *Trans. Moscow Math. Soc.* **19**, 197 (1968).
14. A. M. Monin and I. Yaglom, *Statistical Fluid Mechanics*. MIT Press, Cambridge, MA (1975).
15. G. K. Batchelor, The effect of homogeneous turbulence in materials lines and surfaces, *Proc. R. Soc. Lond.*, **A218**, 349–366 (1952).
16. J. D. Farmer, E. Ott and J. A. Yorke, The dimension of chaotic attractors, *Physica D* **7**, 153–180 (1983).
17. W. J. Cocke, Turbulent hydrodynamic line stretching: Consequences of isotropy, *Phys. Fluids* **12**, 2488–2492 (1969).

18. W. J. Cocke, Turbulent hydrodynamic line stretching: The random walk limit, *Phys. Fluids* **14**, 1624–1628 (1971).
19. S. A. Orszag, Comments on: turbulent hydrodynamic line stretching: consequences of isotropy, *Phys. Fluids* **13**, 2203–2204 (1970).
20. G. K. Batchelor and A. A. Townsend, Turbulent diffusion, in *Surveys in Dynamics*, edited by G. K. Batchelor and R. M. Davies, pp. 352–399. Cambridge University Press, Cambridge (1956).
21. A. A. Townsend, The diffusion of heat spots in isotopic turbulence, *Proc. R. Soc. Lond.* **A209**, 418–430 (1951).
22. W. T. Ashurst, A. R. Kerstein, R. M. Kerr and C. H. Gibson, Alignment of vorticity and scalar gradient with strain rate in simulated Navier–Stokes turbulence, *Phys. Fluids* **30**, 2343–2353 (1986).
23. S. S. Girimaji and S. B. Pope, Material element deformation in isotropic turbulence, *J. Fluid Mech.* **220**, 427–458 (1990).
24. P. Vieillefosse, Local interaction between vorticity and shear in a perfect incompressible fluid, *J. Physique* **43**, 837–842 (1982).
25. P. Vieillefosse, International motion of small element of fluid in an inviscid flow, *Physica A* **125**, 150–162 (1984).
26. A. Majda, Vorticity, turbulence and acoustics in fluid flow, *SIAM Review*, **3**, 349–388 (1991).
27. E. Dresselhaus and M. Tabor, The stretching and alignment of material elements in general flow fields, *J. Fluid Mech.* **236**, 415–444 (1991).
28. Z.-S. She, E. Jackson and S. A. Orszag, Structure and dynamics of homogeneous turbulence: Models and simulations, *Proc. Roy. Soc.* **A434**, pp. 101–124 (1991).
29. E. Dresselhaus and M. Tabor, The persistence of strain in dynamical systems, *J. Phys. A: Math Gen.* **22**, 971–984 (1989).
30. C. Truesdell, *The Kinematics of Vorticity*. Indiana University Press (1954).
31. F. C. Frank and M. R. Mackley, Localized flow birefringence of polyethylene oxide solutions in a two roll mill, *J. Polymer Sci.* **A2**(14), 1121–1131 (1976).
32. R. Betchov, An inequality concerning the production of vorticity in isotropic turbulence, *J. Fluid Mech.* **1**, 497–504 (1956).
33. M. Nelkin and M. Tabor, Unpublished work (1993).
34. E. Dresselhaus, Material element stretching and alignment in turbulence, Ph.D. thesis, Columbia Univeristy, New York (1991).
35. I. Klapper and L. S. Young, Bounds on the fast dynamo growth rate involving topological entropy (1993). Preprint.
36. P. H. Roberts, *An Introduction to Magnetohydrodynamics*. Longmans Green, London (1967).
37. P. H. Roberts and A. M. Soward, Dynamo theory, *Ann. Rev. Fluid Mech.* **24**, 459–512 (1992).
38. I. Klapper and M. Tabor, Current sheet formation in two-dimensional incompressible MHD turbulence from a Lagrangian point of view, *Geophys. Astrophys. Fluid Dynamics*, **73**, 109–122 (1993).
39. A. Brandenberg, R. L. Jennings, A. Nordlund, XX Rieutord, R. F. Stein and I. Tuominen, Magnetic structures in a dynamo simulation, *J. Fluid Mech.* (1994). In press.

APPENDIX

Suppose that at some instant of time, here labelled as $t = 0$, the first two eigenvalues of $\mathbf{S}$ are equal, i.e. $s_1 = s_2 = s$. We choose a coordinate system that diagonalizes $\mathbf{S}$ at this instant, i.e.

$$\mathbf{S}(0) = \begin{pmatrix} s & 0 & 0 \\ 0 & s & 0 \\ 0 & 0 & -2s \end{pmatrix}. \tag{A.1}$$

We denote the eigenvectors of $\mathbf{S}(0)$ as $\mathbf{x}_i$, $i = 1, 2, 3$. We next consider the form of $\mathrm{d}\mathbf{S}(0)/\mathrm{d}t$ which we write as

$$\frac{\mathrm{d}\mathbf{S}(0)}{\mathrm{d}t} = \begin{pmatrix} a & b & * \\ b & d & * \\ * & * & * \end{pmatrix}, \tag{A.2}$$

where * denotes some unspecified element value. We are interested in the symmetric, upper left, 2×2 block which can be diagonalized by some orthogonal matrix $\mathbf{m}$ with elements $m_{ij}(i, j = 1, 2)$. Thus changing coordinates by means of the transformation

$$\mathbf{M} = \begin{pmatrix} m_{11} & m_{12} & 0 \\ m_{21} & m_{22} & 0 \\ 0 & 0 & 1 \end{pmatrix} \tag{A.3}$$

simultaneously puts $\mathbf{S}(0)$ and $\mathrm{d}\mathbf{S}(0)/\mathrm{d}t$ in the forms

$$\mathbf{S}(0) = \begin{pmatrix} s & 0 & 0 \\ 0 & s & 0 \\ 0 & 0 & -2s \end{pmatrix} \tag{A.4}$$

$$\frac{d\mathbf{S}(0)}{dt} = \begin{pmatrix} e & 0 & * \\ 0 & f & * \\ * & * & * \end{pmatrix}. \tag{A.5}$$

We note that this is only possible in general in the degenerate case $s_1 = s_2 = s$ since then the upper-left portion of $\mathbf{S}(0)$ can be written as $s\mathbf{I}$, where $\mathbf{I}$ is the 2×2 identity matrix.

In (A.5) we assume a nondegeneracy condition $e \neq f$ and that $d\mathbf{S}/dt$ is sufficiently smooth (i.e. Lipschitz continuous) so that $\mathbf{S}(t)$ is continuous. Thus

$$\frac{d\mathbf{S}(\Delta t)}{dt} = \begin{pmatrix} e & 0 & * \\ 0 & f & * \\ * & * & * \end{pmatrix} + O(\Delta t), \tag{A.6}$$

and hence that

$$\mathbf{S}(\Delta t) = \begin{pmatrix} s & 0 & 0 \\ 0 & s & 0 \\ 0 & 0 & -2s \end{pmatrix} + \Delta t \begin{pmatrix} e & 0 & * \\ 0 & f & * \\ * & * & * \end{pmatrix} + O(\Delta t^2). \tag{A.7}$$

For the eigenvalues of $\mathbf{S}(\Delta t)$, s_i, $i = 1, 2, 3$, it is clear, because of the nondegeneracy condition $e \neq f$, that $s_1 - s_2 = O(\Delta t)$ and not $O(\Delta t^2)$. Furthermore, the differences between the eigenvectors of $\mathbf{S}(\Delta t)$ and $\mathbf{S}(0)$, namely $|\xi_i - \mathbf{x}_i|$, are, at worst, $O(\Delta t)$. Thus

$$\frac{\left(\xi_2, \dfrac{d}{dt}\mathbf{S}\xi_1\right)}{s_1 - s_2} = \frac{O(\Delta t)}{O(\Delta t)}, \tag{A.8}$$

which is finite in the limit $\Delta t \to 0$. The same argument can also be applied to the two-dimensional case.

Comparisons of Mixing in Chaotic and Turbulent Flows

K. B. SOUTHERLAND, R. D. FREDERIKSEN and W. J. A. DAHM*

Gas Dynamics Laboratories, Department of Aerospace Engineering, The University of Michigan, Ann Arbor, MI 48109-2140, USA

and

D. R. DOWLING

Walter E. Lay Automotive Laboratories, Department of Mechanical Engineering and Applied Mechanics, The University of Michigan, Ann Arbor, MI 48109-2125, USA

Abstract—Results are presented from an experimental investigation comparing geometric scaling properties created by the mixing of dynamically passive tracers in chaotic flows with those resulting at the small scales of fully developed turbulent flows. The low Reynolds number, two-dimensional, time-periodic, closed flow between eccentric rotating cylinders is taken as the archetypal chaotic flow. The turbulent flow for comparison is the high Reynolds number, three-dimensional, unsteady, open flow in the self-similar far field of a steady axisymmetric jet. For each flow, the concentration field $\zeta(\mathbf{x}, t)$ resulting from the mixing of a conserved scalar quantity is used to measure scaling properties of the support set on which the corresponding scalar energy dissipation rate field $(ReSc)^{-1}\nabla\zeta \cdot \nabla\zeta(\mathbf{x}, t)$ is concentrated. The distributions of dissipation layer separations obtained for both flows are found to be identical. Contrary to central limit arguments for multiplicative quantities, the ensemble-averaged distributions in both flows have a -3 power law scaling for all but the smallest separations; classical log-normal scaling for multiplicative processes is found only in regions having undergone extensive stretching and folding. A statistical assessment of the fractal scaling properties based on one-dimensional intersections with the dissipation support set demonstrates that the chaotic flow at this stage of development approaches a global fractal dimension only in these same regions. Unlike previous studies of the fractal scaling of scalar isosurfaces in turbulent flows, the results for the turbulent flow presented here show no strong evidence for global fractal scaling in the dissipation support set.

1. INTRODUCTION

Mixing in turbulent flows represents one of the most widely encountered and technically important classes of problems in the fluid sciences. Yet no broadly applicable theory for treating practical problems involving turbulent mixing currently exists. It is known that the quasi-deterministic large scale structures in a turbulent flow control the entrainment properties of the flow, and that these large structures differ from one flow to another. It is believed that the structure and dynamics of the small scales in fully-developed turbulent flows are quasi-universal, satisfying Reynolds number asymptotics and displaying certain universal statistical scaling properties. However, it is only recently that the fully three-dimensional, time-varying, small scale structure of mixing in turbulent flows has been accessible to direct experimental study [1]. An example of such a measurement is shown in

*Corresponding author.

Fig. 1, where the concentration field $\zeta(\mathbf{x}, t)$ of a dynamically passive, conserved scalar quantity being mixed by the underlying turbulent flow is shown in a small volume element of the flow. The mixing of such a scalar can be usefully quantified in terms of the scalar energy dissipation rate field $(ReSc)^{-1}\nabla\zeta \cdot \nabla\zeta(\mathbf{x}, t)$, also shown for the same volume element in Fig. 1. Of primary interest for the present purposes are the highly folded sheet-like structures into which the scalar dissipation field is formed by the underlying three-dimensional time-varying turbulent flow field. It has recently become possible to extract the underlying time-varying vector velocity field $\mathbf{u}(\mathbf{x}, t)$ from such four-dimensional scalar mixing measurements in turbulent flows [2, 3]. This may allow detailed studies of the mixing process at the small scales of turbulent flows; however, at present the geometric methods and mathematical tools for describing the structure and dynamics of this mixing process in turbulent flows are still at a relatively early stage of development.

By comparison, considerable progress has been made over the past ten years in the development of experiments and analytical methods for understanding, describing, and even predicting certain features of fluid mixing in closed, two-dimensional, time-periodic chaotic flows and three-dimensional, spatially-periodic, steady flows [4]. Some notable examples of the former class of flows include the blinking vortex flow [5, 6], the cavity flow [4, 7, 8], and the journal bearing flow [4, 7, 9] shown in Fig. 2. Examples of the latter include the ABC flow [10] and the EHAM flow [11]. Each of these flows, while outwardly simple, produces a periodic stretching and folding of material elements that rapidly manifests itself as an exponential growth in the total amount of intermaterial area available for molecular diffusion. This repeated stretching and folding leads to the formation of highly layered regions of material striations having a wide distribution of separations. With increasing time, these layered regions are further stretched and folded, forming yet smaller striations from initially larger ones via an apparently self-similar multiplicative process.

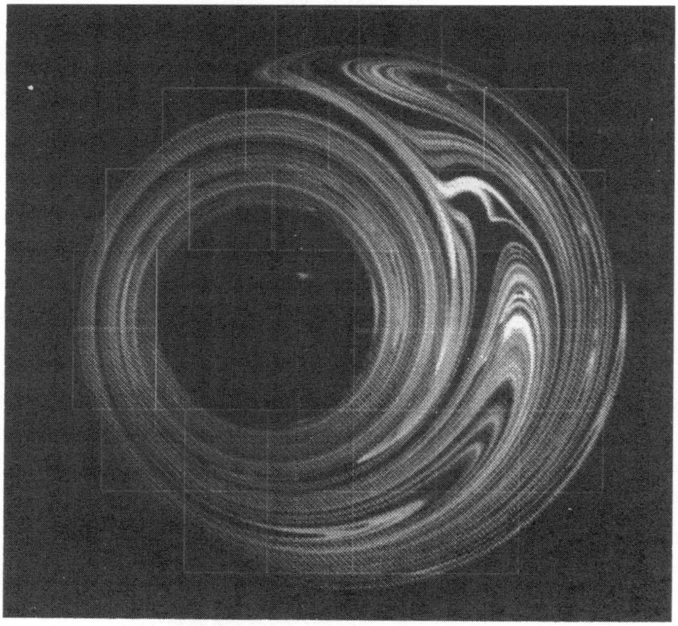

Fig. 2. The scalar field $\zeta(\mathbf{x}, t)$ produced by mixing in the chaotic, two-dimensional, time-periodic eccentric cylinder flow of Swanson and Ottino [12], shown for $\Omega = -3$ and $\vartheta = 4\pi$ at a relatively early period. Note the lamellar structure of the scalar field and compare this with Fig. 1. Shown also are the window boundaries into which the negative is divided for imaging analysis. For reference the boxes are numbered from left to right and top to bottom.

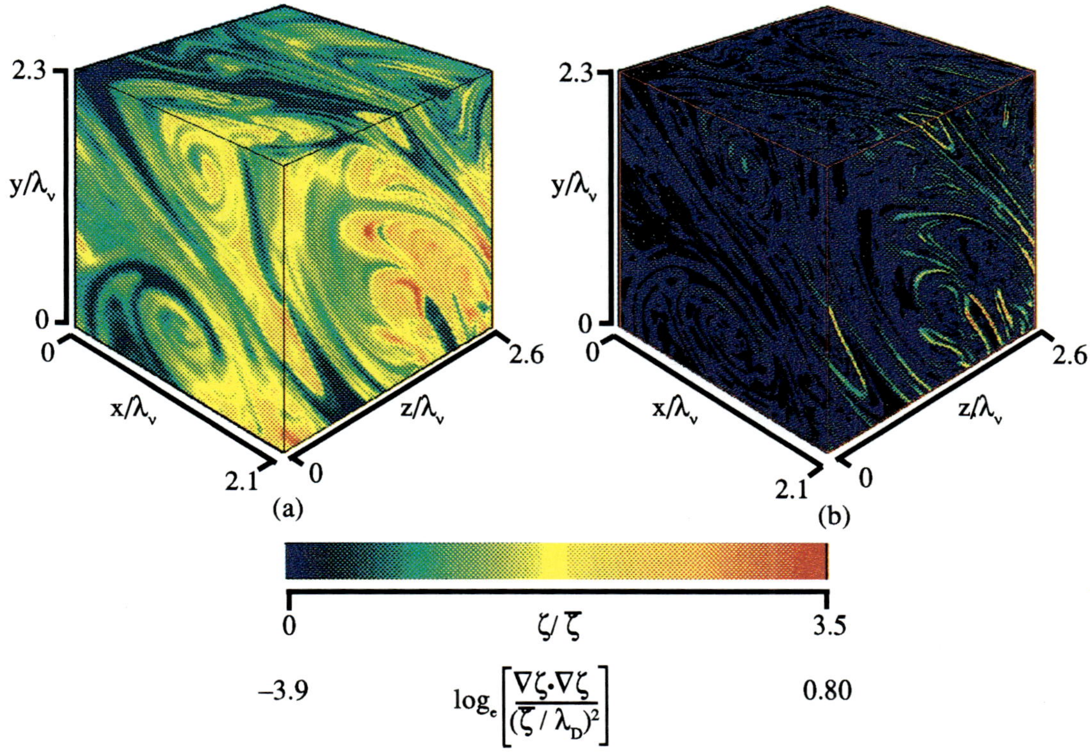

Fig. 1. Experimental data detailing the fluid mixing process at each of over 16 million points in a small three-dimensional spatial data volume taken from a four-dimensional spatio–temporal data space consisting of over 3 billion individual data points obtained in the self-similar far field of an axisymmetric turbulent jet. The volume dimensions are given in terms of the finest velocity gradient length scale λ_ν, proportional to the classical Kolmogorov scale. Shown are: (a) the conserved scalar concentration field $\zeta(\mathbf{x}, t)$, and (b) the associated scalar energy dissipation rate $(ReSc)^{-1}\nabla\zeta\cdot\nabla\zeta(\mathbf{x}, t)$. Note the highly convoluted layers in the dissipation field, created by the continual stretching and folding action of the three-dimensional time-varying flow field. For the case shown, $Re \approx 3700$ and $(ReSc) \approx 8 \times 10^6$.

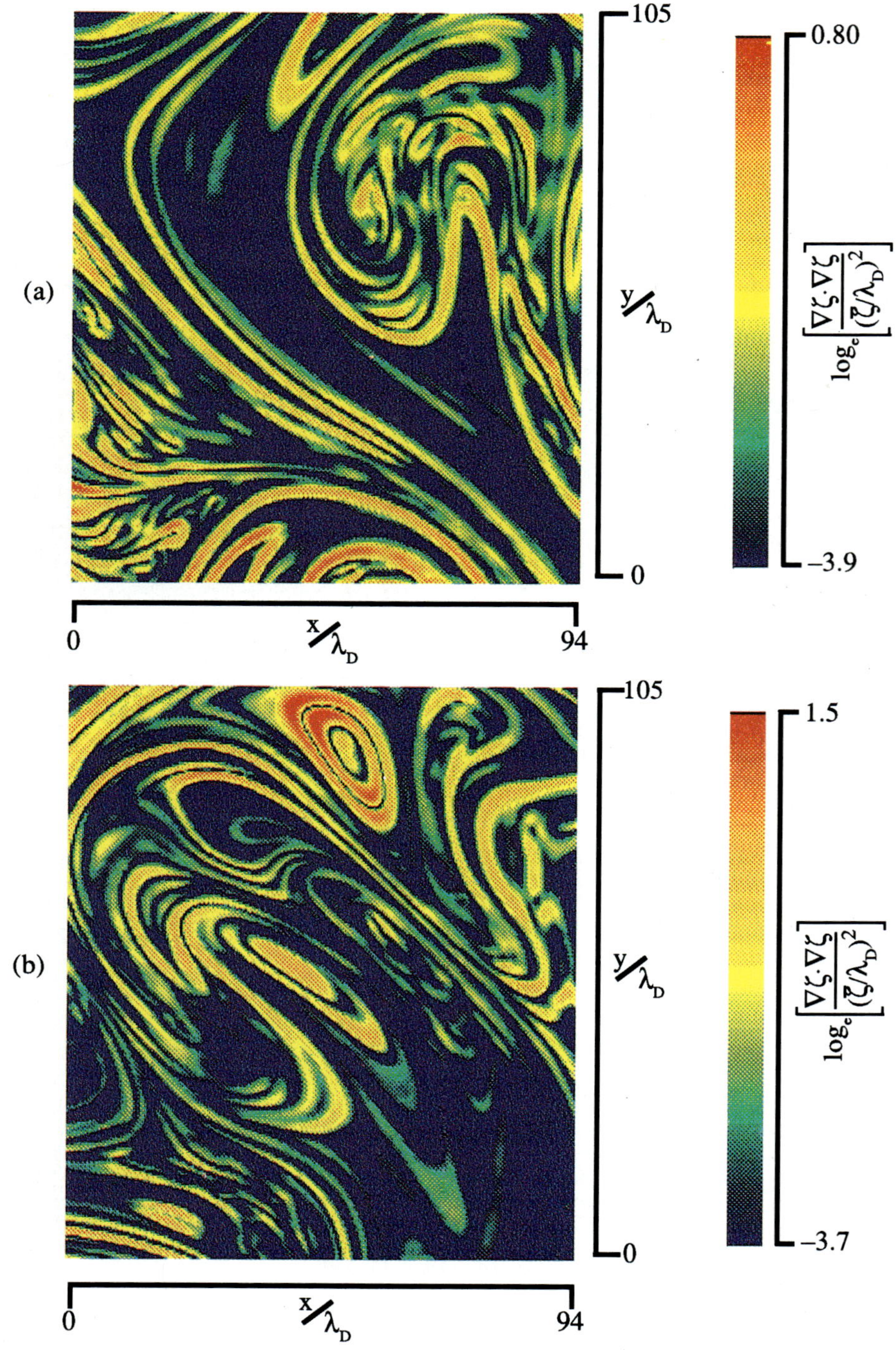

Fig. 3. Examples of the three-dimensional scalar energy dissipation rate field $(ReSc)^{-1}\nabla\zeta\cdot\nabla\zeta(\mathbf{x}, t)$ displayed as two-dimensional cuts from different three-dimensional spatial data volumes of the type shown in Fig. 1. Notice the highly convoluted nature of the layers forming the dissipation field.

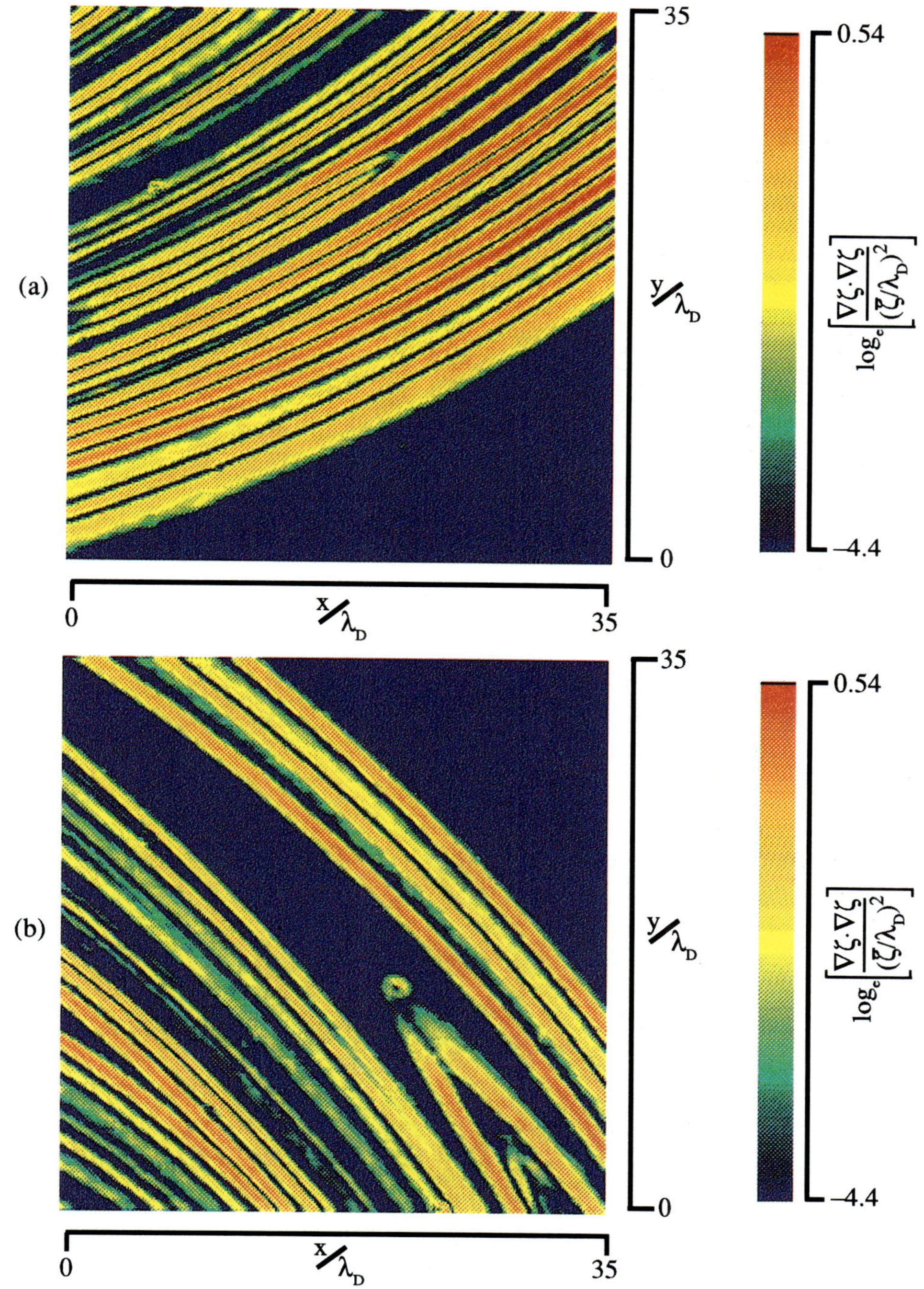

Fig. 4. The logarithm of the scalar energy dissipation rate field $\log_e[\nabla\zeta \cdot \nabla\zeta(\mathbf{x}, t)]$ for two different regions of the journal bearing flow of Fig. 2. (a) Example of a parallel region of the flow where the layers are essentially all aligned uniformly in the same direction. This is the resulting scalar dissipation field obtained from box 31 in Fig. 1. (b) Two-dimensional scalar dissipation field $\log_e[\nabla\zeta \cdot \nabla\zeta(\mathbf{x}, t)]$ calculated in box 7, an example of an intermediate region or one in which the layers show some deviation from a strictly parallel formation.

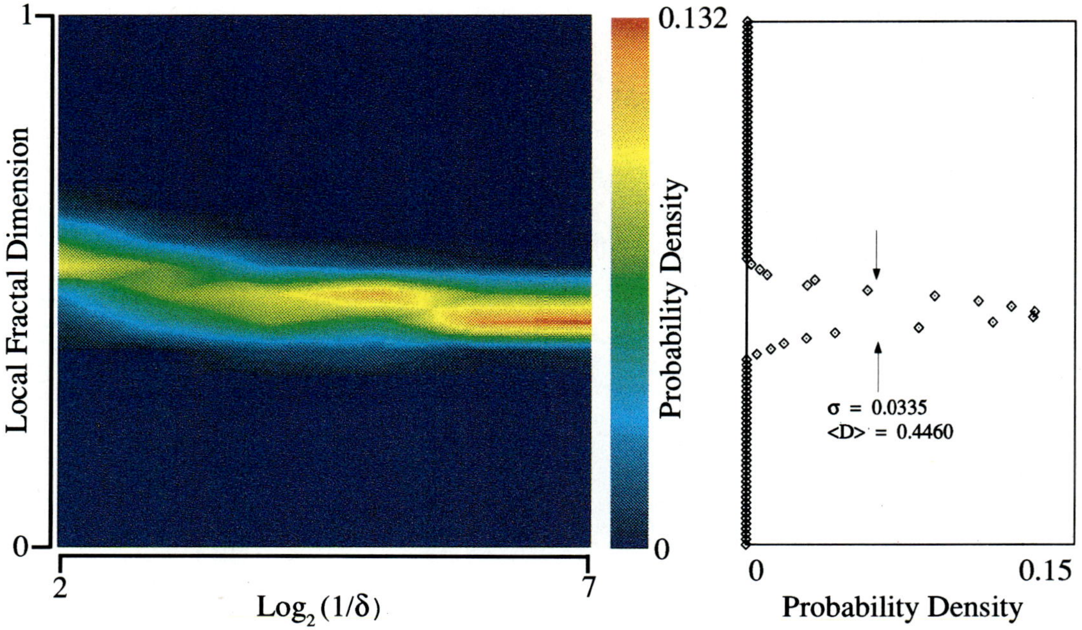

Fig. 18. The probability density plot of all of the local fractal dimension values from all 8000 test cases of digitized irregular cantor sets with dimension = 0.40. A cumulative probability density curve for all of the local fractal dimension values in the range $4 \leq \log_2(1/\delta) \leq 7$ is shown beside it. These test cases serve as a 'yardstick' with which to measure the fractal behavior of the experimental data.

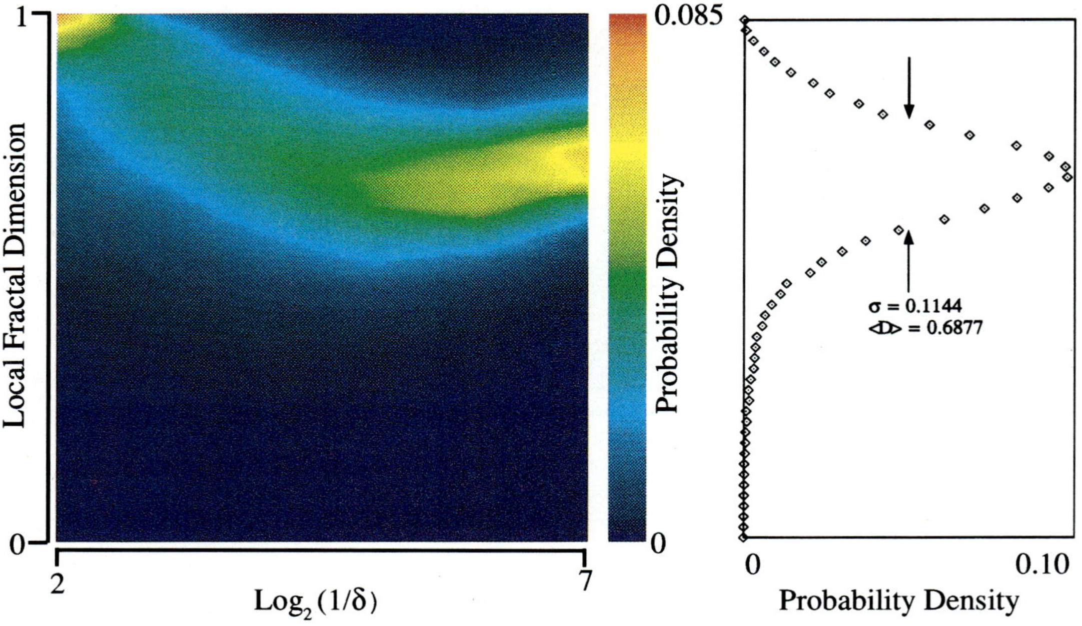

Fig. 22. The probability density plot of all of the local fractal dimension values from the chaotically advected data set. A cumulative probability density curve for all of the local fractal dimension values in the range $4 \leq \log_2(1/\delta) \leq 7$ is shown beside it. From this plot it is obvious that the whole field is not described by one single fractal dimension.

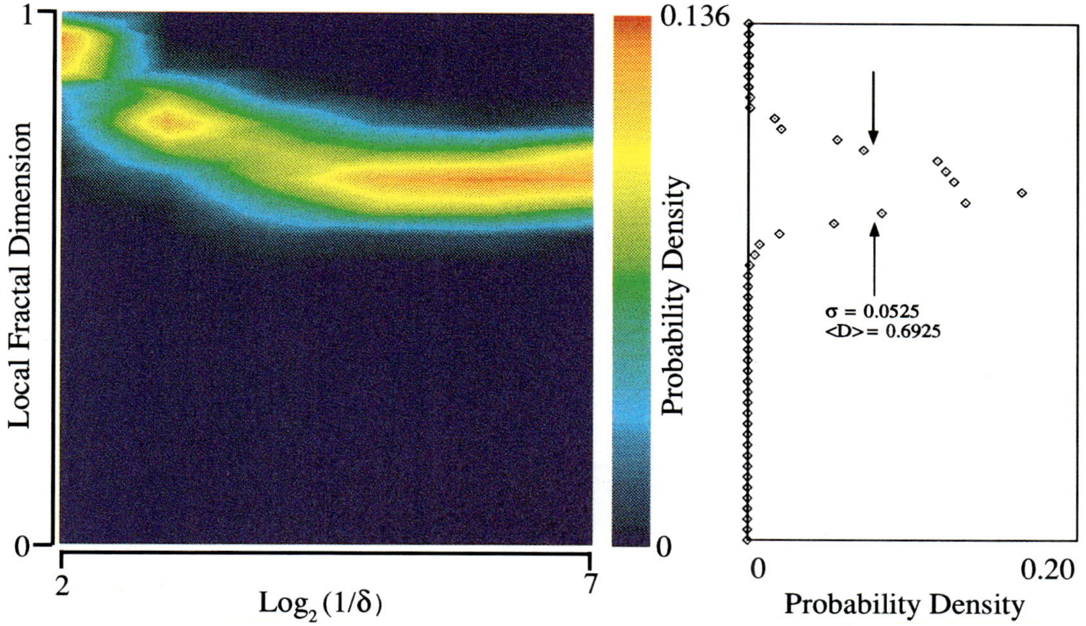

Fig. 23. The probability density plot of all of the local fractal dimension values from a parallel region of the chaotically-advected data set. A cumulative probability density curve for all of the local fractal dimension values in the range $4 \leq \log_2(1/\delta) \leq 7$ is shown beside it. It shows nearly fractal behavior but the standard deviation in the local fractal dimension values is still approximately 2 times the value obtained from the test cases.

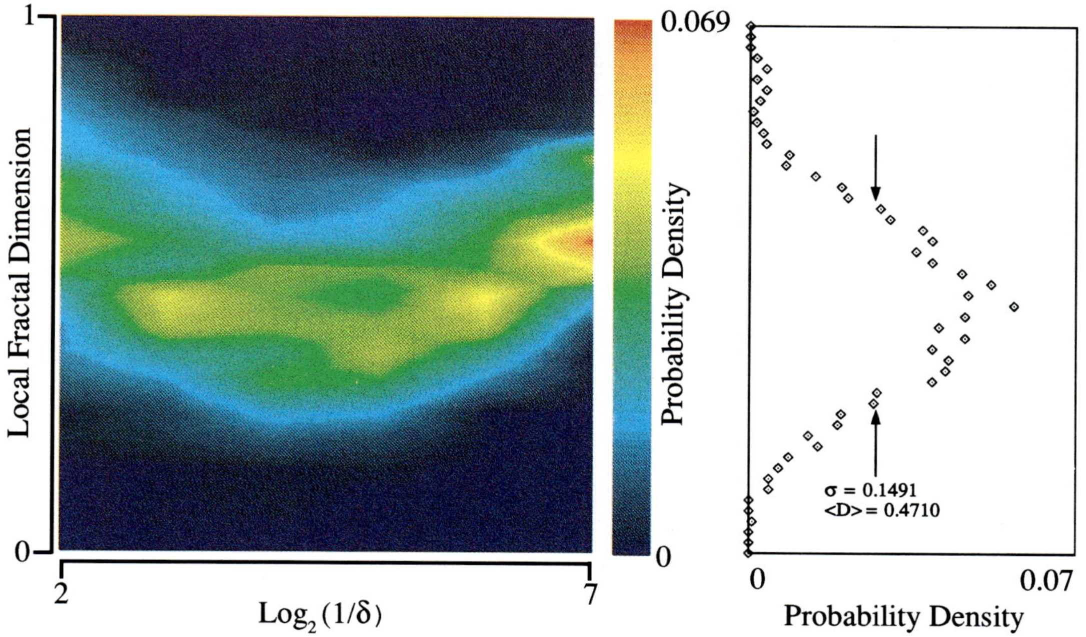

Fig. 24. The probability density plot of all of the local fractal dimension values from a non parallel region of the chaotically-advected data set. A cumulative probability density curve for all of the local fractal dimension values in the range $4 \leq \log_2(1/\delta) \leq 7$ is shown beside it. This region shows absolutely no fractal behavior.

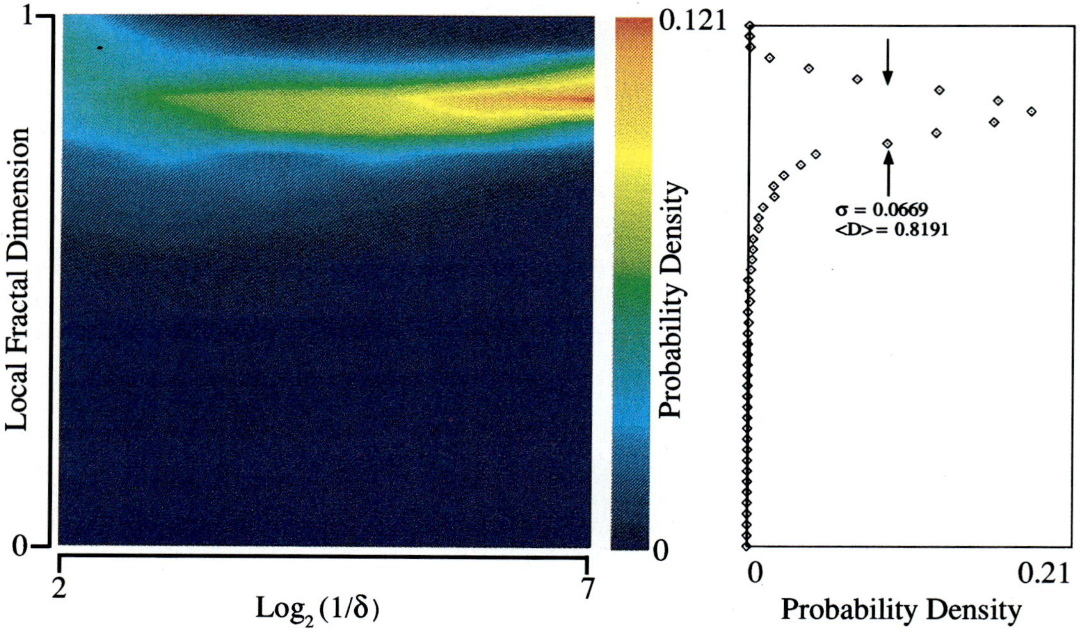

Fig. 27. The probability density of the local fractal dimension values from a typical region of the turbulent data set. Beside it is a cumulative probability density curve for all of the local fractal dimension values in the range $4 \leq \log_2(1/\delta) \leq 7$. This region shows slight fractal behavior with a standard deviation in the local fractal dimensions that is approximately 3 times the value obtained from the test cases.

Given the obvious qualitative similarities in the highly convoluted layered structures seen in such chaotic flows and in turbulent flows like that in Fig. 1, one might hope that the analytical tools that have been fruitfully applied to chaotic mixing might be applicable to turbulent mixing as well.

The widely held view is that turbulent flows are far more complex than these lower-dimensional and periodic chaotic flows, and that insights drawn from chaotic mixing might therefore have limited applicability to turbulent mixing. Indeed, when viewed on relatively large scales, the hydrodynamics of turbulent flows are undeniably more complex. However, at the small scales it becomes more difficult to argue for fundamental differences between these two types of flows. In particular, when turbulent mixing is viewed in a Lagrangian frame, the underlying hydrodynamics at the smallest velocity gradient length scales reduce to a locally uniform strain flow with time-varying principal axes, which leads to a continual stretching and reorientation of material elements in the flow. On slightly larger length scales, differential rotation rates resulting from vorticity gradients fold the drawn-out material elements back onto themselves in a manner reminiscent of the Smale horseshoe in the chaotic flow examples. The picture that results is one of turbulent mixing as a repeated stretching and folding of material elements by time-varying strain-rate and vorticity fields, much as in the chaotic flows, albeit in three dimensions rather than two, and without the simplicity of time-periodicity in the strain-rate eigenvector orientations. Moreover, unlike the closed chaotic mixing systems referred to above, a key feature of many turbulent flows is the continual entrainment of fluid into the mixing region—increasingly more fluid participates in the Lagrangian mixing process as the flow evolves in time. In fact, it is precisely the large fluid entrainment rates that make turbulent flows attractive for most practical applications involving fluid mixing. Keeping in mind these differences, it is tempting to imagine that the underlying similarities in the mixing dynamics at the small scales in a turbulent flow and in a chaotic flow may lead to similar mixing patterns. The tools developed for describing and predicting the mixing properties of chaotic flows might then be productively applicable to turbulent flows as well. Such flows would be relatively simple at the smallest scales, and demonstrate the complexity traditionally associated with turbulence only in the aggregate. This view of turbulent mixing built up from the small scales naturally suggests certain universal scalings and Reynolds number asymptotics traditionally associated with fluid turbulence.

1.1. *Present work*

Within the context of the above discussion, the objective here is to compare various scaling properties of the mixing process at the small scales of high Reynolds number, three-dimensional, unsteady, open turbulent flows with those in low Reynolds number, two-dimensional, time-periodic, closed chaotic flows. To this end the experimental analyses are focused on two flows. We take the creeping flow between eccentric cylinders executing a discontinuous periodic rotation schedule as a canonical representation of the fundamental mixing patterns produced by the closed chaotic flow systems referred to above. The open turbulent flow in the self-similar far field of an axisymmetric jet at high Reynolds number is taken to represent the quasi-universal mixing properties at the small scales of turbulent flows. In each of these flows, measurements of the mixing of a dynamically passive, conserved scalar field $\zeta(\mathbf{x}, t)$ by the underlying flow field are used to analyze geometric scaling properties of the support set on which the associated scalar energy dissipation rate field $\nabla\zeta \cdot \nabla\zeta(\mathbf{x}, t)$ is concentrated.

For both flows we examine two measures of the dissipation support set geometry. The first closely resembles the 'striation thickness distribution' of Ottino [4]. In practice, the

STD is difficult to measure accurately, especially in a system with significant diffusion. For this reason we instead work with the set of points forming the convoluted surface on which the local centers of the scalar dissipation layers reside. From this surface we determine the local one-sided surface-normal distance to the next adjacent dissipation layer. Repeating this over the entire surface produces the distribution of dissipation layer separation distances, which can be readily determined in practice even in highly diffusive systems. We then compare the distributions obtained in the chaotic and turbulent flows as a relatively sensitive measure of the geometry of the dissipation field at the small scales.

Second, we characterize the dissipation support geometry by examining the possible fractal scaling properties of the set of points within which the scalar dissipation rate exceeds a threshold level. Prior attempts at assessing the fractal properties of turbulent mixing have focused largely on the geometry of isoscalar surfaces, and have found that the results depend strongly on the chosen isoscalar value. In contrast, owing to the highly localized layer-like sheets on which the scalar dissipation field is concentrated (see Fig. 1), the geometric scaling properties of the dissipation field are comparatively insensitive to the choice of dissipation threshold. We assess the applicability of a fractal description of the dissipation support for the mixing process in both the chaotic and turbulent flows.

2. EXPERIMENTS

2.1. *Turbulent flow scalar fields*

Conserved scalar fields $\zeta(\mathbf{x}, t)$ as shown in Fig. 1 were measured in the self-similar far field of an axisymmetric turbulent jet at outer scale Reynolds numbers $Re_\delta \equiv (u\delta/\nu)$ as high as 6000, and with resolution finer than the local strain-limited molecular diffusion scale λ_D. Since the imaged volume is small in comparison with the local outer flow scale δ, and comparable to the inner scale λ_ν of the turbulence, the resulting small scale structure of the scalar field will be essentially independent of both the Reynolds number and the particular turbulent shear flow, and should be generic to large Schmidt number $Sc \equiv (\nu/D)$ mixing in high Reynolds number turbulent flows. Details of the measurement technique can be found in ref. [1]. Briefly, we measure the aqueous concentration of a laser fluorescent dye carried by one of the fluids in a small three-dimensional spatial volume located 235 jet momentum diameters (1.15 m) downstream of the jet source and 13 cm off the jet centerline, well into the self-similar far field of the flow. The dye mixture fraction is a conserved scalar with $Sc \approx 2075$. This is measured in time throughout the volume by stimulating laser-induced fluorescence from dye-containing fluid from a collimated laser beam and imaging this onto a high-speed, planar, photodetector array. The beam is swept in a raster fashion through the desired volume in the flow, and the resulting laser-induced fluorescence intensity is collected onto the 256×256 pixel imaging array, which can be driven at continuous rates in excess of 140 data planes per second. The fluorescence data from the array is serially acquired at 8-bits true digital depth, and continuously written in real time to a 3.1 GB high-speed parallel transfer disk rank capable of accommodating more than 50 000 such 256^2 spatial data planes. The spatial separation between adjacent points within each data plane, and between adjacent data planes within each data volume, is smaller than the local strain-limited molecular diffusion length scale λ_D of the scalar field, where $(\lambda_\nu/\delta) \approx 11.2 \cdot Re_\delta^{-3/4}$ and $\lambda_D \approx \lambda_\nu \cdot Sc^{-1/2}$. The scales λ_ν and λ_D are respectively proportional to the classical Kolmogorov and Batchelor scales (which are defined solely on dimensional grounds) but are roughly a factor of 6 larger. Similarly, the temporal separation between adjacent data planes within each data volume and, depending on the number of spatial planes, between the same data plane in successive data volumes, is shorter than the local molecular diffusion scale advection time λ_D/u. This resolution, together with the high signal quality attained, allows accurate differentiation of the measured conserved scalar data in all three space dimensions and in time to determine the components of the true local instantaneous scalar gradient vector field $\nabla\zeta(\mathbf{x}, t)$ at every point in the data space.

Figure 3 shows the scalar dissipation rate data in two typical spatial data planes. Note all quantities are normalized by the inner scales λ_D and $\langle\zeta\rangle$ of the underlying turbulent flow. Each $\nabla\zeta\cdot\nabla\zeta(\mathbf{x}, t)$ plane is obtained by direct linear central differencing of the scalar field data in the three adjacent scalar planes; no explicit smoothing or filtering is used in evaluating the three components of the scalar gradient vector field $\nabla\zeta(\mathbf{x}, t)$. Owing to the wide range of dissipation rates (see Fig. 1(b)), we show $\log_e \nabla\zeta\cdot\nabla\zeta(\mathbf{x}, t)$ in Fig. 3 to allow the structure at low dissipation rates to be readily seen. Notice that essentially all of the molecular mixing occurs in thin sheet-like scalar dissipation layers.

2.2. *Chaotic flow scalar fields*

Measurements of the scalar field $\zeta(\mathbf{x}, t)$ in chaotic mixing are obtained by digitizing photonegatives provided by J. M. Ottino of his experiments on the flow between eccentric rotating cylinders—the 'journal bearing flow'. Details of the experimental apparatus and flow can be found in ref. [12]. Briefly, two cylinders of radii R_1 and R_2, with centers displaced by ε, are respectively rotated in a discontinuous time-periodic fashion through angular displacements ϑ_1 and ϑ_2 during each period. At Reynolds numbers sufficiently low for the creeping flow approximation, the annular flow field is determined by the two geometric parameters (R_1/R_2) and (ε/R_2), held fixed at 1/3 and 3/10 respectively, and the two displacement angles, which can be expressed as $\Omega \equiv (\vartheta_1/\vartheta_2)$ and $\vartheta \equiv \vartheta_2$. For the case considered here, $\Omega = -3$, and $\vartheta = 4\pi$. The fluid in the annulus is glycerine, a region of which is initially marked with fluorescent dye. Owing to the low diffusivity of the dye in glycerine, the Peclét number ($ReSc$) is large even though $Re \to 0$. The flow is allowed to mix this tracer for a given number of rotation periods, after which the dye pattern in the annulus is photographed. The scalar field in the resulting 3 in. × 5 in. photonegatives is digitized on a light table using the same imaging array described above. The negative is divided into individual windows, as indicated in Fig. 2, and each window digitized to 256^2 resolution and 8-bits signal depth. The repeated stretching and folding of the dyed fluid by strain and differential rotation can be seen to lead to a highly layered (lamellar) structure in the scalar field $\zeta(\mathbf{x}, t)$. The basic features of this field are established after just a few periods; further iterations principally increase the fine structure in an apparently self-similar fashion. Within each measurement window, the inherently two-dimensional scalar energy-dissipation rate field $\nabla\zeta\cdot\nabla\zeta(\mathbf{x}, t)$ is computed using linear central differences on this discretized field. Two typical examples of the resulting scalar dissipation fields are shown in Fig. 4.

There are difficulties inherent in this approach that may influence the results. These include errors in the photonegatives themselves, as discussed in ref. [12], among which parallax effects and the nonlinear film response are likely to be the most prominent. Other nonidealities are introduced by our processing of these negatives, e.g. minor flaws in the negative quality occasionally produce spots in the dissipation field. These imperfections appear to be sufficiently infrequent as not to significantly affect the results obtained. A potentially more serious limitation stems from the restricted spatial resolution achievable by indirect imaging measurements from photonegatives. With increasing number of periods, the 'striations' become thinner and eventually cannot be accurately resolved, limiting the maximum period n for which accurate measurements can be obtained. Results of Muzzio *et al.* [13] suggest that, for relatively low values of Ω and ϑ, full self-similarity in the distribution of stretching values may not be achieved until $n \approx 10$. The extent of similarity, however, appears likely to depend on the total stretching that has occurred, while at the same time the striation thicknesses that need to be resolved decrease as the total stretching increases. The central issue, then, is whether or not the striations can still be adequately resolved at a stage in the development of the flow for which self-similarity is reached. Comparing Fig. 1(b) of ref. [13] ($n = 10$ for $\Omega = 3$ and $\vartheta = 2\pi$) with our Fig. 2 suggests a roughly similar total stretching, while their Fig. 4 indicates that this stage of development is indeed marginally sufficient for self-similarity.

3. DISSIPATION LAYER SEPARATIONS

The detailed structure of the mixing patterns that result from the repeated stretching, reorientation, and folding of material elements in the chaotic journal-bearing flow have been discussed in terms of the striation thickness distribution (STD) by Ottino [8, 9]. The STD is, however, notoriously difficult to measure, in part for the reasons outlined above. Consequently, numerical simulations of mixing in chaotic flows (e.g. refs [13–15]) have focused instead on the distribution of stretching values (DSV), which is closely related to the STD and can be readily obtained if access to the underlying velocity field is available. For two-dimensional, time-periodic, creeping flows of the type classically used in studies of chaotic mixing, the velocity field is known, and the numerical studies referred to above have used this to demonstrate remarkable self-similarity and scaling properties of the DSV. In turbulent flows, however, access to the fully three-dimensional, time-varying velocity field is only recently beginning to become available [2], and thus routine access to the DSV is not yet possible. A comparison of the mixing patterns in chaotic and turbulent flows must, therefore, return to measures of the spatial structure of the scalar field. Owing to the inherently sheet-like structure of the scalar dissipation field, we will compare the distribution of dissipation layer separations. Once the effects of initial conditions have become relatively unimportant, this layer separation distribution will be closely related to both the STD and the DSV. Unlike the STD, however, the compact support of the scalar dissipation field allows layer separation distributions to be somewhat more easily measured and, unlike the DSV, this can be obtained in both the chaotic and turbulent flows. The procedure we use for determining these distributions is described in Section 3.1; results for the chaotic and turbulent flows follow in Sections 3.2 and 3.3.

3.1. *Algorithms*

Determining the dissipation layer separation distribution first requires that the set of points containing the dissipation layer centers must be found within each two-dimensional scalar dissipation plane from the chaotic flow experiments, as shown in Fig. 4, and each three-dimensional scalar dissipation volume from the turbulent flow experiments as shown in Figs 1 and 3. We use an algorithm for automatically finding the dissipation layer centers; however, as is often the case in 'machine vision' applications, this is more difficult to do well than it might appear. Extensive tests conducted with various algorithms revealed features necessary for reliably identifying the layer centers in the algorithm that was finally used. This algorithm is based principally on the scalar gradient vector field $\nabla\zeta(\mathbf{x}, t)$. Briefly, it begins by thresholding the gradient magnitude field (dissipation rate field) at some chosen level. Next, it identifies the edges of the regions in which the dissipation rate is above the threshold value. To do this, for each point the local unit normal vector for the

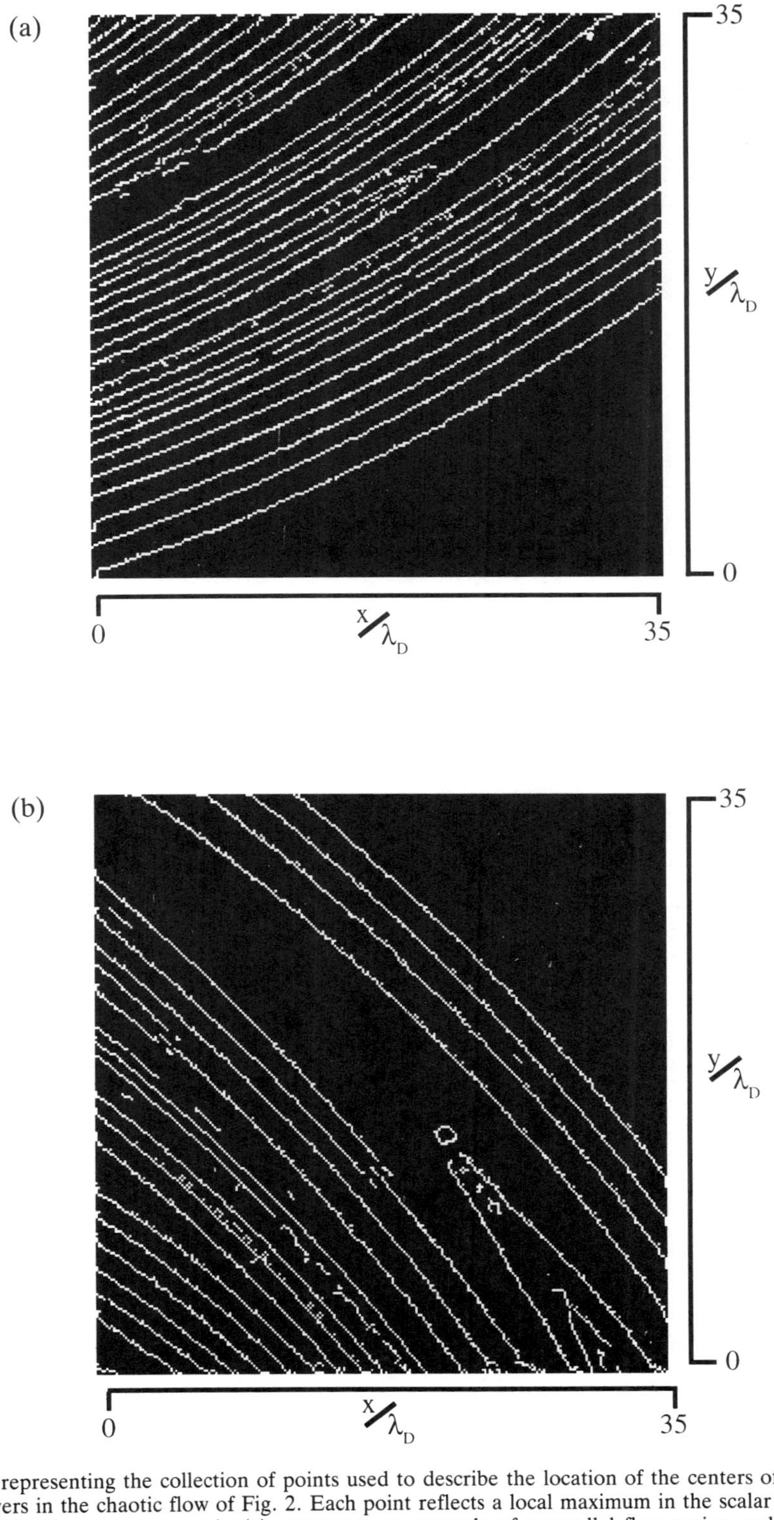

Fig. 5. Fields representing the collection of points used to describe the location of the centers of the scalar energy dissipation layers in the chaotic flow of Fig. 2. Each point reflects a local maximum in the scalar energy dissipation rate fields of Fig. 4 where once again (a) represents an example of a parallel flow region and (b) represents an intermediate region. Note the cleanliness of the layer definitions resulting from the application of the center finding algorithm to this two-dimensional flow. Layer separation PDFs are obtained by measuring the distances between adjacent layers at each point defining the layers.

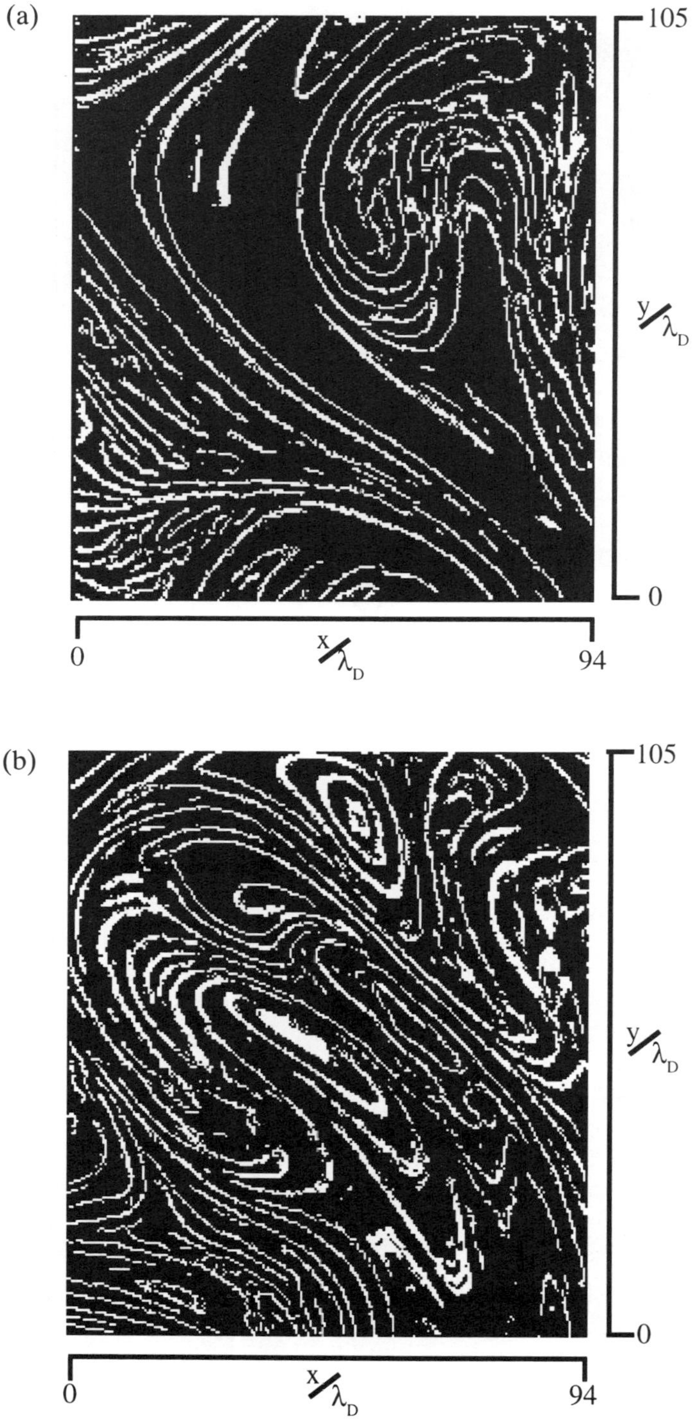

Fig. 6. Fields representing two-dimensional cuts of the collection of points used to describe the location of the centers of the scalar energy dissipation layers in the fully three-dimensional turbulent shear flow of which Fig. 1 is an example. Each point reflects a local maximum in the scalar energy dissipation rate fields of Fig. 3 (a) and (b). Note that the intersections of the three-dimensional surface defining the center of a layer with the two-dimensional cut plane allows adjacent pixels within the plane to be identified with the surface wherever it passes through the plane at more of a grazing incidence. A thick layer will appear wherever the sheet normal direction is pointing essentially perpendicular to the cut plane.

layer is determined based on the local gradient vector information. The edges of the layer are then found by marching in both directions along this unit normal direction until one of two criteria are met: (i) the scalar dissipation at the candidate edge point drops below the chosen threshold level, or (ii) the dot product between the scalar gradient unit vector at the original point and that at the candidate edge point changes sign. The resulting edge locations for the dissipation layers are then used to identify candidate midpoints representing the dissipation layer centers. A further check is made among these candidate dissipation layer center points to insure single pixel thickness for the resulting surface before finally admitting a pixel to the set of layer center points.

Figure 5 shows the resulting dissipation layer centers identified by this algorithm for the two data planes in Fig. 4. Note that most of the layer centers are accurately determined, though there are instances where very weak dissipation layers (the color scale in Fig. 4 is logarithmic) produce either 'holes' or 'erroneous layers'. Generally speaking, algorithms capable of reliably detecting very weak layers produce many erroneous layers. Conversely, attempts to reduce the number of erroneous layers tend to miss the weaker dissipation layers. The final algorithm used is a compromise between these two competing interests, representing roughly an 'optimum' among the various identification strategies attempted. Obvious means for further reducing the number of small spurious layers in the chaotic flow in Fig. 5 tend to remove valid layers in the turbulent flow, owing to the more convoluted nature of the dissipation field in Fig. 3. Examples of the dissipation layer centers for these two planes are shown in Fig. 6, which show that most of the layer centers are accurately identified. The more highly folded character of the dissipation layers in the turbulent flow, together with the three-dimensional nature of the data, make identification of individual layers even conceptually more difficult than for the nearly parallel layers in the two-dimensional chaotic flow data. Notice also that, unlike the inherently two-dimensional surface of layer centers in the chaotic flow, intersections of the three-dimensional surface resulting from the turbulent flow data with any two-dimensional plane, as in Fig. 6, allows adjacent pixels within the plane to be identified with the surface wherever it passes through the plane at a relatively shallow angle.

The influence of imperfections that survive the layer center detection algorithm is considerably reduced by the algorithm used to determine the layer separation distances. This layer separation algorithm begins with the surface of dissipation layer centers, and aims to compute the surface-normal distance in either two or three dimensions from any point on the surface to the next neighboring part of the surface, in a manner that is relatively insensitive to small holes or spurious layers. At each point on the surface, the local surface-normal axis is determined from the scalar gradient vector orientation at neighboring points. The algorithm marches along this axis, staying on one side of the surface (to avoid double-counting), until the next intersection with the surface. The search direction along this axis is accordingly determined by which of the four quadrants (2D) or eight octants (3D) $\nabla\zeta(\mathbf{x}, t)$ points in. To reduce the effects of imperfections in the surface, the search for the next intersection is conducted within a 15° half-angle wedge (2D) or cone (3D) centered on the local surface-normal axis.

Tests conducted with synthetically generated surfaces having varying levels of imperfection verify that the resulting distribution of layer separations can be accurately obtained. An example is shown in Fig. 7(a). In this case, a synthetic scalar field was specified as $\zeta(\mathbf{x}) = \sin(x^2)$ over $0 < x < 3\pi$. From the associated scalar dissipation field, the resulting ideal cumulative distribution of layer separations can be obtained analytically. This agrees well with the distribution obtained from the procedure described above, as shown in Fig. 8. (Since the probability density for this discrete test case is not continuous, we instead compare the cumulative distributions. For the continuous separation distances in the

(b)

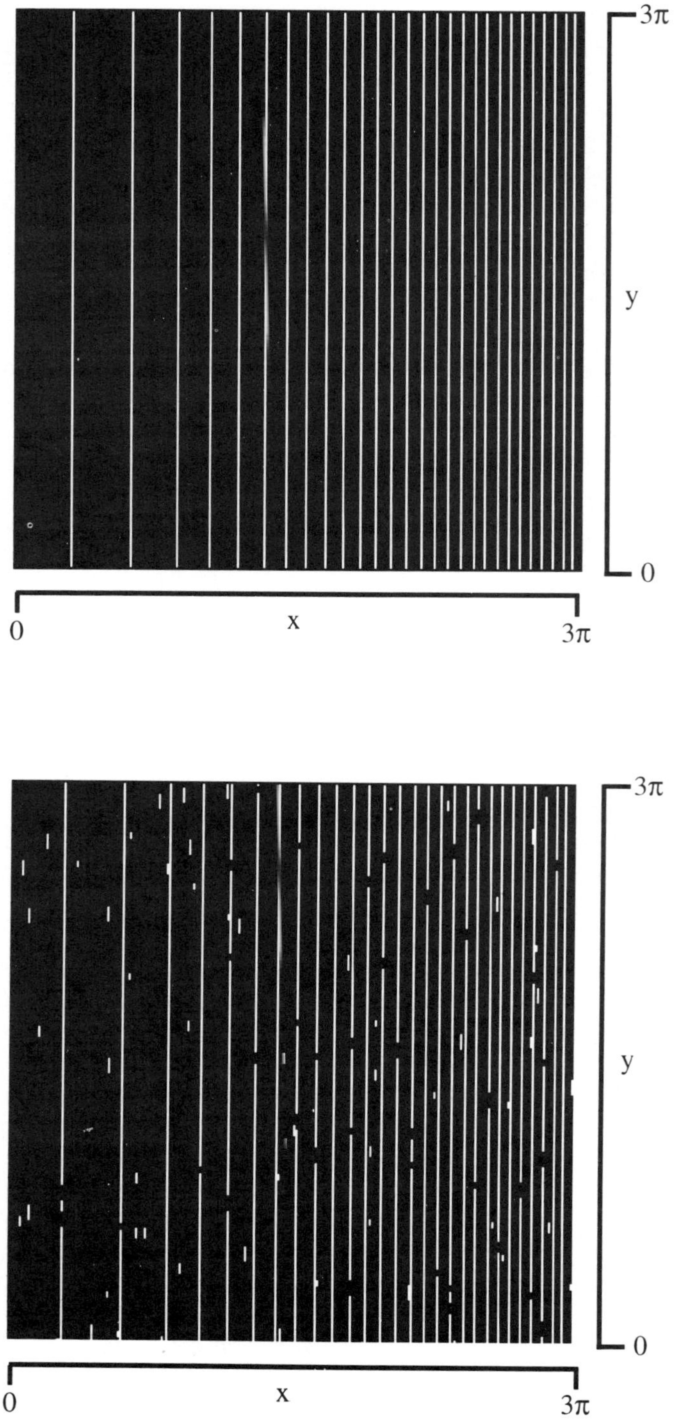

Fig. 7. Layer center definition fields for a test case that was formulated in terms of a conserved scalar field $\zeta(\mathbf{x}, t)$ that varied as $\sin(x^2)$. The scalar energy dissipation field in this case varies as $\nabla\zeta \cdot \nabla\zeta(\mathbf{x}, t) = 4x^2 \cos(x^2)$ so that the magnitude as well as the frequency of the layers vary with the coordinate x. Shown in (a) is the clean version of this field resulting from our layer maximum finding algorithm and in (b) is the field in (a) made noisy by the addition of intentional holes and extraneous layers in a random manner. Both were subjected to a comparison measurement of their individual layer separation distance cumulative distribution functions which can be seen in Fig. 8.

chaotic and turbulent flow cases, we will compare the derivative of the cumulative distributions, namely the probability densities.) In Fig. 7(b), a random pattern of holes and extraneous layers, of various sizes and locations, were added to mimic similar effects anticipated in the measured data. The imperfections in this case are meant to qualitatively match those present in the layer centers in Figs. 5 and 6. The layer separation distribution resulting from this imperfect layer center field is also shown in Fig. 8. Except for very small separations, the resulting distribution still follows the ideal distribution reasonably well, and continuous derivatives (probability densities) of both sets of data are similar for all but the smallest layer separations.

3.2. *Chaotic flow results*

The probability density of dissipation layer separations is computed separately for each of the 31 windows identified in Fig. 2 via the above procedure. The layer separation distribution resulting from ensemble statistics over the entire set of windows covering the flow is shown in Fig. 9. This result is roughly analogous to the distribution of stretching values $F_n(\lambda)$ given in Fig. 3 of ref. [13], where a power law scaling can also be identified in the logarithmic axes, and where evidence of self-similarity can be seen. However, large stretching values in their case correspond to small layer separations in our results and, for a given range of layer separations, the corresponding range of stretching values is much wider. While Fig. 9 gives the statistics for the entire ensemble of windows in Fig. 2, the distributions of layer separations in individual windows turn out to be very different. For this reason, it proves useful to broadly classify the dissipation field structure in various regions of Fig. 2 into three different types. The first of these are termed ‘parallel’ regions, referring to windows in which the dissipation layers are all essentially straight and very nearly parallel to one another. The second are termed ‘nonparallel’ regions, where the layers are not straight or generally not parallel to each other. Finally, windows whose overall dissipation layer structure contains features of both are termed ‘intermediate’.

In Fig. 10 we show the distribution of layer separations for all the parallel regions only.

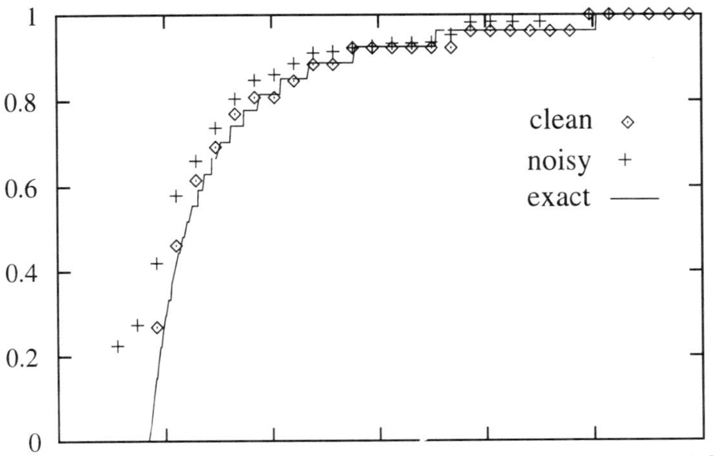

Fig. 8. Layer separation distance cumulative distribution functions for the test cases in Fig. 7. Shown as a solid line is the exact result for the cumulative distribution function for layers defined by $\nabla\zeta \cdot \nabla\zeta(\mathbf{x}, t) = 4x^2 \cos(x^2)$ over the range from 0 to 3π obtained analytically. Overlaid with good agreement are the results obtained from our nearest-neighbor distance-finding algorithm for both the clean and noisy fields of Fig. 7.

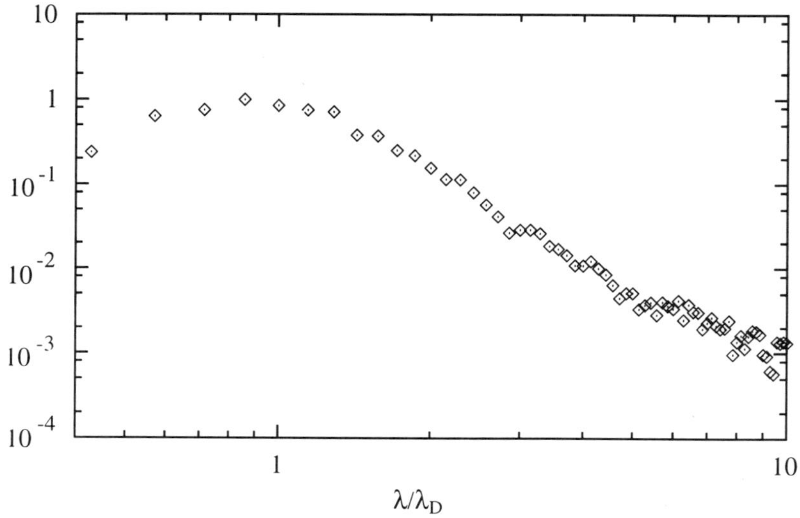

Fig. 9. Composite layer separation PDF for the chaotic flow between two eccentric cylinders as shown in Fig. 2. Result is formulated from the nearest neighbor calculation performed on the scalar energy dissipation rate field for *all* regions mapped out by the boxes in Fig. 2 regardless of the predominant layer arrangement within the box.

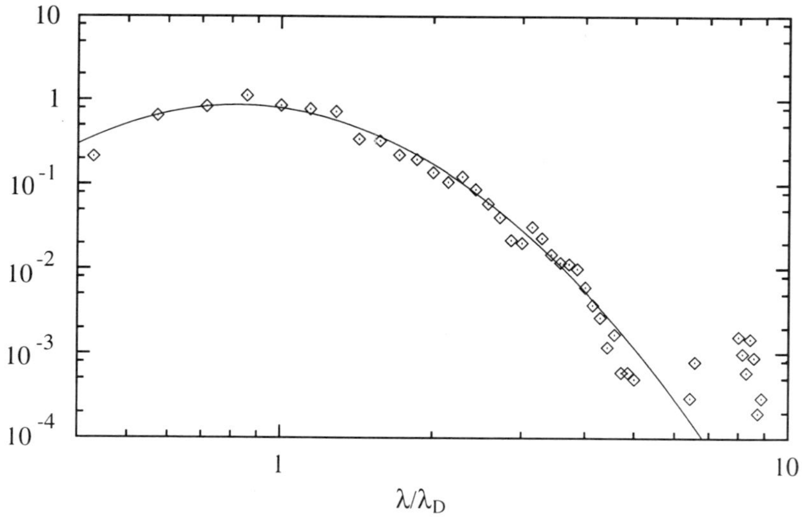

Fig. 10. Layer separation PDF for the parallel layer regions of the chaotic flow displayed in Fig. 2. Here the nearest-neighbor calculation was restricted to those regions of the flow, divided by the boxes in Fig. 2, comprising predominantly parallel arrangements to the scalar dissipation layers. Shown for comparison is a lognormal curve fitted with the same first and second moments of the resulting PDF.

The result can be seen to be distinctly different from that obtained over the entire flow field in Fig. 9, especially for large layer separations. Also shown in this figure is a lognormal distribution having the same first two moments, which can be seen to provide a good fit for all except the very largest layer separations. Such a lognormal form is expected from classical central limit arguments for multiplicative processes (e.g. [13, 14, 16]) provided that the number of independent multiplicative steps producing the layer separations is sufficiently large. One way to meet this provision is to require the flow to have undergone a large number of periods. Muzzio *et al.* [13, 14] show that, when collecting unconditional statistics over the entire flow, the highest stretching values (smallest layer separations) show lognormal scaling even after just a few repetitions (see their Fig. 3(a)). In a flow that has nonchaotic 'islands', however, departures from complete lognormality remain apparent in their DSVs at the lowest stretching values (largest layer separations), regardless of how many repetitions the flow has undergone. These departures from lognormal scaling can be attributed to locally low stretching values that lead to the persistence of large voids in the dissipation support. By collecting unconditional statistics, the continuing influence of these voids will be seen in both distributions of stretching values and layer separations. Following this reasoning, we might expect the lognormal scaling to hold at the smallest layer separations even after just a few periods. Indeed, by collecting conditional statistics as in Fig. 10, and thereby presumably avoiding areas of the flow strongly influenced by the voids, lognormal scaling can be seen over essentially the entire range of layer separations. The 'nonparallel' regions of the scalar field would then presumably be those most strongly influenced by the low strain rates in the vicinity of the voids, where the low stretching values lead to a slower reduction of the dissipation layer separations and a slower reorientation of the dissipation layers by the principal strain axes.

To test this, the individual distributions resulting for the parallel (no voids), nonparallel (voids), and intermediate regions are shown in Fig. 11. These reveal two distinct trends. First, in all three cases, the smallest separations follow a nearly lognormal distribution. For the parallel regions, this lognormality is preserved for all but the largest separations (up to $\lambda/\lambda_D \approx 6$). For the intermediate regions, departures from the lognormal distribution begin at $\lambda/\lambda_D \approx 4$, while in the nonparallel regions the lognormal form holds only up to $\lambda/\lambda_D \approx 3$. Second, as would be expected, large layer separations are rare in the parallel regions, and relatively prominent in the nonparallel regions. The closed nature of this flow suggests precisely these features. Since the total area of the surface of dissipation layer centers initially increases with time as the fluid is stretched and folded, the finite volume occupied by the flow requires a continual reduction in the average layer separation distance. With

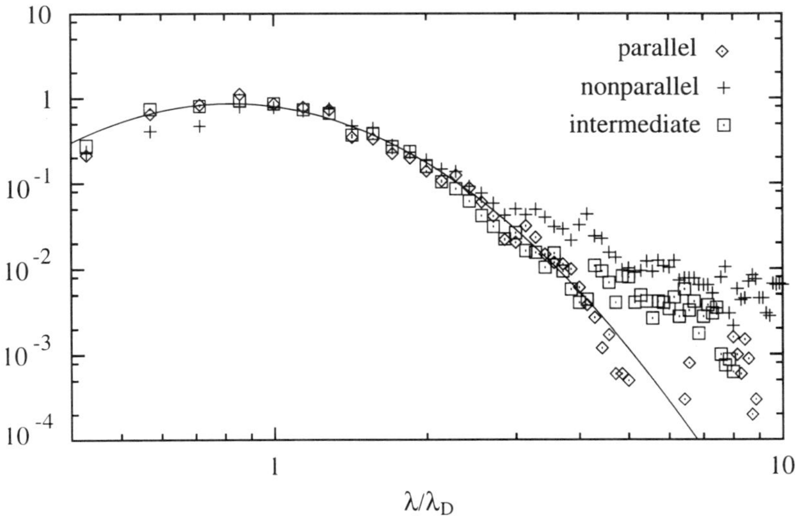

Fig. 11. Layer separation PDFs for the three distinct regions of the chaotic, two-dimensional, time-periodic, eccentric cylinder flow of Fig. 2 defined as parallel, nonparallel, and intermediate. Shown for comparison is a lognormal curve fitted with the same first and second moments of the PDF of layer separations for the parallel regions. Note that the deviation from this lognormal curve occurs at decreasing values of the distance separating neighboring layers as we move from the PDF for the parallel regions through the PDF for the intermediate regions to the PDF for the nonparallel regions.

increasing time, large separations must become increasingly rare (though only at a slow rate set by processes near the voids) and smaller separations must become increasingly dominant. In each type of region, the smallest layer separations presumably result from areas in the flow that have experienced the largest cumulative stretching, while the largest layer separations correspond to the lowest total strain. As the number of periods increases ($n \rightarrow \infty$), the distribution will tend toward lognormality. These distinctly different features near and away from voids are evident in our relatively early time results. Here, the parallel regions show strong evidence of lognormality, and are reminiscent of a later time when the space-filling tendency of the stretching and folding process require the existence of many more parallel layers than nonparallel ones. The fraction of the volume occupied by regions containing nearly parallel layers will then increase with the number of periods, due to this stretching and folding. The distribution for the whole flow field after a large number of periods will asymptotically approach that of just the parallel regions of the flow at the earlier times.

The persistent ($n \rightarrow \infty$) influence of voids on mixing in chaotic flows with islands has been recognized previously for distributions of stretching values and Lyapunov exponents (e.g. [17, 18]). The influence of initial voids has a similar effect in the distribution of dissipation layer separations. However, stretching values and Lyapunov exponents associated with the mixing process are not readily accessible in turbulent flows, while the dissipation layer separation distribution can be determined in a turbulent flow following precisely the same procedure as was done here for a chaotic flow. In the following section, we present results obtained from our turbulent flow measurements, and compare these with the results in Figs 9–11.

3.3. *Turbulent flow results*

Probability densities of dissipation layer separations were computed for each of 29 individual three-dimensional 256^3 spatial data volumes of the type shown in Fig. 1 from our turbulent flow measurements at $Re_\delta \approx 3700$. These volumes were equally spaced in time, and spanned slightly more than 2.5 outer scale turnover times (δ/u) of the flow. The size of each volume was roughly $(2\lambda_v)^3$, where $(\lambda_v/\delta) \approx 11.2 \cdot Re_\delta^{-3/4}$ is the local strain limited velocity gradient length scale in the turbulent flow, and $Re_\delta \equiv (u\delta/\nu)$ is the local outer scale Reynolds number based on the length and velocity scales δ and u characterizing the local mean shear in the flow. The scalar field within any volume thus represents the signature of the mixing process at the small scales of the flow. The dissipation layer separation distribution resulting from ensemble statistics over the entire set of spatial data volumes is shown in Fig. 12. A roughly -3 power law scaling can be seen in the distribution of layer separations over this range of length scales, reminiscent of the power law scale similarities typically found in high Reynolds number turbulent flows.

Although the ensemble-averaged distribution exhibits this -3 power law scaling, the individual dissipation layer separation distributions vary considerably from one data volume to the next, as shown in Fig. 13. The -3 scaling appears to result only from averaging over many such individual uncorrelated volumes, but is not evident in any single volume. However, while the distribution varies significantly at large layer separations, for small separations the results for all volumes are nearly the same. Though perhaps only coincidentally, there are interesting similarities evident when Fig. 13 is compared with Fig. 11 from the chaotic flow results. In both cases, the distributions for small separations are essentially invariant and show a roughly lognormal scaling, while for large separations the distributions are very different. In the chaotic flow, the multiplicative processes that lead to rapid generation of small separations with lognormal scaling in regions of high

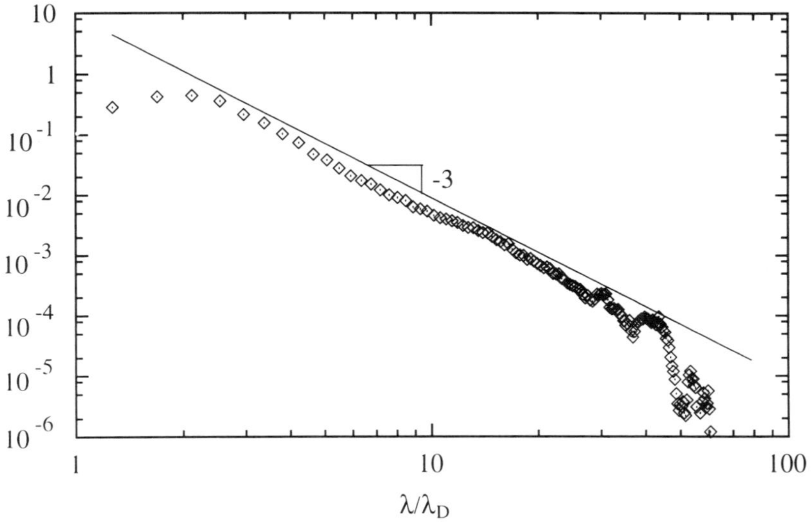

Fig. 12. Composite layer separation PDF for the fully three-dimensional turbulent shear flow data with $Re \approx 3700$. The probability density function is calculated from 29 temporally spaced volumes of which Fig. 1 is an example using the three-dimensional algorithm for finding scalar energy dissipation layer separation distances. Notice the prevalent -3 slope covering nearly all range of separations in this log–log plot.

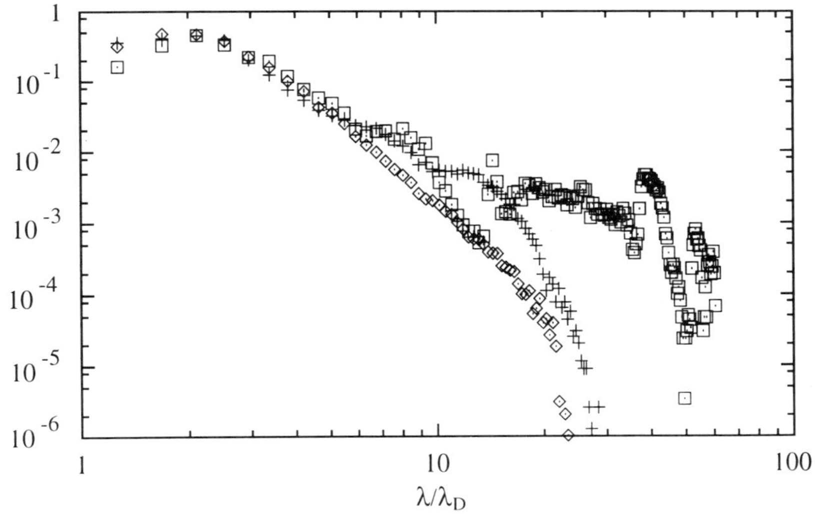

Fig. 13. Layer separation PDFs for three different volumes of the three-dimensional turbulent shear flow data of which Fig. 1 is an example. Notice how the PDFs match fairly well at small separations and can vary quite radically at large separations. This is simply consistent with the shape of the PDF which indicates that the statistics of the more prevalent smaller separation distances will settle down much faster than those of the more infrequent larger separations which will require a longer time average to obtain accurate results.

strain (large stretching values) are fairly well understood. Moreover, in the chaotic flow the presence of low stretching values leads to locally strong departures from lognormal scaling for large separations. In view of the apparent similarities in Figs 11 and 13, it might be tempting to speculate that similar dynamical processes may control the scaling characteristics of the mixing process at these scales in turbulent flows. However, there are important differences between the closed chaotic flow and the open turbulent flows. In particular, the closed chaotic flow initially requires that repeated iterations must lead to reductions in the dissipation layer separations. In the open turbulent flow, however, ambient fluid is continually entrained into the mixing region, creating voids in the dissipation field which are subsequently broken down by the stretching and folding process. The large separations in the turbulent flow result principally from this entrainment process, and the -3 scaling in the ensemble-averaged distribution is merely the result of the distribution of large voids created by the entrainment and their subsequent breakdown ('cascade') process. In the chaotic flow, the breakdown of the initial voids might be expected to occur by a different process, leading to a different scaling of the dissipation layer separations.

Despite these arguments, the similarities in Figs 11 and 13 are sufficiently striking to motivate a comparison of the ensemble-averaged distributions in Figs 9 and 12. The result is shown in Fig. 14, and reveals that the scaling of the layer separation distributions in the chaotic flow and the turbulent flow agree remarkably well. Following the reasoning given above, the apparently common lognormal scaling for small λ is not surprising. Moreover, any similarity in the value of λ at which the crossover from lognormal scaling to the -3 power law scaling might appear to occur is likely only coincidental, since in the chaotic flow this crossover location will depend on the iteration number n (see Fig. 3(b) of ref. [13]). Nevertheless, based on the above arguments there is no apparently obvious reason why the two flows should have the same scaling at large λ values. In the chaotic flow the scaling of the layer separation distribution in this range of length scales appears to be dominated by the persistence of slow processes near the voids. In the turbulent flow, if these large separations are in fact dominated by the entrainment process, then there would seem to be little prospect of any dynamical similarity at these scales between these two types of flows. On the other hand, given the relatively small size of cubes like those in Fig. 1 relative to the local outer scale δ, it seems likely that initially large voids created by the entrainment process have already undergone many periods of a repeated breakdown process, and thus might be more reflective of the dynamics of the small scales of turbulence than of the entrainment process itself. If that is the case, then the similar scaling in Fig. 14 for large values of λ leaves open the possibility of some dynamical similarity between the

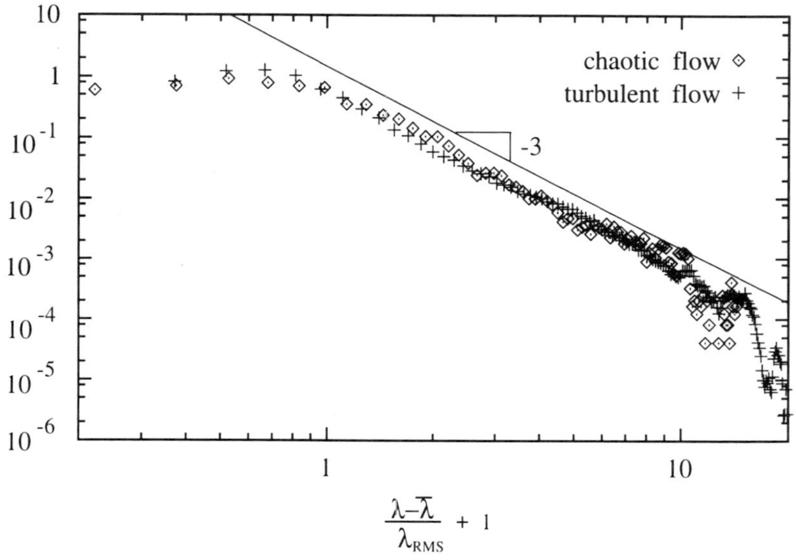

Fig. 14. Overlay of composite PDFs of layer separations for both the chaotic flow of Fig. 2. and the turbulent flow of Fig. 1. Results are presented as shown in terms of their respective first and second moments revealing a common shape between the PDFs of the two-dimensional chaotic flow and the three-dimensional turbulent flow. Notice again the -3 slope prevalent throughout nearly the entire range of scales.

mixing produced by the small scales of turbulent flows and by the chaotic flow. This possibility cannot be argued for strongly on the basis of the results presented here. For the present time we simply note that the similarities evident in Fig. 14 in the mixing signatures of these chaotic and turbulent flows are interesting, and may only be coincidental, but are certainly unexpected.

4. FRACTAL SCALING PROPERTIES

The suggestion of potential fractal scaling in various geometric features of high Reynolds number turbulent flows can be traced back to the oldest descriptions of turbulence. Some of the earliest direct experimental evidence for fractal scaling of mixing in turbulent flows was reported by Sreenivasan and Meneveau [19] from single-point measurements of a conserved scalar field (temperature) in the turbulent wake of a slightly heated body. These findings eventually led to further evidence of fractal geometric scaling in isosurfaces of a conserved scalar quantity from two-dimensional imaging measurements in the self-similar far field of an axisymmetric turbulent jet [20, 21]. Subsequently, Miller and Dimotakis [22] conducted highly-resolved single-point measurements of conserved scalar mixing in a turbulent jet, and reported finding fractal scaling for only a very narrow range of isoscalar values, with distinctly non-fractal scaling for other threshold values. More recently, Lane-Serff [23] has reported two-dimensional imaging measurements of scalar mixing in turbulent jets that appear to support the suggestion that geometric scaling properties of isosurfaces in turbulent mixing can depend significantly on the chosen isoscalar value. Though it is possible that the geometry that results from mixing in turbulent flows might actually differ fundamentally for differing isoscalar surfaces, this runs counter both to intuition and classical scaling concepts for turbulent flows. Instead, it could be argued that the apparent differences in geometric properties of various isoscalar surfaces might be more directly an artifact of the experiments than of the flow, since a small change in the isoscalar value typically produces a rather significant change in the geometry of the isoscalar contour.

By contrast, owing to its compact nature, the geometry of the support of the scalar dissipation field in Fig. 1 should be relatively insensitive to the choice of dissipation threshold level. This suggests an assessment of the potential fractal scaling properties of mixing in turbulent flows from the perspective of the scalar dissipation field rather than the scalar field. In this section, we present a comparison of the applicability of fractal scaling concepts for the geometry of the scalar dissipation field in chaotic and turbulent flows. Although data of the type in Figs 1 and 2 offer access to the fully three-dimensional scalar dissipation rate field, this comes at the expense of a limited spatial dynamic range in the data. Unlike long time-series data, spatial data of this type offers only a limited number of pixels in any given direction from which to assess a potential fractal scaling. This in turn requires novel methods for assessing the geometric scaling properties on which to base a judgement of the applicability of fractal scaling concepts. Accordingly, in Section 4.1 we describe the statistical box counting scheme used here to calculate the local scaling dimension of linear intersections through the three-dimensional dissipation fields. Following this, Sections 4.2 and 4.3, respectively, present results obtained for tne geometric scaling properties of the scalar dissipation support in the chaotic and turbulent flows.

4.1. *Fractal measurement method*

In both the chaotic and turbulent flows, we compute the local scaling dimensions of one-dimensional intersections with the support set on which the three-dimensional scalar dissipation field lies above some chosen threshold value. Owing to the inherently short record lengths available for tiling, a statistical assessment of the geometric scaling properties was used. Briefly, for any given one-dimensional intersection, the tiling method used for computing the local scaling properties is a variation of the conventional box counting scheme. The tile size δ at different tiling levels was varied by factors of two. The tiling procedure used several different initial tile locations (offsets) to remove bias that results from a fixed starting point. The two partial tiles created at both ends of the domain as a result of the offsets were counted together as one tile. This tiling algorithm for one-dimensional intersections was tested on a wide range of both regular and irregular Cantor sets of various known fractal dimensions to assess its suitability for accurately dealing with short record lengths. In each case, the true fractal dimension D was computed via

$$\sum_{i=1}^{n} c_i^D = 1$$

where c_i is the length ratio of the ith segment in the generator to the total length of the initiator, and n is the total number of segments in the generator. The total number of tiles required to cover the set at any scale δ is $N(\delta)$. Typical results obtained for dense Cantor sets with true fractal dimensions of 0.8, 0.6, 0.4, and 0.2 are given in Fig. 15. In each case, the figure shows the local fractal dimension (LFD) obtained at any tile size δ, defined as $-d\log_2 N(\delta)/d\log_2 \delta$, giving a sensitive measure of the local scaling properties. The LFD curves in Fig. 15 were computed via linear central differences of $\log_2 N(\delta)/\log_2 \delta$ curves of the type shown in Fig. 16. Note that simply least squares fitting a straight line through data such as in Fig. 16 *assumes* that fractal scaling holds for the given set, while LFD curves like those in Fig. 15 provide a means for independently assessing the validity of a fractal scaling for the set.

The LFD curves in Fig. 15 demonstrate a number of features essential for this study. First, it should be noted that, for both regular and irregular Cantor sets, for $D \geq 0.4$ the box-counting algorithm accurately converges to the true dimension for any one realization of the fractal as the tile size $\delta \to 0$. In other words, if a sufficiently large number of data points are contained in any one-dimensional intersection, then the present algorithm is capable of accurately determining the fractal dimension of that intersection. However, for low fractal dimensions, this convergence can be seen to become very slow. For $D = 0.2$, for example, 2^{17} points are evidently not sufficient to obtain convergence to a confident assessment of the true dimension. For such low values of D, the fractal set is so sparce that the number of tiles $N(\delta)$ remains relatively small even when the tile size δ is made very small. As a consequence, the ± 1 tile ambiguity inherent in any tiling scheme leads to a non-trivial fractional change in $N(\delta)$, and thus upon differentiation to a large change in the LFD (see Fig. 16). This will remain the case until $\delta \to 0$ as $D \to 0$. This limit is generally more forgiving for irregular Cantor sets than for regular sets, but for any minimum tile size δ there is still a limit to the minimum fractal dimension D that can be reliably determined. A further phenomenon that makes determining low values of D less accurate is the fact that, since low-dimensional sets occupy so little space, the same number of tiles $N(\delta)$ may suffice to cover them for more than one tile size δ. This would produce an LFD value of 0 for that value of δ. If infinitely many offsets were used, this effect could be removed, but any finite number of offsets will cause such imperfections to appear in the LFD curves of low-dimensional fractals.

The results in Fig. 15 also show that, whereas determining the fractal dimension via a tiling scheme is relatively straightforward for arbitrarily long records, for the comparatively short 2^7-point scalar dissipation records available from our imaging measurements of the

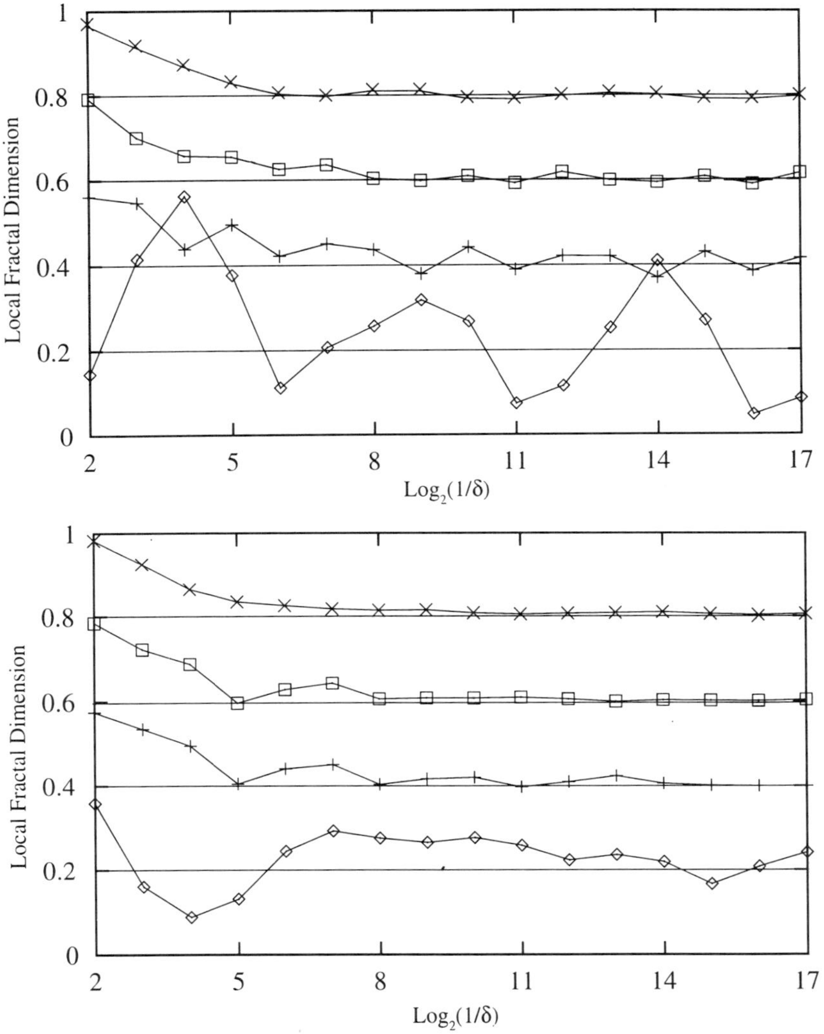

Fig. 15. Local fractal dimension curves for regular and irregular cantor sets with varying fractal dimensions, measured with the box-counting algorithm. Notice that agreement decreases dramatically with decreasing fractal dimension. In the measurements done on the chaotically advected and turbulent data sets only values in the range $2 \leqslant \log_2(1/\delta) \leqslant 7$ will be available. (a) Regular cantor sets; (b) irregular cantor sets.

chaotic and turbulent flows, finding the fractal dimension is anything but straightforward. In particular, any direct assessment of the dimension on the basis of LFD plots such as those in Fig. 15 up to $\log_2 \delta \leqslant -7$ is out of the question for all but the highest values of D. For $\log_2 \delta \leqslant -4$, it is evident that the tiles are too coarse to provide any useful scaling information. For our short records, this leaves the range $-4 \leqslant \log_2 \delta \leqslant -7$ for assessing the potential fractal scaling properties of the dissipation support set, and for determining its dimension if it is fractal.

However, we can recognize that the fluctuations in the LFD curves in Fig. 15 for any given value of D result from idiosynchracies of that one particular realization of the Cantor set. Since in general each realization of the fractal will be similar only in its scaling properties and will differ in its precise details, we can anticipate that the fluctuations will differ from one realization to another. This motivates a statistical approach as indicated in

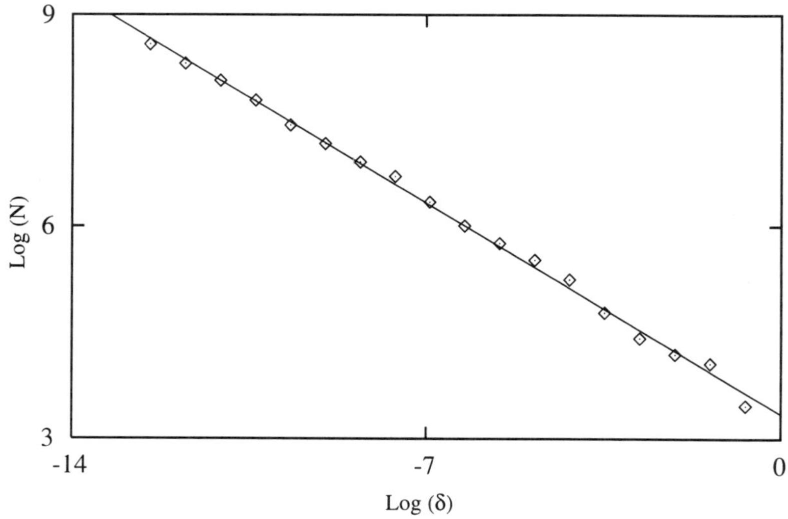

Fig. 16. A plot of the logarithm of the number of cells vs the logarithm of the cell size of an irregular cantor set with $D = 0.40$. Comparing this with the local fractal dimension curve in Fig. 15(b) shows that the local fractal dimension curves are visually more sensitive to non-fractal behavior.

Figs 17 and 18. In particular, for *many* realizations of any one fractal with given dimension D, we can collect statistics of the LFD value over the range $-4 \leqslant \log_2 \delta \leqslant -7$ to construct the probability density of LFD values. Note that, if it were possible to take $\delta \to 0$, the variance σ of this distribution would also tend to zero, and a direct assessment of D would be possible. Constrained to the range of δ achievable with our short record lengths, we can nevertheless determine $\langle D \rangle$ and $\sigma^2(\langle D \rangle)$ as indicated from a large number of independent realizations of known fractal sets over this range of δ. The resulting $\sigma^2(\langle D \rangle)$ then provides the measure by which a given set can be determined to be fractal or non-fractal. For a given set, its $\langle D \rangle$ and σ^2 can be determined as in Figs 17 and 18, and then compared against the ideal $\sigma(\langle D \rangle)$ generated from perfect fractals using irregular Cantor sets. If the measured variance is comparable to the ideal variance for the measured value of $\langle D \rangle$, then the set 'is as fractal as a true fractal is' over this range of tile sizes δ.

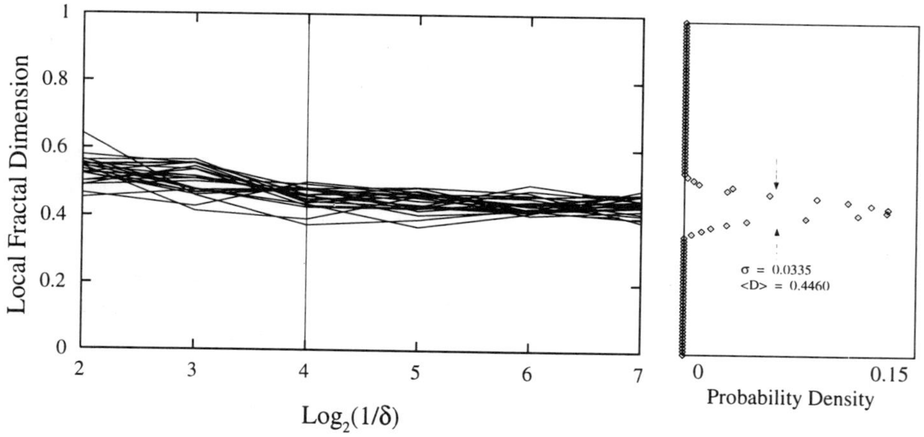

Fig. 17. Sample local fractal dimension curves from among the 8000 irregular cantor sets generated as test cases (dimension = 0.40) along with a cumulative probability density curve for all of the local fractal dimension values in the range $4 \leqslant \log_2 (1/\delta) \leqslant 7$.

It thus becomes essential to accurately tabulate $\sigma^2(\langle D\rangle)$ for genuine fractal sets. To do this, the tiling algorithm described above was repeated for 8000 individual independent realizations of irregular Cantor sets each having the same fractal dimension D. This was repeated for each of nine different fractal dimensions between 0.1 and 0.9 (Table 1).

Note that the slight increase in $\sigma^2(\langle D\rangle)$ for $D = 0.9$ is probably due to the difficulty in applying the generator for high-dimensional Cantor sets until the appropriate resolution is reached. In order to make a set with high dimension and the appropriate resolution in a reasonable number of generator applications, a limit had to be placed on the maximum c_i. This restricted the details of these high-dimensional sets, which could affect the convergence of the LFD statistics. Note also that, in addition to the variance $\sigma^2(\langle D\rangle)$, local probability density plots of the type shown in Fig. 18 provide an additional and somewhat more detailed signature of a true fractal, against which analogous results for a given set can be compared to assess the applicability of fractal concepts to its scaling properties.

An assessment of the possible effects of noise in the original scalar field measurements on the scaling properties of the resulting scalar dissipation support set was conducted via the following procedure. First, an approximate value of the r.m.s. noise in the dissipation fields was obtained by measuring the r.m.s. noise found in uniform regions of the original scalar fields and applying classical noise estimation techniques. Next, a Gaussian noise distribution with the same r.m.s. value was added to the scalar energy dissipation rate fields, which were then thresholded and compared with the original thresholded scalar energy dissipation rate fields. By comparing the original fields to the noisy fields, probability density functions were calculated giving the likelihood of a given pixel being turned on or off due to noise, depending on the distance from the nearest point on the original noise-free set. This procedure was used to add noise to an irregular Cantor set and LFD curves generated as described above. The results in Fig. 19 show that even noise levels much higher than those in either set of experimental data appear to have little effect on the measured LFD curves.

4.2. *Chaotic flow results*

In the chaotic flow, LFD curves were generated for each non-empty row and column in the 256×256 spatial data planes in the two-dimensional scalar dissipation field. Statistics were collected in each window shown in Fig. 2. There were many individual rows and columns that exhibited fractal scaling properties, i.e. for which $\sigma(\langle D\rangle)$ was within the values in Table 1 over the range $-4 \leqslant \log_2 \delta \leqslant -7$ (see Fig. 20). However, many other one-dimensional intersections displayed a scaling that could not be remotely termed fractal

Table 1. Statistics of known fractal sets collected over the range $-4 \leqslant \log_2 \delta \leqslant -7$ for many different realizations via the procedure outlined in Figs 17 and 18.

D	$\langle D\rangle$	$\sigma^2(\langle D\rangle)$
0.1	0.199	0.065
0.2	0.233	0.064
0.3	0.333	0.052
0.4	0.446	0.036
0.5	0.543	0.026
0.6	0.638	0.023
0.7	0.733	0.021
0.8	0.830	0.020
0.9	0.930	0.021

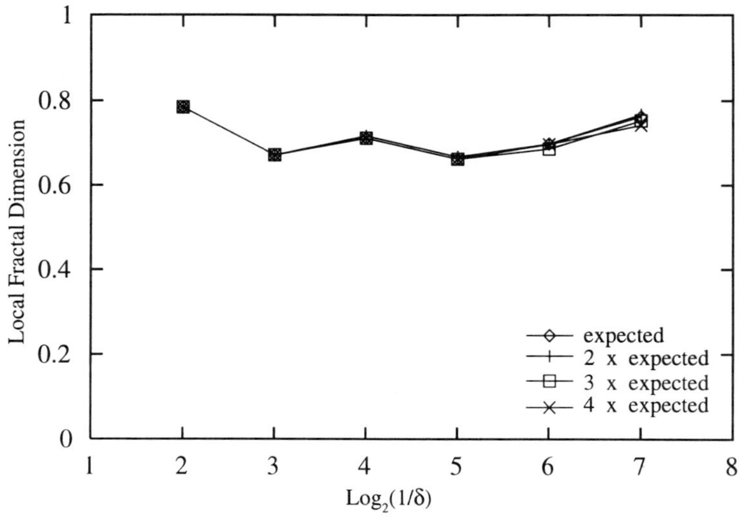

Fig. 19. The effect of differing amounts of noise on the local fractal dimension curve of an irregular cantor set that has been digitized to 256 pixels. Noise does not appear to be much of a factor in influencing the local fractal dimension, even at levels much higher than those present in the experimental data.

(see Fig. 20). Notice that the differences are not as strong in the log $N(\delta)$ vs log δ curves in Fig. 20(a), but show up clearly in the LFD curves in Fig. 20(b). However, none of the data within an entire window produced $\sigma^2(\langle D\rangle)$ values that fell even within a factor of two of the values given in Table 1. Moreover, changing the dissipation threshold level over the range $1.0 \leqslant \chi/\langle\chi\rangle \leqslant 2.5$, as shown in Fig. 21, did not produce any change in this result – no window produced a geometric scaling that came within a factor of two of the criterion in Table 1 and Fig. 18 for declaring a set as being fractal.

Figure 22 shows the local corresponding local probability density plot analogous to Fig. 18 for the ensemble-averaged statistics over all 31 windows in Fig. 2. As the figure indicates, these ensemble statistics can be confidently declared as being non-fractal.

(a)

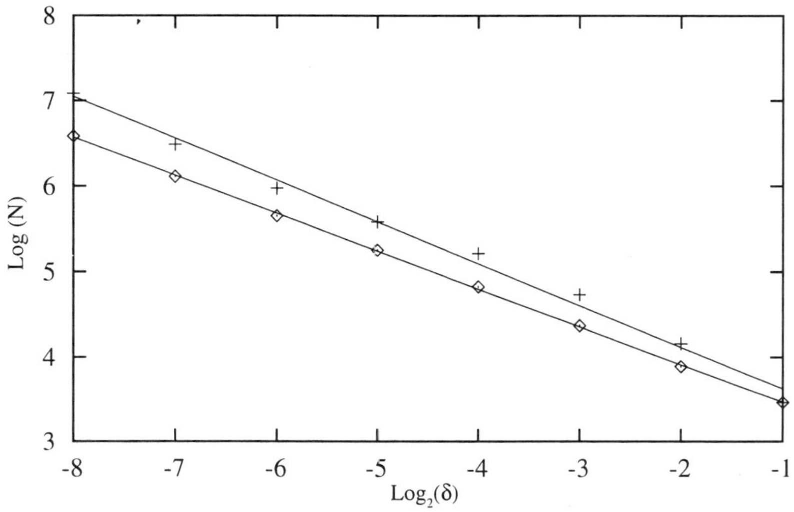

Fig. 20(a).

(b)

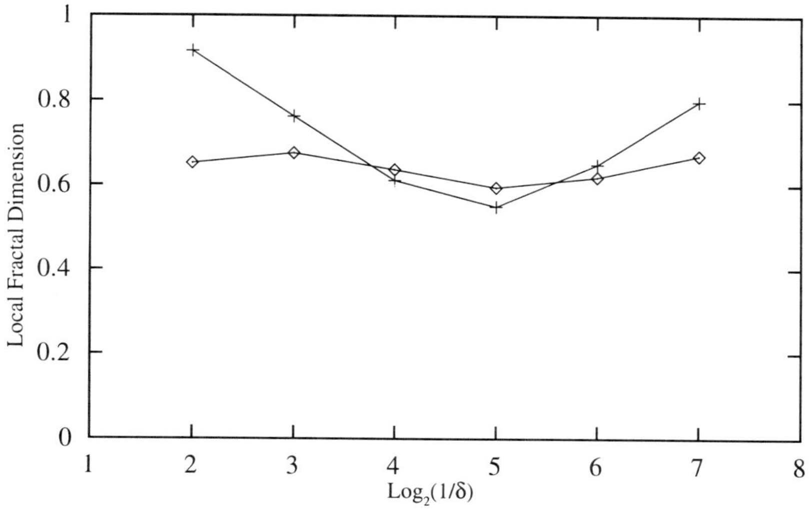

Fig. 20. Local fractal dimension curves and $\log(N)$ vs $\log_2(\delta)$ plots for two data samples from the chaotically advected data set. One data sample appears to be quite fractal while the other shows no sign of being fractal. Notice that the local fractal dimension curves accentuate any deviations from fractal behavior. (a) $\text{Log}(N)$ vs $\log_2(\delta)$ plot. (b) Local fractal dimension curves.

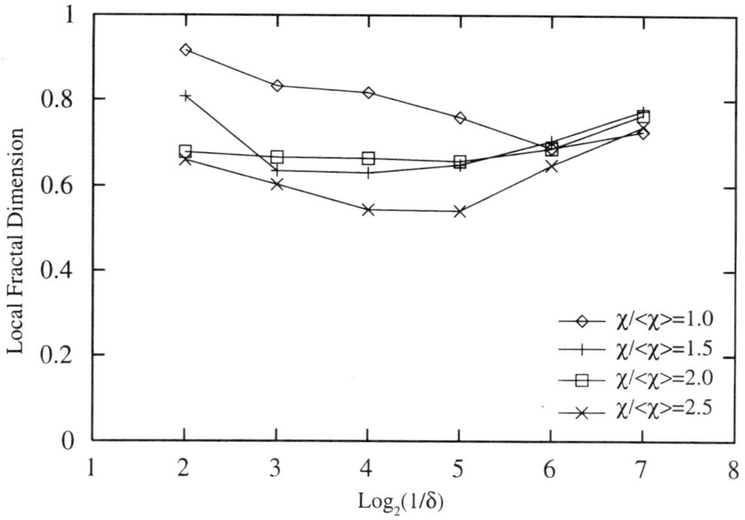

Fig. 21. Threshold level effects on a representative data sample from the chaotically advected data set are shown above. It appears that in general the results are relatively threshold-independent considering that the threshold value varies by a factor of 2.5.

However, within 'parallel' regions of the flow, much tighter LFD probability distributions were found. An example is shown in Fig. 23, corresponding to a region in which essentially all the dissipation layers were parallel, and which produced an essentially lognormal layer separation distribution in Section 3. Even in this case, however, the $\sigma^2(\langle D \rangle)$ value lies well above that given in Table 1 for the same $\langle D \rangle$. This can be seen by comparing Figs 18 and 23, and keeping in mind that the width of the LFD distributions in Table 1 decrease with increasing D. Lastly, Fig. 24 shows an LFD map for a window corresponding to a 'non parallel' region of the flow. In this case, the variance is nearly five times that given in

Table 1, and the scaling can be confidently said to be non-fractal. Generally speaking, the most nearly fractal regions of the flow seemed to coincide with the areas in which the layers were parallel, while the clearly non-fractal regions coincided with intermediate and non parallel regions. However, at this stage of development of the flow, no regions led to a scaling sufficient to declare the dissipation support as being globally fractal.

4.3. *Turbulent flow results*

For the turbulent flow, analogous LFD curves were generated for each non-empty row and column in the 256^3 spatial data volumes in the three-dimensional scalar dissipation field. As was the case in the chaotic flow results, many individual one-dimensional intersections of the turbulent flow data exhibited nearly fractal scaling (e.g. see Fig. 25).

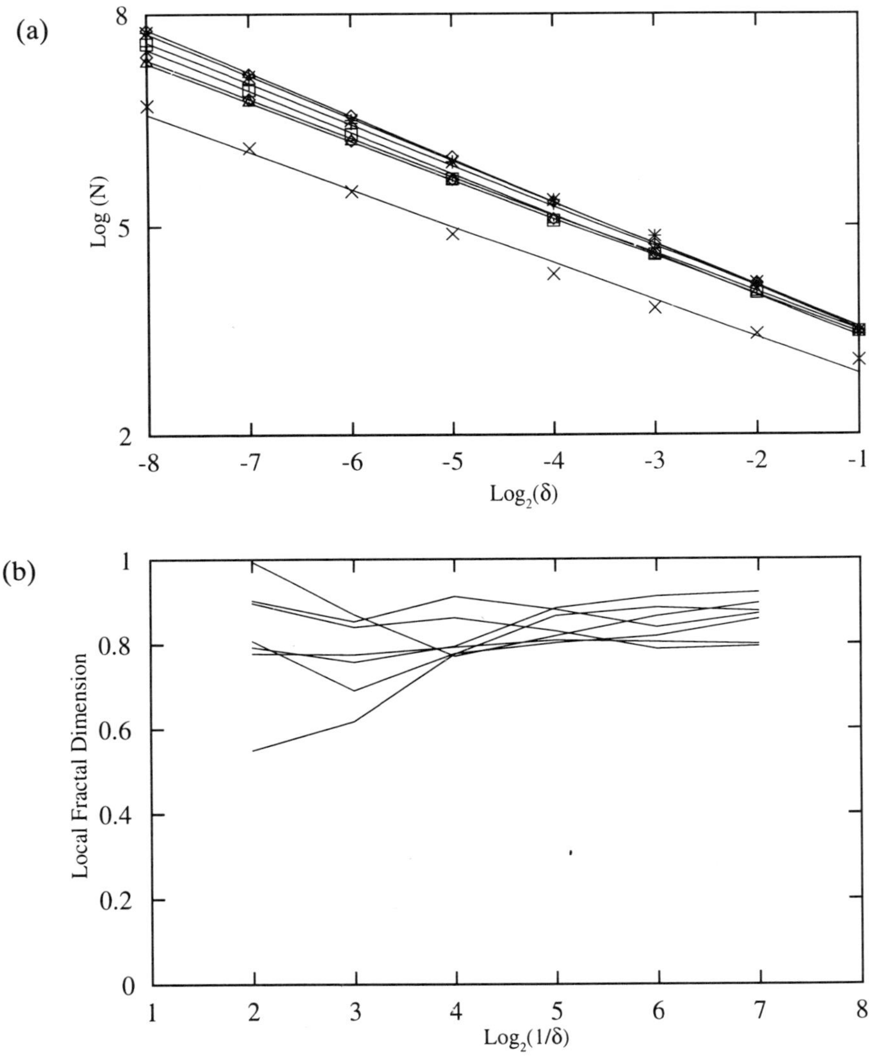

Fig. 25. Local fractal dimension curves and $\log(N)$ vs $\log_2(\delta)$ plots for seven data samples from the turbulent data set. Some data samples once again appear to be quite fractal, while others show no sign of being fractal. (a) $\text{Log}(N)$ vs $\log_2(\delta)$ plot; (b) local fractal dimension curves.

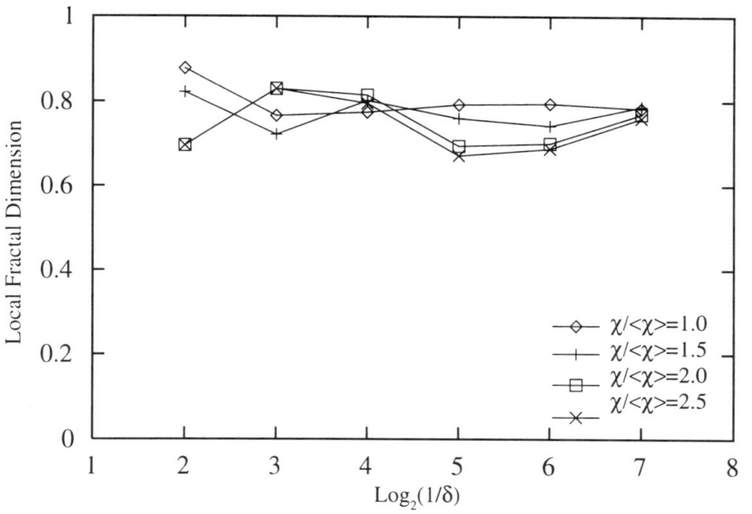

Fig. 26. Threshold level effects on a representative data sample from the turbulent data set are shown above. It appears that, in general, the results are relatively threshold-independent considering that the threshold value varies by a factor of 2.5.

Moreover, the LFD curves for the turbulent flow were found to be largely independent of the $\chi/\langle\chi\rangle$ threshold values over the range from 1.0–2.5, as shown for typical cases in Fig. 26. Nevertheless, when local probability density plots were constructed for typical regions in the turbulent flow data (e.g. Fig. 27), the resulting $\sigma^2(\langle D \rangle)$ values were all well above those in Table 1. Even for the most nearly fractal regions, the measured $\sigma^2(\langle D \rangle)$ based on these one-dimensional intersections was still approximately three times higher than in Table 1. Generally speaking, in the turbulent flow data there was much less variation in the results among different regions. This was presumably due to the fact that there were no large regions in the data where all the scalar energy dissipation layers were all parallel to each other. Throughout most of the turbulent flow data, the dissipation layers were highly convoluted and tended to fold back on themselves many times in a single region. Collectively, the present results do not support a contention that the scalar dissipation rate field in turbulent flows display a global fractal scaling over the range of length scales accessible by our data.

5. CONCLUSIONS

The results presented here have compared two rather sensitive statistical geometric signatures of the mixing process at the small scales of high Reynolds number: three-dimensional, highly unsteady, open turbulent flows with those in low Reynolds number and two-dimensional, time-periodic, closed chaotic flows. In each of these flows, measurements of the mixing of a dynamically passive, conserved scalar field $\zeta(\mathbf{x}, t)$ by the underlying flow field were used to analyze geometric scaling properties of the support set on which the associated scalar energy dissipation rate field $\nabla\zeta \cdot \nabla\zeta(\mathbf{x}, t)$ was concentrated. The results from the present one-dimensional intersections are not consistent with a global fractal description, having a single fractal dimension, for the geometric scaling properties of the scalar dissipation support set in this turbulent flow. In the chaotic flow, however, the present results are essentially consistent with a fractal description of the scaling in those regions of the flow that have undergone sufficient stretching and folding. Perhaps most interesting is the observation that the distribution of dissipation layer separations in both the chaotic and turbulent flows follow the same scaling over a wide range of length scales. This suggests the possibility that similar dynamical features may play a role in the mixing process in both the chaotic and the turbulent flows. It may even be possible that the classical power law scaling characteristics of the small scales in turbulent mixing may be usefully approached from this novel perspective.

Despite the attraction of working in terms of the full three-dimensional scalar dissipation rate field rather than the scalar field itself (for both practical and dynamical reasons), the results obtained here must be viewed cautiously. At a minimum, the Reynolds numbers in the turbulent flow data are large enough for a highly unsteady and apparently 'turbulent' mixing process to result, but are still well below the range of values traditionally considered safe for statements of high Reynolds number asymptotic behavior. However, our attention here has been focused on the small scales of the turbulent mixing process. If these are governed largely by the local dynamics of the continual stretching and folding of material elements that results from the time-varying strain rate and vorticity in the underlying velocity field, then traditional requirements for Reynolds number asymptotics, such as an extensive inertial range, may be relatively unimportant. While it would certainly be interesting to look for additional Reynolds number scaling and asymptotic limits in measures of the mixing dynamics such as those considered here, such high Reynolds number studies are not possible with presently available experimental techniques. Moreover, in view of the dynamical similarities of the stretching and folding processes at the small scales of turbulent flows and in chaotic flows, it may be more important to have access to fully three-dimensional dissipation field data, as we have done here, than to consider lower-dimensional estimates of the mixing process at higher Reynolds number values.

Acknowledgements—This work is supporting the program in Turbulence Structure and Control at the Air Force Office of Scientific Research (AFOSR) through AFOSR Grant No. F49620-92-J-0025. Dr. J. M. McMichael is the program manager. We are grateful to Julio Ottino for providing us with the original photonegatives of his eccentric-cylinder flow experiments, and to Lester Su for assistance in development of the layer center detection algorithm.

REFERENCES

1. W. J. A. Dahm, K. B. Southerland and K. A. Buch, Direct, high resolution, four-dimensional measurements of the fine scale structure of $Sc \gg 1$ molecular mixing in turbulent flows, *Phys. Fluids* **A3**, 1115–1127 (1991).
2. W. J. A. Dahm, L. K. Su and K. B. Southerland, A scalar imaging velocimetry technique for four-dimensional velocity field measurements in turbulent flows, *Phys. Fluids* **A4**, 2191–2206 (1992).
3. L. K. Su and W. J. A. Dahm, Scalar imaging velocimetry measurements of the small scale structure and dynamics of fluid turbulence, *Phys. Fluids* **A**, submitted (1993).
4. J. M. Ottino, *The Kinematics of Mixing: Stretching, Chaos, and Transport*. Cambridge University Press, Cambridge (1989).
5. H. Aref, Integrable, chaotic, and turbulent vortex motion in two-dimensional flows, *Ann. Rev. Fluid Mech.* **15**, 345–389 (1983).
6. H. Aref, Stirring by chaotic advection, *J. Fluid Mech.* **143**, 1–21 (1984).
7. J. M. Ottino, C. W. Leong, H. Rising and P. D. Swanson, Morphological structures produced by mixing in chaotic flows, *Nature Lond.* **333**, 419–425 (1988).
8. J. M. Ottino, The mixing of fluids, *Sci. Am.* **260**, 56–67 (1989).
9. J. M. Ottino, Mixing, chaotic advection, and turbulence, *Ann. Rev. Fluid Mech.* **22**, 207–253 (1990).
10. T. Dombre, U. Frisch, J. M. Greene, M. Hénon, A. Mehr and A. M. Soward, Chaotic streamlines in the ABC flows, *J. Fluid Mech.* **167**, 353–391 (1986).
11. H. A. Kusch and J. M. Ottino, Experiments on mixing in continuous chaotic flows, *J. Fluid Mech.* **236**, 319–348 (1992).
12. P. D. Swanson and J. M. Ottino, A comparative computational and experimental study of chaotic mixing in viscous fluids, *J. Fluid Mech.* **213**, 227–249 (1990).
13. F. J. Muzzio, P. D. Swanson and J. M. Ottino, The statistics of stretching and stirring in chaotic flows, *Phys. Fluids* **A3**, 822–831 (1991).

14. F. J. Muzzio, P. D. Swanson and J. M. Ottino, Mixing distributions produced by multiplicative stretching in chaotic flows, *Int. J. Bifurcations & Chaos* **2**, 37–50 (1992).
15. F. J. Muzzio, C. Meneveau, P. D. Swanson and J. M. Ottino, Scaling and multifractal properties of mixing in chaotic flows, *Phys. Fluids* **A4**, 1439–1456 (1992).
16. S. Redner, Random multiplicative processes: An elementary tutorial, *Am. J. Phys.* **58**, 267–273 (1990).
17. E. Ott and T. M. Antonsen, Fractal measures of passively convected vector fields and scalar gradients in chaotic fluid flows, *Phys. Rev.* **A39**, 3660–3671 (1989).
18. E. Ott and T. M. Antonsen, The spectrum of fractal dimensions of passively convected scalar gradients in chaotic fluid flows, *Phys. Fluids* **A3**, 1417–1430 (1991).
19. K. R. Sreenivasan and C. Meneveau, The fractal facets of turbulence, *J. Fluid Mech.* **173**, 357–386 (1986).
20. K. R. Sreenivasan, R. Ramshankar and C. Meneveau, Mixing, entrainment, and fractal dimension of surfaces in turbulent flow, *Proc. R. Soc. Lond.* **A421**, 79–108 (1989).
21. K. R. Sreenivasan, Fractals and multifractals in fluid turbulence, *Ann. Rev. Fluid Mech.* **23**, 539–600 (1991).
22. P. L. Miller and P. E. Dimotakis, Stochastic geometric properties of scalar interfaces in turbulent flows, *Phys. Fluids* **A3**, 168–177 (1991).
23. G. F. Lane-Serff, Investigation of fractal structure of jets and plumes, *J. Fluid Mech.* **249**, 521–534 (1993).

Tracer Microstructure in the Large-eddy Dominated Regime

R. T. PIERREHUMBERT

Department of Geophysical Sciences, University of Chicago, Chicago, IL 60637, USA

Abstract—We have re-examined the inertial range behavior of a passive scalar which is advected by a large-scale velocity field causing a cascade of tracer variance to small scales, where it is dissipated by diffusion. This has been done within the context of an idealized model based on mixing by a 2D area-preserving map alternating with a weak diffusion step; the model is a special case of the general advection–diffusion problem. Both freely decaying and forced equilibrium systems were considered. Our main interest in this concerns the validity of Batchelor's theory predicting a k^{-1} tracer variance spectrum, but the tracer microstructure has been diagnosed in terms of concentration probability distribution functions, generalized dimensions of the dissipation field, structure functions, and cancellation exponents. 2D simulations carried out at 1024^2 resolution show that in the decaying case the evolution settles into a 'fractal eigenmode' in which the variance decays exponentially with time at a rate dependent on the Lyapunov exponent but independent of the diffusion coefficient. Although the concentration pattern is self-similar with time, the power spectrum is not algebraic. Concentration PDFs have exponential tails. The dissipation field is not multifractal, and formally has $D_q = 2$ for all q. The convergence of the squared-gradient PDFs under coarse-graining indicates some underlying fractal behavior, however, and we have introduced the notion of 'fractal degree of freedom' systems to describe such entities. Cancellation exponents and structure functions were also considered, and have a self-similarity which is compatible with a non-intermittent behavior of the dissipation field. These matters have also been addressed for the equilibrium case. The main difference is that the power spectrum of concentration variance in equilibrium exhibits a power-law inertial range, though it is steeper than k^{-1}, but not as steep as k^{-2}. Other features are similar to the decaying case. Very high resolution simulations of the undiffused problem indicate that a k^{-1} spectrum is approached asymptotically, but only at resolutions corresponding to $10^6 \times 10^6$. The passive scalar behavior is compared and contrasted with the behavior of scalar pseudo-vorticity (an 'active scalar') in the family of generalized 2D turbulence models introduced elsewhere in this issue.

1. INTRODUCTION

There are two great spectra believed to prevail for passive scalars under the action of turbulent or random velocity fields. The first is the $k^{-5/3}$ Obukhov–Corrsin spectrum prevailing in inertial-range homogeneous isotropic turbulence when the viscous cutoff of the velocity field is comparable to or smaller than the diffusive cutoff of the tracer field. In this case, the tracer microstructure is created by advection by the $k^{-5/3}$ velocity field, which has infinite shears and can be expected to lead to crinkling of tracer isosurfaces even at short times. This spectrum appears to be uncontroversial and robust, and is readily observable in natural and laboratory experiments.

The other spectrum is the Batchelor k^{-1} spectrum [1], which is expected to prevail when the tracer diffusivity is so small that the smallest scales in the tracer field are smaller than the viscous cutoff of the velocity field. In this case, the tracer microstructure is created by long-term rearrangement by a smooth velocity field which, in Batchelor's theory is characterized by a single dominant timescale. To the best of our knowledge, this spectrum has never been unambiguously observed in laboratory experiments. The same problem has bedeviled the related k^{-1} enstrophy spectrum of 2D turbulence. One begins to suspect that there may be something fundamentally wrong with Batchelor's argument.

The essence here is the nature of mixing by an organized velocity field, and as such comes under the general heading of 'chaotic advection.' If Batchelor's argument is to work at all, it should work for chaotic mixing by simple flow fields. At the same time, the nature of microstructure created by chaotic mixing and small-scale diffusion is of considerable interest in its own right, and has many applications to naturally occurring flows [2–5]. There is also a great deal of current interest in more sophisticated measures of tracer variability than the spectrum: fractals, multifractals, probability distribution functions (PDFs) and cancellation exponents, e.g. refs [6, 7].

The general problem of mixing by an organized, smooth velocity field can be qualitatively captured by the action of area-preserving maps in two dimensions. Our overriding theme is to show how such maps with an interposed dissipative step can be used to probe the critical questions in a computationally tractable way. The model problem is described in Section 2. Freely decaying behavior is explored in Section 3, which includes a lengthy digression on the general subject of structure functions, cancellation exponents and generalized dimensions. Statistical equilibrium solutions are described in Section 4. Some extremely high resolution spectra for the undiffused case are given in Section 5. Our principal conclusions are discussed in Section 6.

2. DESCRIPTION OF THE MAP

In this paper we explore mixing induced by the map

$$x_{n+1} = x_n + a\sin(y_n + \phi_n) \qquad (\mathrm{mod}\ 2\pi) \tag{1a}$$

$$y_{n+1} = y_n + \cos(x_{n+1} + \phi_n) \qquad (\mathrm{mod}\ 2\pi) \tag{1b}$$

where ϕ_n is a random variable uniformly distributed over $[0, 2\pi]$. The randomization of the phase is introduced so as to break up invariant tori. Such tori are known to provoke very complicated scaling behavior in the stretching statistics, via the tendency of orbits to stick near tori or to become trapped behind cantori, e.g. ref. [8]. Conventional wisdom has it that such tori are uncharacteristic of real fluid turbulence, where the aperiodic flows make it more difficult for invariant structures to exist. In light of the prevalence of coherent structures in turbulence, we are not convinced of the validity of this assumption, but we adopt the randomized map (1) anyway, because its statistics are easier to study.

Because of the complete scrambling of the phase, each individual orbit undergoes an uncorrelated random walk on the torus. However, because the same ϕ_n is used at each iteration throughout the phase space, interesting spatial correlations nevertheless arise from (1). The most important characterization of the map is through its stretching statistics. The Jacobian of the map is

$$J = \begin{pmatrix} 1 & a\cos(y_n + \phi) \\ -\sin(x_{n+1} + \phi) & [1 - a\cos(y_n + \phi)\sin(x_{n+1} + \phi)] \end{pmatrix} \tag{2}$$

which has eigenvalues

$$e^{\lambda} = \gamma \pm \sqrt{(\gamma^2 - 1)}, \qquad \text{where} \qquad \gamma = 1 - \tfrac{1}{2}a\cos(y_n + \phi)\sin(x_{n+1} + \phi) \tag{3}$$

where λ is the exponential stretching rate over one iteration. Note $\gamma \in [1 - \frac{1}{2}a, 1 + \frac{1}{2}a]$. For any phase and any value of a, there is always a part of the domain near which the trajectories separate exponentially. Because of averaging over the spectrum of strains, the long-time Lyapunov exponent is considerably lower than the maximum given by (3). For maps without invariant tori it is known that the finite-time (n-iteration) Lyapunov exponents have a Gaussian PDF, with width decreasing like $n^{-1/2}$ [9]. We have verified this

Table 1. Finite time Lyapunov exponents

Iterations	Mean (λ)	Std. Dev (λ)
10	0.57	0.20
20	0.53	0.14
40	0.50	0.10

property directly for the map (1), by accumulating multiplications of J over ensembles of orbits. For $a = 4$, which is the parameter used throughout the rest of this paper, the mean and standard deviation of the exponential stretching rates for various n are given in Table 1.

There have been innumerable studies of mixing and tracer microstructure due to maps such as (1). The chief novel interest in the present work comes from incorporating the effects of weak diffusion, which removes tracer variance as it cascades to small scales. Within the context of discrete-time maps on the plane, this is accomplished in the following manner. We define a continuous concentration field $c(x, y)$ on the plane, and spatially discretize it to a regular grid so $c_{ij} = c(x_i, y_j)$. The map (1) is first used to rearrange $c(x, y)$, through mapping (x, y) and re-interpolating to the grid. This iteration, which yields a new $c(x, y)$ is followed by the diffusion step:

$$c_{ij} = (1 - D)c_{ij} + \tfrac{1}{4}D(c_{(i+1)j} + c_{i(j+1)} + c_{(i+1)j} + c_{i(j-1)}) \tag{4}$$

where $0 < D < 1$. On each iteration, this damps out the $2\Delta x$ wave by a factor $(1 - D)$. A generally similar approach has been employed by Ott *et al.* [10], and Du and Ott [11] to study diffusive effects on fast magnetic dynamos.

3. FREELY DECAYING EVOLUTION

Because of the introduction of dissipation, the variance of the tracer field $c_n(x, y)$ will decay to zero with time. The first property of interest is the rate of decay, and how the rate scales with diffusivity D. The variance decay for $D = 0.5$ and $D = 1$ is shown in Fig. 1; the initial condition for this case is $c(x, y) = \cos(x)\cos(y)$, which would not perceptibly

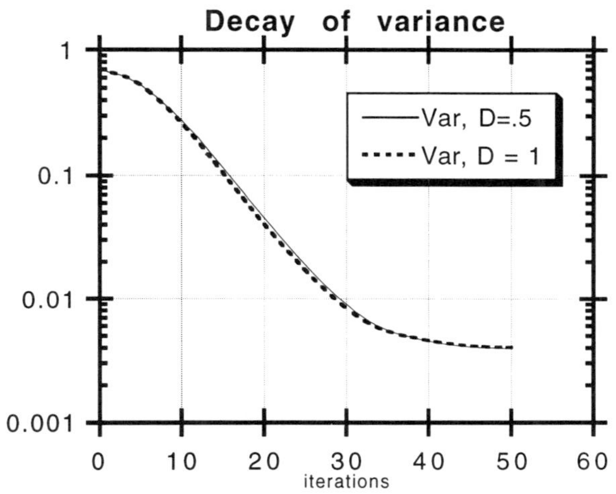

Fig. 1. Decay of variance for mixing by randomized sine–sine map, with two different values of diffusion D.

decay over the time period shown, were it not for the exponential amplification of gradients caused by the chaotic mixing.

At long times, the variance reaches a small nonzero asymptotic value. This is an artifact of the numerical method, arising from a small variance source associated with interpolation errors in going between the map (1) and the discrete grid on which the diffusive step is carried out. Apart from this, the variance decay is *independent of* D, and has an exponential stage at intermediate times. We conjecture that the exponential stage would continue indefinitely were it not for discretization error; verification of this conjecture must await formulation of a more accurate discrete advection–diffusion scheme.

The diffusion-independent decay stage shows that the simple map can reproduce the classical inertial-range cascade to small scales, where variance can be dissipated. The independence of D, of course, comes from the fact that as D gets smaller, the dissipation range simply moves to smaller scales. Further light can be shed on the decay properties through examination of the variance equation

$$\frac{\mathrm{d}}{\mathrm{d}t}\langle c^2\rangle = -\kappa\langle|\nabla c|^2\rangle, \tag{5}$$

where κ is the molecular diffusivity. If λ is the stretching rate experienced by a fluid element (roughly the typical Lyapunov exponent), then balancing strain against diffusion yields a filament scale $L = \sqrt{(\kappa/\lambda)}$, whence the gradient scales like $\delta c/L$, where δc is the typical concentration gradient across a filament. Then, if δc can be estimated as $\delta c = b\sqrt{\langle c^2\rangle}$ the variance decay rate becomes simply $b^2\lambda$, which is indeed independant of diffusivity. In the numerical experiments, the variance decays approximately like $\exp(-0.178n)$. Using $\lambda = 0.5$ from the Lyapunov exponent calculation, we infer $b = 0.6$. The fact that b is order unity indicates that the concentration values are well 'scrambled.' The typical jump across a filament is similar to what would be obtained by randomly picking values from a distribution covering the range prevailing over the entire domain. It can easily be shown that these results are not altered if the stretching rate is characterized by a distribution of different λ instead of a unique value. As long as δc is independent of λ, fluctuations in λ only require replacing λ in the above formulae by its mean value.

The exponential decay suggests that, following a transient adjustment stage, the concentration pattern settles into an 'eigenmode.' The amplitude decays, but the pattern is (in some statistical sense) time-invariant. A snapshot of the concentration eigenmode at $n = 25$ is shown in Fig. 2. One is tempted to invoke the term 'fractal,' but it would perhaps be less prejudicial at this point to refer to it as a 'strange eigenmode.' The indication is, at any rate, that this 'eigenmode' has arbitrarily fine scales, that are limited only by the presence of diffusion. The behavior is similar to the 'horrible eigenvalue problems' discussed by Bayley [12]. There is a close analogy with the less exotic example of critical-line singularity in the shear instability problem. In that case, as viscosity approaches zero the eigenmodes show arbitrarily fine structure near the line where Doppler-shifted phase speed vanishes. Perhaps the 'strange eigenmodes' can profitably be viewed as a situation in which the 'critical line,' which is really just a form of resonance, becomes space-filling.

To flesh out the notion of a 'strange eigenmode' we need to quantify in what sense the 'eigenmode' is time-independent, by studying statistical characterizations of the spatial structure of the concentration field. The power spectrum is one of the most familiar characterizations of the spatial structure of a field. Power spectra at $n = 10$, 20 and 30 are shown on a log–linear plot in Fig. 3. All fields have been normalized to have unit variance, and the wavenumbers have been normalized such that $k = 512$ corresponds to the $2\Delta x$ wave. Notably, the spectra collapse onto the same curve for wavenumbers > 10, indicating

Fig. 2. Snapshot of concentration pattern for the decaying mode after 25 iterations.

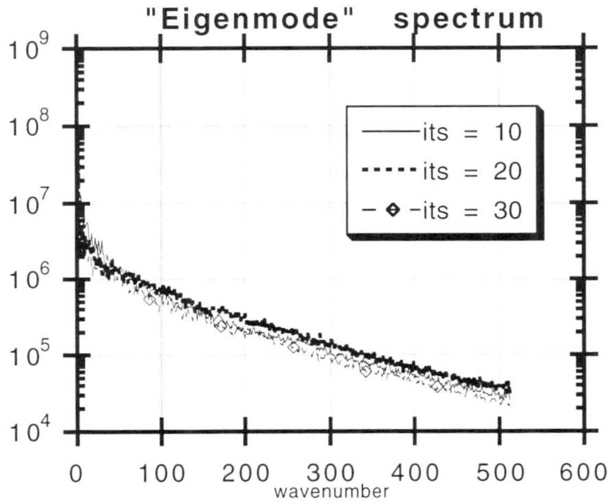

Fig. 3. Concentration power spectra of the decaying mode at $n = 10$, 20, 30 and 250 (note linear–log axes).

self-similar behavior in time. Further, the spectra do not have a power-law form (k^{-1} or otherwise) even at large scales that are not directly affected by dissipation. Rather, the spectra show exponential rolloff at constant rate for wavenumbers > 10. Thus, the spectral behavior supports the notion that a (strange) eigenmode emerges, but also shows that Batchelor's argument leading to a k^{-1} spectrum fails when where is not a sufficient resupply of variance from large scales. In fact, the possibility of a breakdown of this sort was hinted at in ref. [1]; it was remarked there that the k^{-1} spectrum implies unbounded variance in the limit of small dissipation, whereas the variance is globally bounded for many physical means of maintaining concentration variance (e.g. introducing dye at a fixed concentration). Batchelor noted that something different must happen when the supply of variance ('θ^2-stuff,' as he called it) from large scales is insufficient to fill out the k^{-1} spectrum. Up to now, there have been no suggestions as to just what happens when the breakdown takes place. We conjecture that as the global limit of 'θ^2-stuff' becomes much smaller than that needed to fill out the k^{-1} spectrum to the dissipation range, the exponential behavior characteristic of the eigenmode takes over. We shall see in Sections 4 and 5, however, that there appear to be other factors at work that also tend to yield spectra steeper than k^{-1}.

The concentration PDF provides more detailed information about the tracer variability, though by itself it contains no information about spatial scales. Following the experimental work on temperature PDFs in turbulent convection (see ref. [13] and references therein) and the theoretical work by Sinai and Yakhot [7] and by Pumir *et al*. [14], there has been considerable interest in the mechanisms leading to non-Gaussian PDFs. In particular, the appearance of exponential rather than Gaussian tails implies a relatively high probability of extreme events—parcels surviving a long time without mixing much with their environment. PDFs of concentration at $n = 25$ are shown in Fig. 4(a). The PDF is self-similar for $n = 10$ through 30, which further supports the notion that the decaying field is an eigenmode. The PDF is unimodal, indicating efficient mixing between high and low concentrations present in the initial field. While the profile is Gaussian for small to intermediate fluctuations, the large-fluctuation tail is markedly exponential. The decaying 'eigenmode' we have studied is a perfect match for the assumptions of Sinai and Yakhot [7]; these authors have shown that non-Gaussian tails arise from correlations between the concentration field and its gradient, and that the advection–diffusion equation can generate such correlations on its own. We believe the decaying mode of the advection–diffusion

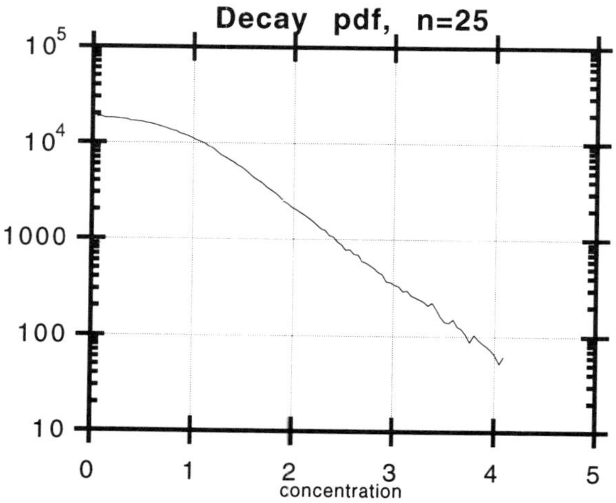

Fig. 4(a). PDF of concentrations for the decaying mode at $n = 25$ (linear–log plot). The PDF is symmetric about $c = 0$, so only positive concentrations are shown.

map is the simplest example to date of a system in which the Sinai–Yakhot theory works. In the next section, however, we will see that the theory can break down in the forced/dissipated equilibrium situation.

The concentration field itself has bounded variations and therefore cannot show multifractal scaling. Indeed, a calculation of the behavior of the concentration PDFs under various degrees of spatial smoothing revealed no interesting scaling behavior. The gradient fields, however, can have singularities (at long time, in the limit of small diffusion), and therefore can conceivably show multifractal scaling. The PDFs of gradients also contain information about the spatial arrangement of the concentration fluctuations. To address this question, we normalized the $n = 25$ field to have unit variance, computed the squared-gradient of this field, subjected the resulting field to various degrees of smoothing, and computed the PDFs. Results are shown in Fig. 4(b). These curves are fit almost exactly by the lognormal distribution:

$$P(g) = \frac{1}{g\sigma\sqrt{\pi}} \exp\left[-\left(\frac{\log(g/g_1)}{\sigma}\right)^2\right], \tag{6a}$$

where g is the squared-gradient. The parameters of the fit are given in Table 2.

The qth moment of the distribution (6a) is

$$\zeta_q = g_1{}^q \exp(\sigma^2 q^2/4). \tag{6b}$$

The behavior of g_1 and σ under smoothing thus determine the scaling properties of the gradient field. If we assume the log of the gradient to be a random field of some sort

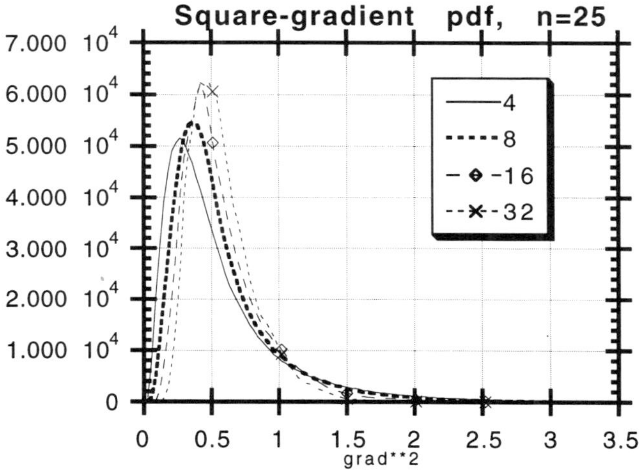

Fig. 4(b). Squared-gradient PDF as function of smoothing for $n = 25$ in the decaying case. Abscissa is fixed for all smoothings, and has not been renormalized.

Table 2.

Box size	g_1	σ
4	2.32	1.03
8	2.13	0.80
16	2.00	0.67
32	1.9	0.57
64	1.9	0.24
128	1.9	0.16

(related to the distribution of Lyapunov exponents), then the scaling of σ gives the effective number of degrees of freedom within a box of side m. If the field were 2D white noise, then the number of degrees of freedom is m^2, so we would have $\sigma = m^{-1}$. If the field were constant on each of an array of space-filling thin filaments, but randomly distributed across filaments, the number of degrees of freedom would be m^1, and so we would have $\sigma = m^{-0.5}$. Other values of the exponent would be indicative of a 'fractal number of degrees of freedom.' However, power-law scaling of σ (with g_1 fixed) would *not* correspond to a multifractal. For the field to be multifractal, ζ_q must be a power-law in m, the exponent of which determines the generalized dimension D_q for the squared-gradient measure. According to (6b), then, multifractal scaling requires $\sigma^2 = -a \log(m/M)$. If $\sigma^2 = m^{-\gamma}$ instead, the scaling of ζ_q treated as a power-law is asymptotically trivial, and we have $D_q = D$, where D is the dimension of the whole space. It is interesting to note that we obtain a non-intermittent field ($D_q = D$) in the limit of a random field which is nowhere-differentiable as well as in the limit a smooth field. Smooth fields and random fields are opposite extremes of non-intermittency, with multifractals lying in between. Clearly, when $D_q = D$, the multifractal formalism fails to characterize the geometry of the field. On the other hand, the exponent γ does contain important geometric information. We introduce the term 'fractal degree of freedom' (FDF) field for random fields that have a power law for σ having non-integer γ.

For the data in Table 2, the width parameter exhibits the scaling $\sigma \approx m^{-0.29}$, and is decidedly different from the behavior required of a multifractal. Thus, the squared-gradient field for this advection–diffusion problem appears to be an FDF random field rather than a multifractal. Since $\gamma = 0.58$, the spatial structure of concentration gradients exhibits more long-range order than even the filamentary model would imply, as filaments yield $\gamma = 1$. This suggests considerable correlations of gradient values amongst bundles of filaments, which retard the convergence of the coarse-grained gradient to its mean value.

Multifractal scaling of gradients of undiffused tracers has been predicted by Varosi *et al.* [15] but, apart from the nondiffusive limitation, there are some important differences between the approach of Varosi *et al.* and that employed here. In their theoretical formula linking the generalized dimension spectrum to the PDF of Lyapunov exponents, Varosi *et al.* employ a variable box-size cover designed to resolve each filament individually. Boxes in the cover do not contain an ensemble of filaments. This makes it possible to compute multifractal properties of the field in terms of the PDF of Lyapunov exponents, but in the process, all information about the spatial arrangement of the filaments is lost; it makes no difference whether the filaments are straight or curled, or whether the high-gradient filaments are segregated or randomly distributed. Our fixed-box calculation, in contrast, retains the geometric information. Most analysis of experimental data would use a cover that is more like the uniform-sized cover we employ in our analysis. In such a case, we would predict that the squared-gradient field is an FDF random field rather than a multifractal, for concentration fluctuations dominated by advection having a characteristic timescale. This comment notwithstanding, the relation of our results to results of the type of Varosi *et al.* requires further exploration.

The scaling properties of squared-gradient fields provide a lot of interesting information about the arrangement of gradients, but lose all information relating to the sign of the gradients; these characterizations would be invariant under multiplication of the underlying *signed* gradient by an arbitrary field taking on the values +1 or −1. Worse still, in two dimensions we can rearrange not only the *sign* of the gradients but also the *directions* without altering the scaling properties of the squared-gradient fields. The lost information is of great importance to many physical applications, because it makes a great deal of difference whether gradients of like sign tend to congregate, or whether alternating signs occur in close proximity so that in the average they tend to cancel out. The problem, of course, arises from the fact that the operations of smoothing and taking the square do not commute.

Interest in matters related to sign-singularity and restoration of the missing information has arisen in a number of contexts recently. The classical way of characterizing the variability of a field $\theta(\mathbf{r})$ is through its set of structure functions:

$$Z_L(q) = L^{-q}\langle|\theta(\mathbf{r} + L\hat{\mathbf{e}}) - \theta(\mathbf{r})|^q\rangle \tag{7}$$

where L is a length and $\hat{\mathbf{e}}$ is a unit vector whose direction is immaterial in a statistically isotropic situation. Angle brackets represent a spatial average over the entire domain. The description in terms of $Z(q, L)$ is equivalent to a description in terms of the PDF $\Pi_L(g)$ of the mean gradient

$$g = \frac{|\theta(\mathbf{r} + L\hat{\mathbf{e}}) - \theta(\mathbf{r})|}{L} \tag{8}$$

since the PDF can be recovered from the set of moments and vice versa provided that the tails of the PDF are steeper than algebraic. The structure functions retain information about the sign of the bare gradient, because the averaging over L is performed *before* the absolute value is taken. Scaling laws for velocity structure functions have long played a key role in characterization of turbulence, and through application of Kolmogorov's second hypothesis can be linked to the generalized dimensions of the energy dissipation field (see ref. [16] and references therein). If $Z_L(q)$ has a power-law behavior in L, the exponents behave very much like generalized dimensions, and the relation between Z and Π obeys a thermodynamic formalism essentially identical to that relating D_q and $f(\alpha)$ for multifractals. This has been exploited in analysis of turbulence data by Du *et al*. [17] for analysis of magnetic fields and turbulent velocity fields, and by Marshak *et al*. [18] and Davis *et al*. [19] in characterization of a variety of cloud-related atmospheric quantities. It has recently become clear that the structure functions are in fact intimately related to the generalized dimensions D_q of the gradient-magnitude field $|\nabla\theta|$, provided one additional bit of information—the *cancellation exponent* is specified. Cancellation exponents were first introduced by Ott *et al*. [10] and Du and Ott [11], and have been subsequently explored in more depth by Du *et al*. [17] and by Vainshtein *et al*. [20]. Our discussion in the following is based loosely on these works, except that we have made a few straightforward changes to allow us to deal with concentration gradients rather than magnetic fields or vorticity fields.

Let $\bar{c}(\mathbf{r})$ be the concentration field smoothed spatially over lengthscale L. We imagine the smoothing to be carried out by spatially averaging over a shape of size L (e.g. a square or a circle) centered on $\mathbf{r}$. Now define the moments

$$Z_L(q) = \langle|\nabla\bar{c}|^q\rangle. \tag{9}$$

In one dimension, this is identical to the structure function (7). In two or more dimensions, it is similar to, but not identical to, a structure function. The difference arises because the mean gradient g in (8) results from integrating the gradient along a line, whereas the 2D smoothing operator in (9) implies

$$\nabla\bar{c} = \frac{1}{A}\int_\Gamma c(\mathbf{r})(-\hat{x}\,\mathrm{d}y + \hat{y}\,\mathrm{d}x), \tag{10}$$

where A is the area of the smoothing region and Γ is its boundary. For a square-averaging region, this yields the spatial difference between the one-dimensional *averages* of c along opposing edges of the square, rather than between the values of c itself. The extra average is what keeps Z_L in (9) from being identical to the classical structure function. In some cases, it may be possible to relate the two structure functions by employing auxiliary assumptions about the field. Henceforth, by 'structure function' we will mean (9). The corresponding PDF of the mean gradient will be referred to as the 'structure PDF.'

If $Z_L(q)$ has a power-law dependence on q, then the spectrum of cancellation exponents κ_q is defined by

$$Z_L(q) \approx L^{-\kappa_q-(q-1)D}. \tag{11}$$

This formula differs somewhat in appearance from that used by Du *et al*. [17], because we have defined Z_L in terms of an average instead of a sum, but it is in fact equivalent. $Z_L(2)$ gives the scaling of the smoothed variance field, and so κ_2 is directly related to the power spectrum. Specifically, if the power spectrum of c has the scaling $k^{-\beta}$, then $Z_L(2) \approx L^{\beta-3}$ whence

$$\kappa_2 = 3 - \beta - D. \tag{12}$$

It is a very remarkable result, due to Du *et al*. [17], that the κ_q are not all independent, and in fact can be determined given knowledge of κ_1 and the generalized dimensions D_q for the gradient-magnitude field. This means the κ_1 alone is sufficient to characterize the cancellation, and to permit reconstruction of structure functions from knowledge of D_q. The remarkable relationship is:

$$\kappa_q = q\kappa_1 - (q-1)D_q \tag{13}$$

whence

$$Z_L(q) \approx L^{-q\kappa_1-(q-1)(D-D_q)}. \tag{14}$$

This permits all the cancellation exponents to be obtained in terms of D_q and the familiar power spectrum. It is interesting to note that the fractional integration approach in Tessier *et al*. [21] characterizes cloud-field variability by a D_q for the gradient-magnitude field and a power spectrum H for underlying field; the implication of (14) is that the fractionally integrated fields obtained from their procedure should have the correct structure function, despite the fact that only the behavior of the variance scaling H is built into the construction.

From (12) and (13) it follows that $\kappa_1 = \frac{1}{2}(D_2 - D + 3 - \beta)$. The generalized dimension D_2 measures the intermittency of the dissipation field. For the Batchelor spectrum, κ_1 ranges from 1 for space-filling dissipation ($D_2 = D = 2$) to 0 for dissipation concentrated on isolated points ($D_2 = 0$). The Batchelor spectrum is thus compatible with both intermittent and non-intermittent dissipation fields, because differences in the intermittency can be offset by different degrees of cancellation. Intermittency is reflected in the behavior of the structure PDF, however. If the $|\nabla c|$ field is monofractal, with $D_q = D'$ for all q, then (14) implies that the corresponding structure PDF is *self-similar* under rescaling of its argument by $L^{-\kappa_1-(D-D')}$. In the limit of a space-filling gradient field, so $D' = D$, the renormalization factor is $Z_L(1)$ itself. Such behavior can be referred to as 'Kolmogorovian,' by analogy with a similar hypothesis made concerning the relation between velocity structure functions and the energy dissipation field in the K41 homogeneous turbulence phenomenology. On the other hand, if $|\nabla c|$ is a multifractal field the PDFs will not be self-similar.

We can estimate $Z_L(2)$ in terms of random field models of the signed quantity ∇c, much as we did for the squared-gradient field. Consider the two-dimensional case. If ∇c is white noise distributed on filaments, then $Z_L(2) \approx L^{-1}$, whence $\beta = 2$. This result provides an alternate derivation of the Saffman k^{-2} spectrum [22], which arises from uncorrelated jumps of the value of a scalar across smooth curves. If ∇c is 2D white noise, then

$Z_L(2) \approx L^{-2}$, and we have the Batchelor spectrum $\beta = 1$. This provides a geometric interpretation of the Batchelor spectrum, as well as of spectra with steepness intermediate between the Batchelor and Saffman spectra. Intermediate spectra correspond to gradient structures that are more folded than filaments, yet not so much crinkled as to produce 2D white noise.

Because the concentration power spectrum for the decaying tracer case does not exhibit a well-defined power-law, one cannot unambiguously compute a cancellation exponent for this case. One can nonetheless compute the structure PDF and examine its scaling properties under averaging of c. In Fig. 4(c) we show the structure PDFs for various m after 25 iterations. For each m, the argument of the PDF has been rescaled by the mean gradient $\langle|\nabla\bar{c}|\rangle$. The most important result is that the rescaled PDFs are almost exactly self-similar under variation of averaging length m. This is what would be expected for a non-intermittent gradient magnitude field with $D_q = D$ for all q. A further interesting feature is that the most probable gradient is nonzero. This demonstrates the existence of long-range correlations in the fields. If the smoothed gradient were behaving like the difference of two samples randomly chosen from the concentration PDF, the most probable gradient would instead be zero.

4. EQUILIBRIUM PROBLEMS

The preceding has provided a characterization of the tracer variability for the freely decaying case, and we have seen that the overall picture may be described in terms of a 'strange eigenmode.' In many situations of physical interest, however, the tracer variance is maintained against small-scale dissipation by large-scale forcing. In the present section, we will see how the introduction of such forcing alters the picture obtained in the decaying case.

We employ the following forcing protocol to maintain variance. Within a strip of width 2ε centered on the line $y = \pi/2$, we reset the concentration to $+1$ at each iteration; the corresponding strip centered on $y = -\pi/2$ is reset to -1. This kind of forcing is adopted rather than the more common (in numerical experiments only!) additive forcing $c \rightarrow c + F$

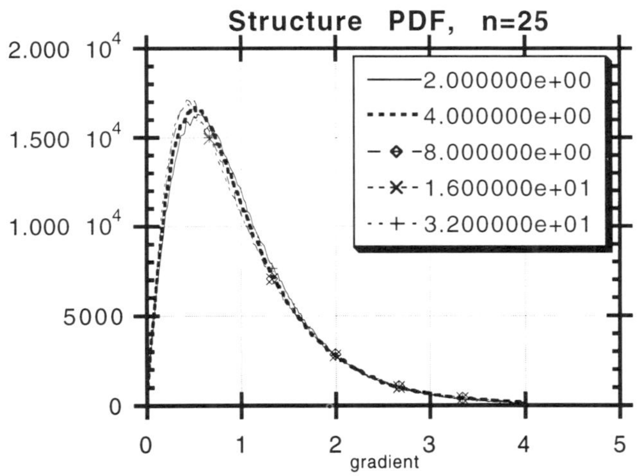

Fig. 4(c). Rescaled PDFs of magnitude of the gradient computed *after* various degrees of smoothing of the concentration field. Abscissa has been rescaled by the r.m.s. gradient for each curve. Note that this is the PDF for the gradient, rather than for the *squared*-gradient. The squared-gradient PDF peaks at zero gradient instead of a finite value.

because we believe it to be more representative of the means by which concentration variance is maintained in most physical experiments. In particular, the 'resetting' forcing assures that the concentrations will remain in a bounded range $c \in [-1, 1]$, so that the variance remains globally bounded. This is similar to the situation in which the tracer represents dye concentration, and the dye is introduced at fixed concentration from a nozzle, or by diffusion from a solid source at a boundary (whereupon concentration is maintained at saturation in a diffusive boundary layer). It is also similar to the situation prevailing for temperature in convection experiments. Fluid parcels in that case are reset to either the temperature T_1 of the hot lower boundary or T_2 of the cold upper boundary only when they are processed through the thin thermal-boundary layers near each plate. The PDF theory described by Sinai and Yakhot [7] neglects the dynamics of variance maintenance associated with this process, and so it will be of particular interest to see what concentration PDF emerges in the equilibrium case.

It should be noted that our forcing protocol destroys some small-scale variance within the thin strips, as well as maintaining the large-scale variance. Variance is free to cascade to small scales outside the strips, however. The rate at which the forcing protocol removes variance depends on the frequency with which parcels are processed through the forcing strips. Assuming ergodicity, a parcel has a high probability of passing through one of the strips after $\pi/2\varepsilon$ iterations. For small ε, this leaves ample time for exponential enhancement of gradients outside the strips. Similar considerations are of physical interest in the case of diffusive thermal or concentration boundary layers in laboratory experiments.

We turn first to the spectra, which have settled into an equilibrium after 25 iterates. The spectra for $n = 25$, 50, 250 and 500 with $\varepsilon = 0.1$ are shown in Fig. 5. Apart from some continued evolution in the dissipation range, the spectra have converged by $n = 25$. These calculations were carried out with dissipation $D = 0.5$. Notably, there is a clear power-law range. Equally notably, however, the spectrum is steeper than the k^{-1} Batchelor spectrum. The best fit power-law for $k > 20$ is $k^{-1.6}$. It thus appears that though this is a situation where Batchelor's argument ought to apply precisely, the tracer spectrum shows anomalous steepness. This phenomenon appears to be robust, extending well beyond passive tracers and mixing by iterated maps; anomalous steepness has been found in a family of fully nonlinear 2D turbulence models [22] in the range where large-eddy advection should dominate. In particular, the spectral shape for $\alpha = 2$ is $k^{-1.4}$ and for $\alpha = 3$ is $k^{-1.3}$. Is this anomalous steepness due to insufficient separation between the dissipation range and the

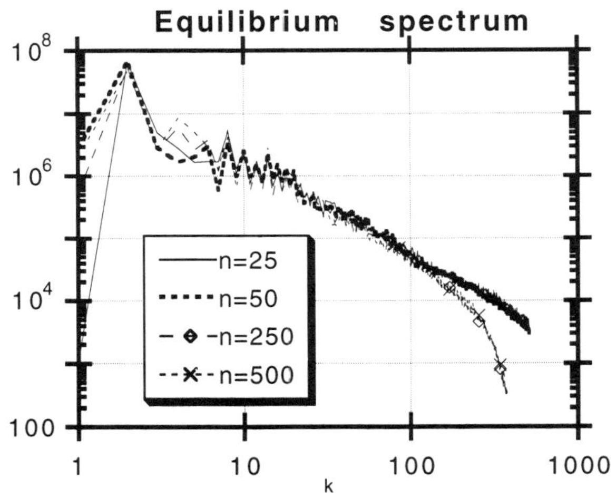

Fig. 5. Spectra for equilibrium advection–diffusion, forced with $\varepsilon = 0.1$ strips. Various times.

injection range? In Section 5 we will present some evidence that this is indeed the case. However, it will be seen that truly enormous resolutions are needed to approach the Batchelor spectrum. It seems to us a more important matter to resolve the question of how dissipation determines the anomalously steep spectrum. We raise the question, but do not yet have an answer.

The concentration PDF for the equilibrium case is shown in Fig. 6. There is a spike at $|c| = 1$, arising from the forcing in the strips; the plotting of this has been suppressed to make the rest of the PDF clearer. The PDFs take longer to reach equilibrium than do the spectra, particularly for small concentrations. For $|c| > 0.2$ the PDF has converged by $n = 50$, but the smaller concentrations continue to evolve gradually up to about $n = 250$. We cannot at this time discount the possibility that the slow evolution of the small concentration values is influenced by numerical error in our mapping scheme. At $n = 250$, the PDF is very different from that prevailing in the decay case. Rather than being Gaussian, or even exponential, the PDF is characterized by a *linear* decay from its peak at $c = 0$, tailing off to a uniform distribution for $0.67 < |c| < 1$. We do not suggest that all forcings will produce such dramatic discrepancies with the decay PDFs, but our results certainly show that the nature of the forcing cannot be swept under the rug when considering equilibrium concentration PDFs. Incorporating forcing into the theory of concentration PDFs will be an interesting and challenging task for the future.

A counterpoint to these results is provided by analysis of the pseudo-vorticity PDFs for the family of turbulence models simulated in Pierrehumbert *et al.* [23]. These are also equilibrium runs, but rather than the tracer (in this case pseudo-vorticity) values being reset to a fixed value at each time-step, only the large-scale projection is re-set, and even then the phase is randomized. This system results in PDFs (Fig. 7(a)) that are more like those prevailing in the decaying tracer case. The PDFs appear to have non-Gaussian tails. Rather remarkably, there does not seem to be any great difference in the PDFs between the spectrally local case (dominated by small-scale advection) $\alpha = 1$, and the highly nonlocal case $\alpha = 3$ dominated by large-scale advection. The marginally local case $\alpha = 2$ does show some peculiar and erratic behavior in the tails.

Returning for the moment to the passive tracer case, we note that although the concentration fields show markedly different behavior from the decay case, the *gradient* PDFs (not shown) exhibit essentially the same behavior as that prevailing in the decay case. They are lognormal, and self-similar under scaling. At $n = 25$ the width of the

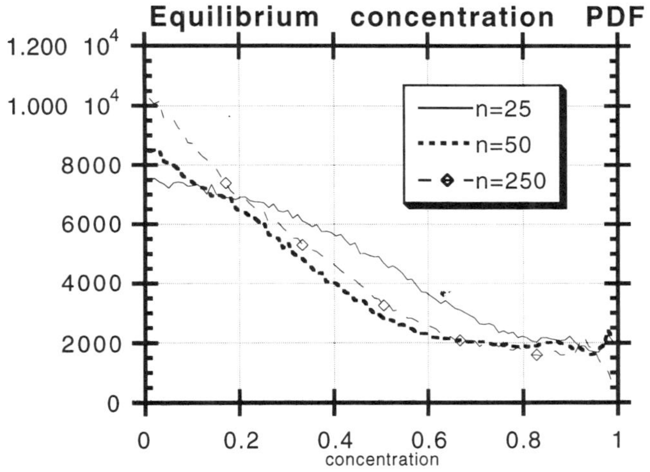

Fig. 6. Concentration PDF for equilibrium case.

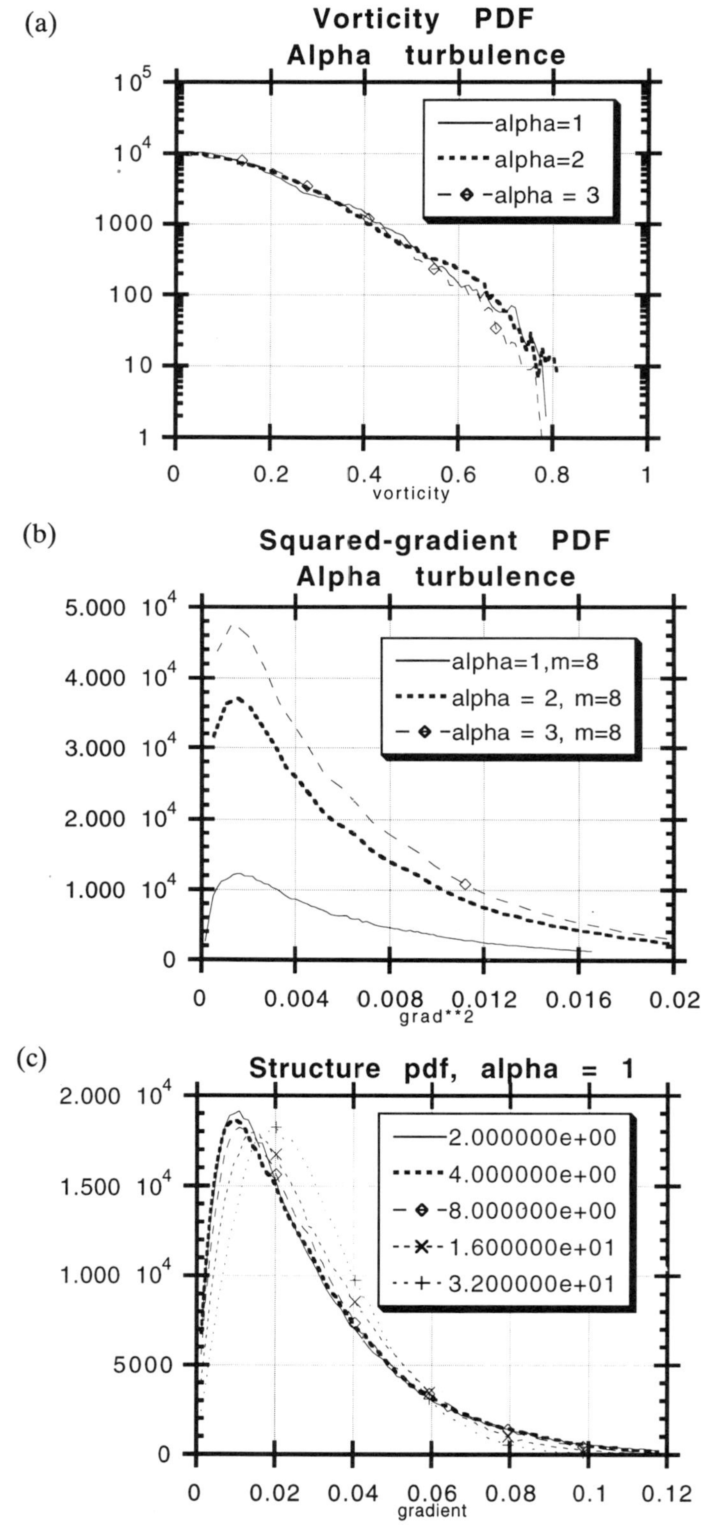

Fig. 7. PDF for 2D α-turbulence ($\alpha = 2$ is conventional Navier–Stokes). (a) Pseudo-vorticity PDF for various α. (b) PDF of grad^2 field of pseudo-vorticity for various α, with smoothing $m = 8$. (c) Structure PDF (as in Fig. 4(c) but for pseudo-vorticity instead of concentration) for $\alpha = 1$.

distribution, σ, scales like $m^{-0.37}$, with box-size which is slightly steeper than the scaling law in the decay case. The 'structure PDFs' (i.e. with the squared-gradient computed *after* smoothing) also show similar shapes and scaling to those in the decay case.

The squared-gradient PDFs for the turbulence models of Pierrehumbert *et al*. [23] show behavior that is quite similar to those we have seen for the passive tracer case. Some representative curves are shown in Fig. 7(b). The width-scaling law for the gradient PDF is $m^{-0.34}$ for $\alpha = 1$, $m^{-0.38}$ for $\alpha = 2$, and $m^{-0.32}$ for $\alpha = 3$. To within the errors of the estimate, these can all be considered the same, and the indication is that the gradient field is an FDF random field rather than an intermittent multifractal. In this case, however, we cannot discard the possibility of multifractality with as much confidence as we did in the passive tracer case. The extent of the scaling range is smaller for the turbulence models, and so the logarithmic behavior required for multifractality can arguably fit the data as well. Considerably higher resolutions will be needed to resolve this question. The substantial independence of α indicates that although $\alpha = 1$ is formally more local spectrally than the other cases, and although the tracer filaments curdle rather than remaining lamellar (see ref. [23]), the curdling process is evidently intermittent enough that the gradient field over most of the domain is still controlled by large-scale processes.

The structure PDFs are shown in Fig. 7(c) for $\alpha = 1$. These are rather different from what we have seen for the passive tracers. Rather than collapsing onto a single curve, the rescaled PDFs systematically change shape as m is increased, and only collapse between $m = 16$ and $m = 32$. This is indicative of a substantially different kind of long-range order than that prevailing in the passive scalar advection problem. Other values of α behave similarly. This result is somewhat at odds with the previous indications that the squared-gradient field is FDF rather than multifractal, and therefore formally has $D_q = D$. Such a system should have self-similar ('Kolmogorovian') structure PDFs. However, this prediction is dependent on the link between κ_q and D_q and other asymptotic relations. Since the scaling associated with $D_q = D$ is only asymptotically valid at large m, the behavior of the structure PDFs may just result from the failure of asymptoticity. Again, much higher resolutions are needed to resolve the question.

The fact that none of the PDFs we have discussed is able to distinguish the obviously different tracer morphology prevailing between $\alpha = 1$ and $\alpha = 2$ (see Fig. 2 in ref. [23]) reveals that these tools still leave much to be desired.

5. ULTRA-HIGH RESOLUTION UNDIFFUSED SOLUTIONS

The question naturally arises as to whether the anomalous spectral steepness in the equilibrium case is simply due to insufficient resolution. With a full 2D calculation, it is unlikely that sufficient resolution will be available definitively to answer the question anytime soon. However, for *undamped* exactly conservative tracers, it is possible to determine the spectrum at resolutions several order of magnitudes higher. This is so because in the absence of diffusion, the concentration value on a particle is independent of that on neighboring particles. Thus, we can compute the equilibrium concentration profile $c(x)$ along a given line by running trajectories backwards for each point on the line, and determining whether they originate in the $+1$ forcing strip or the -1 forcing strip. Isotropy can be probed by doing the calculation for lines of various orientations.

It should be noted that even in the absence of small-scale damping, the forced problem as we have outlined it has a shortwave cutoff scale, which arises in the following manner. For forcing strips of total area A, the probability of a particle avoiding the forcing region for n iterations is $(1 - A/(4\pi^2))^n$, which becomes very small for $n \gg 4\pi^2/A$. Thus, a

Table 3.

ε	Inner scale
0.50000	0.043214
0.25000	0.00093373
0.10000	3.0141×10^{-8}
0.0100000	10^{-69}

'typical' age of a parcel following its most recent resetting is $4\pi^2/A$ iterations. If λ is the typical Lyapunov exponent, the scale generated in this time from forcing strips of thickness ε is $\varepsilon \exp(-\lambda 4\pi^2/A)$. With $\lambda = 0.5$ and the forcing geometry we have adopted, this inner length scale is given in Table 3. The inner scale is exceedingly small (smaller in fact than the radius of an electron, for $\varepsilon = 0.01$), but it is in fact present.

Spectra computed along the lines $x = \pi$ and $y = \pi$ were found to be practically identical at small scales, confirming that there is sufficient isotropy for spectra along a single line to be indicative of the full behavior. Spectra for the former line at various ε are shown in Fig. 8(a). These were computed based on 1 048 576 points, and with such an enormous range of scales the power-law scaling cannot be held in dispute. Given that the inner cutoff scale is only marginally resolved for $\varepsilon = 0.1$, and is orders of magnitude below the sampling interval for $\varepsilon = 0.01$ and $\varepsilon = 0.001$, error due to aliasing is a prime concern. For $\varepsilon = 0.001$, the spectrum is entirely flat; because of undersampling, all information about spatial correlation is lost and the signal is indistinguishable from white noise. For $\varepsilon = 0.01$, the white noise spectrum due to aliasing is visible at short wavelengths, but there is indication of a power-law spectrum at somewhat longer waves. The spectral slope of this range is approximately $k^{-0.6}$. Given that the power in this spectrum is highly divergent, however, one must be suspicious that it, too, is an artifact of aliasing. This is confirmed by examining spectra computed from signals subsampled by a factor of 16 and 64, and also by recomputing spectra for the full 1 048 576 points arranged on a segment of length 0.001 the original one. Both spectra (not shown) indicate that the $\varepsilon = 0.01$ spectra in Fig. 8(a) are strongly influenced by unresolved small scales.

Finally, for $\varepsilon = 0.1$, a k^{-1} Batchelor spectrum is obtained. Given that k^{-1} is divergent, one must still be wary of aliasing in this case. However, the fact that the inner scale is only slightly shorter than the sampling interval partly allays this fear. Further, in Fig. 8(b) we compare the spectra for the full record with spectra computed from subsampling by a factor of 16 and 64. It appears that the aliasing error reliably moves to short waves as the sampling increases. From this we conclude that the Batchelor spectrum is indeed realizable given an extensive enough inertial range. The results suggest that extremely high resolutions are needed to recover the spectrum, however.

As ε is increased beyond 0.1, the inner scale increases and moves into the range of the resolved scales, compressing the 'inertial' range. When this happens, the spectra begin to be influenced by the detailed physics determining the statistics of the inner scale. In fact, there is not really a unique 'inner scale' but rather a spectrum of 'inner scales' associated with the spread of Lyapunov exponents and the spread of times taken before an orbit encounters a resetting region. In the limit of an inner scale that is sharply bounded below, one would recover the k^{-2} jump spectrum as in the theory of Saffman [22]. Small-scale events are sufficiently probable to render the spectra shallower than Saffman's theory, but steeper than k^{-1}. For $\varepsilon = 0.25$, the spectral shape is $k^{-1.25}$, and for $\varepsilon = 0.5$ the shape is $k^{-1.52}$.

(a)

(b)

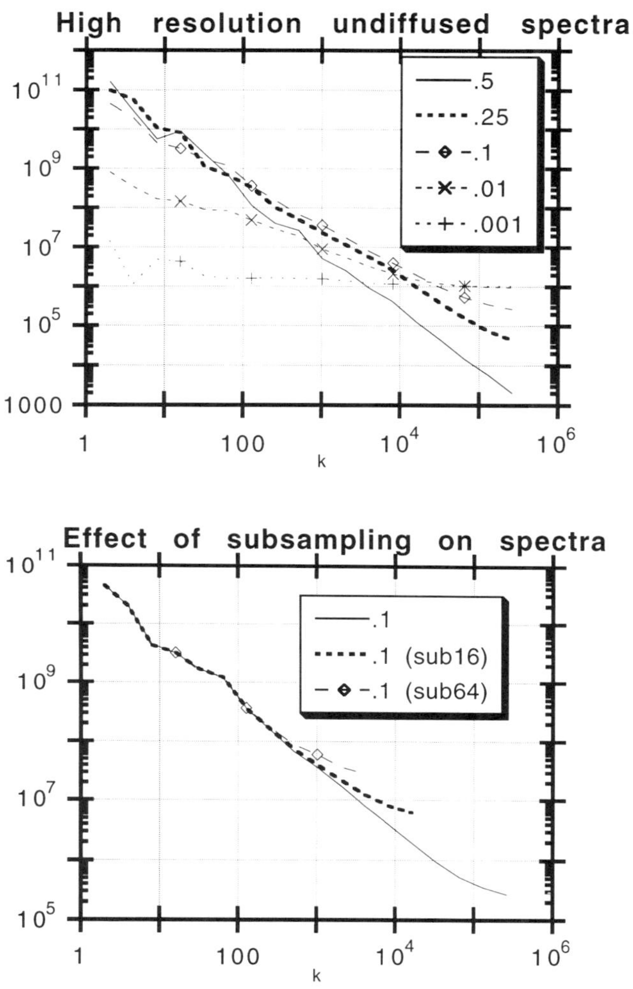

Fig. 8. (a) High-resolution (1 048 576 point spectra) for undamped equilibrium case with various forcing widths ε. (b) Effect of subsampling for $\varepsilon = 0.1$, showing localization of aliasing errors at high wavenumbers.

6. CONCLUSIONS

We have outlined a general computationally tractable approach for addressing a range of problems concerning tracer microstructure created by large-scale advection and small-scale diffusion, through the use of iterated maps. Using this technique, we have studied spectra, concentration PDFs, multifractal scaling of squared-gradients, and cancellation exponents. Results for the idealized system have been compared and contrasted with similar diagnostics applied to a family of fully nonlinear two-dimensional turbulence models. While more questions have been raised than have been resolved, a few clear points have at least emerged.

For the freely decaying advection–diffusion problem, there is evidence that the evolution settles into a ‘strange eigenmode’ which has arbitrarily fine-scaled variability limited only by diffusion, but which is statistically self-similar in time. The variance decays exponentially in time, with a rate governed by the dominant Lyapunov exponent, and independent of diffusivity. The concentration power spectrum for this eigenmode is not algebraic, but rather exhibits a broad exponential rolloff. The concentration PDFs are self-similar in time, and are approximately Gaussian for small concentrations but have exponential tails for extreme concentration excursions. These PDFs appear to conform well to the theory of Sinai and Yakhot [7]. It is clear from these results that a Batchelor k^{-1} spectrum is not the long-term solution for the freely evolving case in the presence of dissipation. The 'strange eigenmode' is the correct end-state.

A very different picture is obtained when the variance is maintained against decay by periodically reintroducing unit-magnitude concentration in certain strips at large scales. In this case, the spectrum is indeed algebraic, though it is significantly steeper than k^{-1}. It is significant that, at similar resolutions, the pseudo-vorticity spectra for the large-eddy dominated cases $\alpha = 2$ and $\alpha = 3$ in the 2D turbulence models of Pierrehumbert *et al.* [23] show the same kind of anomalous steepness. We suggest that steep spectra of this sort are characteristic of mixing by large-scale advection and small-scale diffusion, at least until truly enormous resolutions (perhaps $10^6 \times 10^6$) in 2D. An undiffused simulation at the latter extremely high resolution indeed shows a k^{-1} spectrum when aliasing problems are surmounted and when the non diffusive concentration microscale (which we have defined above) is sufficiently far removed from the inertial range. There is as yet no quantitative theory of the diffusively influenced steep spectra prevailing in the lower resolution calculations.

The PDFs of concentration in the equilibrium case are markedly different from those in the freely decaying case. Apart from a spike at the concentration value introduced by the forcing, these distributions are much more uniform than those in the decaying problem. This indicates that theories of the type of Sinai and Yakhot do not always work in the presence of variance maintenance by large-scale forcing. In contrast, though, the equilibrium pseudo-vorticity PDFs for the turbulence models of Pierrehumbert *et al.* [23] do show a behavior that is more like the Sinai–Yakhot theory, though the range of Gaussian behavior is broader and the exponential tails are less pronounced.

A robust result we have found throughout all the models we have examined—decaying tracer, equilibrium tracer and 'α-turbulence' (Pierrehumbert *et al.* [23])—is that the squared-gradient field is a 'fractal degree of freedom' random field rather than a multifractal. We have argued that the scaling of the width of the lognormal distribution of the squared gradient under smoothing is anomalous, and reflects the underlying geometry. The lognormal behavior of the squared-gradient PDF is closely related to the Gaussian distribution of Lyapunov exponents, which is known for the advection–diffusion problem, and which we conjecture for the α-turbulence problems.

We have also made a foray into the subject of cancellation exponents, by examining the PDFs of the magnitude of the gradient computed *after* smoothing is first applied to the concentration (or pseudo-vorticity) field. For both the decaying and equilibrium passive tracer case, the concentration structure PDFs for various degrees of smoothing are 'Kolmogorovian,' in the sense that they are self-similar under rescaling by the mean gradient. It was argued that this is indicative of a non-intermittent squared-gradient field. We gave a concrete realization of the k^{-1} Batchelor spectrum in terms of a non-intermittent space-filling random gradient model. However, it was also shown that the Batchelor spectrum is perfectly compatible with intermittent models as well, because of the leeway provided by varying the degree of cancellation between positive and negative gradient regions.

The structure PDFs are less self-similar for the α-turbulence cases and suggest that (relative to the typical gradient at each scale) weak gradients are more probable at small scales then they are at large scales, as if localized weak-gradient regions are systematically eaten up by neighboring same-signed high gradients when the field is smoothed. Most of our results concerning α-turbulence should be regarded as tentative, though, owing to a marginally resolved inertial range.

Under what situations would behavior of the sort we have seen above be expected to prevail? It comes down to a question of whether the advecting field has in some sense a dominant timescale, and the existence of such a timescale can be probed by examination of the finite-time Lyapunov exponents for the trajectory problem arising from a given velocity field. For the iterated map we have studied, these exponents are well-behaved by construction. For a turbulent velocity field with small scales, however, the Lyapunov exponents could be highly dependent on the dissipative cutoff. Notably, for a $k^{-5/3}$ energy spectrum, the associated velocity field has divergent gradients. In principle, then, the Lyapunov exponents could become divergent as resolution is increased. This need not happen, however, if the correlation time of high-gradient regions is sufficiently short. Indeed, this appears to happen for atmospheric winds simulated by a general circulation model, for which case the trajectory problem has a well-defined exponent [5]. The velocity field for the $\alpha = 1$ turbulence model discussed in Pierrehumbert *et al.* [23] also has divergent gradients. While we have not computed Lyapunov exponents yet for this case, the PDFs of the squared-gradient field are strongly suggestive of behavior similar to that of the iterated map. Coming to an understanding of finite-time Lyapunov exponent distributions for trajectory problems arising from turbulent velocity fields is, in any event, of prime importance. The $\alpha = 1$ model provides a good forum for probing the relative role of large-scale and small-scale advection, and would be a good starting point for such a study. Simulations of passive tracer mixing by the velocity field of this model would provide invaluable complementary information.

Acknowledgements—The advection/diffusion calculation reported here grew out of a stimulating discussion with Leo Kadanoff and Itamar Procaccia. Discussions with Samuel Vainshtein and Yunson Du were indispensible in clarifying the connection between cancellation exponents, structure functions, and generalized dimensions. My thinking about structure functions has also been much influenced by discussions with Alexander Marshak and Anthony Davis. The results in Section 5 were stimulated by exploratory calculations carried out by Yannis Antoniades. This work was supported by the National Science Foundation, under grant ATM 89-20589.

REFERENCES

1. G. K. Batchelor, Small-scale variation of convected quantities like temperature in turbulent fluid—Part 1: General discussion and the case of small conductivity, *J. Fluid Mech.* **5**, 113–133 (1959).
2. R. T. Pierrehumbert, Chaotic mixing of tracer and vorticity by modulated travelling Rossby waves, *Geophys. Astrophys. Fluid Dynamics* **58**, 285–319 (1991).
3. R. T. Pierrehumbert, Large-scale horizontal mixing in planetary atmospheres, *Phys. Fluids* **A3**, 1250–1260 (1991).
4. R. T. Pierrehumbert, Spectra of tracer distributions: A geometric approach, in *Nonlinear Phenomena in Atmospheres and Oceans*, edited by R. Pierrehumbert and G. Carnevale. Springer, New York (1992).
5. R. T. Pierrehumbert and H. Yang, Global chaotic mixing on isentropic surfaces, *J. Atmos. Sci.* **50**, 2462–2480 (1993).
6. R. R. Prasad, C. Meneveau and K. R. Sreenivasan, Multifractal nature of the dissipation field of passive scalars in fully turbulent flows, *Phys. Rev. Lett.* **61**, 74–77 (1988).
7. Ya. G. Sinai and V. Yakhot, Limiting probability distributions of a passive scalar in a random velocity field, *Phys. Rev. Lett.* **63**, 1962–1964 (1989).
8. T. Horita, H. Hiroki, I. Ishizaki and H. Mori, Long-time correlations and expansion-rate spectra of chaos in hamiltonian systems, *Prog. Theor. Phys.* **83**, 1065–1070 (1990).
9. P. Grassberger, R. Badii and A. Politi, Scaling laws for invariant measures on hyperbolic and nonhyperbolic attractors, *J. Stat. Phys.* **51**, 135–178 (1988).
10. E. Ott, Y. Du, K. R. Sreenivasan, A. Juneja and A. K. Suri, Sign-singular measures: Fast magnetic dynamos, and high Reynolds number fluid turbulence, *Phys. Rev. Lett.* **69**, 2654–2656 (1992).
11. Y. Du and E. Ott, Fractal dimensions of fast dynamo magnetic fields, *Physica D*, **67**, 387–417 (1993).
12. B. J. Bayley, Infinitely conducting dynamos and other horrible eigenproblems, in *Nonlinear Phenomena in Atmospheres and Oceans*, edited by R. Pierrehumbert and G. Carnevale. Springer, New York (1992).
13. E. Ching, Probability densities of turbulent temperature fluctuations, *Phys. Rev. Lett.* **70**, 283–286 (1993).
14. A. Pumir, B. Schraiman and E. D. Siggia, *Phys. Rev. Lett.* **66**, 2984–2986 (1991).

15. F. Varosi, T. M. Antonsen and E. Ott, The spectrum of fractal dimensions of passively convected scalar gradients in chaotic fluid flows, *Phys. Fluids* **A3**, 1017-1028 (1991).
16. E. Aurell, U. Frisch, J. Lutsko and M. Vergassola, On the multifractal properties of the energy dissipation derived from turbulence data, *J. Fluid Mech.* **238**, 467–486 (1992).
17. Y. Du, T. Tel and E. Ott, Characterization of sign-singular measures, *Physica D*, in press (1994).
18. A. Marshak, A. Davis, R. Cahalan and W. Wiscombe, Bounded cascade models as non-stationary multifractals. *Phys. Rev. E*, **49**, 55–79 (1994).
19. A. Davis, A. Marshak, W. Wiscombe and R. Cahalan, Multifractal characterizations of non-stationarity and intermittency in geophysical fields, observed, retrieved or simulated, *J. Geophys. Res.* **99**, 8055–8072 (1994).
20. S. I. Vainshtein, Y. Du and K. R. Sreenivasan, Sign-singular measure and its association with turbulent scalings, *Phys. Rev. E*, **49**, 2521–2523.
21. Y. Tessier, S. Lovejoy and D. Schertzer, Universal multifractals: Theory and observations for rain and clouds, *J. Appl. Meteor.* **32**, 223–250 (1993).
22. P. G. Saffman, On the spectrum and decay of random two-dimensional vorticity distributions at large Reynolds number, *Stud. Appl. Math.* **50**, 377–383 (1971).
23. R. T. Pierrehumbert, I. Held and K. Swanson, Spectra of local and nonlocal turbulence in two dimensions. *Chaos, Solitons & Fractals*, to appear (1994).
24. T. M. Antonson and E. Ott, Fractal measures of passively convected vector fields and scalar gradients in chaotic fluid flow, *Phys. Rev.* **A39**, 3660–3671 (1989).

Spectra of Local and Nonlocal Two-dimensional Turbulence

RAYMOND T. PIERREHUMBERT

Department of Geophysical Sciences, University of Chicago, Chicago, IL 60637, USA

and

ISAAC M. HELD

Geophysical Fluid Dynamics Laboratory/NOAA, Princeton University, Princeton, NJ 08540, USA

and

KYLE L. SWANSON

Department of Geophysical Sciences, University of Chicago, Chicago, IL 60637, USA

Abstract—We propose a family of two-dimensional incompressible fluid models indexed by a parameter $\alpha \varepsilon [0,\infty]$, and discuss the spectral scaling properties for homogeneous, isotropic turbulence in these models. The family includes two physically realizable members. It is shown that the enstrophy cascade is spectrally local for $\alpha < 2$, but becomes dominated by nonlocal interactions for $\alpha > 2$. Numerical simulations indicate that the spectral slopes are systematically steeper than those predicted by the local scaling argument.

Incompressible fluid turbulence in two dimensions is of interest because of its applications in geophysical problems [1] and because of its role as a computationally tractable testbed for theories. Spectrally local similarity theories of the inertial range spectrum based on Kolmogorov's phenomenology [2, 3] often provide a starting point for such studies. The enstrophy cascading inertial range in 2D turbulence has a pathology in this regard; assuming the local similarity spectrum, the aggregate straining effects of eddies larger than a given scale L just fail to be dominated by the effects of eddies with scales similar to L. Hence, the enstrophy range of conventional 2D turbulence is precisely on the threshold of locality [3, 4]. In this note we examine a family of modified 2D fluid equations, which span the range between spectrally local and strongly nonlocal behavior.

The equation for advection of a conserved scalar q in a velocity field with streamfunction ψ is:

$$\partial_t q + J(\psi, q) = 0 \tag{1a}$$

where $J(A, B) = \partial_x A \partial_y B - \partial_x B \partial_y A$. Advection of even a passive scalar decoupled from ψ embraces a rich variety of phenomena [5, 6], but the 2D hydrodynamic equations are distinguished by coupling between q and ψ, which renders the equation nonlinear. The form of the coupling determines the degree of locality. We consider the family of couplings defined in Fourier transform space by

$$\hat{\psi}(\mathbf{k}) = -|\mathbf{k}|^{-\alpha}\hat{q}(\mathbf{k}), \tag{1b}$$

where $\mathbf{k}$ is the two-dimensional wavenumber vector, with magnitude $k = |\mathbf{k}|$. Essentially, $\psi(x, y)$ is a smoothed form of $q(x, y)$; as α is made larger, the smoothing becomes more extreme, the advecting field becomes more decoupled from the small-scale structure of the advected field, and the problem becomes more spectrally nonlocal. For $\alpha = 2$, q is vorticity and the equations are the conventional 2D Euler equations. For $\alpha = 1$, the equations are also physically realizable; they govern the evolution of three-dimensional stratified rapidly rotating flow with zero potential vorticity in a semi-infinite domain, in which case $q(x, y)$ is identified with the temperature field on the (flat) lower boundary [7]. In terms of the geostrophic streamfunction $\psi(x, y, z, t)$, the boundary temperature is $\partial\psi/\partial z$ at $z = 0$. ψ satisfies a three-dimensional Laplace equation, and the problem is closed by imposing decay conditions at $z \to \infty$. This leads to the spectral space expression (1b) with $\alpha = 1$.

Equations (1a, b) have two quadratic conserved quantities, namely, mean energy and enstrophy, defined, respectively, by:

$$\bar{E} = -\frac{1}{A}\iint \psi q \,\mathrm{d}x\,\mathrm{d}y \quad \text{and} \quad \bar{G} = \frac{1}{A}\iint q^2 \,\mathrm{d}x\,\mathrm{d}y \tag{2}$$

in which A is the area of the domain. The power spectra $E(k)$ and $G(k)$ are defined so that

$$\bar{E} = \int_0^\infty E\,\mathrm{d}k \quad \text{and} \quad \bar{G} = \int_0^\infty G\,\mathrm{d}k \tag{3}$$

with $G = k^\alpha E = k^{2(\alpha-1)}V$, where V is the power spectrum of the velocity field. If $\alpha = 2$, $E = V$; if $\alpha = 1$, $G = V$. The spectral fluxes are defined by $\partial_t E = -\partial_k F_{\mathrm{E}}$ and $\partial_t G = -\partial_k F_G$. In equilibrium, both F_E and F_G must be constant.

The dimensions of F_E/F_G are $L^{2\alpha}$. The fundamental tenet of the Kolmogorov–Kraichnan scale analysis is that the only available length scale is the local eddy scale k^{-1}. Hence, since the fluxes must be independant of k, only one of F_E and F_G can be nonzero. Nonzero F_E yields the *energy cascading* spectrum; since the dimensions of E are $L^{5-\alpha}T^{-2}$ and the dimensions of F_E are $L^{4-\alpha}T^{-3}$, dimensional analysis implies

$$E = a(\alpha)(F_E)^{2/3}k^{\gamma-\alpha}, \qquad G = a(\alpha)(F_E)^{2/3}k^{\gamma} \tag{4}$$

with $\gamma = -\frac{7}{3} + \frac{4}{3}\alpha$. This yields the conventional $k^{-5/3}$ energy spectrum for $\alpha = 2$. Following similar reasoning, the spectrum in the enstrophy cascading range is

$$E = a(\alpha)(F_G)^{2/3}k^{\gamma-\alpha}, \qquad G = a(\alpha)(F_G)^{2/3}k^{\gamma} \tag{5}$$

with $\gamma = -\frac{7}{3} + \frac{2}{3}\alpha$. For $\alpha = 2$, this yields the customary k^{-1} enstrophy spectrum, while for $\alpha = 1$, the enstrophy spectrum is $k^{-5/3}$. The enstrophy spectrum is integrable at shortwaves whenever $\alpha < 2$, but diverges like $k^{(2/3)\alpha}$ for large α. The spectral shape for $\alpha = 1$ was first obtained ref. [7], and has been reproduced in EDQNM closure model simulations [8].

Following Kraichnan [3, 4, 5, 9] the strain rate due to eddies with scales between $1/k$ and $2/k$ is $k\surd(kV) = k^{2-\alpha}\surd(kG)$. Substituting (5) for the spectrum, we find that the enstrophy cascade is dominated by local strain for $\alpha < 2$ and dominated by large eddy strain for $\alpha > 2$. Notably, the k^{-1} local scaling spectrum for $\alpha = 2$ is the same as the passive scalar spectrum predicted [6, 10] for the strongly nonlocal case in which straining is dominated by large eddies with a fixed timescale. Thus, the spectral behavior shades continuously over to the nonlocal scaling as α approaches 2 from below. This is because the conserved quantity q is coincidentally the straining rate for $\alpha = 2$. Strong deviations from the predictions of local scaling are expected for $\alpha > 2$.

Making use of (4), the strain-rate scaling for the energy cascading range is found to be

$k^{4/3-\alpha/3}$, whence, as is well known, the cascade is already local for $\alpha = 2$. The energy cascading range does not become spectrally nonlocal until α exceeds 4.

Advection of a tracer by a time-varying large-scale flow typically produces filaments whose thickness decreases exponentially with time [5, 6]. The stability properties of such filaments shed further light on the locality of the enstrophy cascade. Consider a filament with profile $q(y) = q_{\max} f(y/L)$, which is to be thought of as a local representation of a curved filament which is varying slowly in time compared to the growth of instabilities upon it. If (1a, b) is linearized about this profile, it follows from dimensional analysis that the wavelength of maximum instability scales with L, and that the maximum growth rate of instabilities scales with $q_{\max} L^{\alpha-2}$. Large-scale straining with rate ε stabilizes instabilities with growth rates less than $O(\varepsilon)$ (see the discussion in ref. [11] for $\alpha = 2$). Thus, in the presence of large eddy-strain of a given magnitude, filaments always succumb to a secondary instability for $\alpha < 2$, instantiating the classical 'large swirls have smaller swirls' cascade. For $\alpha > 2$ the large-scale strain always suppresses the instability. In the marginal case $\alpha = 2$, whether or not the filament undergoes secondary rollup depends on the initial condition, via $q_{\max}/\varepsilon$.

The scaling of the instability suggests the possibility of finite-time singularities for $\alpha < 2$. Filament thickness decreases exponentially in time at the local strain-rate, but the local strain is roughly proportional to the growth rate of the filament instability (which creates the strain). Substituting the instability scaling yields $\mathrm{d}L/\mathrm{d}t = -aq_{\max} L^{\alpha-1}$. Thus, a filament with initial width L_0 collapses to zero thickness at time $(L_0^{2-\alpha})/((2-\alpha)aq_{\max})$. This agrees with the result derived in ref. [8] for $\alpha = 1$. As $\alpha \to 2^-$, the blowup time diverges and we recover global regularity.

We carried out a series of numerical simulations of 1(a, b) with $\alpha = 0.5$, 1, 2 and 3. The simulations were carried out using a de-aliased spectral model with 512^2 resolution and ∇^8 dissipation. It was forced by holding the vorticity amplitudes for modes with wavenumbers between 6 and 8 fixed, with random, time-dependent phases. The enstrophy spectra are shown in Fig. 1, and the spectral slopes calculated in the wavenumber range $10 < k < 100$ are summarized in Table 1. These do not represent true equilibrium results, as the inverse energy cascade has not proceeded to completion; the slopes in the enstrophy cascading range have nonetheless converged. A calculation carried out at 256^2 resolution revealed essentially the same inertial range slopes.

As predicted by local similarity theory, the spectra become steeper as α is decreased. However, the simulated spectra are systematically steeper than the similarity prediction. The difficulties in obtaining the local similarity spectrum for conventional 2D turbulence

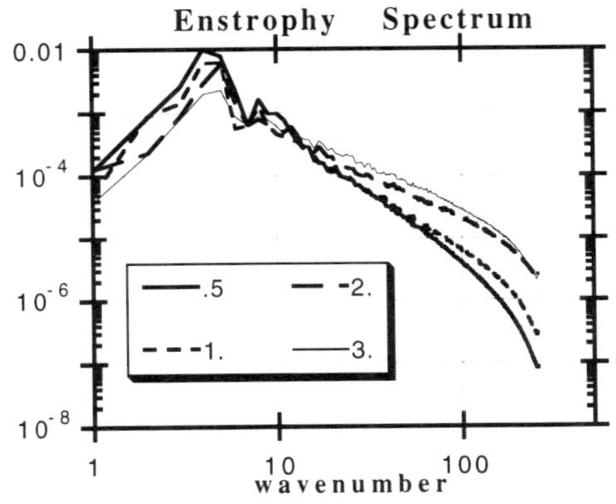

Fig. 1. Power spectra for $\alpha = 0.5$, 1, 2 and 3.

Fig. 2. Vorticity patterns for $\alpha = 1$ (left panel) and 2 (right panel).

Table 1.

α	Slope (actual)	Slope (theory)
0.5	−2.2	−2.0
1	−1.9	−1.7
2	−1.4	−1.0
3	−1.3	−0.33

have been discussed extensively in the literature (e.g. ref. [12]). The generalized case we have studied exhibits the same anomaly, despite the unambiguous spectral locality for $\alpha < 2$. The EDQNM model [8] does not yield an anomalously steep spectrum for $\alpha = 1$; this may reflect a shortcoming in the closure scheme, or it may be indicative of insufficient resolution of the inertial range in our calculation. Intermittency corrections [13], or coherent vortex formation [14] may also come into play in steepening the spectra. As α is increased beyond 2, the local similarity theory predicts that the slope continues to become more shallow. In reality, the slope for $\alpha = 3$ is nearly the same as that for $\alpha = 2$, suggesting a dominance of the enstrophy cascade by spectrally nonlocal large eddy advection, and a crossover to Batchelor-type passive scalar behavior.

Images of the vorticity fields for $\alpha = 1$ and 2 are shown in Fig. 2, and show the decreasing importance of secondary instability and small vortex formation as α increases. For $\alpha = 1$, one sees a mix of small vortices, 'curdled' filaments and undisturbed filaments. The vortices arising from instability of the filaments are nearly two orders of magnitude smaller than the primary vortices at the injection range. This suggests that the finite-time collapse discussed above emerges from a sequence of large, discrete spectral jumps, of which our calculation has resolved only the first. For $\alpha = 2$ the scene is dominated by large vortices, large spirals, and smooth filaments. Given the marginal locality in this case, the ability of the filaments to resist strain is probably dependent on the details of the forcing. An example showing the coexistence of passively strained filaments with filaments exhibiting strong rollup in conventional 2D turbulence is given in ref. [15].

It appears that the anomalous spectral steepness commonly found in two-dimensional turbulence is not closely related to spectral nonlocality. On the one hand, $\alpha = 2$ behaves as if the cascade were dominated by large-scale advection, for which Batchelor's argument gives a k^{-1} enstrophy spectrum. Thus, Batchelor's phenomenology is failing; the reasons for the failure can be most clearly studied in the context of large eddy passive advection problems such as Ottino [5] or Pierrehumbert [6]. On the other hand, the $\alpha < 2$ cases exhibit anomalous steepness despite spectral locality; the reasons for this breakdown can be most clearly studied in the $\alpha = 1$ model, for which the issue is disentangled from the marginal locality bedeviling $\alpha = 2$.

REFERENCES

1. P. B. Rhines, Geostrophic turbulence, *Ann. Rev. Fluid Mech.* **11**, 401–442 (1979).
2. A. N. Kolmogorov, The local structure of turbulence in incompressible fluid at very high Reynolds number, *Dokl. Akad. Sci. USSR*, **30**, 299–303 (1941).
3. R. Kraichnan, Eddy viscosity in two and three dimensions, *J. Atmos. Sci.* **33**, 1521–1536 (1976).
4. R. Kraichnan, Inertial-range transfer in two- and three-dimensional turbulence, *J. Fluid Mech.* **47**, 525–535 (1971).
5. J. Ottino, *The Kinematics of Mixing: Stretching, Chaos and Transport*. Cambridge University Press, Cambridge (1989).
6. R. T. Pierrehumbert, Chaotic mixing of tracer and vorticity by modulated travelling Rossby waves, *Geophys. Astrophys. Fluid Dynamics* **58**, 285–319 (1991).
7. W. Blumen, Uniform potential vorticity flow—Part I: Theory of wave interactions and two-dimensional turbulence, *J. Atmos. Sci.* **35**, 774–783 (1978).

8. J.-M. Hoyer and R. Sadourny, *J. Atmos. Sci.* **39**, 707–721 (1982).
9. R. Kraichnan, Inertial ranges in two-dimensional turbulence, *Phys. Fluids* **10**, 1417–1423 (1967).
10. G. K. Batchelor, Small-scale variation of convected quantities like temperature in turbulent fluid—Part 1; General discussion and the case of small conductivity, *J. Fluid Mech.* **5**, 113–133 (1959).
11. D. G. Dritschel, P. H. Haynes, M. N. Juckes and T. G. Shepherd, The stability of a two-dimensional vorticity filament under uniform strain, *J. Fluid Mech.* **230**, 647–665 (1991).
12. M. E. Brachet, M. Meneguzzi and P. L. Sulem, Small-scale dynamics of high Reynolds number two-dimensional turbulence, *Phys. Rev. Lett.* **57**, 683–686 (1986).
13. C. Basdevant, B. Legras, R. Sadourny and M. Beland, A study of barotropic model flows: Intermittency, waves and predictability, *J. Atmos. Sci.* **38**, 2305–2326 (1981).
14. J. Herring and R. Kraichnan, *J. Fluid Mech.* **153**, 229–242 (1985).
15. M. N. Juckes and M. E. McIntyre, A high resolution, one-layer model of breaking planetary waves in the stratosphere, *Nature Lond.* **328**, 590–596 (1987).

Author Index

Acharya, N. 213, 222, 225, 227–30
Acrivos, A. 194, 196
Afanasyev, Y. D. 235, 244–5, 265
Antonsen, T. M. Jr. 6, 72, 93, 95, 119–20, 125, 127, 135, 137, 149, 332, 345, 354, 366
Aref, H. 1–4, 6, 9, 29, 73, 98, 117, 119, 125, 149, 185–6, 192, 196–7, 209, 212–13, 230, 235–8, 245–7, 252, 263, 265–7, 285, 287, 309, 314, 344
Arnold, V. I. 2, 4, 158, 167, 209, 212, 230, 290, 309
Arter, W. 2, 4
Ashurst. W. T. 297, 303, 306–7, 310
Auerbach, F. 261, 266
Aurell, E. 355, 366

Badii, R. 93, 95, 118, 120, 126, 135, 137, 149, 348, 365
Bajer, K. 4, 151–2, 163, 165, 167
Balanchandar, S. 6, 93, 117, 125, 149
Ball, R. 234, 264
Basdevant, C. 235, 244–5, 263, 265, 371–2
Batchelor, G. K. 73, 117, 152, 167, 190, 234, 237, 265, 290, 294–5, 297, 301, 309–10, 347, 352, 365, 371–2
Bayley, B. J. 189, 196, 350, 365
Beerens, S. P. 267
Behringer, R. F. 6, 117
Beigie, D. 3, 5–6, 14–15, 21, 23, 37, 45, 51, 57, 59, 79, 89–90, 114, 117–18, 125, 135, 137, 149, 274, 285
Beland, M. 371–2
Benettin, G. 293, 309
Bensimon, D. 45, 118
Benzi, R. 93, 118, 263, 266
Berlin, D. 6, 120
Bernardin, M. P. 151, 167
Berry, M. V. 9, 118
Bertozzi, A. L. 275, 285
Betchev, R. 302, 310
Biferale, L. 93, 118
Blumen, W. 368, 371
Brachet, M. E. 371–2
Brandenberg. A. 308, 310
Bretherton, F. P. 171, 183
Buch, K. A. 118, 322, 344
Burton, T. D. 186, 196

Cahalan, R. 355, 366
Camassa, R. 6, 14–15, 18, 30, 45, 110, 115, 118
Carrier, G. F. 118–19
Cary, J. 151, 167
Castelain, C. 229, 231
Chaiken, J. 118, 125, 149, 212, 230, 287, 309
Chandrasekhar, S. 186, 196
Chang, H.-C. 4, 211, 213–14, 218–19, 222, 227–31
Chaplygin, S. A. 245, 265
Chen, S. H. 213, 230
Chevray, R. 118, 125, 149, 212, 230, 287, 309
Chien, W. L. 118, 125, 149, 212, 230
Ching, E. S. C. 93, 118, 365
Christiensen, J. P. 264, 266
Clarke, J. 6, 118
Cocke, W. J. 295, 309–10
Corrsin, S. 169, 182
Couder, Y. 235, 244–5, 263, 265

Dahm, W. J. A. 3, 118, 313–14, 322, 324, 344
Davis, A. 355, 365–6
de Swart, H. E. 270, 285
de Vriend, H. J. 270, 285
Dean, W. R. 222, 231
Demekhin, E. 214
Dimotakis, P. E. 73, 118–19, 335, 345
Dombre, T. 6, 118, 187, 196, 293, 309, 314, 344
Dresselhaus, E. 6, 73, 118, 298–301, 305–6, 310, 305–6, 310
Dritschel, D. G. 369, 372
Du, Y. 349, 355–6, 365–6

Easton, R. W. 45, 118
Eckhardt, B. 252, 266
Eckmann, D. M. 197, 209
Edwards, M. F. 211, 230
Escudier, M. P. 34–6, 118

Farmer, J. D. 294, 309
Feder, J. 98, 118
Feingold, M. 29, 118
Fendell, F. E. 118–19
Fenichel, N. 21, 118
Ferziger, J. H. 6, 14, 34–5, 120
Filonienko, N. N. 151, 167
Finn, J. M. 6, 29, 93, 118–19
Fischer, H. B. *et al.* 285
Flieril, G. R. 235, 265
Flór, J. B. 235, 244–6, 255, 259, 265
Franjione, J. G. 127–8, 149, 212, 230
Frank, F. C. 300, 310
Frederiksen, R. D. 3, 313
Frisch, J. M. 6, 118, 187, 196, 290, 309, 314, 344, 355, 366
Fung, J. C. H. 144, 149

Galgani, L. 293, 309
Gañán-Calvo, A. M. 169, 182
Gharib, M. 6, 118
Ghosh, S. 218–19, 230–1
Gibson, C. H. 297, 303, 306–7, 310
Giorgilli, A. 293, 309
Girimaji, S. S. 297, 303, 310
Goldhirsch, I. 293, 309
Gollub, J. P. 6, 30, 93, 118, 120
Goryachev, D. N. 252, 266
Govers, W. S. S. 246, 265
Gradshteyn, I. S. 276, 279–80, 285
Grassberger, P. 93, 95, 118, 126, 135, 137, 149, 348, 365
Greated, C. A. 199, 209
Greene, J. M. 6, 118, 187, 196, 290, 309, 314, 344
Gröbli, W. 238, 246–7, 252, 258, 265
Grotberg, J. B. 197, 209
Guckenheimer, J. 14, 16, 58, 118, 158, 167, 175, 183, 273, 285
Gurzhi, A. A. 6, 28, 119, 235, 248–50, 255–6, 264–5

Hairer, E. 238, 265
Hata, H. 93, 118, 126, 135, 137, 149
Haynes, P. H. 369, 372
Held, I. M. 359, 361, 364–7
Helmholtz, H. 3, 233, 237, 240, 261, 264
Hénon, M. 2, 4, 6, 118, 137, 187, 196, 212, 230, 290, 309, 314, 344
Herring, J. 371–2
Hicks, W. M. 234, 238, 265
Hiroki, H. 348, 365
Hirsch, M. W. 118
Holmes, P. 6, 13–14, 16, 29–30, 58, 118, 120, 158, 167, 175, 183, 218–19, 231
Holmes, W. T. 119
Hopfinger, E. J. 235, 265
Horita, T. 93, 118, 126, 135, 137, 149, 348, 365
Hoyer, J.-M. 368–9, 371–2
Hulscher, S. J. M. H. 270, 285
Hydon, P. E. 4, 197, 199, 205, 210

Ide, K. 14, 119
Ishizaki, R. 93, 118, 348, 365
Itzykson, C. 167

Jackson, E. 298, 305–6, 310
Jeffery, G. B. 183
Jennings, R. L. 308, 310
Jones, S. W. 4, 6, 73, 117, 119, 125, 144, 149, 185, 191, 196–7, 205, 209–10, 213, 230
Juckes, M. N. 369, 371–2
Juneja, A. 349, 355, 365

Kadanoff, L.P. 29, 45, 118, 365
Kailasnath, P. 93, 119
Kamm, R. D. 197–8, 209
Kang, I. S. 119
Kaper, T. J. 14, 119
Karniadakis, G. E. 197, 209
Kelvin, (see Thomson, J. J.)
Kerr, R. M. 297, 303, 306–7, 310
Kerst, D. W. 151, 167
Kerstein, A. R. 297, 303, 306–7, 310
Khakhar, D. V. 6, 119, 125, 127, 149, 212, 230
Kimmel, E. 197, 209
Kirchhoff, G. 237, 246, 265
Klafter, J. 126, 149, 201, 205, 210
Klapper, I. 4, 287, 307–8, 310
Kloosterziel, R. C. 235, 265
Knobloch, E. 6, 120, 218, 231
Kolmogorov, A. N. 367, 371
Konovaljuk, T. P. 6, 28, 119, 235, 247–50, 255–6, 265–6
Konstantinov, M. Y. 6, 28, 119, 235, 247–50, 255–6, 258, 265–6
Kopelevich, D. 214
Kraichnan, R. H. 263, 266, 367–8, 371–2
Kruskal. M. 152, 167

Kusch, H. A. 212, 230

Lamb, H. 234, 237, 245–6, 265
Lane, B. 6, 118
Lane-Serff, G. F. 335, 345
Lasheras, J. C. 169, 182
Lau, Y.T. 6, 29, 119
Le Guer, Y. 213, 229–31
Leal, L. G. 4, 119, 169–74, 177–8, 180, 182–3
Lee, C. S. 213, 230
Legras, B 263, 266, 371–2
Leibovich, S. 6, 30, 119, 190
Leonard, A. 3, 5–6, 14–15, 19, 21, 23, 34–5, 45, 51, 57, 59, 117–21, 125, 135, 137, 149, 234, 238, 243, 265, 274–5, 285, 287, 309
Leong, C. W. 6, 119, 125, 127–8, 146, 149
Lichtenberg, A. J. 159, 167, 201, 210, 275, 285
Lieberman, M. A. 159, 167, 201, 210, 275, 285
Ling, F. H. 126–8, 149, 212, 230
Littlejohn, R. G. 151, 167
Liu, M. 3, 125, 135–6, 149
Lochak, P. 158, 167
Lopez, J. M. 6, 30, 34–6, 119
Lorenz, E. 2, 4
Love, A. E. H. 261, 266
Lovejoy, S. 356, 366
Lutsko, J. 355, 366

Maas, L. R. M. 269–70, 285
MacKay, R. S. 29, 45, 58, 119
Mackley, M. R. 300, 310
Majda, A. 298, 310
Mallier, R. 169, 182
Marble, F. E. 118–19
Marshak, A. 355, 365–6
Matsui, T. 34–5, 120
Maxey, M. R. 169, 182
Maxwell, J. C. 233, 236, 246, 264–5
McIntyre, M. E. 371–2
McLaughlin, J. B. 169, 182, 186, 196
McWilliams, J. C. 235, 245, 263–6
Mehr, A. 6, 118, 187, 196, 314, 344
Meiburg, E. 29, 119
Meiss, J. D. 45, 57–8, 119
Meleshko, V. V. 4, 6, 28, 119, 233, 235, 245, 247–9, 255–6, 258, 264–6
Melinkov, V. K. 5, 37, 119
Meneguzzi, M. 93, 120, 371–2
Meneveau, C. 6, 119–20, 125–6, 135–6, 144, 149, 324, 335, 345, 348, 365
Mesquita, O. N. 6, 118
Meunier, C. 158, 167
Meyer, K. R. 26, 119
Meyers, S. D. 6, 117
Mezic, I. 14–15, 29, 119
Miller, P. L. 116, 119, 335, 345
Milliken, W. J. 170–1, 174, 177, 182
Moffatt, H. K. 152, 160, 163, 165, 167
Monin, A. M. 294, 297, 309
Montgomery, D. 263, 266
Moon, F. C. 119
Mori, H. 93, 118, 126, 135, 137, 149, 348, 365
Mullin, T. 199, 209
Muzzio, F. J. 3, 6, 93, 119, 125–7, 135–7, 144, 149, 324, 330, 334, 344–5
Myers, M. R. 273, 285

Nadim, A. 125, 149, 160, 165, 167
Nelkin, M. 303, 310
Newton, P. K. 29, 119
Nex, F. H. 163, 167
Nguyen Duc, J.-M. 235, 244-5, 265
Nordlund, A. 308, 310
N_rset, S. P. 238, 265
Northrup, E. P. 242-3, 265

Olbricht, W. L. 171, 183
Onsager, L. 261, 266
Orlandi, P. 235, 255, 257, 263, 265
Ornstein, L. S. 196
Orszag, S. A. 93, 120, 189, 196, 293, 295, 298, 305-6, 309-10
Oseledec, V. I. 293, 309
Oshima, Y. 264, 266
Ott, E. 6, 45, 58, 72, 93, 95, 118-19, 125, 127, 135, 137, 149, 294, 309, 332, 345, 349, 354-6, 365-6
Ottino, J. M. 6, 14, 30, 36, 38, 93, 118–20, 125–8, 135–7, 144, 149, 169, 182, 212, 230, 236, 265, 267, 271, 285, 287, 309, 314, 321, 323–4, 330, 334, 344–5, 367–9, 371
Ou, J. J. 213, 230

Paladin, G. 93, 118
Pasmanter, R. 267, 269, 285
Pedley, T. J. 198, 209
Peerhossaini, H. 213, 229–31

Percival, I. C. 45, 58, 119, 201, 210
Perry, A. D. 6, 30, 34–6, 119
Peskin, R. L. 3, 125, 127, 135–6, 149
Pierrehumbert, R. T. 4, 6, 93, 119, 135, 137, 149, 359, 361, 364–7, 369, 371
Piro, O. 29, 118
Poincaré, H. 12, 119, 238, 265
Politi, A. 93, 95, 118, 126, 135, 137, 149, 348, 365
Pollak, E. 93, 120, 126, 135, 137, 149
Pope, S. B. 297, 303, 310
Prandtl, L. 190, 196
Prasad, R. R. 116, 119-20, 348, 365
Price, T. 252, 266
Proctor, M. R. E. 163, 167
Pugh, C. C. 118
Pulliam, T. H. 6, 14, 30, 36, 38, 120
Pullin, D. I. 264, 266
Pumir, A. 352, 365

Quint, N. 246, 266

Rallison, J. M. 171, 183
Ramshanker, R. 6, 120, 335, 345
Rankine, W. J. M. 236, 265
Redner, S. 330, 345
Reusch, E. 234, 265
Reynolds, O. 234, 236, 265
Rhines, P. B. 190, 196, 217, 231, 367, 371
Ridderinkhof, H. 267, 269, 284–5
Rieutord, X. 308, 310
Rising, H. 6, 118–19, 125, 149, 212, 230
Roberts, P. H. 307, 310
Robertson, J. M. 238, 265
Robinson, A. R. 234, 265
Robinson, C. 120
Rogers, W. B. 234, 264
Rom-Kedar, V. 6, 14–15, 45, 58, 72, 119–21, 125, 149, 238, 243, 265, 273, 275, 285, 287, 309
Rott, N. 246, 266
Ryrie, S. C. 274–5, 283, 285
Ryzhik, I. W. 276, 279–80, 285

Sadourny, R. 368–9, 371–2
Saffman, P. G. 186, 196, 233, 237, 264, 356, 358, 362, 366
Sagdeev, R. Z. 151, 167
Salam, F. M. A. 226, 231
Santangelo, P. 263, 266
Sastry, S. S. 226, 231
Sawyers, D. R. 213, 230
Schertzer, D. 356, 366
Scheurle, J. 120
Schmidt, G. 126–7, 149
Schraiman, B. 352, 365
Scorer, R. 234, 265
Sell, G. R. 26, 119
Sen, M. 4, 211, 213, 218–19, 222, 227–31
Sepúlveda, M. A. 93, 120, 126, 135, 137, 149
Shapiro, A. H. 197, 209
Shariff, K. 6, 14, 30, 34–6, 38, 120, 234, 265
Sharp, M. K. 197, 209
She, Z.-S. 93, 120, 298, 305–6, 310
Shepherd, T. G. 369, 372
Shlessinger, M. F. 126, 149, 201, 205, 210
Shub, M. 118
Siggia, E. D. 352, 365
Sinai, Ya. G. 348, 352–3, 358, 364–5
Sirovich, L. 93, 117
Solomon, T. H. 6, 30, 93, 118, 120
Sommeria, J. 235, 244–5, 265
Southerland, K. B. 3, 118, 313–14, 322, 324, 344
Soward, A. M. 6, 118, 187, 196, 314, 344
Sreenivasan, K. R. 93, 119–20, 335, 345, 349, 355, 365
Stein, R. F. 308, 310
Stern, M. E. 235, 265
Stock, D. E. 186, 196
Stoffer, D. 28, 120
Stolovitzky, G. 93, 119
Stone, E. 120
Stone, H. A. 125, 149, 160, 165, 167
Strelcyn, J.-M. 283, 309
Strogatz, S. H. 125, 149, 160, 165, 167
Stuart, J. T. 2, 4, 287, 309
Su, L. K. 314, 324, 344
Sulem, P.-L. 293, 309, 371–2
Suri, A. K. 349, 355, 365
Swanson, K. L. 359, 361, 364–7
Swanson, P. D. 6, 93, 119, 125–6, 135–7, 144, 149, 314, 323–4, 330, 334, 344–5
Swinney, H. L. 6, 117
Szeri, A. J. 4, 169–74, 177–8, 180, 182–3

Tabor, M. 2, 4, 6, 73, 118, 125, 149, 212, 230, 267, 273, 285, 287–9, 298–301, 303, 305, 308–10
Tait, P. G. 234, 265
Tan, Q. M. 118, 125, 149, 212, 230, 287, 309
Tataronis, J. A. 151, 167

Taylor, G. I. 193–4, 196
Taylor, T. D. 194, 196
Tel, T. 355-6, 366
Temme, N. M. 269–70, 285
Tessier, Y. 356, 366
Thomman, H. 246, 266
Thomas, O. M. 6, 119, 197, 209, 213, 230
Thomson J. J. (Kelvin) 233–4, 261, 264, 266
Thomson, W. 233–4, 238, 240, 264
Tio, K. K. 169, 182
Tjahjadi, M. 169, 182
Townsend, A. A. 296–7, 301, 310
Truesdell, C. 2, 300, 310
Tsuboi, K. 264, 266
Tuominen, I. 308, 310

Uhlenbeck, G. E. 186, 196

Vainshtein, S. I. 355, 365–6
van den Berg, J. C. 246, 266
Van Dyke, M. 36-8, 120
van Heijst, G. J. F. 4, 233, 235, 244–6, 255, 257, 259, 263–6
van Sluis, R. 246, 265
Városi, F. 93, 95, 120, 127, 135, 137, 149, 354, 366
Vassilicos, J. C. 146, 149
Velasco Fuentes, O. U. 264, 266
Vergassola, M. 93, 118, 355, 366
Vieillefosse, P. 298, 310
Vincent, A. 93, 120
Voropayev, S. I. 235, 244–5, 265
Vulpiani, A. 93, 118

Wang, L.-P. 186, 196
Wang, M. C. 186, 196
Wanner, G. 238, 265
Weiss, J. B. 6, 120, 218, 231
Welander, P. 3–4, 236, 261, 263, 265, 274, 285
Whitehead, J. A. 235, 265
Whiteman, K. J. 151, 167
Wicklin, F. J. 273, 285
Wiggins, S. 3, 5–6, 18–19, 21, 23, 29–30, 37, 45, 51, 57–9, 110, 115, 117–21, 125, 135, 137, 149, 166–7, 170–1, 182, 238, 243, 265, 267, 271, 273–5, 285, 287, 309
Wiscombe. W. 355, 366
Worfolk, P. A. 273, 285

Yaglom, I. 294, 297, 309
Yakhot, V. 189, 196, 348, 352–3, 358, 364–5
Yamada, H. 34–5, 120
Yang, H. 348, 365
Yorke, J. A. 294, 309
Young, L. S. 307–8, 310
Young, W. R. 190, 194, 196, 205, 210, 217, 231

Zabusky, N. J. 6, 14, 34, 120, 235, 245, 263–6
Zaslavsky, G. M. 126, 149, 151, 167, 201, 205, 210
Zimmerman, J. T. F. 267, 269–70, 278, 282, 284–5
Zuber, J.-B. 167

Subject Index

ABC flow 188, 189, 314
Alignment 287-311
Aperiodic
flows 3, 125-149
systems 3, 126-128
Atmosphere 234, 235
of vortex pair 238, 239-244
Attractor 13, 169-183

Batchelor spectrum 294, 295, 322, 347, 352, 356, 357, 358, 359, 362, 364, 371
Beltrami flow 189

Chaotic advection 1, 2, 3, 4, 267, 287
and diffusion 185-196, 205-209
and heat transfer 211-230
in a curved pipe 197-210
in perspective 1-4
templates for 5-124
Chaotic mixing 8, 12, 148, 211, 212-213, 221, 225, 230, 271-278, 280-283, 313-345
Chaotic stirring 8, 46, 267-285
Chaplygin-Lamb dipole 235, 244-246, 251, 255
Convective heat transfer, enhancement of 228-229
Curved pipe 197-210

Dispersion 185, 186-9, 195
Dynamical system 1, 3, 9, 14, 107, 137, 152, 175, 77, 182, 209, 267, 284, 287, 288-298, 303-306

Eulerian 3, 185, 197, 245, 264, 289, 290, 298, 303, 304, 368
current field 236, 268-271
representation 2, 215, 235, 236

Hamiltonian 6, 23, 31, 45, 199, 224, 225, 237, 271, 287, 289
canonical equations 152-156
formulation 4, 151-167, 240
system 3, 7, 8, 151, 153, 159, 165, 166, 198, 218, 237, 238, 267, 268, 271, 173, 275-283

Heat transfer 4, 14, 211-230
Helmholtz equation 3, 233, 237, 261
Horseshoe map 14, 28, 58, 59, 65, 70, 98

Interaction
of chaotic advection and diffusion 185-196
of vortex pairs 245, 246-261
of vortex pair with monopole 246-252

Kinematics 3, 152-156, 169, 185, 235, 239, 252, 255, 256, 258, 264, 268, 284, 288, 303-306, 308
Kolmogorov-Arnold-Moser (KAM) theory 9, 15, 157, 193, 195-196, 198, 200, 201, 205, 219, 225, 289, 303, 322, 355, 356, 367

Lagrangian representation 2, 3, 185, 187, 197, 199, 214, 215, 236, 237, 267, 278, 279, 282, 287, 288, 289, 296, 297, 305, 321
Lobe dynamics 3, 5, 6-15, 17, 18, 22, 24, 35, 45-52, 58, 72, 273-274, 275, 278
Lyapunov exponent 14, 21, 60, 73, 93, 94-96, 98, 99, 101, 106, 116, 135, 137, 191, 225, 228, 243, 288, 291-294, 295, 296, 297, 298, 332, 348, 350, 354, 362, 363, 364, 365

Manifold 240, 273, 276
geometry 13, 14, 37-45
invariant 5-124, 273
Markov model 58
Melnikov's method 21, 28, 29, 30, 37-45, 52-56, 218, 219, 221, 275-276, 277, 278, 283
Microstructure 347-66
Mixing 2, 6, 7, 9, 14, 29, 54, 60, 107-117, 125-149
advection and mixing 12, 107-110
chaotic 8, 12, 148, 211, 212-213, 221, 225, 230, 271-278, 280-283, 313-345
control of 108, 115-116
experiments 322-324
fractal scaling properties of 116, 322, 335-343
lobe boundary evolution and 60-73, 110-115

Orientation 88, 169-183, 321

Poincaré map 8, 10, 11, 16, 17, 19, 20, 21, 25, 31, 34, 36, 40, 42, 46, 48, 49, 42, 58, 88, 157, 170, 172, 174-181, 224, 225, 226, 228, 243, 283
Prandtl-Batchelor theorem 190

Resonance 210-215, 209

Scalar 70, 93, 110, 116, 157, 186, 193, 230, 258, 264, 288, 293, 306, 308, 313, 314, 321, 322, 323-324, 327, 331, 332, 334, 335, 339, 342, 343, 344, 347, 361, 367, 368, 371
Smale-Birkhoff theorem 58, 321
Spectrum 347, 352, 356, 357, 362, 363, 367-372
Stirring 9, 107, 135, 185, 193, 255
  chaotic 46, 267-285
Streamlines 46, 74, 75, 76, 77, 78, 81, 82, 96, 151-167, 188, 191, 211, 212, 213, 216, 218-219, 240
Stretching 5, 6, 7, 12, 14, 29, 58-107, 108, 111, 114, 115-116, 128-136, 170, 173, 178-179, 181, 288-298, 305, 306, 321, 343

Tidal areas 267-285
Tracer 128-136, 185, 186, 188, 190, 191, 193, 194, 195, 235, 241, 246, 250, 347-372
Transport 5, 6, 8, 12, 14, 29, 35, 45-58, 60, 73, 74, 89, 107, 108, 110, 115-116, 185 186, 201, 211, 212
Turbulence 2, 30, 211, 235, 246, 287-311, 313-345, 367-372
Turnstile 5, 14, 45-52, 58, 60, 74, 89, 90, 91, 110, 274, 275, 280, 281, 282

Velocity field 1, 6, 7, 10, 14, 15-31, 39, 46, 49, 52, 54, 58, 62, 70, 125, 127, 212, 267, 271, 278, 289, 295, 347, 365
  steady 151-167, 170, 172-173
  turbulent 9, 15, 30
  2D single-frequency 15-18, 40-41
  2D multi-frequency 13, 18-26, 42-3
  2D aperiodic 13, 26-29, 44-45
  3D 6, 13, 15, 29, 40, 151-167
Vortex 29, 33, 235, 305, 306
  dipole 4, 234, 235, 239-246
  dynamics 3, 238, 261
  pair 10, 235, 238, 239-244, 252, 255, 258, 264
  point 235, 236-238, 245, 251, 255, 256, 258, 261, 263, 264
  structures 233-266
Vorticity 180, 244, 245, 246, 263, 264, 269, 270, 288, 289, 290, 297, 298, 302, 303-306, 321, 355, 364, 368

Wadden Sea 271, 284